DIE GRUNDLEHREN DER
MATHEMATISCHEN WISSENSCHAFTEN
IN EINZELDARSTELLUNGEN MIT BESONDERER BERÜCKSICHTIGUNG DER ANWENDUNGSGEBIETE

HERAUSGEGEBEN VON
R. GRAMMEL · E. HOPF · H. HOPF · F. RELLICH
F. K. SCHMIDT · B. L. VAN DER WAERDEN

BAND LXXV

CONTINUOUS TRANSFORMATIONS IN ANALYSIS

WITH AN INTRODUCTION TO ALGEBRAIC TOPOLOGY

BY

T. RADO

AND

P. V. REICHELDERFER

SPRINGER-VERLAG
BERLIN · GÖTTINGEN · HEIDELBERG
1955

CONTINUOUS TRANSFORMATIONS IN ANALYSIS

WITH AN INTRODUCTION TO ALGEBRAIC TOPOLOGY

BY

T. RADO
RESEARCH PROFESSOR OF MATHEMATICS
IN THE OHIO STATE UNIVERSITY

AND

P. V. REICHELDERFER
PROFESSOR OF MATHEMATICS IN THE OHIO STATE UNIVERSITY

MIT 53 TEXTABBILDUNGEN

SPRINGER-VERLAG
BERLIN · GÖTTINGEN · HEIDELBERG
1955

ISBN-13: 978-3-642-85991-5 e-ISBN-13: 978-3-642-85989-2
DOI: 10.1007/978-3-642-85989-2
Softcover reprint of the hardcover 1st edition 1955

Introduction.

The general objective of this treatise is to give a systematic presentation of some of the topological and measure-theoretical foundations of the theory of real-valued functions of several real variables, with particular emphasis upon a line of thought initiated by BANACH, GEÖCZE, LEBESGUE, TONELLI, and VITALI. To indicate a basic feature in this line of thought, let us consider a real-valued continuous function $f(u)$ of the single real variable u. Such a function may be thought of as defining a *continuous transformation* T under which $x=f(u)$ is the image of u. About thirty years ago, BANACH and VITALI observed that the fundamental concepts of *bounded variation, absolute continuity*, and *derivative* admit of fruitful geometrical descriptions in terms of the transformation $T: x=f(u)$ associated with the function $f(u)$. They further noticed that these geometrical descriptions remain meaningful for a continuous transformation T in Euclidean n-space R^n, where T is given by a system of equations of the form

$$x^1 = f_1(u^1, \ldots, u^n), \ldots, \quad x^n = f_n(u^1, \ldots, u^n),$$

and n is an arbitrary positive integer. Accordingly, these geometrical descriptions can be used to define, for continuous transformations in Euclidean n-space R^n, *n-dimensional concepts of bounded variation and absolute continuity*, and to introduce a *generalized Jacobian* without reference to partial derivatives. These ideas were further developed, generalized, and modified by many mathematicians, and significant applications were made in Calculus of Variations and related fields along the lines initiated by GEÖCZE, LEBESGUE, and TONELLI. In their fundamental aspects, these researches constitute a study of continuous transformations in Euclidean n-space R^n in terms of Point Set Theory, Algebraic Topology, and the modern Theory of Functions of Real Variables. The purpose of this volume is to give a systematic account of some of the basic concepts, methods, and results in this general area.

The development of this program involves a study of several individual topics which may deserve mention here. The n-dimensional concepts of bounded variation, absolute continuity and generalized Jacobians, already referred to above, are discussed thoroughly from various points of view in Part IV. The topological index, a concept of general utility in Mathematics, is studied in detail in §§ **II.2, II.3** for the case of Euclidean n-space, and in §§ **VI.1, VI.2** the important case of the plane is given special consideration. The transformation of multiple integrals, a troublesome issue even under the very restrictive assumptions customary in classical areas of Analysis, is discussed in several forms and in extreme generality in various sections of Parts IV, V, VI. Local linear approximations in Euclidean n-space are discussed in § **V.2** in terms of the various types of total differentials. Since this treatise has been designed for use by mathematicians primarily interested in Analysis, it did not seem appropriate to assume that the reader is familiar with Algebraic Topology to the extent necessary for our purposes. Accordingly, a self-contained exposition of the required background material in this field is given in §§ **I.4, I.5, I.6, I.7, II.1.** This exposition covers, in a very detailed manner, the basic algebraic apparatus needed in general cohomology theory, as well as some of the fundamental topics in the topology of Euclidean n-space R^n. Thus this exposition may serve as a first introduction to Algebraic Topology, especially since a definite effort has been made to prepare the reader for the study of the excellent comprehensive treatises in this field by ALEXANDROFF-HOPF, LEFSCHETZ, EILENBERG-STEENROD (see the *Bibliography*).

While our main concern is the discussion of a general theory applicable in Euclidean n-space, we included a detailed study of the special cases $n = 1$ and $n = 2$. The case $n = 2$ is considered in Part VI, with due emphasis upon the special features which are present in this case and are lost in the general case $n > 2$. The case $n = 1$, discussed in § **V.1**, furnishes a concrete picture of the general abstract concepts involved, and the study of this case yields instructive geometrical interpretations for many of the basic ideas and theorems in the modern theory of real-valued functions of a single real variable.

The method of presentation in this volume is based upon the assumption that the reader has progressed beyond the stage of basic training in the theory of Functions of Real Variables, Point Set Theory, and Group Theory. The background material in these fields is merely summarized in §§ **I.1, I.2, I.3, III.1.** For the convenience of the reader, an *Index* of the terms and symbols used in the book has been added. The *Bibliography* contains only a list of treatises directly related to our subject. References to literature concerning individual topics are given in footnotes throughout the text, with no attempt at completeness.

The drawings for the diagrams were prepared by our colleague F. W. NIEDENFUHR. We take pleasure in thanking him for his help.

Columbus, Ohio. T. RADO
January, 1955. P. V. REICHELDERFER.

Table of Contents.

Part I. Background in topology.

§ I.1. Survey of general topology.

I.1.1. Sets and mappings. If X is a set, then $x \in X$ means that x is an element of X, while $y \notin X$ means that y is not an element of X. A set which has a single element x is denoted by (x). On logical grounds, it is necessary to distinguish between an object x and the set (x) consisting of the single object x. However, as a matter of notational convenience we shall use frequently the same symbol for an object x and the set consisting of the single object x. It is also convenient to use the concept of the *empty set* which has no element. The empty set will be denoted by the symbol $\emptyset$.

A set may be finite or infinite. A set X is termed *countable* if there exists a one-to-one correspondence between X and a set consisting of some or all of the positive integers $1, 2, \ldots,$ or if X is the empty set. Thus the elements of a non-empty countable set X can be arranged into a (finite or infinite) sequence $x_1, x_2, \ldots$.

If $S(x)$ designates a certain statement relating to the object x, then we write

$$E = \{x \mid S(x)\}$$

to state that E is the set of those objects x for which the statement $S(x)$ holds. For example, if A and B are two given sets, then their *union* $A \cup B$ consists of those objects x which belong to at least one of the sets A and B, and their *intersection* $A \cap B$ consists of those objects x which belong to both A and B. These definitions may be stated in the form of the following equations.

$$A \cup B = \{x \mid x \in A \quad \text{or} \quad x \in B\},$$
$$A \cap B = \{x \mid x \in A \text{ and } x \in B\}.$$

The concepts of union and intersection apply to any number of sets. If F is a family of sets, then

$$\bigcup_{X \in \mathsf{F}} X = \{x \mid x \in X \text{ for some } X \in \mathsf{F}\},$$

$$\bigcap_{X \in \mathsf{F}} X = \{x \mid x \in X \text{ for every } X \in \mathsf{F}\}.$$

Two sets A, B are said to be *disjoint* if $A \cap B = \emptyset$.

If every element of a set A belongs to a set B, then A is termed a *subset* of B, and we write $A \subset B$ or equivalently $B \supset A$. If A is not a subset of B, then we write $A \not\subset B$. As a matter of convenience, the empty set is considered a subset of every set A.

If A is a subset of X, then X and A are said to constitute a *pair* (X, A). If X is any set, then $(X, \emptyset)$ is a pair. If (X, A) and (Y, B) are two pairs such that $X \subset Y$ and $A \subset B$, then we write $(X, A) \subset (Y, B)$.

If $S \subset X$, then the *complement* $\complement_X S$ of S with respect to X is the set of those elements of X which do not belong to S. In symbols

$$\complement_X S = \{x \mid x \in X,\ x \notin S\}.$$

If S is also a subset of some other set Y, then $\complement_Y S$ and $\complement_X S$ are different sets. If we consider, in a certain situation, only subsets S of a fixed set X, then we write $\complement S$ instead of $\complement_X S$ as a matter of convenience.

The *difference* $A - B$ of two sets A and B is defined by the formula

$$A - B = \{x \mid x \in A,\ x \notin B\}.$$

Consider a fixed set X and its subsets (thus $\complement$ will stand for complement with respect to X). The symbols $\cup$, $\cap$, $\complement$, $-$, introduced above, give rise to a number of identities involving subsets of X. Some examples follow.

$$\complement\complement A = A,\quad A - B = A \cap \complement B.$$
$$A \cup A = A,\quad A \cap A = A,\quad A \cup B = B \cup A,\quad A \cap B = B \cap A.$$
$$A \cup (B \cup C) = (A \cup B) \cup C,\quad A \cap (B \cap C) = (A \cap B) \cap C.$$
$$A \cap (B \cup C) = (A \cap B) \cup (A \cap C),\quad A \cup (B \cap C) = (A \cup B) \cap (A \cup C).$$
$$\complement(A \cup B) = \complement A \cap \complement B,\quad \complement(A \cap B) = \complement A \cup \complement B.$$

Similar identities hold in situations involving an arbitrary number of subsets of X. For example, if F is any family of subsets of X, then

$$\complement \cup S = \cap \complement S,\quad S \in \mathsf{F},$$
$$\complement \cap S = \cup \complement S,\quad S \in \mathsf{F}.$$

Furthermore, there exist certain logical relationships between the symbols already introduced. For example, $A \subset B$ if and only if $A \cup B = B$. Similarly, $A \subset B$ if and only if $A \cap B = A$.

If S is a subset of X, then the function $c(x, S)$ defined by the formula

$$c(x, S) = \begin{cases} 1 & \text{if } x \in S, \\ 0 & \text{if } x \in \complement S, \end{cases}$$

is termed the *characteristic function* of S in X. For any two subsets A, B of X we have the identity

$$c(x, A \cup B) + c(x, A \cap B) = c(x, A) + c(x, B).$$

If X_1, X_2 are two sets, then their *Cartesian product* $X_1 \times X_2$ is the set of all ordered pairs (x_1, x_2) such that $x_1 \in X_1$, $x_2 \in X_2$. For example, if (X, A) is a pair of sets, and $I: 0 \leq u \leq 1$ is the *unit interval* on the real number-line, then one has the Cartesian products $X \times I$, $A \times I$, $X \times u$, $A \times u$ (where u is any element of I), and the pairs

$$(X \times I,\ A \times I), \quad (X \times u,\ A \times u),$$

as well as the inclusion

$$(X \times u,\ A \times u) \subset (X \times I,\ A \times I).$$

Given two sets X and Y, a *mapping* (or *transformation*) $f: X \to Y$ from X into Y is a rule which assigns to each $x \in X$ a unique element $y = fx \in Y$ as its *image*.

Given a mapping $f: X \to Y$ and a subset $A \neq \emptyset$ of X, we denote by $f|A$ the mapping from A into Y which agrees with f on A. Thus

$$(f|A)\, x = fx \quad \text{for} \quad x \in A.$$

The mapping $f|A$ is referred to as f *cut down* to A or also as the *restriction* of f to A.

Given a mapping $f: X \to Y$ and a subset A of X, the image fA of A under f is the set of the images fx of the elements of A. If B is a subset of Y, then $f^{-1}B$ is the set of those elements x of X for which $fx \in B$. In particular, if y_0 is an element of Y, then $f^{-1}y_0$ is the set of those elements x of X for which $fx = y_0$.

Given a mapping $f: X \to Y$, f is said to be *onto* if $fX = Y$, and f is said to be *one-to-one* if $x_1 \neq x_2$ implies that $fx_1 \neq fx_2$. If f is both onto and one-to-one, then f is termed *bi-unique*. If f is bi-unique, then for each $y \in Y$ the set $f^{-1}y$ consists of a single point of X. Thus in this case we have at our disposal the mapping $f^{-1}: Y \to X$, the *inverse* of f.

The following set of identities, relating to an arbitrary mapping $f: X \to Y$, will be constantly used. If F is any family of subsets of X, then

$$f \cup A = \cup f A, \quad A \in \mathsf{F},$$
$$f \cap A \subset \cap f A, \quad A \in \mathsf{F}.$$

If A_1, A_2 are subsets of X such that $A_1 \subset A_2$, then $fA_1 \subset fA_2$. If Φ is any family of subsets of Y, then

$$f^{-1} \cup B = \cup f^{-1} B, \quad f^{-1} \cap B = \cap f^{-1} B, \quad B \in \Phi.$$

If B_1, B_2 are any two subsets of Y, then

$$f^{-1}(B_2 - B_1) = f^{-1} B_2 - f^{-1} B_1 .$$

If A, B are subsets of X and Y respectively, then

$$f^{-1} f A \supset A , \quad f f^{-1} B \subset B .$$

Consider now two mappings $f : X \to Y$, $g : Y \to Z$. Then the *product mapping* $gf : X \to Z$ is defined by the formula

$$(g f)\, x = g\,(f\, x), \quad x \in X .$$

Note that a product of mappings is to be read *from the right to the left.*

Given a mapping $f : X \to Y$, consider a subset A of X and a subset B of Y. If $fA \subset B$, then f is said to be a mapping from the pair (X, A) into the pair (Y, B), and one writes

$$f : (X, A) \to (Y, B) .$$

If f is a mapping from (X, A) into (Y, B) and g is a mapping from (Y, B) into (Z, C) then clearly gf is a mapping from (X, A) into (Z, C).

If X is any set, then the *identity mapping* $i : X \to X$ on X is defined by $i\,x = x$ for $x \in X$. If A is a subset of X, then the mapping $i|A : A \to X$ is termed the *inclusion mapping* from A into X.

Consider now two pairs (X, A), (Y, B) such that $(X, A) \subset (Y, B)$, and let j be the identity mapping on Y. Then the mapping

$$j\,|\,X : (X, A) \to (Y, B)$$

is termed the *inclusion mapping* from (X, A) into (Y, B).

If one has to consider several pairs of sets and certain mappings from these pairs, it is helpful to set up a *mapping diagram.* We give two examples.

Let (X, A) be a pair of sets, and let $I : 0 \leqq u \leqq 1$ be the unit interval on the real number-line. The following diagram will be referred to later on as the *first homotopy diagram.*

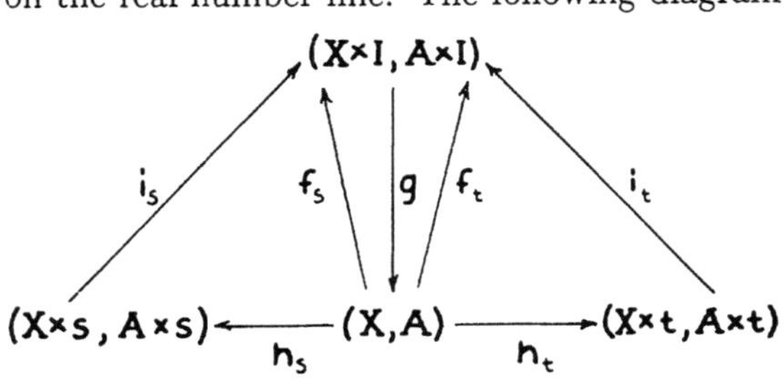

Fig. 1.

In this diagram, s and t denote two real numbers in I. The mappings i_s, i_t are the inclusion mappings corresponding to the inclusions

$$(X \times s, A \times s) \subset (X \times I, A \times I), \quad (X \times t, A \times t) \subset (X \times I, A \times I) .$$

The mappings h_s, h_t are defined by the formulas

$$h_s x = (x, s), \quad h_t x = (x, t), \quad x \in X.$$

The mappings f_s, f_t are defined by the formulas

$$f_s x = (x, s), \quad f_t x = (x, t), \quad x \in X.$$

The mapping g is defined by the formula $g(x, u) = x$, $(x, u) \in (X \times I)$. Clearly h_s and h_t are bi-unique, f_s and f_t are one-to-one, and g is onto. Furthermore

$$f_s = i_s h_s, \quad f_t = i_t h_t, \quad g f_s = g f_t = i,$$

where i is the identity mapping on X.

The diagram in Fig. 2 will be referred to as the *second homotopy diagram*.

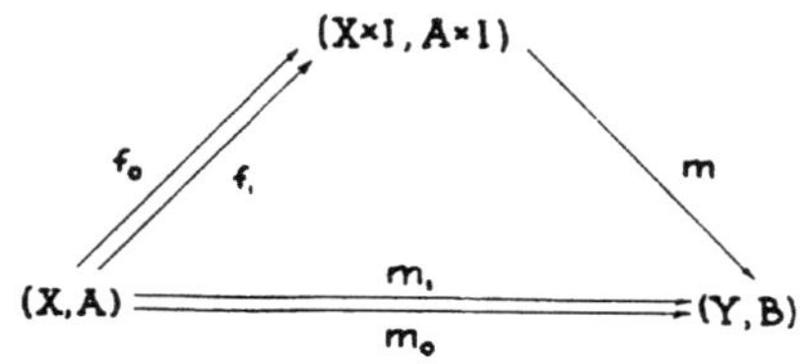

Fig. 2.

In this diagram, the mappings f_0 and f_1 have the same meaning as in the first homotopy diagram (corresponding to $s=0$, $t=1$). The mapping m is an arbitrary mapping from $(X \times I,\ A \times I)$ into (Y, B), and the mappings m_0 and m_1 are defined by the formulas

$$m_0 x = m(x, 0), \quad m_1 x = m(x, 1), \quad x \in X.$$

Clearly $m_0 = m f_0$, $m_1 = m f_1$.

I.1.2. Preliminary comments on real-valued functions. Let $X \neq \emptyset$ be an arbitrary set. A real-valued function $f(x)$ on X arises if with every element x of X there is associated a unique real number $f(x)$. The characteristic function $c(x, S)$ of a set $S \subset X$ (see **I.1.1**) is an example. It is convenient to extend the concept of a real-valued function by permitting such a function to assume the values ∞ and $-\infty$ also. If a real-valued function $f(x)$ on X does not assume infinite values, then $f(x)$ is termed *finite-valued*.

Given a mapping $T: X \to Y$ (see **I.1.1**) and a set $S \subset X$, the *crude multiplicity function* $N(y, T, S)$ is defined, for $y \in Y$, as the number of elements of the set $S \cap T^{-1} y$. Thus $N(y, T, S)$ may be equal to ∞. The finite values of $N(y, T, S)$ are non-negative integers.

While the use of ∞ and $-\infty$ is a matter of convenience, it is also true that one has to provide certain obvious but at times cumbersone explanations in this connection. The following agreements will be used.

$$\infty + \infty = \infty, \quad -\infty + (-\infty) = -\infty, \quad a \cdot \infty = \infty \text{ if } 0 < a \leqq \infty,$$
$$a \cdot \infty = -\infty \text{ if } -\infty \leqq a < 0, \quad a + \infty = \infty \text{ if } -\infty < a < \infty.$$

However, no meaning is assigned to $\infty - \infty$ and $0 \cdot \infty$.

Let $f(x)$ be a real-valued function on a set X, such that the finite values of f are non-negative and f does not assume the value $-\infty$ (however, f may assume the value ∞). We shall then say that f is non-negative on X, and we shall write $f(x) \geqq 0$, $x \in X$. Given a real-valued, non-negative function $f(x)$ on X and a set $E \subset X$, we shall have to consider later on the *sum of the values* of $f(x)$ on E, to be denoted by $\sum(E, f)$. Thus

$$\sum(E, f) = \sum_{x \in E} f(x).$$

The following comments will serve to clarify the intended meaning of the symbol $\sum(E, f)$.

(i) If $f(x) = \infty$ for some $x \in E$, then $\sum(E, f) = \infty$.

(ii) If $f(x) \equiv 0$ on E, then $\sum(E, f) = 0$.

(iii) If $f(x)$ is finite-valued on E and vanishes on E except for a finite number of distinct elements $x_1, \ldots, x_M$ of E, then

$$\sum(E, f) = f(x_1) + \cdots + f(x_M).$$

(iv) If $f(x)$ is finite-valued on E and vanishes on E except for an infinite sequence of distinct elements $x_1, \ldots, x_k, \ldots$ of E, then

$$\sum(E, f) = \sum_{k=1}^{\infty} f(x_k)$$

if the series on the right converges, and $\sum(E, f) = \infty$ if this series diverges.

(v) If the set of those elements $x \in E$ where $f(x) > 0$ is non-countable, then $\sum(E, f) = \infty$.

(vi) If $E = \emptyset$, then $\sum(E, f) = 0$.

The sum $\sum(E, f)$ has various obvious properties, some of which will be listed presently.

(a) If f, F are real-valued, non-negative functions on X and $f \leqq F$ on the set $E \subset X$, then

$$\sum(E, f) \leqq \sum(E, F).$$

(b) If f is a real-valued, non-negative function on X and c is a positive constant, then

$$\sum(E, cf) = c \sum(E, f), \quad E \subset X.$$

(c) If $\{f_j\}$ is a (finite or infinite) sequence of real-valued, non-negative functions on X, then

$$\sum\Big(E, \sum_j f_j\Big) = \sum_j \sum(E, f_j), \quad E \subset X.$$

(d) If f is a real-valued, non-negative function on X and $\{f_j\}$ is a sequence of real-valued, non-negative functions on X such that one has on a certain set $E \subset X$ the relations

$$f_1 \leqq f_2 \leqq \cdots, \quad f_j \to f,$$

then

$$\sum (E, f_j) \to \sum (E, f).$$

Let us recall a few definitions relating to sets of real numbers. If S is a set of real numbers and B is a (finite) real number such that $x \leqq B$ for $x \in S$, then B is termed an *upper bound* for S. A set S of real numbers is said to be *bounded from above* if there exists an upper bound for S. A fundamental property of the real number system is expressed by the statement that if a set S of real numbers is bounded from above, then among its upper bounds there exists a smallest one. This smallest upper bound is called the *least upper bound* of S and is denoted by l.u.b. S. Clearly the relation l.u.b. $S \in S$ holds if and only if S contains a largest element (which is then equal to l.u.b. S). As a matter of convenience, we agree to set l.u.b. $S = \infty$ if S is not bounded from above. The *greatest lower bound* gr.l.b. S of a set S of real numbers is defined in a similar manner.

Consider now a real-valued, finite-valued function $f(x)$ on a set X. Then the values assumed by $f(x)$ on X constitute a set V of real numbers. The least upper bound and the greatest lower bound of $f(x)$ on X are defined then by the formulas

$$\underset{x \in X}{\text{l.u.b.}}\, f(x) = \text{l.u.b.}\, V, \quad \underset{x \in X}{\text{gr.l.b.}}\, f(x) = \text{gr.l.b.}\, V.$$

These quantities may be equal to ∞ and $-\infty$ respectively.

If $f(x)$ is a real-valued function on X which takes on the value ∞, then we agree to consider ∞ as its least upper bound. Similarly, if $f(x)$ takes on the value $-\infty$, then we agree to consider $-\infty$ as its greatest lower bound.

If $f(x)$ is a real-valued, finite-valued function on a set X and $E \neq \emptyset$ is a subset of X, then the *oscillation* $\omega(f, E)$ of f on E is defined as the least upper bound of the set of the numbers $|f(x_2) - f(x_1)|$ corresponding to all pairs of elements x_1, x_2 of E.

If $\{x_n\}$ is an infinite sequence of real numbers, then the symbols

$$\limsup_{n \to \infty} x_n, \quad \liminf_{n \to \infty} x_n$$

denote the *limit superior* and the *limit inferior* of the sequence respectively in the usual sense. Let us recall merely that the relation

$$\limsup_{n \to \infty} x_n = \infty$$

holds if and only if for every real number K there exists an integer n such that $x_n > K$.

I.1.3. Topological spaces. *Definition 1.* A *topological space* X is a set of elements (to be termed the *points* of X) in which a family Ω of subsets of X (to be termed the *open sets* of X) is assigned in such a manner that the following holds.

(i) $\emptyset \in \Omega$, $X \in \Omega$.

(ii) If $O_1, \ldots, O_m$ is any *finite* system of open sets of X, then $\bigcap O_j$, $j = 1, \ldots, m$, is also an open set.

(iii) If F is any (finite or infinite) family of open sets of X, then $\bigcup O$, $O \in \mathsf{F}$, is also an open set.

Definition 2. Let X be a topological space. A subset F of X is termed *closed* if $\complement F$ is open (where $\complement$ stands for complement relative to X).

Definition 3. Let X be a topological space. If A is a subset of X, then the *closure* of A (to be denoted by $\bar{A}$) is the intersection of all those closed sets of X which contain A.

Definition 4. Let X be a topological space. If A is a subset of X, then the *interior* of A (to be denoted by int A) is the union of all those open sets of X which are contained in A.

Definition 5. Let X be a topological space. If A is a subset of X, then the *frontier* of A (to be denoted by fr A) is defined by the formula

$$fr\, A = \bar{A} \cap \overline{\complement A}.$$

Definition 6. Let X be a topological space. If A is a subset of X and $\mathfrak{U}$ is a family of subsets of X such that

$$A \subset \bigcup U, \quad U \in \mathfrak{U},$$

then $\mathfrak{U}$ is termed a *covering* of A. If $\mathfrak{U}'$ is a sub-family of $\mathfrak{U}$ such that $\mathfrak{U}'$ is also a covering of A, then $\mathfrak{U}'$ is referred to as a *sub-covering*. A covering $\mathfrak{U}$ of A is *finite* if $\mathfrak{U}$ is a finite family. A covering $\mathfrak{U}$ of A is *open* (*closed*) if every set $U \in \mathfrak{U}$ is open (closed).

Definition 7. Let X be a topological space, and let $\mathfrak{U}$, $\mathfrak{V}$ be two open coverings of X. If for every $U \in \mathfrak{U}$ there exists a $V \in \mathfrak{V}$ such that $U \subset V$, then $\mathfrak{U}$ is termed a *refinement* of $\mathfrak{V}$.

Definition 8. Let X be a topological space and $\mathfrak{U}$ an open covering of X. If $U \in \mathfrak{U}$, then the *star* of U relative to $\mathfrak{U}$ (to be denoted by $St_{\mathfrak{U}}\, U$) is the set defined by the formula

$$St_{\mathfrak{U}}\, U = \bigcup U', \quad U' \in \mathfrak{U}, \quad U' \cap U \neq \emptyset.$$

Note that this set is open as a consequence of (iii) in definition 1. Evidently the sets $St_{\mathfrak{U}}\, U$, corresponding to all the sets $U \in \mathfrak{U}$, constitute an open covering of X. The open covering so derived from $\mathfrak{U}$ will be denoted by $\mathfrak{U}^*$.

Definition 9. Let X be a topological space. If $\mathfrak{U}, \mathfrak{V}$ are two open coverings of X such that $\mathfrak{U}^*$ is a refinement of $\mathfrak{V}$ then $\mathfrak{U}$ is termed a *star refinement* of $\mathfrak{V}$.

Definition 10. Let X be a topological space, S a subset of X, and A, B subsets of S. Then A and B are said to constitute a *partition* of S (in symbols, $S = A|B$) provided that

$$A \cup B = S, \quad A \cap B = \emptyset, \quad A \neq \emptyset, \quad B \neq \emptyset, \quad \bar{A} \cap B = \emptyset, \quad A \cap \bar{B} = \emptyset.$$

Definition 11. Let X be a topological space. A subset S of X is *disconnected* if there exists some partition $S = A|B$. If there exists no partition of S, then S is *connected.*

Definition 12. Let X be a topological space. If E is a subset of X and F is a connected subset of E, then F is termed a *maximal connected subset* of E if F is not a proper subset of any connected subset of E. A non-empty maximal connected subset of a set $E \subset X$ is termed a *component* of E.

Definition 13. Let X be a topological space. A non-empty, open, connected subset of X is termed a *domain* in X.

Definition 14. Let X be a topological space. A subset S of X is *compact* if every open covering of S contains a finite sub-covering.

Definition 15. Let X be a topological space. A non-empty, compact, connected subset of X is termed a *continuum.* If a continuum does not reduce to a single point, then it is termed *non-degenerate.*

Definition 16. A HAUSDORFF *space* is a topological space X which satisfies the following additional condition (the so-called *separation axiom*): if x_1, x_2 are any two distinct points of X, then there exist open sets O_1, O_2 such that $x_1 \in O_1$, $x_2 \in O_2$, $O_1 \cap O_2 = \emptyset$.

Definition 17. A topological space X is *normal* if for every pair of disjoint closed sets F_1, F_2 there exist open sets O_1, O_2 such that $F_1 \subset O_1$, $F_2 \subset O_2$, $O_1 \cap O_2 = \emptyset$.

Definition 18. A topological space X is *fully normal* if every open covering of X admits of a star refinement.

Definition 19. A topological space X is *locally connected* if for every open set O it is true that the components of O are open.

Definition 20. Let X be a topological space. A subset E of X is *dense* in X if $\bar{E} = X$.

Definition 21. A topological space is *separable* if it contains a countable dense subset.

Definition 22. Let X be a topological space. A *subspace* A of X is a non-empty subset of X with the following assignment of the family

Ω_A of the open sets of A: a subset of A belongs to Ω_A if and only if it is of the form $O \cap A$, where O is open in X.

Note that a subset S of a subspace A of X may be open in A without being open in X. In a general way, the meaning of several topological concepts depends upon the topological space relative to which such concepts are considered. Given a topological space X and a subspace A of X, as well as a subset $S \subset A$, to avoid ambiguity we shall write $fr_A S$, $\mathrm{int}_A S$, $\complement_A S$, and so forth, to indicate that the concepts involved are considered relative to the subspace A. On the other hand, it is easy to see that a set $S \subset A$ is compact relative to A if and only if it is compact relative to X, and similarly S is connected relative to A if and only if it is connected relative to X.

Definition 23. Let X_1, X_2 be topological spaces. Then their *topological product* is the Cartesian product $X_1 \times X_2$ with the following assignment of open sets for $X_1 \times X_2$. Let ω be the family of those subsets of $X_1 \times X_2$ which are of the form $O_1 \times O_2$, where O_1 is open in X_1 and O_2 is open in X_2. A subset S of $X_1 \times X_2$ is then, by definition, open in $X_1 \times X_2$ if and only if S is the union of a family of sets belonging to ω [it is immediate that the conditions (i), (ii), (iii) in definition 1 are satisfied]. The topological product of X_1, X_2 is denoted by the same symbol $X_1 \times X_2$ as the Cartesian product (it will always be clear from the context which one of the two products is meant).

Definition 24. A *pair* (X, A) of topological spaces consists of a topological space X and a subspace A of X. As a matter of convenience, one uses also on certain occasions the symbol $(X, \emptyset)$ to refer to a single topological space X.

Definition 25. Let S be a topological space and (X_0, A_0) a pair of subspaces of S such that X_0 and A_0 are closed in S. Furthermore, let F be a family of pairs (X, A) of subspaces of S such that X and A are closed in S. Then we shall write $\mathsf{F} \Rightarrow (X_0, A_0)$ as an abbreviation for the following set of conditions.

(i) $(X_0, A_0) \subset (X, A)$ for $(X, A) \in \mathsf{F}$.

(ii) If $(X_1, A_1) \in \mathsf{F}$, $(X_2, A_2) \in \mathsf{F}$, then there exists a pair $(X_3, A_3) \in \mathsf{F}$ such that

$$(X_3, A_3) \subset (X_1 \cap X_2,\ A_1 \cap A_2).$$

(iii) If O and U are any two open sets of S such that $X_0 \subset O$, $A_0 \subset U$, then there exists a pair $(X, A) \in \mathsf{F}$ such that $X \subset O$, $A \subset U$.

Definition 26. Let X be a topological space. A sequence $\{x_n\}$ of points of X converges to a point x of X (in symbols, $x_n \to x$ or $x = \lim x_n$) if for every open set O containing x there exists an integer $N = N(O)$ such that $x_n \in O$ for $n > N$.

Clearly, if X is a HAUSDORFF space, then a sequence $\{x_n\}$ cannot converge to two distinct points.

Definition 27. Given a topological space X, let $f(x)$ be a real-valued, finite-valued function defined on a subset E of X. Then $f(x)$ is said to be continuous (relative to E) at a point $x_0 \in E$ if for every real number $\varepsilon > 0$ there exists an open set $O = O(\varepsilon)$ containing x_0 such that $|f(x) - f(x_0)| < \varepsilon$ for $x \in E \cap O$. If $f(x)$ is continuous (relative to E) at every point of E, then $f(x)$ is said to be continuous on E (relative to E).

Definition 28. Given a topological space X and a real-valued function $f(x)$ on X, $f(x)$ is said to be *lower semi-continuous* at a point x_0 of X if for every real number $a < f(x_0)$ there exists an open set $O = O(a)$ such that $x_0 \in O$ and $f(x) > a$ for $x \in O$. If $f(x)$ is lower semi-continuous at every point of X, then $f(x)$ is said to be lower semi-continuous on X.

The concepts introduced so far give rise to a number of theorems. We shall list presently a series of such theorems (which will be used later on) as exercises for the reader. The proofs may be found in the treatises on Topology listed in the Bibliography, or else the proofs are easy consequences of the definitions involved.

Exercise 1. Let A be a subset of the topological space X. Then the closure $\bar{A}$ of A is closed, and $A \subset \bar{A}$. Furthermore, if A, B are any two subsets of X, then

$$\overline{A \cup B} = \bar{A} \cup \bar{B}, \quad \overline{A \cap B} \subset \bar{A} \cap \bar{B}.$$

Exercise 2. Let A be a subset of the topological space X. Then a point $x \in X$ belongs to $\bar{A}$ if and only if every open set O containing x intersects A.

Exercise 3. If O is an open subset of the topological space X, then $fr\,O \subset \complement O$.

Exercise 4. Let $\{O_j\}$ be a (finite or infinite) sequence of pairwise disjoint open sets in a topological space X. Set $\bigcup O_j = O$. Then $fr\,O_j \subset fr\,O \subset \complement O$.

Exercise 5. Let O and F be subsets of a topological space X such that O is open, F is closed, and $fr\,O \subset F$. Then $\bar{O} - F$ is open.

Exercise 6. Let X and Y be closed subsets of a topological space S such that $X \supset Y \neq \emptyset$, $X - Y \neq \emptyset$. Then $X - Y$ is open if and only if $fr\,X \subset Y$.

Exercise 7. In a topological space X let O_1, O_2 be open sets such that $O_1 \subset O_2$. Put $\bar{O}_2 - O_1 = F_{12}$. Then

$$\bar{O}_2 \supset F_{12} \supset fr\,O_2, \quad \bar{O}_2 - F_{12} = O_1, \quad \bar{O}_2 = \bar{O}_1 \cup F_{12}, \quad fr\,O_1 = \bar{O}_1 \cap F_{12}.$$

Exercise 8. Let U and A be subsets of the topological space X such that $U \subset A$. Then $\bar{U} \subset \operatorname{int} A$ if and only if $(\operatorname{int} A) \cup (\operatorname{int} \complement U) = X$.

Exercise 9. In a topological space the union of a *finite* system of compact sets is compact.

Exercise 10. Let A be a closed subset of the fully normal topological space X. Then A (as a subspace of X) is fully normal.

Exercise 11. Let $E \neq \emptyset$ be a subset of the topological space X. Then the following statements hold concerning the components of E.

(i) If C_1, C_2 are distinct components of E, then $C_1 \cap C_2 = \emptyset$.

(ii) If F is a non-empty connected subset of E, then F is contained in a (unique) component of E.

(iii) E is the union of its components.

(iv) If C is a component of E and F is a connected set such that $C \subset F \subset E$, then $C = F$.

(v) If E is closed, then every component of E is closed.

Exercise 12. If the subsets A, B of the topological space X constitute a partition of X (see definition 10), then A and B are both open and closed.

Exercise 13. Let A be a subset of the connected topological space X such that A is both open and closed. Then either $A = \emptyset$ or $A = X$.

Exercise 14. In a topological space X, let O_1, O_2 be non-empty, disjoint open sets. Then O_1 and O_2 constitute a partition of $O_1 \cup O_2$ (see definition 10).

Exercise 15. Let O_1, O_2, O_3 be open subsets of the topological space X, such that O_1, O_2 are non-empty and disjoint, and O_3 does not intersect both O_1 and O_2. Then $O_1 \cup O_2 \cup O_3$ is disconnected.

Exercise 16. Let E and F be subsets of a topological space X such that E is connected and $E \subset F \subset \overline{E}$. Then F is connected.

Exercise 17. Let E and Γ be subsets of a topological space X such that Γ is connected. If Γ intersects both E and $\complement E$, then it also intersects $fr\,E$.

Exercise 18. If E is a non-empty proper subset of the connected topological space X, then $fr\,E \neq \emptyset$.

Exercise 19. Let X, Y be non-empty, closed subsets of the locally connected HAUSDORFF space S, such that $X \supset Y$ and $X - Y$ is non-empty and connected. Then $X - Y$ is contained in a unique component O_1 of $\complement Y$. Let O_2 be the union of those components of $\complement Y$ which do not contain $X - Y$ (if no such components exist, then put $O_2 = \emptyset$). Set $X^* = X \cup O_2$, $Y^* = Y \cup O_2$. Then the following holds.

(i) X^* and Y^* are closed.

(ii) $X^* = X \cup Y^*$, $Y = X \cap Y^*$, $S - Y^* = O_1$.

(iii) $S - X^*$ is empty or non-empty according as $X - Y$ is open or not open.

(iv) $S - X^*$ is empty or non-empty according as Y does or does not contain $fr\,X$.

Exercise 20. In a HAUSDORFF space X, a single point x (considered as a subset of X) is closed, and $\complement x$ is open.

Exercise 21. Let X be a HAUSDORFF space which consists of precisely two distinct points a and b. Then the family of open sets of X consists of the sets $\emptyset$, X, a, b.

Exercise 22. If A is a closed subset of the compact topological space X, then A is compact.

Exercise 23. If F is a compact subset of the HAUSDORFF space X, then F is closed.

Exercise 24. In a HAUSDORFF space X, let $\{G_j\}$ be a sequence of non-empty compact sets such that $G_1 \supset G_2 \supset \cdots$. Set $E = \bigcap G_j$. Then the following holds.

(i) E is non-empty and compact.

(ii) If O is an open set containing E, then $O \supset G_j$ for j sufficiently large.

Exercise 25. Let F and O be subsets of the normal topological space X such that F is closed, O is open, and $F \subset O$. Then there exists an open set G such that $F \subset G$, $\overline{G} \subset O$.

Exercise 26. Let U and A be subsets of the normal topological space X such that U is open, A is closed, and $U \subset A$. Let F be the family of all the pairs of the form $(X - V, A - V)$, where $\overline{V} \subset U$ and V is open. Then (see definition 25)

$$\mathsf{F} \Rightarrow (X - U, A - U).$$

Exercise 27. Let Y be a non-empty, compact subset of the HAUSDORFF space X, and let $\{F_n\}$ be a sequence of compact subsets of X such that $F_1 > F_2 > \cdots$, $\bigcap F_n = Y$. Denote by F the family of the pairs (X, F_n), $n = 1, 2, \ldots$. Then (see definition 25)

$$F \Rightarrow (X, Y).$$

Exercise 28. Let O be a non-empty open subset of the locally connected, separable topological space X. Then the components of O constitute a countable family of pair-wise disjoint domains.

Exercise 29. A compact HAUSDORFF space is normal.

Exercise 30. Let $\{G_j\}$ be a sequence of continua in the HAUSDORFF space X such that $G_1 > G_2 > \cdots$. Then $\bigcap G_j$ is a continuum.

Exercise 31. Let X, Y be non-empty, compact subsets of the connected HAUSDORFF space S, such that $X > Y$ and $X - Y = O$ is open and non-empty. Then the following holds.

(i) $\overline{O}$ and $fr\,O$ are compact.

(ii) $fr\,O \neq \emptyset$, $\overline{O} - fr\,O = O$.

(iii) $X = \overline{O} \cup Y$, $fr\,O = \overline{O} \cap Y$.

Exercise 32. Let F be a closed subset of the locally connected topological space X. If D is a component of $\mathsf{C}F$, then $fr\,D < F$.

Ecercise 33. Let D be a component of an open subset O of the locally connected topological space X. Then

$$fr\,D < fr\,O < \mathsf{C}\,O.$$

Exercise 34. Let D_0, D_1 be non-empty domains in the connected topological space X, such that $\overline{D}_0 < D_1$ and $fr\,D_0$ is connected. Then $D_1 - D_0$ is connected.

Exercise 35. Let F be a closed subset of the compact, locally connected HAUSDORFF space X, such that $\mathsf{C}F$ is connected. Let O be an open set containing F. Then there exists an open set U such that $F < U < O$ and $\mathsf{C}\,U$ is connected.

Exercise 36. If X_1, X_2 are compact HAUSDORFF spaces, then their topological product $X_1 \times X_2$ is also a compact HAUSDORFF space.

Exercise 37. Let $f(x)$ be a real-valued, finite-valued, continuous function defined on a connected subset C of the topological space X. Then the following holds.

(i) If $f(x)$ takes on the values a, b on C, then it also takes on every value between a and b.

(ii) If the set V of the values taken on by $f(x)$ is countable, then $f(x)$ is constant on C.

Exercise 38. If $f(x_1, \ldots, x_n)$ is a real-valued, finite-valued, continuous function of the variable points $x_1, \ldots, x_n$ on the compact topological space X, then $f(x_1, \ldots, x_n)$ attains a maximum and a minimum on X.

Exercise 39. If $f(x)$ is a real-valued, finite-valued continuous function on the compact topological space X, then there exist two points x_1, x_2 such $|f(x_2) - f(x_1)|$ is equal to the oscillation $\omega(f, X)$ of $f(x)$ on X.

Exercise 40. Let A be a compact subspace of the compact HAUSDORFF space X. Consider the topological products $X \times I$, $A \times I$, where I is the unit interval $0 \leq u \leq 1$ on the real number-line. Given $t \in I$, let O, U be open subsets of $X \times I$ such that $O > X \times t$, $U > A \times t$. Then there exists an $\varepsilon > 0$ (which depends upon t, O, U) such that

$$(X \times s,\ A \times s) < (\overline{O}, \overline{U}) \quad \text{for} \quad s \in I, \quad |s - t| < \varepsilon.$$

Exercise 41. Let (X, Y) be a pair of compact HAUSDORFF spaces such that $\emptyset \neq Y \neq X$. Let $\complement Y$ be represented as the union of a (finite or infinite) sequence of pair-wise disjoint open sets $O_1, \ldots, O_n, \ldots$. Set $Y_n = \complement O_n, Y_n^* = Y \cup O_n, F_n = \bigcap Y_k$, $k = 1, \ldots, n$. Then the following holds.

(i) Y_n, Y_n^*, F_n are non-empty and compact.
(ii) $Y_n > Y$, $Y_n^* > Y$, $Y_m > Y_n^*$ if $m \neq n$, $F_1 > F_2 > \cdots$, $Y_n > F_n$, $\bar{O}_n < Y_n^*$, $fr O_n < Y$.
(iii) $Y = \bigcap F_n$, $X = Y_n \cup \bar{O}_n$, $fr O_n = Y_n \cap \bar{O}_n$.
(iv) $\{(X, F_n)\} \Rightarrow (X, Y)$ (see definition 25).

Exercise 42. In a HAUSDORFF space S, let (X_1, Y_1), (X_2, Y_2) be two pairs of compact subsets such that (*a*) $(X_2, Y_2) < (X_1, Y_1)$ and (*b*) $X_i - Y_i$ is connected and open, $i = 1, 2$. Then either $X_1 - Y_1$ and $X_2 - Y_2$ are disjoint, or else $X_1 - Y_1$ is a subset of $X_2 - Y_2$.

Exercise 43. If X is a fully normal topological space, then X is normal.

I.1.4. Metric spaces. *Definition 1.* A metric space arises if in a set X of elements (to be termed the *points* of the metric space) there is assigned a *distance* $d(x_1, x_2)$ for every pair of points x_1, x_2 of X, such that the following holds. (i) $d(x_1, x_2)$ is real-valued, finite, and non-negative. (ii) $d(x_1, x_2) = 0$ if and only if $x_1 = x_2$. (iii) $d(x_1, x_2) = d(x_2, x_1)$. (iv) If x_1, x_2, x_3 are any three points of X, then one has the *triangle inequality*

$$d(x_1, x_3) \leq d(x_1, x_2) + d(x_2, x_3).$$

Definition 2. Let X be a metric space, x a point of X, and r a positive real number. Then the set

$$\Delta_r(x) = \{x' \mid x' \in X, \ d(x, x') < r\}$$

is termed the *open spherical neighborhood* of x with radius r. As a matter of convenience, we shall also use the symbol $\Delta(x, r)$ to refer to this spherical neighborhood.

Definition 3. The *natural topology* of a metric space X is determined by the following assignment of the family of the open sets of X: A set $E < X$ is termed open if and only if for every point $x \in E$ there exists a real number $r = r(x) > 0$ such that $\Delta_r(x) < E$. It is immediate that X is then a HAUSDORFF space.

Definition 4. Let E be a subset of the metric space X. Then the *diameter* of E, to be denoted by δE, is defined by the formula

$$\delta E = \text{l.u.b.}\, d(x_1, x_2), \quad x_1 \in E, \ x_2 \in E.$$

It is convenient to agree that the diameter of the empty set is equal to zero.

Definition 5. A subset E of the metric space X is *bounded* if $\delta E < \infty$.

Definition 6. A metric space X is *boundedly compact* if every bounded closed subset of X is compact.

Definition 7. Let E be a non-empty subset and x a point of the metric space X. Then the *écart* of x from E, to be denoted by $e(x, E)$, is defined by the formula

$$e(x, E) = \text{gr.l.b.}\, d(x, y), \quad y \in E.$$

Definition 8. Let $\{x_j\}$ be a sequence of points in the metric space X. Then $\{x_j\}$ is termed a *Cauchy sequence* if for every $\varepsilon > 0$ there exists an integer $n = n(\varepsilon)$ such that $d(x_j, x_k) < \varepsilon$ if $j, k > n$.

Definition 9. A metric space X is *complete* if every CAUCHY sequence in X is convergent.

Definition 10. Let X be a metric space. A family F of subsets of X is termed *completely additive* if the following holds. (i) Every open set of X belongs to F. (ii) If $E \in \mathsf{F}$, then $\complement E \in \mathsf{F}$. (iii) If $E_1, E_2, \ldots$ is any (finite or infinite) sequence of sets of F, then their union also belongs to F.

Clearly, those subsets of X which belong to *every* completely additive family of subsets of X constitute a completely additive family. This observation justifies the next definition.

Definition 11. The completely additive family of those subsets of X which belong to *every* completely additive family in X is termed the *Borel class* in X. A set which belongs to this class is termed a *Borel set* in X.

As an immediate consequence of this definition it follows that every open set and every closed set in X is a BOREL set, and that the complement of a BOREL set in X is again a BOREL set. Furthermore, countable unions and countable intersections of BOREL sets are again BOREL sets.

Definition 12. Let X be a metric space. A *determining system* $\mathfrak{A}$ of closed sets in X arises if with every finite sequence of positive integers $n_1, \ldots, n_k$ there is associated a closed subset $F(n_1, \ldots, n_k)$ of X. If $\nu = (n_1, \ldots, n_k, \ldots)$ is an infinite sequence of positive integers, then we put

$$F_\nu = \bigcap_{k=1}^{\infty} F(n_1, \ldots, n_k), \quad \mathsf{A}(\mathfrak{A}) = \bigcup_{\nu} F_\nu,$$

where in the second formula the union is taken with respect to all infinite sequences ν of positive integers. The subsets $\mathsf{A}(\mathfrak{A})$ of X which are obtained in this manner are termed analytic sets in X.

Definition 13. Let X be a metric space. A determining system $\mathfrak{A}$ of closed subsets $F(n_1, \ldots, n_k)$ in X is termed *regular* if always

$$F(n_1, \ldots, n_k) \supset F(n_1, \ldots, n_k, n_{k+1}).$$

Definition 14. Let $f(x)$ be a real or complex-valued, finite-valued function on the metric space X. Then $f(x)$ is said to be *uniformly continuous* on X if for every $\varepsilon > 0$ there exists an $\eta = \eta(\varepsilon) > 0$ such that $|f(x_2) - f(x_1)| < \varepsilon$ whenever $d(x_1, x_2) < \eta$.

We proceed to list some theorems, relating to the various concepts introduced in this section, in the form of exercises (the proofs may be found in the treatises on Topology listed in the Bibliography, or else the proofs are easy consequences of the definitions involved).

Exercise 1. If $f(x)$ is a real or complex-valued continuous function on the compact metric space X, then $f(x)$ is uniformly continuous on X.

Exercise 2. Let x be a point and E a non-empty compact subset of the metric space X. Then there exists a point $y \in E$ such that $d(x, y) = e(x, E)$ (see definition 7).

Exercise 3. In the metric space X, let E be a non-empty compact set and x a point such that $x \notin E$. Then $e(x, E) > 0$.

Exercise 4. A metric space is fully normal.

Exercise 5. If X is a boundedly compact metric space, then X is complete.

Exercise 6. Let $\mathfrak{U}$ be an open covering of the compact subset F of the metric space X. Then there exists a number $\eta > 0$ such that the following holds: if S is any subset of X such that $\delta S < \eta$ and $S \cap F \neq \emptyset$, then S is contained in some set $U \in \mathfrak{U}$.

Exercise 7. In a complete, separable metric space X the following holds for Borel sets and analytic sets.

(i) Every Borel set is analytic.

(ii) If $\mathfrak{A}$ is a determining system of closed sets, then there exists a determining system $\mathfrak{A}'$ of closed sets $F'(n_1, \ldots, n_k)$ such that $\mathsf{A}(\mathfrak{A}') = \mathsf{A}(\mathfrak{A})$, $\mathfrak{A}'$ is regular, and for every infinite sequence $n_1, n_2, \ldots$ of positive integers one has

$$\delta F'(n_1, \ldots, n_k) \to 0 \quad \text{for} \quad k \to \infty.$$

I.1.5. Continuous transformations. *Definition 1.* Let $f: X \to Y$ be a transformation from a topological space X into a topological space Y. Then f is termed continuous at a point $x_0 \in X$ if for every open set O_Y in Y such that $f x_0 \in O_Y$ there exists an open set O_X in X such that $x_0 \in O_X$ and $f O_X \subset O_Y$. If f is continuous at every point of X, then f is said to be continuous on X.

Definition 2. Let $f: (X, A) \to (Y, B)$ be a transformation, where (X, A) and (Y, B) are pairs of topological spaces. Then f is a transformation from X into Y such that $f A \subset B$ (see **I.1.1**), and the statement that f is continuous is to be interpreted in the sense of definition 1.

Definition 3. A transformation $f: X \to Y$, where X and Y are topological spaces, is termed a *homeomorphism* from X onto Y if it is biunique, and if furthermore f is continuous on X and f^{-1} is continuous on Y. If there exists a homeomorphism from the topological space X onto the topological space Y, then X and Y are said to be *homeomorphic*.

Definition 4. Given two metric spaces X, Y, a transformation $T: X \to Y$, and a non-empty subset E of X, the *oscillation* $\omega(T, E)$ of T on E is defined by the formula

$$\omega(T, E) = \delta T E.$$

Definition 5. Given two metric spaces X, Y, two transformations $T_1: X \to Y$, $T_2: X \to Y$, and a non-empty subset E of X, the deviation of T_2 from T_1 on E is measured by the quantity

$$\varrho(T_1, T_2, E) = \text{l.u.b.}\, d_Y(T_1 x, T_2 x), \quad x \in E,$$

where d_Y denotes distance in Y.

Definition 6. Let A be a subspace of the topological space X and let $f: X \to A$ be a continuous transformation from X into A. If $fx = x$ for $x \in A$, then f is termed a *retraction* from X onto A. If such a retraction exists, then A is termed a *retract* of X.

Definition 7. Let (X, A) and (Y, B) be pairs of topological spaces, and let m_0, m_1 be two continuous transformations from (X, A) into (Y, B). Then a continuous transformation

$$m: (X \times I, \ A \times I) \to (Y, B),$$

where I is the unit interval $0 \leq u \leq 1$ on the real number-line, is termed a *homotopy* connecting m_0 and m_1 provided that

$$m_0 x = m(x, 0), \quad m_1 x = m(x, 1), \quad x \in X.$$

If such a homotopy exists, then the continuous transformations m_0, m_1 are termed *homotopic*.

Definition 8. Let A be a subspace of the topological space X. Then A is termed a *deformation retract* of X if there exists a continuous transformation $f: (X, A) \to (X, A)$ such that (a) f is a retraction from X onto A and (b) f is homotopic to the identity transformation $i: (X, A) \to (X, A)$.

We shall now list, in the form of exercises, a series of theorems relating to the preceding concepts. The proofs may be found in the treatises on Topology listed in the Bibliography or else the proofs are easy consequences of the definitions involved.

Exercise 1. Let $i: X \to X$ be the identity transformation in the topological space X. Then i is continuous on X.

Exercise 2. Let A be a subspace of the topological space X. Then all the transformations occurring in the *first homotopy diagram* (see **I.1.1**) are continuous, and the transformations denoted there by h_s and h_t are homeomorphisms onto.

Exercise 3. Let $T: X \to Y$ be a continuous transformation from the topological space X into the topological space Y. If F is a closed subset of Y, then $T^{-1}F$ is closed in X.

Exercise 4. Given a topological space X and a Hausdorff space Y, let $T: X \to Y$ be a continuous transformation from X into Y. Then for every point $y \in Y$ the set $T^{-1}y$ is closed in X.

Exercise 5. Let $T: X \to Y$ be a continuous transformation from the topological space X into the topological space Y. If $\mathfrak{U}_Y$ is an open covering of Y, then the sets $T^{-1}U_Y$, corresponding to the sets $U_Y \in \mathfrak{U}_Y$, constitute an open covering of X.

Exercise 6. Let X, Y be topological spaces. Consider an open subset O_X^0 of X and a continuous transformation $T: O_X^0 \to Y$. Assume that T is one-to-one and carries open sets in O_X^0 into open sets in Y. Then T is a homeomorphism from O_X^0 onto $T O_X^0$.

Exercise 7. Given two HAUSDORFF spaces X and Y, consider a continuous transformation $T: E \to Y$ from a compact subset E of X into Y. Assume that T is one-to-one. Then T is a homeomorphism from E onto TE.

Exercise 8. Given two topological spaces X and Y, consider a continuous transformation $T: X \to Y$. If S is a compact subset of X, then TS is a compact subset of Y. Similarly, if S is a connected subset of X, then TS is a connected subset of Y.

Exercise 9. Given a continuous transformation $T: X \to Y$ from the HAUSDORFF space X into the HAUSDORFF space Y, let E be a subset of X such that $\overline{E}$ is compact. Then $T\overline{E} = \overline{TE}$.

Exercise 10. Given two topological spaces X and Y, let $h: X \to Y$ be a homeomorphism from X onto Y. If S is any subset of X, then $h\, fr\, S = fr\, h\, S$, $h\overline{S} = \overline{hS}$, $h \complement S = \complement h S$, $h \operatorname{int} S = \operatorname{int} h S$.

Exercise 11. Let Y be a metric space and X a compact metric space. If $T: X \to Y$ is a continuous transformation from X into Y, then T is uniformly continuous on X in the following sense: for every $\varepsilon > 0$ there exists an $\eta = \eta(\varepsilon) > 0$ such that if the distance (in X) of two points x_1, x_2 is less than η, then the distance (in Y) of the image points Tx_1, Tx_2 is less than ε.

Exercise 12. Let T_1, T_2 be two continuous transformations from the metric space X into the metric space Y, and let E be a non-empty compact subset of X. Then there exists a point $x_0 \in E$ such that (see definition 5)

$$\varrho(T_1, T_2, E) = d_Y(T_1 x_0, T_2 x_0),$$

where d_Y stands for distance in Y.

Occasionally there arise situations where one has to consider *discontinuous* transformations in connection with topological spaces. We shall now discuss in some detail an important example.

Modification theorem. Let there be given a fully normal topological space X, an open covering $\mathfrak{U}$ of X, and a non-empty closed subset Y of X. Then there exists an open set $O \supset Y$, a star refinement $\mathfrak{V}$ of $\mathfrak{U}$, and a (generally discontinuous) transformation $f: X \to X$, such that $f\overline{O} \subset Y$ and $fV \subset St_{\mathfrak{V}} V$ for every $V \in \mathfrak{V}$.

Proof. In view of **I.1.3**, definition 18 we can select a star refinement $\mathfrak{V}$ of $\mathfrak{U}$. Denote by O' the union of all those sets $V \in \mathfrak{V}$ which intersect Y. Then clearly $Y \subset O'$ and O' is open. By **I.1.3**, exercises 43 and 25 there follows the existence of an open set O such that

$$Y \subset O, \quad \overline{O} \subset O'. \tag{1}$$

We now define a (generally discontinuous) transformation $f: X \to X$ in the following steps.

$$fx = x \quad \text{if} \quad x \in \complement O'. \tag{2}$$

If $x \in O'$, then by the definition of O' we can select a set $V(x) \in \mathfrak{V}$ such that $x \in V(x)$, $V(x) \cap Y \neq \emptyset$. Accordingly, we can choose a point in $V(x) \cap Y$ as the image fx of x. Then

$$x \in V(x) \in \mathfrak{V}, \quad fx \in V(x) \cap Y \quad \text{for} \quad x \in O'. \tag{3}$$

We assert that $\mathfrak{V}$, O, f satisfy the requirements of the theorem. Indeed, (3) implies that $fx \in Y$ if $x \in O'$, and hence $fO' \subset Y$. In view of (1) it follows that $f\overline{O} \subset Y$. There remains to show that

$$fV \subset St_{\mathfrak{V}} V \quad \text{if} \quad V \in \mathfrak{V}. \tag{4}$$

Select a set $V \in \mathfrak{V}$ and consider a point $x \in V$. If $x \in \complement O'$, then $fx = x \in V \subset St_{\mathfrak{V}} V$. If $x \in O'$, then in view of (3) we have $fx \in V(x) \subset St_{\mathfrak{V}} V$, since $V(x)$ contains the point $x \in V$ and hence $V \cap V(x) \neq \emptyset$. Thus $fx \in St_{\mathfrak{V}} V$ in either case. Since x was an arbitrary point of V, (4) follows and the theorem is proved.

§ I.2. Survey of Euclidean spaces.

I.2.1. Preliminaries. For each positive integer n, Euclidean n-space R^n is the metric space defined as follows. The points of R^n are n-term sequences $(x^1, \ldots, x^n)$ of real numbers. Of course, the superscripts are indices and *not* exponents. If

$$x = (x^1, \ldots, x^n), \quad y = (y^1, \ldots, y^n)$$

are two points of R^n, then their distance $d(x, y)$ is defined by the formula

$$d(x, y) = \left[\sum_{j=1}^{n} (x^j - y^j)^2\right]^{\frac{1}{2}}.$$

Clearly the conditions required in the definition of a metric space are satisfied (see **I.1.4**, definition 1). As a matter of convenience, one also considers a zero-dimensional Euclidean space R^0 which consists of the real number zero alone.

If $x = (x^1, \ldots, x^n)$ is a point of R^n, $n \geq 1$, then $x^1, \ldots, x^n$ are termed the coordinates of x. It is convenient to use vector notation in R^n to condense formulas. A vector v in R^n has n components $v^1, \ldots, v^n$, and one writes $v = (v^1, \ldots, v^n)$ if it is desired to display the components of v. Thus points of R^n and vectors in R^n are identified by assigning an n-term sequence of real numbers. In the sequel, the terms point in R^n and vector in R^n will be used interchangeably. The reader is assumed to be familiar with the fundamental properties of vectors, and we merely state here certain agreements concerning terminology. If

$$v_1 = (v_1^1, \ldots, v_1^n), \quad v_2 = (v_2^1, \ldots, v_2^n)$$

are two vectors in R^n, then their sum and difference are given by the formulas

$$v_1 \pm v_2 = (v_1^1 \pm v_2^1, \ldots, v_1^n \pm v_2^n).$$

The scalar product $v_1 \cdot v_2$ is given by the formula

$$v_1 \cdot v_2 = \sum_{j=1}^{n} v_1^j v_2^j.$$

If $v_1 = v_2 = v$, then one writes v^2 instead of $v \cdot v$. Thus the length $\|v\|$ of a vector v is given by the formula $\|v\| = (v^2)^{\frac{1}{2}}$.

If $v = (v^1, \ldots, v^n)$ is a vector and c is a real number, then cv is the vector given by the formula $cv = (cv^1, \ldots, cv^n)$.

If $x = (x^1, \ldots, x^n)$, $y = (y^1, \ldots, y^n)$ are two points in R^n, then in vector notation one has $d(x, y) = \|x - y\|$. Thus, in particular, $\|x\|$ is the distance of the point x from the origin $(0, \ldots, 0)$.

We shall now list, in the form of exercises, various elementary theorems about R^n. The proofs may be found, for instance, in the ALEXANDROFF-HOPF treatise (see the *Bibliography*), or else the proofs are easy consequences of the definitions involved.

Exercise 1. Since R^n is a metric space, for each point $x \in R^n$ and each real number $r > 0$ one has the open spherical neighborhood $\Delta_r(x)$, in the sense of **I.1.4**, definition 2. In the present case, $\Delta_r(x)$ and its closure $\bar{\Delta}_r(x)$ are given by the formulas

$$\Delta_r(x) = \{x' \mid x' \in R^n, \quad \|x - x'\| < r\},$$
$$\bar{\Delta}_r(x) = \{x' \mid x' \in R^n, \quad \|x - x'\| \leq r\}.$$

Furthermore, the following holds. (i) $\Delta_r(x)$, $\bar{\Delta}_r(x)$ and (*if* $n \geq 2$) $fr\,\Delta_r(x)$ are connected. (ii) $\Delta_r(x) = \operatorname{int} \bar{\Delta}_r(x)$. (iii) $x \in \bar{\Delta}_r(x) - fr\bar{\Delta}_r(x)$. (iv) $\bar{\Delta}_r(x)$ and $fr\bar{\Delta}_r(x)$ are compact. (v) $fr\,\Delta_r(x) = fr\,\bar{\Delta}_r(x)$.

Exercise 2. In R^n, $n \geq 1$, take a point x and a real number r such that $0 < r < 1$. Put

$$A_r(x) = \left\{x' \mid x' \in R^n, \quad \|x - x'\| \leq \frac{1}{r}\right\},$$

$$B_r(x) = \left\{x' \mid x' \in R^n, \quad r \leq \|x - x'\| \leq \frac{1}{r}\right\}.$$

Then the following holds.

(i) $A_r(x) \supset B_r(x)$, $A_r(x) - B_r(x) = \Delta_r(x)$.
(ii) $A_r(x)$ and $B_r(x)$ are compact.
(iii) $A_r(x) - B_r(x)$ is a domain (see **I.1.3**, definition 13).

Exercise 3. If x_0 is a point of R^n, E is a subset of R^n, and $\Delta_r(x_0) \cap E \neq \emptyset$ for every $r > 0$, then $x_0 \in \bar{E}$.

Exercise 4. R^n is separable, connected, locally connected, and boundedly compact (see **I.1.3**, definitions 21, 11, 19, and **I.1.4**, definition 6).

Exercise 5. A subset E of R^n is termed *isolated* if for every point $x \in E$ there exists an open set $O(x)$ containing x such that $E \cap O(x) = x$. An isolated set in R^n is countable.

Exercise 6. Let F be a non-empty compact subset of R^n, $n \geq 1$. Then $\complement F$ has precisely one unbounded component if $n > 1$, and precisely two unbounded components if $n = 1$.

Exercise 7. If E is a non-empty compact subset of R^n and T is a continuous transformation from E into R^n, then T is *bounded* in the sense that there exists a (finite) positive constant M such that $\| Tx \| < M$ for $x \in E$.

Exercise 8. In R^1, a continuum (see **I.1.3**, definition 15) is a set consisting of all those real numbers x which satisfy the condition $a \leq x \leq b$, where a, b are any two real numbers such that $a \leq b$. The most general bounded domain in R^1 is a set consisting of all those real numbers x which satisfy the condition $a < x < b$, where a, b are any two real numbers such that $a < b$. The frontier of the domain is then the set consisting of a and b.

Exercise 9. Let there be given in R^n a finite system of vectors $v_j = (v_j^1, \ldots, v_j^n)$, $j = 1, \ldots, m$. Then these vectors are said to be linearly dependent if there exist real numbers $c_1, \ldots, c_m$ such that at least one of $c_1, \ldots, c_m$ is different from zero and $c_1 v_1 + \cdots + c_m v_m = 0$. Otherwise $v_1, \ldots, v_m$ are said to be linearly independent. The following holds. (i) A system of n vectors $v_j = (v_j^1, \ldots, v_j^n)$, $j = 1, \ldots, n$, in R^n is linearly independent if and only if the determinant of the matrix (v_j^i) is different from zero. (ii) If the vectors $v_1, \ldots, v_n$ in R^n are linearly independent, then every vector v in R^n can be represented in the form $v = c_1 v_1 + \cdots + c_n v_n$, where $c_1, \ldots, c_n$ are uniquely determined real numbers.

Exercise 10. Let there be given n vectors $v_j = (v_j^1, \ldots, v_j^n)$, $j = 1, \ldots, n$, in R^n, such that $\| v_j \| = 1$, $j = 1, \ldots, n$, and $v_j \cdot v_k = 0$ for $j \neq k$. Then the determinant of the matrix (v_j^i) has the value ± 1.

Exercise 11. Let b_n be a vector in R^n, $n > 1$, such that $\| b_n \| = 1$. Then one can select vectors $b_1, \ldots, b_{n-1}$ in R^n such that $\| b_i \| = 1$, $i = 1, \ldots, n$, and $b_i \cdot b_j = 0$ if $i \neq j$.

Exercise 12. If v_1, v_2 are vectors in R^n, then $| v_1 \cdot v_2 | \leq \| v_1 \| \, \| v_2 \|$.

I.2.2. Elementary figures in R^n. The n-cell E^n in R^n, $n \geq 1$, is defined by the formula

$$E^n = \{x \mid x \in R^n, \ \|x\| \leq 1\}.$$

The 0-cell E^0 in R^0 is defined as $E^0 = R^0$ (thus E^0 consists of a single point).

The n-sphere S^n in R^{n+1}, $n \geq 0$, is defined by the formula

$$S^n = \{x \mid x \in R^{n+1}, \ \|x\| = 1\}.$$

Thus S^0 consists of precisely two points (namely, $+1$ and -1).

For $n \geq 1$, we have in R^{n+1} the following configuration (which will be referred to as the *basic diagram* in R^n). Starting with the $(n+1)$-cell E^{n+1} in R^{n+1}, we consider the following sets as constituting the basic diagram (in describing these sets, $x = (x^1, \ldots, x^{n+1})$ denotes a generic point of R^{n+1}).

$$E^{n+1} = \{x \mid \|x\| \leq 1\}, \quad S^n = \{x \mid \|x\| = 1\},$$
$$E_+^n = \{x \mid \|x\| = 1, \ x^{n+1} \geq 0\}, \quad E_-^n = \{x \mid \|x\| = 1, \ x^{n+1} \leq 0\},$$
$$\hat{S}^{n-1} = \{x \mid \|x\| = 1, \ x^{n+1} = 0\}.$$

If $a_j, b_j, j=1, \ldots, n$, are real numbers such that $a_j < b_j, j=1, \ldots, n$, then the set of those points $x=(x^1, \ldots, x^n)$ in R^n which satisfy the conditions $a_j \leq x^j \leq b_j$, $j=1, \ldots, n$, is called an n-interval in R^n, or briefly an interval in R^n. If $b_j - a_j = h > 0$ for $j=1, \ldots, n$, then the n-interval is called an oriented n-cube of side-length h. Thus the points $x=(x^1, \ldots, n^n)$ of an oriented n-cube $Q \subset R^n$ may be characterized by means of inequalities of the form

$$a_j \leq x^j \leq a_j + h, \quad j = 1, \ldots, n,$$

where $h > 0$ is the side-length of Q. For fixed j, the set of those points of Q for which $x^j = a_j$ is termed an $(n-1)$-dimensional face of Q, and the set of those points of Q for which $x^j = a_j + h$ is also termed an $(n-1)$-dimensional face of Q. Thus the number of $(n-1)$-dimensional faces of Q is equal to $2n$. The point

$$\left(a_1 + \frac{h}{2}, \ldots, a_n + \frac{h}{2}\right)$$

is called the center of Q, and the point $(a_1, \ldots, a_n)$ is called the initial vertex of Q.

In R^n a segment with end-points x_1, x_2 is defined as the set of those points x which are (in vector notation) of the form

$$x = t x_2 + (1 - t) x_1, \quad 0 \leq t \leq 1.$$

A non-empty subset E of R^n is termed convex if $x_1 \in E$, $x_2 \in E$ imply that the segment with end-points x_1, x_2 is contained in E.

Let there be given in R^n a point $x_0 = (x_0^1, \ldots, x_0^n)$ and n linearly independent vectors $v_i = (v_i^1, \ldots, v_i^n)$, $i=1, \ldots, n$. Then the *parallelotope* P with *initial vertex* x_0 and *edge-vectors* $v_1, \ldots, v_n$ is the set of those points $x = (x^1, \ldots, x^n)$ which can be represented in the form

$$x^j = x_0^j + \sum_{i=1}^{n} \mu_i v_i^j, \quad j = 1, \ldots, n,$$

where $\mu_1, \ldots, \mu_n$ are real numbers such that $0 \leq \mu_i \leq 1$, $i=1, \ldots, n$. In vector notation, the points x of P may be represented in the form

$$x = x_0 + \sum_{i=1}^{n} \mu_i v_i, \quad 0 \leq \mu_i \leq 1, \ i = 1, \ldots, n.$$

The point

$$x_0 + \tfrac{1}{2} \sum_{i=1}^{n} v_i$$

is termed the center of the parallelotope.

Consider in R^{n+1}, $n \geq 1$, the n-sphere S^n. Denote by $\hat{R}^n$ the set of those points $x = (x^1, \ldots, x^{n+1})$ of R^{n+1} for which $x^{n+1} = 0$ [thus a generic point x of $\hat{R}^n$ is of the form $(x^1, \ldots, x^n, 0)$]. Denote by N

the point $(0, \ldots, 0, 1)$ (the *north pole* of S^n). If $\hat{x} = (x^1, \ldots, x^n, 0)$ is a point of $\hat{R}^n$, then the line through $\hat{x}$ and N is easily seen to intersect S^n in precisely one point $\tilde{x}$ distinct from N. The transformation $f: \hat{R}^n \to S^n - N$ defined by $f\hat{x} = \tilde{x}$ is easily seen to be a homeomorphism from $\hat{R}^n$ onto $S^n - N$. The inverse f^{-1} of f, which is a homeomorphism from $S^n - N$ onto $\hat{R}^n$, is termed the *stereographic projection* from $S^n - N$ onto $\hat{R}^n$. On denoting by h the transformation from $\hat{R}^n$ onto R^n defined by

$$h(x^1, \ldots, x^n, 0) = (x^1, \ldots, x^n),$$

clearly hf^{-1} is a homeomorphism from $S^n - N$ onto R^n. This homeomorsphism hf^{-1} is frequently referred to as the stereographic projection from $S^n - N$ onto R^n.

We shall list presently a series of elementary theorems in the form of exercises. The proofs may be found in the ALEXANDROFF-HOPF treatise (see the *Bibliography*), or else the proofs follow readily in view of the definitions involved.

Exercise 1. The following holds for the *basic diagram* in R^{n+1} (see above).

(i) All the sets occurring in the diagram are compact.

(ii) $\hat{S}^{n-1}$ is homeomorphic to the $(n-1)$-sphere S^{n-1}.

(iii) E_+^n, E_-^n are homeomorphic to the n-cell E^n.

(iv) $S^n = E_+^n \cup E_-^n$, $\hat{S}^{n-1} = E_+^n \cap E_-^n$.

Exercise 2. Let Q be an oriented n-cube in R^n, $n \geq 1$, of side-length $h > 0$. Let $x_0 = (x_0^1, \ldots, x_0^n)$ be the initial vertex of Q. Then Q consists of those points $x = (x^1, \ldots, x^n)$ which can be represented in the form

$$x^j = x_0^j + \mu_j h, \quad j = 1, \ldots, n,$$

where $\mu_1, \ldots, \mu_n$ are real numbers such that $0 \leq \mu_j \leq 1$, $j = 1, \ldots, n$.

Exercise 3. Let I be an n-interval in R^n. Then I is convex, compact, and connected, and int I is a convex domain.

Exercise 4. In R^n, $n \geq 1$, let P be the parallelotope with center y and edge-vectors $v_1, \ldots, v_n$. The P consists of those points $x \in R^n$ which can be represented in the form

$$x = y + \sum_{i=1}^{n} \lambda_i v_i,$$

where $\lambda_1, \ldots, \lambda_n$ are real numbers such that

$$-\tfrac{1}{2} \leq \lambda_i \leq \tfrac{1}{2}, \quad i = 1, \ldots, n.$$

Exercise 5. The n-cell E^n is convex, compact, connected, and int E^n is a convex domain.

Exercise 6. The n-sphere S^n, $n \geq 1$, is a non-degenerate continuum. Furthermore, S^n is locally connected and separable.

Exercise 7. In R^n, $n \geq 1$, the following holds for the open spherical neighborhoods $\Delta_r(x)$ (see **I.1.4**, definition *2*). (i) $\Delta_r(x)$ is a convex domain. (ii) $\overline{\Delta}_r(x)$ is a convex continuum.

Exercise 8. If O is an open set in R^n, $n \geq 1$, then there exists a sequence $I_1, I_2, \ldots$ of n-intervals such that $I_j \subset O$, $j = 1, 2, \ldots$, and $O = \bigcup \operatorname{int} I_j$.

Exercise 9. In R^n, $n \geq 1$, let F be a compact subset of an open set O. Then there exists a finite system of oriented n-cubes $Q_1, \ldots, Q_m$ such that

$$F \subset \bigcup_{k=1}^{m} Q_k \subset O.$$

Exercise 10. For $n \geq 1$, the stereographic projection (see above) can be used to transfer configurations from R^n to S^n, and *vice versa.* The following statements are useful in this respect.

(i) If $n \geq 2$, and $F \neq \emptyset$ is a compact subset of S^n such that $S^n - F$ is non-empty and connected, then F is homeomorphic to a non-empty compact subset F^* of R^n such that $R^n - F^*$ is connected.

(ii) In R^n, $n \geq 1$, let Y be a non-empty compact set. Then there exists a non-empty compact subset Y^* of S^n such that (a) Y^* is homeomorphic to Y, and (b) on denoting by q the number of components of $S^n - Y^*$ and by k the number of *bounded* components of $R^n - Y$, one has $q = k + 1$.

Exercise 11. In R^n, $n \geq 2$, denote by Q^n the *unit cube* [the set of those points $x = (x^1, \ldots, x^n)$ for which $0 \leq x^i \leq 1$, $i = 1, \ldots, n$]. If F is a non-empty compact set in R^n such that $R^n - F$ is connected, then there exists a non-empty compact subset F^* of $\operatorname{int} Q^n$ such that (a) F^* is homeomorphic to F, and (b) $Q^n - F^*$ is connected.

Exercise 12. Let Q be an oriented n-cube in R^n, $n \geq 1$. Then Q is homeomorphic to the n-cell E^n, and $fr\, Q$ is homeomorphic to the $(n-1)$-sphere S^{n-1}.

Exercise 13. In R^n, $n \geq 1$, the origin $(0, \ldots, 0)$ is a deformation retract of the n-cell E^n.

Exercise 14. Let F be a non-empty compact subset of the one-sphere S^1, such that $S^1 - F$ is non-empty and connected. Then either F consists of a single point, or else F is homeomorphic to the one-cell E^1.

I.2.3. Subdivisions in R^n.

Definition 1. Let n and m be positive integers. Then the subdivision $\Delta(n, m)$ of R^n consists of those oriented n-cubes in R^n which are determined by inequalities of the form

$$\frac{k_j}{2^m} \leq x^j \leq \frac{k_j + 1}{2^m}, \quad j = 1, \ldots, n,$$

where $k_1, \ldots, k_n$ are arbitrary integers.

Definition 2. Let $Q^n: 0 \leq x^j \leq 1$, $j = 1, \ldots, n$, be the unit n-cube in R^n, $n \leq 1$, and let m be a positive integer. Then the subdivision $\Delta^*(n, m)$ of Q^n consists of those cubes of the subdivision $\Delta(n, m)$ of R^n which are contained in Q^n.

Definition 3. Let $p_1, \ldots, p_j, \ldots$ be the sequence of positive primes (thus $p_1 = 2$, $p_2 = 3, \ldots$), and let n be a positive integer. Then the subdivision D_{p_j} of R^n consists of those oriented n-cubes in R^n which are determined by inequalities of the form

$$\frac{k_i}{p_j} \leq x^i \leq \frac{k_i + 1}{p_j}, \quad i = 1, \ldots, n,$$

where $k_1, \ldots, k_n$ are arbitrary integers. Given an oriented n-cube

$$Q: a_i \leq x^i \leq a_i + h, \quad i = 1, \ldots, n, \quad h > 0,$$

in R^n, the subdivision D_{p_j} of Q consists of all those oriented n-cubes which are determined by inequalities of the form

$$a_i + \frac{k_i}{p_j} h \leq x^i \leq a_i + \frac{k_i + 1}{p_j} h, \quad i = 1, \ldots, n,$$

where $k_1, \ldots, k_n$ are integers such that $0 \leq k_i \leq p_j - 1$, $i = 1, \ldots, n$.

Definition 4. If K is a collection of cubes of the subdivision $\Delta^*(n, m)$ (see definition 2), then $|K|$ denotes the union of the cubes $q \in K$.

Definition 5. Two cubes q', q'' of the subdivision $\Delta^*(n, m)$ (see definition 2) are *strongly adjacent* if $q' \neq q''$ and $q' \cap q''$ is a common $(n-1)$-dimensional face of q' and q''.

Definition 6. A finite sequence $q_1, \ldots, q_s$ of $s \geq 2$ cubes of the subdivision $\Delta^*(n, m)$ is termed a *chain* if q_j and q_{j+1} are strongly adjacent, $j = 1, \ldots, s-1$ (see definitions 2 and 5). Thus $q_j \neq q_{j+1}$, but one may have $q_j = q_k$ if $|j - k| \geq 2$.

Definition 7. A collection K of cubes of the subdivision $\Delta^*(n, m)$ is *strongly connected* if (a) K is empty, or (b) K consists of a single cube, or (c) for every pair q', q'' of distinct cubes of K there exists a chain $q_1, \ldots, q_s$ in K such that $q_1 = q'$, $q_s = q''$ (see definitions 2 and 6).

We shall state presently, in the form of exercises, a series of elementary theorems relating to the preceding concepts. The proofs may be found in the Alexandroff-Hopf treatise (see the *Bibliography*), or else the proofs are easily deduced from the definitions involved.

Exercise 1. Let x_0 be a point of R^n, $n \geq 1$, such that no coordinate of x_0 is equal to zero. Then there exists an integer j_0 such that for $j > j_0$ the point x_0 is contained in the interior of a (unique) cube of the subdivision D_{p_j} of R^n.

Exercise 2. Given an oriented n-cube $Q \subset R^n$, $n \geq 1$, and a point $x_0 \in \operatorname{int} Q$, there exists an integer j_0 such that for $j > j_0$ the point x_0 is contained in the interior of a (unique) cube of the subdivision D_{p_j} of Q.

Exercise 3. Let Γ be a finite collection of $s \geq 2$ distinct cubes of the subdivision $\Delta(n, m)$ of R^n, $n \geq 1$. Then the cubes of Γ can be arranged into a sequence $q_1, \ldots, q_s$ in such a manner that on setting

$$Y = q_s \cap \left(\bigcup_{j=1}^{s-1} q_j \right),$$

Y is a proper subset of $fr\, q_s$.

Exercise 4. Let F_1, F_2 be disjoint compact subsets of the unit cube Q^n in R^n, $n \geq 1$. Then there exists an integer m_0 such that for $m > m_0$ no cube of the subdivision $\Delta^*(n, m)$ of Q^n intersects both F_1 and F_2.

Exercise 5. Let O and F be subsets of R^n, $n \geq 1$, such that O is open, F is compact, and $F \subset O$. Denote by F_m the union of all those cubes of the subdivision $\Delta(n, m)$ of R^n which intersect F. Then there exists an integer m_0 such that $F_m \subset O$ for $m > m_0$.

Exercise 6. Consider the subdivision $\Delta^*(n, m)$ of the unit cube Q^n in R^n. Let q', q'' be a pair of strongly adjacent cubes of $\Delta^*(n, m)$. Then there exists a point x_0 such that the following holds.

(a) $x_0 \in fr\, q' \cap fr\, q''$.

(b) If q is a cube of $\Delta^*(n, m)$ such that $x_0 \in q$, then either $q = q'$ or $q = q''$.

Exercise 7. Let Γ be a continuum in the unit cube Q^n of R^n, $n \geq 1$. Denote by K the collection of all those cubes of the subdivision $\Delta^*(n, m)$ of Q^n which intersect Γ. Then K is strongly connected.

Exercise 8. Let K_s be a strongly connected collection of s cubes of $\Delta^*(n, m)$, where $s \geq 2$, and let q^* be a cube of K_s. Then the cubes of K_s can be arranged into a sequence $q_1, \ldots, q_s$ in such a manner that the following holds. (i) q_{s-1} and q_s are strongly adjacent. (ii) The collection K_{s-1} consisting of the cubes $q_1, \ldots, q_{s-1}$ is stiongly connected, and $q^* \in K_{s-1}$.

Exercise 9. Let D be a bounded domain in R^n, $n \geq 1$. Then there exists in R^n a sequence of domains $\{D_j\}$ such that the following holds.

(i) D_j is the interior of the union of a finite number of cubes of the subdivision $\Delta(n, m_j)$ of R^n, where $m_1 < m_2 < \cdots$.

(ii) $fr\, D_j$ is the union of a finite number of $(n-1)$-dimensional faces of cubes of $\Delta(n, m_j)$.

(iii) $\overline{D}_j \subset D_{j+1}$, $j = 1, 2, \ldots$.

(iv) $D = \bigcup D_j$, $j = 1, 2, \ldots$.

§ I.3. Survey of Abelian groups.

I.3.1. Abelian groups, factor groups, direct sums. In the present **§ 1.3.** we collect, for convenient reference, the definitions and facts concerning Abelian groups which will be needed later on. Proofs and further details may be found in the ALEXANDROFF-HOPF treatise listed in the *Bibliography*.

Abelian groups will be written additively (that is, the group operation will be denoted by the symbol $+$). An Abelian group G contains at least one element, namely 0. If G consists of a zero-element alone, then G is termed *trivial*, and one writes $G = 0$. The following special groups will frequently occur in the sequel.

The additive group of integers will play an important role, and will be denoted by I. For each positive integer n, an Abelian group I^n is defined as follows. The elements of I^n are n-term sequences $(k_1, \ldots, k_n)$ of integers. Addition in I^n is defined by the formula

$$(j_1, \ldots, j_n) + (k_1, \ldots, k_n) = (j_1 + k_1, \ldots, j_n + k_n).$$

In particular, one has $\mathsf{I}^1 = \mathsf{I}$. As a matter of convenience, one sets $\mathsf{I}^0 = 0$. A further important group I^ω (where ω denotes the cardinal of the set of positive integers) is defined as follows. The elements of I^ω are those infinite sequences $\{k_j\}$ of integers which contain at most a finite number of non-zero terms. Addition in I^ω is defined by the formula

$$\{k_j\} + \{i_j\} = \{k_j + i_j\}.$$

An Abelian group G is termed infinite cyclic if it contains an element g_0 such that (a) every element of G can be written in the form ng_0, where n is some integer, and (b) one has $ng_0 = 0$ if and only if the integer n is equal to zero. An element g_0 with these properties is termed a generator of the infinite cyclic group G. If g_0 is a generator of the infinite cyclic group G, then $-g_0$ is also a generator, and G possesses no further generators. The additive group I of integers is the prototype of infinite cyclic groups, its generators being $+1$ and -1.

Let K be a subgroup of the Abelian group G. If g is an element of G, then the set of all those elements g' of G for which $g' - g \in K$ is called a coset relative to K, and is denoted by $[g, K]$, or merely by $[g]$ if the subgroup K is thought of as fixed. For given K, the cosets relative to K constitute a group, the factor group G/K, in which addition is defined by the formula

$$[g_1] + [g_2] = [g_1 + g_2].$$

It is easy to see that the sum so defined is independent of the choice of g_1, g_2 in the cosets $[g_1]$, $[g_2]$. The factor group G/K is again an Abelian group.

Let $G_1, \ldots, G_m$ be a finite system of subgroups of the Abelian group G, such that each element $g \in G$ can be represented in one and only one way in the form

$$g = g_1 + \cdots + g_m, \quad g_1 \in G_1, \ldots, g_m \in G_m.$$

Then G is said to be the *direct sum* of its subgroups $G_1, \ldots, G_m$, and one writes

$$G = G_1 \dot{+} \cdots \dot{+} G_m.$$

Let $\{G_j\}$ be an infinite sequence of subgroups of the Abelian group G, such that the following holds: for every element $g \neq 0$ of G there exists a unique (finite) system of integers $k_1, \ldots, k_m$ such that $0 < k_1 < \cdots < k_m$ and

$$g = g_{k_1} + \cdots + g_{k_m}, \quad g_{k_i} \in G_{k_i}, \quad g_{k_i} \neq 0, \quad i = 1, \ldots, m.$$

Then G is said to be the *weak direct sum* of the subgroups $G_1, G_2, \ldots$.

We shall now state, in the form of exercises, a few facts needed in the sequel. The proofs may either be found in the ALEXANDROFF-HOPF treatise (see the *Bibliography*) or can be readily deduced from the definitions involved.

Exercise 1. Let $[g]$ be an element of the factor group G/K. Then $[g] = 0$ if and only if $g \in K$.

Exercise 2. If 0 denotes the subgroup of G consisting of the zero-element of G alone, then $G/0 = G$.

Exercise 3. $G/K = 0$ if and only if $K = G$.

Exercise 4. If g_0 is a generator of the infinite cyclic group G and n_1, n_2 are integers, then $n_1 g_0 = n_2 g_0$ if and only if $n_1 = n_2$.

Exercise 5. Let $n \geqq 2$ be an integer. For each integer $j = 1, \ldots, n$, denote by I_j^n the subgroup of I^n consisting of those elements $(k_1, \ldots, k_n)$ of I^n for which $k_i = 0$ for $i \neq j$. Then I_j^n is infinite cyclic, and I^n is the direct sum of $\mathsf{I}_1^n, \ldots, \mathsf{I}_n^n$.

Exercise 6. For each positive integer i, denote by I_i^ω the subgroup of those elements $\{k_j\}$ of I^ω for which $k_j = 0$ if $j \neq i$. Then I_i^ω is infinite cyclic, and I^ω is the weak direct sum of $\mathsf{I}_1^\omega, \mathsf{I}_2^\omega, \ldots$.

I.3.2. Homomorphisms, isomorphisms. Let $h: G_1 \to G_2$ be a mapping from the Abelian group G_1 into the Abelian group G_2. If

$$h(g_1' + g_1'') = h g_1' + h g_1''$$

for every pair of elements g_1', g_1'' of G_1, then h is termed a homomorphism from G_1 into G_2. If $hG_1 = G_2$, then h is said to be a homomorphism from G_1 *onto* G_2. The set of those elements $g_1 \in G_1$ for which $hg_1 = 0$ is termed the *nucleus* of h. If $hG_1 = 0$ (that is, if every element g_1 of G_1 is carried by h into the zero-element of G_2), then h is called a *zero-homomorphism.* If either $G_1 = 0$ or $G_2 = 0$, then clearly there exists only one homomorphism $h: G_1 \to G_2$, and this unique h is a zero-homomorphism. In these cases, one refers to this unique h as *the trivial zero-homomorphism.* If G is any Abelian group, then the mapping $h: G \to G$ defined by $hg = g$ for $g \in G$ is clearly a homomorphism, termed the identity homomorphism in G. One writes $h = 1$ to state that h is the identity homomorphism.

If the nucleus of the homomorphism $h: G_1 \to G_2$ reduces to the zero-element of G_1, then clearly h is one-to-one. If this is the case, then h is called an isomorphism from G_1 into G_2, and if furthermore $hG_1 = G_2$, then h is called an isomorphism from G_1 *onto* G_2, and one writes $h: G_1 \approx G_2$. Two Abelian groups G_1, G_2 are termed isomorphic if there exists an isomorphism from G_1 onto G_2.

Let G be an infinite cyclic group and $h: G \approx G$ an isomorphism from G onto G. If g_0 is a generator of G, then it is easy to see that either $hg_0 = g_0$ or $hg_0 = -g_0$. In the first case, h is termed *even*, and in the second case h is termed *odd.* It is immediate that these properties of h do not depend upon the choice of the generator g_0.

Let $h: G_1 \to G_2$ be a homomorphism from the Abelian group G_1 into the Abelian group G_2, and let K_1, K_2 be subgroups of G_1, G_2 respectively such that $hK_1 \subset K_2$. For each element g_1 of G_1 denote by $[g_1]_1$ the coset in G_1, relative to K_1, which contains g_1, and let $[g_2]_2$ have an analogous meaning relative to G_2, K_2. Consider an element $[g_1]_1$ of the factor group G_1/K_1. Then $[hg_1]_2$ is an element of the factor group G_2/K_2, and it is easy to see that this element depends only upon the coset $[g_1]_1$ and is independent of the particular element g_1 selected in that coset. Accordingly, we can define a mapping

$$h_*: G_1/K_1 \to G_2/K_2$$

by the formula $h_*[g_1]_1 = [hg_1]_2$, and it is immediate that h_* is a homomorphism. The homomorphism h_* so obtained is said to be induced by the homomorphism $h: G_1 \to G_2$ (which is assumed to satisfy the condition $hK_1 < K_2$).

The proofs of the statements included in the following exercises are either contained in the ALEXANDROFF-HOPF treatise (see the *Bibliography*) or can be readily deduced from the definitions involved.

Exercise 1. If $h: G_1 \to G_2$ is a homomorphism from the Abelian group G_1 into the Abelian group G_2, then the nucleus of h is a subgroup of G_1, and hG_1 is a subgroup of G_2.

Exercise 2. Given three Abelian groups G_1, G_2, G_3, let $h_1: G_1 \to G_2$, $h_2: G_2 \to G_3$ be homomorphisms such that h_2 is an isomorphism into. Then the homomorphism $h = h_2 h_1$ has the same nucleus as h_1.

Exercise 3. Given three Abelian groups G_1, G_2, G_3, let $h_1: G_1 \to G_2$, $h_2: G_2 \to G_3$ be homomorphisms such that $h_2 h_1$ is *onto*. Then h_2 is also onto.

Exercise 4. Consider a homomorphism $h: \mathsf{I} \to \mathsf{I}$, where I is the additive group of integers. If h is onto, then it is an isomorphism onto.

Exercise 5. Given two Abelian groups G_1, G_2, let $h_1: G_1 \to G_2$, $h_2: G_2 \to G_1$ be homomorphisms such that $h_2 h_1$ is the identity homomorphism in G_1. Then one has the relation $G_2 = I_1 \dotplus N_2$ (see **I.3.1**), where N_2 is the nucleus of h_2 and $I_1 = h_1 G_1$.

Exercise 6. Consider two groups I^r, I^s, where r is equal to some non-negative integer or to ω, and similarly s is equal to some non-negative integer or to ω (see **I.3.1**). Then I^r and I^s are isomorphic if and only if $r = s$.

Exercise 7. Let G be an Abelian group which is the direct sum of its subgroups $G_1, \ldots, G_m$. If these subgroups are infinite cyclic, then G is isomorphic to I^m (see **I.3.1**).

Exercise 8. If $\{G_j\}$ is an infinite sequence of infinite cyclic subgroups of an Abelian group G, and G is the weak direct sum of these subgroups, then G is isomorphic to I^ω (see **I.3.1**).

I.3.3. Exactness[1]. Let G, H, K be Abelian groups, and let $g: G \to H$, $h: H \to K$ be homomorphisms. We shall use the diagram

$$G \xrightarrow{g} H \xrightarrow{h} K$$

to refer to this sort of situation, and we shall say that we are given a *three-term sequence* of groups and connecting homomorphisms. The three-term sequence is said to be *exact* if the nucleus of h coincides with gG.

More generally, a diagram of the form

$$\cdots \xrightarrow{h_{j-2}} G_{j-1} \xrightarrow{h_{j-1}} G_j \xrightarrow{h_j} G_{j+1} \xrightarrow{h_{j+1}} \cdots$$

[1] For historical comments and further details, see the EILENBERG-STEENROD treatise listed in the *Bibliography*.

will be used to refer to a situation where we are given a (finite or infinite) sequence of Abelian groups and connecting homomorphisms. Such a sequence is termed *exact* if every three-term sequence

$$G_{j-1} \xrightarrow{h_{j-1}} G_j \xrightarrow{h_j} G_{j+1}$$

contained in it is exact. The EILENBERG-STEENROD treatise (listed in the *Bibliography*) contains a comprehensive study of exact sequences. We shall consider here merely a few elementary facts relating to such sequences.

Lemma 1. If the three-term sequence

$$G_1 \xrightarrow{h_1} G_2 \xrightarrow{h_2} G_3$$

is exact and $G_3 = 0$, then h_1 is onto.

Indeed, the assumption $G_3 = 0$ implies that the nucleus of h_2 coincides with G_2. Since the sequence is exact, it follows that $G_2 = h_1 G_1$, and thus h_1 is onto. The proofs of the following lemmas are equally obvious.

Lemma 2. If the three-term sequence

$$G_1 \xrightarrow{h_1} G_2 \xrightarrow{h_2} G_3$$

is exact and $G_1 = 0$, then h_2 is an isomorphism into.

Lemma 3. If the three-term sequence

$$G_1 \xrightarrow{h_1} G_2 \xrightarrow{h_2} G_3$$

is exact and $G_1 = 0$, $G_3 = 0$, then $G_2 = 0$.

Lemma 4. If the four-term sequence

$$G_1 \xrightarrow{h_1} G_2 \xrightarrow{h_2} G_3 \xrightarrow{h_3} G_4$$

is exact and $G_1 = 0$, $G_4 = 0$, then h_2 is an isomorphism onto.

§ I.4. MAYER complexes[1].

I.4.1. Basic definitions. A MAYER complex

$$M = M(C^p, \delta^p) \tag{1}$$

is a mathematical system consisting of the following objects. (i) For every integer p (positive, negative, or zero) there is given an Abelian

[1] For historical comments and further details concerning this fundamental concept, see the EILENBERG-STEENROD treatise listed in the *Bibliography*. We restrict ourselves to a detailed discussion (designed for a reader unfamiliar with this type of algebraic approach) of those topics which are actually needed in the sequel.

group C^p. (ii) For every integer p, there is given a homomorphism

$$\delta^p : C^p \to C^{p+1} \tag{2}$$

such that

$$\delta^{p+1} \delta^p = 0. \tag{3}$$

Explicitly, (3) means that if c^p is any element of C^p, then $\delta^{p+1} \delta^p c^p$ is the zero-element of C^{p+2}.

Generically, an element of C^p will be termed a p-cochain of the Mayer complex M, and C^p will be termed the group of p-cochains of M. If $c^p \in C^p$, then the $(p+1)$-cochain $\delta^p c^p$ is termed the coboundary of c^p. A p-cochain whose coboundary is zero is termed a p-cocycle. Thus the p-cocycles constitute precisely the nucleus (see **I.3.2**) of the homomorphism δ^p, and hence the p-cocycles form a subgroup of C^p which will be denoted by Z^p. A p-cochain is termed a p-coboundary if it is the coboundary of some $(p-1)$-cochain. Thus the p-coboundaries constitute precisely the image of C^{p-1} in C^p under the homomorphism δ^{p-1}, and hence (see **I.3.2**, exercise 1) the p-coboundaries form a subgroup of C^p which will be denoted by B^p. Thus $c^p \in B^p$ if and only if there exists some $c^{p-1} \in C^{p-1}$ such that $c^p = \delta^{p-1} c^{p-1}$. By (3) it follows that $\delta^p c^p = \delta^p (\delta^{p-1} c^{p-1}) = 0$. Hence every p-coboundary is a p-cocycle. In summary:

$$B^p \subset Z^p \subset C^p. \tag{4}$$

The factor group

$$H^p = Z^p / B^p \tag{5}$$

is termed the p-th cohomology group of the Mayer complex M. As explained in **I.3.1**, the elements of H^p are cosets in Z^p with respect to the subgroup B^p. If z^p is an element of Z^p, then the coset (relative to B^p) containing z^p will be denoted by $[z^p]$ and will be termed the cohomology class of z^p. Thus the generic element of H^p is a cohomology class $[z^p]$, where z^p is a p-cocycle of M. By **I.3.1**, exercise 1, the relation $[z^p] = 0$ holds if and only if $z^p \in B^p$. In other words, the cohomology class $[z^p]$ is the zero element of H^p if and only if the p-cocycle z^p is a p-coboundary. If z_1^p, z_2^p are two p-cocycles, then the relation $[z_1^p] = [z_2^p]$ holds if and only if the difference $z_2^p - z_1^p$ is a p-coboundary.

If clarity requires explicit reference to the complex M, then notations like δ_M^p, c_M^p, $C^p(M)$, $Z^p(M)$, $B^p(M)$, $H^p(M)$, $[z_M^p]_M$ will be used as needed.

I.4.2. Cochain mappings. Let there be given two Mayer complexes

$$K = \left(C^p(K), \delta_K^p\right), \quad L = \left(C^p(L), \delta_L^p\right). \tag{1}$$

Suppose that for every integer p there is also given a homomorphism

$$\lambda^p : C^p(K) \to C^p(L). \tag{2}$$

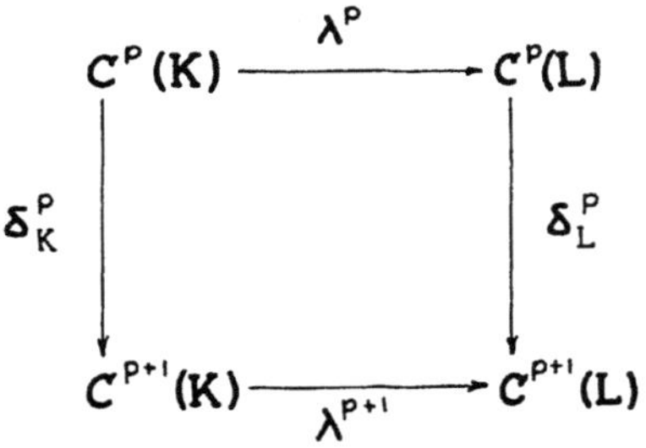

Fig. 3.

Consider the diagram in Fig. 3.

Starting with a p-cochain $c_K^p \in C^p(K)$, we can apply to c_K^p the homomorphisms δ and λ in two different ways, obtaining the $(p+1)$-cochains $\delta_L^p \lambda^p c_K^p$ and $\lambda^{p+1} \delta_K^p c_K^p$ of L. If

$$\delta_L^p \lambda^p c_K^p = \lambda^{p+1} \delta_K^p c_K^p \tag{3}$$

for every integer p and for every p-cochain c_K^p, then the homomorphisms (2) are said to constitute a cochain mapping from K to L. Briefly, we have a cochain mapping if the *vector law* holds in the preceding diagram.

Suppose that the homomorphisms λ^p constitute a cochain mapping. We assert then that

$$\lambda^p Z^p(K) \subset Z^p(L), \tag{4}$$

$$\lambda^p B^p(K) \subset B^p(L). \tag{5}$$

That is, under a cochain mapping cocycles are carried into cocycles and coboundaries are carried into coboundaries. Indeed, consider any cocycle $z_K^p \in Z^p(K)$. Then, in view of (3),

$$\delta_L^p \lambda^p z_K^p = \lambda^{p+1} \delta_K^p z_K^p = 0,$$

since $\delta_K^p z_K^p = 0$. Thus $\lambda^p z_K^p \in Z^p(L)$, and (4) is proved. Similarly, consider any coboundary $b_K^p = \delta_K^{p-1} c_K^{p-1}$. Then

$$\lambda^p b_K^p = \lambda^p \delta_K^{p-1} c_K^{p-1} = \delta_L^{p-1} \lambda^{p-1} c_K^{p-1} \in B^p(L),$$

and (5) is also proved. In view of (4), (5) and **I.4.1** (5), there arises (see **I.3.2**) for each p a homomorphism from $H^p(K)$ into $H^p(L)$, to be denoted by

$$\lambda_*^p : H^p(K) \to H^p(L). \tag{6}$$

The discussion in **I.3.2**, applied to the present situation, yields the following explicit description of the homomorphism λ_*^p. The generic element of $H^p(K)$ is of the form (see **I.4.1**)

$$h_K^p = [z_K^p]_K, \quad z_K^p \in Z^p(K). \tag{7}$$

Then $\lambda^p z_K^p$ is a p-cocycle of L by (4), and hence $\lambda^p z_K^p$ determines a cohomology class $[\lambda^p z_K^p]_L$ in $Z^p(L)$ relative to $B^p(L)$. We have then explicitly

$$\lambda_*^p [z_K^p]_K = [\lambda^p z_K^p]_L. \tag{8}$$

The homomorphisms λ_*^p are said to be induced by the cochain mapping (2).

In addition to the complexes (1) and the cochain mapping (2) let there be given a third MAYER complex

$$M = \left(C^p(M),\, \delta_M^p\right) \tag{9}$$

and a cochain mapping

$$\mu^p : C^p(L) \to C^p(M). \tag{10}$$

We have then the homomorphisms

$$\mu^p \lambda^p : C^p(K) \to C^p(M). \tag{11}$$

On taking any p-cochain $c_K^p \in C^p(K)$, we obtain [since (2) and (10) are cochain mappings] the relation

$$\delta_M^p \mu^p \lambda^p c_K^p = \mu^{p+1} \delta_L^p \lambda^p c_K^p = \mu^{p+1} \lambda^{p+1} \delta_K^p c_K^p,$$

showing that the homomorphisms (11) constitute a cochain mapping from K to M. Briefly, *the product of two cochain mappings is again a cochain mapping.*

The cochain mappings (2), (10), (11) induce homomorphisms λ_*^p, μ_*^p, $(\mu^p \lambda^p)_*$ for the cohomology groups of the complexes K, L, M. There arises the following diagram.

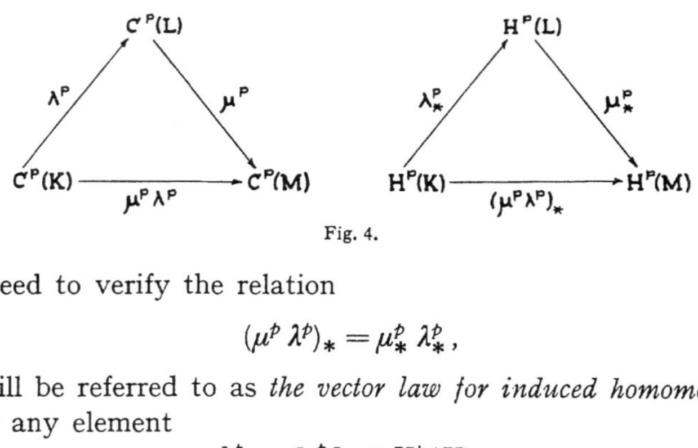

Fig. 4.

We proceed to verify the relation

$$(\mu^p \lambda^p)_* = \mu_*^p \lambda_*^p, \tag{12}$$

which will be referred to as *the vector law for induced homomorphisms.* Consider any element

$$h_K^p = [z_K^p]_K \in H^p(K).$$

In view of (8) we obtain

$$(\mu^p \lambda^p)_* h_K^p = [\mu^p \lambda^p z_K^p]_M = \mu_*^p [\lambda^p z_K^p]_L = \mu_*^p \lambda_*^p [z_K^p]_K = \mu_*^p \lambda_*^p h_K^p,$$

and (12) is proved.

I.4.3. Standard triples of MAYER complexes. Let there be given three MAYER complexes

$$L = \left(C^p(L),\, \delta_L^p\right), \quad K = \left(C^p(K),\, \delta_K^p\right), \quad F = \left(C^p(F),\, \delta_F^p\right), \tag{1}$$

and cochain mappings

$$\eta^p : C^p(L) \to C^p(K), \quad \pi^p : C^p(K) \to C^p(F). \tag{2}$$

It will be convenient to write (2) in the concise form

$$C^p(L) \xrightarrow{\eta^p} C^p(K) \xrightarrow{\pi^p} C^p(F). \tag{3}$$

The complexes L, K, F, jointly with the cochain mappings (2), will be said to constitute a *standard triple* of MAYER complexes if the following conditions hold for every integer p.

(i) The homomorphism η^p is an *isomorphism into*. That is, if $c_L^p \in C^p(L)$ and $\eta^p c_L^p = 0$, then $c_L^p = 0$.

(ii) The homomorphism π^p is *onto*. That is, if $c_F^p \in C^p(F)$, then there exists at least one p-cochain $c_K^p \in C^p(K)$ such that $\pi^p c_K^p = c_F^p$.

(iii) The three-term sequence (3) is exact (see **I.3.3**). That is, if $c_K^p \in C^p(K)$, then $\pi^p c_K^p = 0$ if and only if there exists a cochain $c_L^p \in C^p(L)$ such that $c_K^p = \eta^p c_L^p$.

Observe that a standard triple consists of three MAYER complexes *and* of two cochain mappings, as indicated in (3), with the properties (i), (ii), (iii).

Given a standard triple as in (1) to (3), the cochain mappings in (3) induce homomorphisms (see **I.4.2**)

$$H^p(L) \xrightarrow{\eta_*^p} H^p(K) \xrightarrow{\pi_*^p} H^p(F). \tag{4}$$

Next, it will be shown how the properties (i), (ii), (iii) may be used to introduce, for each integer p, a homomorphism

$$\delta_*^p : H^p(F) \to H^{p+1}(L), \tag{5}$$

termed the *coboundary homomorphism* of the standard triple. A definition for δ_*^p is suggested by an inspection of the following diagram.

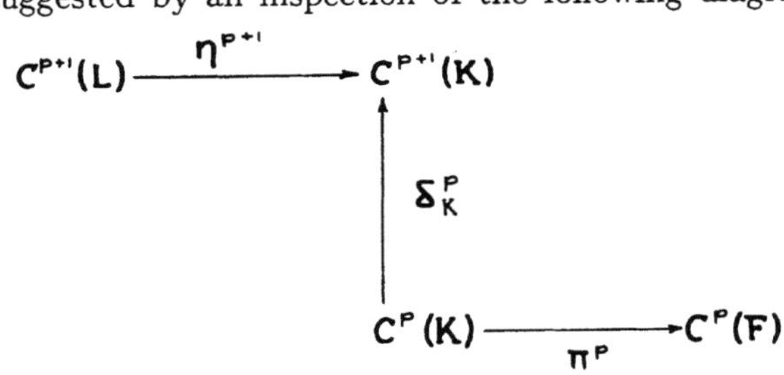

Fig. 5.

The generic element $h_F^p \in H^p(F)$ is of the form (see **I.4.1**)

$$h_F^p = [z_F^p]_F, \quad z_F^p \in Z^p(F). \tag{6}$$

Since π^p is a homomorphism onto by (ii), there exists a cochain $c_K^p \in C^p(K)$ such that

$$\pi^p c_K^p = z_F^p. \tag{7}$$

Then $\delta_K^p c_K^p$ is a coboundary in $B^{p+1}(K)$; since one seeks an element of $H^{p+1}(L)$ to assign as the image of h_F^p in (6) under δ_*^p, one is led to ask whether there is a cocycle $z_L^{p+1} \in Z^{p+1}(L)$ such that

$$\eta^{p+1} z_L^{p+1} = \delta_K^p c_K^p. \tag{8}$$

Note that if such a cocycle z_L^{p+1} exists, then it is unique by (i). We shall find that z_L^{p+1} does exist, and we shall also find that on assigning $[z_L^{p+1}]_L$ as the image of $[z_F^p]_F$ there results a homomorphism from $H^p(F)$ into $H^{p+1}(L)$. To facilitate the discussion, we introduce a binary relation

$$z_L^{p+1} \mathfrak{A} z_F^p, \quad z_L^{p+1} \in Z^{p+1}(L), \quad z_F^p \in Z^p(F), \tag{9}$$

as an abbreviation of the statement that for the cocycles z_F^p, z_L^{p+1} there exists a cochain $c_K^p \in C^p(K)$ such that (7) and (8) hold. The following diagram indicates the groups and homomorphisms that will be used in the course of the discussion.

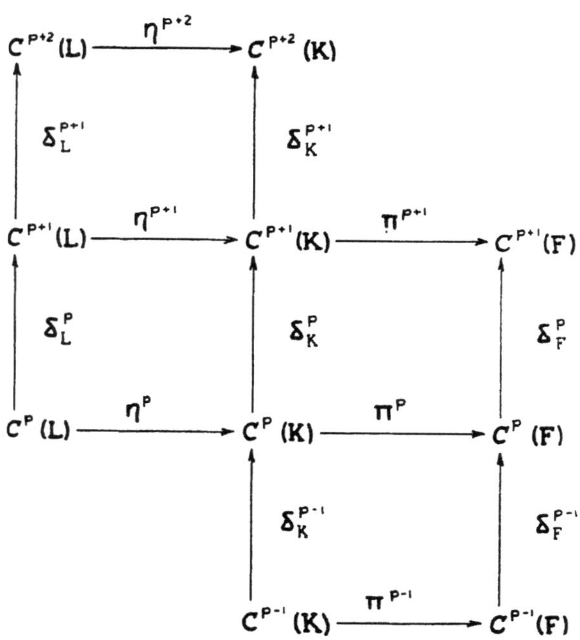

Fig. 6.

Lemma 1. Given $z_F^p \in Z^p(F)$, there exists at least one cocycle $z_L^{p+1} \in Z^{p+1}(L)$ such that (9) holds.

Proof. Since π^p is *onto*, there exists a cochain $c_K^p \in C^p(K)$ such that

$$\pi^p c_K^p = z_F^p. \tag{10}$$

We have then

$$\pi^{p+1} \delta_K^p c_K^p = \delta_F^p \pi^p c_K^p = \delta_F^p z_F^p = 0, \tag{11}$$

since z_F^p is a cocycle. In view of the exactness property (iii) above, (11) implies the existence of a cochain $c_L^{p+1} \in C^{p+1}(L)$ such that

$$\eta^{p+1} c_L^{p+1} = \delta_K^p c_K^p. \tag{12}$$

In view of (9), (10), (12), the lemma will be proved if we show that

$$c_L^{p+1} \in Z^{p+1}(L). \tag{13}$$

Now by (12) and **I.4.1** (3) we have

$$\eta^{p+2} \delta_L^{p+1} c_L^{p+1} = \delta_K^{p+1} \eta^{p+1} c_L^{p+1} = \delta_K^{p+1} \delta_K^p c_K^p = 0.$$

Since η^{p+2} is an isomorphism into, it follows that $\delta_L^{p+1} c_L^{p+1} = 0$, and (13) is verified.

The proofs of the following lemmas 2, 3, 4, 5, 6 are quite similar and are left to the reader.

Lemma 2. If $z_L^{p+1} \mathfrak{A} z_F^p$ and $\bar{z}_L^{p+1} \mathfrak{A} \bar{z}_F^p$, then

$$(z_L^{p+1} \pm \bar{z}_L^{p+1}) \mathfrak{A} (z_F^p \pm \bar{z}_F^p).$$

Lemma 3. If $z_L^{p+1} \in B^{p+1}(L)$ and $z_F^p \in B^p(F)$, then (9) holds.

Lemma 4. If $z_L^{p+1} \mathfrak{A} z_F^p$, and $\bar{z}_F^p$ is a p-cocycle of F such that $\bar{z}_F^p - z_F^p \in B^p(F)$, then $z_L^{p+1} \mathfrak{A} \bar{z}_F^p$.

Lemma 5. If $z_L^{p+1} \mathfrak{A} z_F^p$ and $\bar{z}_L^{p+1}$ is a $(p+1)$-cocycle of L such that $\bar{z}_L^{p+1} - z_L^{p+1} \in B^{p+1}(L)$, then $\bar{z}_L^{p+1} \mathfrak{A} z_F^p$.

Lemma 6. If $z_L^{p+1} \mathfrak{A}\, 0$, then $z_L^{p+1} \in B^{p+1}(L)$.

We are now ready to define the homomorphism δ_*^p [see (5)]. The generic element h_F^p of $H^p(F)$ is of the form (see **I.4.1**)

$$h_F^p = [z_F^p]_F, \quad z_F^p \in Z^p(F). \tag{14}$$

By lemma 1 there exists a cocycle $z_L^{p+1} \in Z^{p+1}(L)$ such that

$$z_L^{p+1} \mathfrak{A} z_F^p. \tag{15}$$

We define

$$\delta_*^p h_F^p = [z_L^{p+1}]_L. \tag{16}$$

To justify this definition, we have to show that $\delta_*^p h_F^p$ is independent of the particular choice of the cocycles z_F^p, z_L^{p+1} [subject to the conditions (14), (15)]. So let $\bar{z}_F^p$, $\bar{z}_L^{p+1}$ be any other pair of cocycles such that

$$h_F^p = [\bar{z}_F^p]_F, \quad \bar{z}_L^{p+1} \mathfrak{A} \bar{z}_F^p. \tag{17}$$

We have to show that

$$[\bar{z}_L^{p+1}]_L = [z_L^{p+1}]_L. \tag{18}$$

Now $\bar{z}_F^p - z_F^p \in B^p(F)$ by (14) and (17), and hence in view of (15) we conclude by lemma 4 that

$$z_L^{p+1} \mathfrak{A} \bar{z}_F^p. \tag{19}$$

From (17) and (19) we obtain, by lemma 2,

$$(\bar{z}_L^{p+1} - z_L^{p+1}) \mathfrak{A} 0. \tag{20}$$

By lemma 6 (applied to the cocycle $\bar{z}_L^{p+1} - z_L^{p+1}$) it follows from (20) that $\bar{z}_L^{p+1} - z_L^{p+1} \in B^{p+1}(L)$, and (18) is proved (see **I.4.1**).

It is thus shown that δ_*^p is a single-valued mapping from $H^p(F)$ into $H^{p+1}(L)$. If $h_F^p, \bar{h}_F^p$ are any two elements of $H^p(F)$, then lemma 2 and the defining formulas (14), (15), (16) yield directly the relation

$$\delta_*^p (h_F^p + \bar{h}_F^p) = \delta_*^p h_F^p + \delta_*^p \bar{h}_F^p,$$

showing that δ_*^p is a homomorphism.

Lemma 7. Given two cocycles $z_F^p \in Z^p(F)$ and $\bar{z}_L^{p+1} \in Z^{p+1}(L)$, the relation

$$\delta_*^p [z_F^p]_F = [\bar{z}_L^{p+1}]_L \tag{21}$$

holds if and only if

$$\bar{z}_L^{p+1} \mathfrak{A} z_F^p. \tag{22}$$

Proof. By the definition of δ_*^p, there exists a cocycle z_L^{p+1} such that

$$z_L^{p+1} \mathfrak{A} z_F^p, \tag{23}$$

$$\delta_*^p [z_F^p]_F = [z_L^{p+1}]_L. \tag{24}$$

Suppose now that (21) holds. Then from (21) and (24) it follows that

$$\bar{z}_L^{p+1} - z_L^{p+1} \in B^{p+1}(L). \tag{25}$$

In view of lemma 5, (23) and (25) imply (22). Conversely, if (22) holds, then (21) follows directly from the definition of δ_*^p.

Jointly with the homomorphisms η_*^p and π_*^p in (4), the homomorphism δ_*^p gives rise to the following sequence (infinite in both directions):

$$\cdots \xrightarrow{\delta_*^{p-1}} H^p(L) \xrightarrow{\eta_*^p} H^p(K) \xrightarrow{\pi_*^p} H^p(F) \xrightarrow{\delta_*^p} H^{p+1}(L) \xrightarrow{\eta_*^{p+1}} \cdots \tag{26}$$

which is called the *cohomology sequence* of the standard triple.

Theorem. The cohomology sequence of a standard triple is exact (see **I.3.3**).

Proof. Let us use the symbols $I(\eta_*^p), I(\pi_*^p), I(\delta_*^p)$ to denote the images of the groups $H^p(L), H^p(K), H^p(F)$ under the homomorphisms

$\eta_*^p, \pi_*^p, \delta_*^p$ respectively. Similarly, let us denote by $N(\pi_*^p), N(\delta_*^p), N(\eta_*^{p+1})$ the nuclei of the homomorphisms $\pi_*^p, \delta_*^p, \eta_*^{p+1}$.

Exactness at $H^p(K)$. Let h_K^p be an element of $H^p(K)$ such that

$$h_K^p \in I(\eta_*^p). \tag{27}$$

Then there exists an element $[z_L^p]_L \in H^p(L)$ such that

$$h_K^p = \eta_*^p [z_L^p]_L = [\eta^p z_L^p]_K.$$

It follows that

$$\pi_*^p h_K^p = \pi_*^p [\eta^p z_L^p]_K = [\pi^p \eta^p z_L^p]_F = 0,$$

since $\pi^p \eta^p z_L^p = 0$ by the exactness property (iii) above. Thus $h_K^p \in N(\pi_*^p)$. Since h_K^p was an arbitrary element satisfying (27), we conclude that

$$I(\eta_*^p) \subset N(\pi_*^p). \tag{28}$$

Consider now an element $h_K^p = [z_K^p]_K \in H^p(K)$, such that

$$h_K^p = [z_K^p]_K \in N(\pi_*^p). \tag{29}$$

Then $\pi_*^p h_K^p = 0$, and hence $\pi^p z_K^p \in B^p(F)$. Hence there exists a cochain $c_F^{p-1} \in C^{p-1}(F)$ such that

$$\pi^p z_K^p = \delta_F^{p-1} c_F^{p-1}. \tag{30}$$

Since π^{p-1} is *onto*, there exists a cochain $c_K^{p-1} \in C^{p-1}(K)$ such that

$$c_F^{p-1} = \pi^{p-1} c_K^{p-1}. \tag{31}$$

From (30) and (31) we conclude that

$$\begin{aligned} \pi^p (z_K^p - \delta_K^{p-1} c_K^{p-1}) &= \pi^p z_K^p - \pi^p \delta_K^{p-1} c_K^{p-1} \\ &= \pi^p z_K^p - \delta_F^{p-1} \pi^{p-1} c_K^{p-1} = \pi^p z_K^p - \delta_F^{p-1} c_F^{p-1} = 0. \end{aligned}$$

By the exactness property (iii) above it follows that $z_K^p - \delta_K^{p-1} c_K^{p-1}$ is the image, under η^p, of an element of $C^p(L)$. We denote this element by z_L^p, anticipating the fact (to be verified in a moment) that this element is a cocycle. We have

$$z_K^p - \delta_K^{p-1} c_K^{p-1} = \eta^p z_L^p. \tag{32}$$

To verify that z_L^p is a cocycle, we calculate [using (32)]:

$$\eta^{p+1} \delta_L^p z_L^p = \delta_K^p \eta^p z_L^p = \delta_K^p z_K^p - \delta_K^p \delta_K^{p-1} c_K^{p-1} = 0 - 0 = 0,$$

since z_K^p is a cocycle. As η^{p+1} is an isomorphism into, it follows that $\delta_L^p z_L^p = 0$, and thus z_L^p is a cocycle. Now since $\delta_K^{p-1} c_K^{p-1} \in B^p(K)$, we have

$$[z_K^p - \delta_K^{p-1} c_K^{p-1}]_K = [z_K^p]_K = h_K^p.$$

From (32) we conclude now that

$$\eta_*^p [z_L^p]_L = [\eta^p z_L^p]_K = [z_K^p - \delta_K^{p-1} c_K^{p-1}]_K = h_K^p.$$

Thus $h_K^p \in I(\eta_*^p)$. Since h_K^p was an arbitrary element satisfying (29), it follows that $N(\pi_*^p) \subset I(\eta_*^p)$. By (28) we conclude that $N(\pi_*^p) = I(\eta_*^p)$, and exactness at $H^p(K)$ is proved.

Exactness at $H^{p+1}(L)$. Take an element $h_L^{p+1} = [z_L^{p+1}]_L$ of $H^{p+1}(L)$ such that $h_L^{p+1} \in I(\delta_*^p)$. Then there exists an element $h_F^p = [z_F^p]_F \in H^p(F)$ such that $h_L^{p+1} = \delta_*^p h_F^p$. By lemma 7 we have then the relation $z_L^{p+1} \mathfrak{A} z_F^p$, and hence [see (7), (8), (9)] there exists a cochain $c_K^p \in C^p(K)$ such that

$$\eta^{p+1} z_L^{p+1} = \delta_K^p c_K^p, \quad \pi^p c_K^p = z_F^p.$$

It follows that

$$\eta_*^{p+1} h_L^{p+1} = [\eta^{p+1} z_L^{p+1}]_K = [\delta_K^p c_K^p]_K = 0,$$

since $\delta_K^p c_K^p \in B^{p+1}(K)$. Thus $h_L^{p+1} \in N(\eta_*^{p+1})$, and it is proved that

$$I(\delta_*^p) \subset N(\eta_*^{p+1}). \tag{33}$$

Consider next an element $h_L^{p+1} = [z_L^{p+1}]_L$ of $H^{p+1}(L)$ such that $h_L^{p+1} \in N(\eta_*^{p+1})$. Then

$$\eta_*^{p+1} h_L^{p+1} = [\eta^{p+1} z_L^{p+1}]_K = 0,$$

and hence there exists a cochain $c_K^p \in C^p(K)$ such that

$$\eta^{p+1} z_L^{p+1} = \delta_K^p c_K^p. \tag{34}$$

We calculate

$$\delta_F^p \pi^p c_K^p = \pi^{p+1} \delta_K^p c_K^p = \pi^{p+1} \eta^{p+1} z_L^{p+1} = 0,$$

by the exactness property (iii) above. Thus $\pi^p c_K^p$ is a cocycle of F:

$$\pi^p c_K^p = z_F^p \in Z^p(F). \tag{35}$$

From (34) and (35) it follows [see (7), (8)] that $z_L^{p+1} \mathfrak{A} z_F^p$. Hence, by lemma 7, $h_L^{p+1} = \delta_*^p [z_F^p]_F$, and it is thus proved that

$$N(\eta_*^{p+1}) \subset I(\delta_*^p). \tag{36}$$

From (33) and (36) it follows that $I(\delta_*^p) = N(\eta_*^{p+1})$, and exactness at $H^{p+1}(L)$ is proved.

Exactness at $H^p(F)$. Take an element $h_F^p \in H^p(F)$ such that $h_F^p \in I(\pi_*^p)$. Then there exists an element $[z_K^p]_K \in H^p(K)$ such that

$$h_F^p = \pi_*^p [z_K^p]_K = [\pi^p z_K^p]_F.$$

Now $\delta_*^p h_F^p$ is [see (14), (15), (16)] of the form

$$\delta_*^p h_F^p = [z_L^{p+1}]_L, \quad z_L^{p+1} \mathfrak{A} \pi^p z_K^p. \tag{37}$$

Accordingly, there exists a cochain $c_K^p \in C^p(K)$ such that

$$\eta^{p+1} z_L^{p+1} = \delta_K^p c_K^p, \quad \pi^p c_K^p = \pi^p z_K^p. \tag{38}$$

Thus $\pi^p(c_K^p - z_K^p) = 0$. By the exactness property (iii) above, there follows the existence of a cochain $c_L^p \in C^p(L)$ such that

$$c_K^p - z_K^p = \eta^p c_L^p. \tag{39}$$

From (38), (39) we infer that

$$\begin{aligned}\eta^{p+1}(z_L^{p+1} - \delta_L^p c_L^p) &= \eta^{p+1} z_L^{p+1} - \eta^{p+1} \delta_L^p c_L^p \\ &= \delta_K^p c_K^p - \delta_K^p \eta^p c_L^p = \delta_K^p c_K^p - (\delta_K^p c_K^p - \delta_K^p z_K^p) = 0,\end{aligned}$$

since z_K^p is a cocycle. Since η^{p+1} is an isomorphism into, it follows that $z_L^{p+1} - \delta_L^p c_L^p = 0$ and hence $z_L^{p+1} \in B^{p+1}(L)$. In view of (37) it follows that $\delta_*^p h_F^p = 0$, and thus $h_F^p \in N(\delta_*^p)$. This shows that

$$I(\pi_*^p) \subset N(\delta_*^p). \tag{40}$$

Consider now an element $[z_F^p]_F \in H^p(F)$ such that $[z_F^p]_F \in N(\delta_*^p)$. Now [see (14), (15), (16)]

$$\delta_*^p [z_F^p]_F = [z_L^{p+1}]_L, \tag{41}$$

and there exists a cochain $c_K^p \in C^p(K)$ such that

$$\eta^{p+1} z_L^{p+1} = \delta_K^p c_K^p, \quad \pi^p c_K^p = z_F^p. \tag{42}$$

Since $\delta_*^p [z_F^p]_F = 0$ by assumption, we conclude from (41) that $z_L^{p+1} \in B^{p+1}(L)$. Hence there exists a cochain $c_L^p \in C^p(L)$ such that

$$z_L^{p+1} = \delta_L^p c_L^p. \tag{43}$$

From (42) and (43) we obtain

$$\delta_K^p(c_K^p - \eta^p c_L^p) = \delta_K^p c_K^p - \eta^{p+1} \delta_L^p c_L^p = \delta_K^p c_K^p - \eta^{p+1} z_L^{p+1} = 0.$$

Thus $c_K^p - \eta^p c_L^p$ is a cocycle of K:

$$c_K^p - \eta^p c_L^p = z_K^p \in Z^p(K). \tag{44}$$

In view of (44), (42) we have then

$$\pi_*^p [z_K^p]_K = [\pi^p z_K^p]_F = [\pi^p c_K^p - \pi^p \eta^p c_L^p]_F = [\pi^p c_K^p]_F = [z_F^p]_F,$$

since $\pi^p \eta^p c_L^p = 0$ by the exactness property (iii) above. Thus $[z_F^p]_F \in I(\pi_*^p)$, and hence

$$N(\delta_*^p) \subset I(\pi_*^p). \tag{45}$$

From (40) and (45) it follows that $I(\pi_*^p) = N(\delta_*^p)$, and the proof is complete.

I.4.4. Mappings for standard triples. To avoid prohibitive notational complications in the sequel, we make the following agreements. Given two MAYER complexes K, L and a cochain mapping from K

to L as in **I.4.2**, we shall write

$$K \xrightarrow{\lambda} L \quad \text{or} \quad \lambda : K \to L$$

as an abbreviation for the explicit formulas (1), (2), (3) in **I.4.2**. Similarly, the formulas **I.4.3** (1), (2), (3), relating to a standard triple, will be abbreviated by writing

$$L \xrightarrow{\eta} K \xrightarrow{\pi} F.$$

Let there be given two standard triples

$$T_1 : L_1 \xrightarrow{\eta_1} K_1 \xrightarrow{\pi_1} F_1 \tag{1}$$

$$T_2 : L_2 \xrightarrow{\eta_2} K_2 \xrightarrow{\pi_2} F_2 \tag{2}$$

and three cochain mappings

$$L_1 \xrightarrow{l} L_2, \quad K_1 \xrightarrow{k} K_2, \quad F_1 \xrightarrow{f} F_2. \tag{3}$$

For each integer p we have then the following diagram.

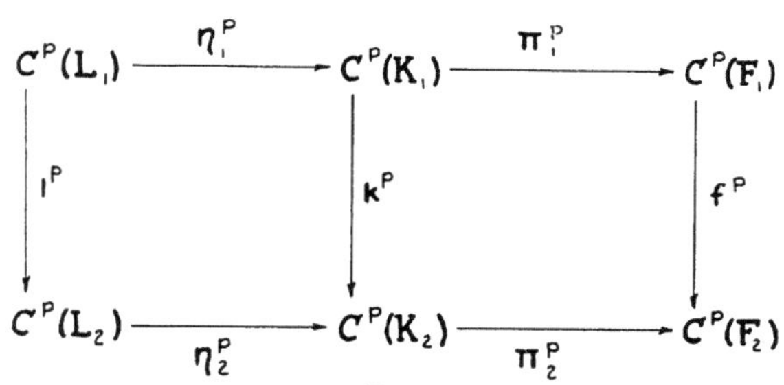

(4)

Fig. 7.

The cochain mappings l, k, f will be said to constitute a mapping from the standard triple (1) to the standard triple (2) provided that the *vector law* holds in each box of the preceding diagram (for every integer p). Explicitly, this condition means that

$$k^p \eta_1^p = \eta_2^p l^p, \quad f^p \pi_1^p = \pi_2^p k^p. \tag{5}$$

Let there be given a third standard triple

$$T_3 : L_3 \xrightarrow{\eta_3} K_3 \xrightarrow{\pi_3} F_3,$$

and cochain mappings

$$L_2 \xrightarrow{\lambda} L_3, \quad K_2 \xrightarrow{\varkappa} K_3, \quad F_2 \xrightarrow{\varphi} F_3.$$

By **I.4.2** we have then also the cochain mappings

$$L_1 \xrightarrow{\lambda l} L_3, \quad K_1 \xrightarrow{\varkappa k} K_3, \quad F_1 \xrightarrow{\varphi f} F_3.$$

The following statement is an immediate consequence of the preceding definitions.

Theorem 1 (product law for triple mappings). If l, k, f constitute a mapping from T_1 to T_2 and $\lambda, \varkappa, \varphi$ constitute a mapping from T_2 to T_3, then $\lambda l, \varkappa k, \varphi f$ constitute a mapping from T_1 to T_3.

Returning to the standard triples T_1, T_2 in (1), (2), suppose that the cochain mappings (3) constitute a mapping from T_1 to T_2. There arises the following diagram.

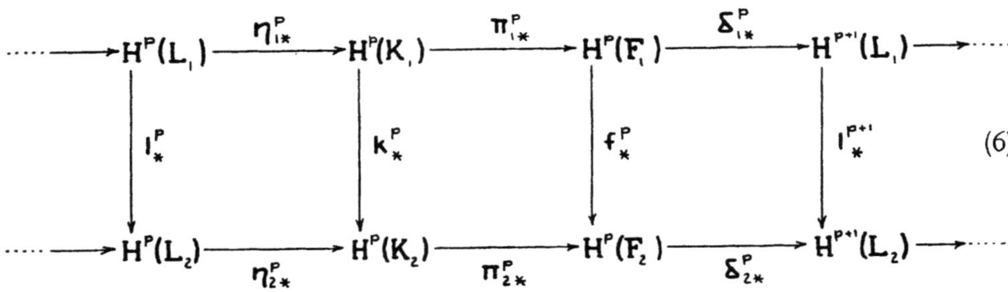

(6)

Fig. 8.

This diagram consists of the cohomology sequences (see **I.4.3**) of the standard triples (1) and (2), connected by homomorphisms (indicated by vertical arrows) which are induced (see **I.4.2**) by the cochain mappings (3).

Theorem 2 (vector law for cohomology sequences). The vector law holds in each box of diagram (6). Explicitly: for every integer p we have the relations

$$k^p_* \eta^p_{1*} = \eta^p_{2*} l^p_*, \tag{7}$$

$$f^p_* \pi^p_{1*} = \pi^p_{2*} k^p_*, \tag{8}$$

$$l^{p+1}_* \delta^p_{1*} = \delta^p_{2*} f^p_*. \tag{9}$$

Proof. In view of (5), the relations (7) and (8) follow directly from the vector law for induced homomorphisms (see **I.4.2**). To prove (9), take any element

$$h^p_{F_1} = [z^p_{F_1}]_{F_1} \in H^p(F_1). \tag{10}$$

By the definition of the homomorphism δ^p_{1*} [see **I.4.3** (14), (15), (16), (9)], we have then a cocycle

$$z^{p+1}_{L_1} \in Z^{p+1}(L_1)$$

and a cochain

$$c^p_{K_1} \in C^p(K_1)$$

such that

$$\delta^p_{1*} h^p_{F_1} = [z^{p+1}_{L_1}]_{L_1}, \quad \eta^{p+1}_1 z^{p+1}_{L_1} = \delta^p_{K_1} c^p_{K_1}, \quad \pi^p_1 c^p_{K_1} = z^p_{F_1}. \tag{11}$$

By (11) and the relations (5), applied for p and $p+1$, we obtain

$$\pi_2^p k^p c_{K_1}^p = f^p \pi_1^p c_{K_1}^p = f^p z_{F_1}^p, \tag{12}$$

$$\eta_2^{p+1} l^{p+1} z_{L_1}^{p+1} = k^{p+1} \eta_1^{p+1} z_{L_1}^{p+1} = k^{p+1} \delta_{K_1}^p c_{K_1}^p = \delta_{K_2}^p k^p c_{K_1}^p. \tag{13}$$

Since l, k, f are cochain mappings, we have (see **I.4.2**).

$$l^{p+1} z_{L_1}^{p+1} \in Z^{p+1}(L_2), \quad f^p z_{F_1}^p \in Z^p(F_2), \quad k^p c_{K_1}^p \in C^p(K_2).$$

Accordingly, applying **I.4.3** (9) to the standard triple (2), the relations (12) and (13) mean that

$$l^{p+1} z_{L_1}^{p+1} \mathfrak{A} f^p z_{F_1}^p.$$

Hence, by lemma 7 in **I.4.3**,

$$\delta_{2*}^p [f^p z_{F_1}^p]_{F_2} = [l^{p+1} z_{L_1}^{p+1}]_{L_2}. \tag{14}$$

Observe now that

$$[f^p z_{F_1}^p]_{F_2} = f_*^p [z_{F_1}^p]_{F_1} = f_*^p h_{F_1}^p, \tag{15}$$

$$[l^{p+1} z_{L_1}^{p+1}]_{L_2} = l_*^{p+1} [z_{L_1}^{p+1}]_{L_1} = l_*^{p+1} \delta_{1*}^p h_{F_1}^p, \tag{16}$$

by (10) and (11). From (14), (15), (16) we conclude that

$$\delta_{2*}^p f_*^p h_{F_1}^p = l_*^{p+1} \delta_{1*}^p h_{F_1}^p,$$

and (9) is proved.

I.4.5. Homotopy. Let there be given two MAYER complexes

$$L = (C^p(L), \delta_L^p), \quad K = (C^p(K), \delta_K^p),$$

and two cochain mappings

$$l_1 : L \to K, \quad l_2 : L \to K. \tag{1}$$

By the agreement stated in **I.4.4**, l_1 and l_2 are abbreviations for homomorphisms

$$l_1^p : C^p(L) \to C^p(K), \quad l_2^p : C^p(L) \to C^p(K),$$

satisfying the condition **I.4.2** (3). The cochain mappings l_1, l_2 induce (see **I.4.2**) homomorphisms

$$l_{1*}^p : H^p(L) \to H^p(K), \quad l_{2*}^p : H^p(L) \to H^p(K).$$

Our present objective is to state a relationship between the cochain mappings l_1 and l_2 which implies that $l_{1*}^p = l_{2*}^p$ for every integer p.

The cochain mappings l_1, l_2 are termed *homotopic*, in symbols $l_1 \simeq l_2$, provided that there exists homomorphisms

$$D^p : C^p(L) \to C^{p-1}(K), \quad p = 0, \pm 1, \pm 2, \ldots, \tag{2}$$

such that for every cochain $c_L^p \in C^p(L)$ the following relation holds:

$$\delta_K^{p-1} D^p c_L^p + D^{p+1} \delta_L^p c_L^p = l_2^p c_L^p - l_1^p c_L^p. \tag{3}$$

The homomorphisms D^p are then said to constitute a *homotopy operator* for the pair of cochain mappings l_1, l_2, and (3) is called the *homotopy identity*.

Theorem. If $l_1 \simeq l_2$, then $l_{1*}^p = l_{2*}^p$ for every integer p.

Proof. Take any element

$$h_L^p = [z_L^p]_L \in H^p(L).$$

Since z_L^p is a cocycle and hence $\delta_L^p z_L^p = 0$, the substitution $c_L^p = z_L^p$ in (3) yields

$$l_2^p z_L^p - l_1^p z_L^p = \delta_K^{p-1} D^p z_L^p \in B^p(K).$$

Hence (see **I.4.1, I.4.2**)

$$[l_1^p z_L^p]_K = [l_2^p z_L^p]_K,$$

and consequently

$$l_{1*}^p [z_L^p]_L = [l_1^p z_L^p]_K = [l_2^p z_L^p]_K = l_{2*}^p [z_L^p]_L.$$

Since $[z_L^p]_L$ was an arbitrary element of $H^p(L)$, the theorem is proved.

I.4.6. Subcomplexes. Let there be given two MAYER complexes

$$L = \left(C^p(L), \delta_L^p\right), \quad K = \left(C^p(K), \delta_K^p\right). \tag{1}$$

Then L is termed a *subcomplex* of K, in symbols $L \subset K$, if the following two conditions hold for every integer p.

(i) $C^p(L)$ is a subgroup of $C^p(K)$.

(ii) $\delta_L^p = \delta_K^p \mid C^p(L)$ (see **I.1.1**).

Lemma 1. Suppose that $L \subset K$. For each integer p, let

$$\eta^p : C^p(L) \to C^p(K)$$

denote the identity homomorphism (that is, $\eta^p c^p = c^p$ for every cochain $c^p \in C^p(L)$. Then the homomorphisms η^p constitute a cochain mapping from L to K.

Proof. If c^p is any element of $C^p(L)$, then by condition (ii) above we have

$$\delta_K^p \eta^p c^p = \delta_K^p c^p = \delta_L^p c^p = \eta^{p+1} \delta_L^p c^p.$$

Thus the condition **I.4.2** (3) is satisfied, and the lemma is proved.

Lemma 2. Given a MAYER complex $K = \left(C^p(K), \delta_K^p\right)$, let there be assigned for each integer p a subgroup G_p of $C^p(K)$, such that $\delta_K^p G_p \subset G_{p+1}$ for every p. Then the subgroups G_p and the homomorphisms $\delta_K^p \mid G_p$ constitute a subcomplex of K.

This is a direct consequence of the definition of a subcomplex.

Lemma 3. Let there be given two MAYER complexes

$$K_1 = \left(C^p(K_1), \delta^p_{K_1}\right), \quad K_2 = \left(C^p(K_2), \delta^p_{K_2}\right),$$

and a cochain mapping (see **I.4.4** for the notation)

$$\lambda : K_1 \to K_2. \tag{2}$$

Furthermore, let

$$L_1 = \left(C^p(L_1), \delta^p_{L_1}\right), \quad L_2 = \left(C^p(L_2), \delta^p_{L_2}\right)$$

be subcomplexes of K_1, K_2 respectively, such that $\lambda^p C^p(L_1) \subset C^p(L_2)$ for every p. Then the homomorphisms (see **I.1.1**)

$$\mu^p = \lambda^p \mid C^p(L_1) : C^p(L_1) \to C^p(L_2)$$

constitute a cochain mapping from L_1 to L_2.

Proof. Take any cochain $c^p \in C^p(L_1)$. Since (2) is a cochain mapping, we obtain

$$\delta^p_{L_2} \mu^p c^p = \delta^p_{K_2} \lambda^p c^p = \lambda^{p+1} \delta^p_{K_1} c^p = \mu^{p+1} \delta^p_{L_1} c^p.$$

Thus the condition **I.4.2** (3) is satisfied, and the lemma is proved.

§ I.5. Formal complexes[1].

I.5.1. *p*-functions. Let $X \neq \emptyset$ be an arbitrary set and $p \geq 0$ an integer. A *p-function* $c^p(x_0, \ldots, x_p)$ for X is an integral-valued function of $p+1$ variable points $x_0, \ldots, x_p$ of X. Thus, for example, a 0-function $c^0(x_0)$ is an integral-valued function defined for all points $x_0 \in X$, a 1-function $c^1(x_0, x_1)$ is an integral-valued function defined for every choice of the points $x_0, x_1 \in X$, and so forth. The sum of two p-functions c^p and d^p, to be denoted by $c^p + d^p$, is the p-function defined by the formula

$$(c^p + d^p)(x_0, \ldots, x_p) = c^p(x_0, \ldots, x_p) + d^p(x_0, \ldots, x_p). \tag{1}$$

Since c^p and d^p are integral-valued, clearly $c^p + d^p$ is also integral-valued. It is obvious that for fixed $p \geq 0$ the p-functions for X form an Abelian group with addition defined by (1). This Abelian group will be denoted by $C^p_F(X)$. The subscript F stands for *formal*, and in turn the term *formal* is used in the present § **I.5** to emphasize the fact that the concepts to be discussed here relate to un-topologized sets, and hence are purely formal in character.

[1] These complexes are frequently referred to as *abstract simplicial complexes*. For historical comments and further details concerning this subject, see the treatises by ALEXANDROFF-HOPF, LEFSCHETZ, and EILENBERG-STEENROD listed in the *Bibliography*. We restrict ourselves to topics actually needed in the sequel. The presentation is designed for a reader unfamiliar with this subject.

Given a 0-function $c^0 \in C_F^0(X)$ and any two points $x_0, x_1 \in X$, the expression $c^0(x_1) - c^0(x_0)$ may be thought of as representing the increment of c^0 over the pair of points x_0, x_1. The same expression may be thought of as a (manifestly integral-valued) function of the variable points $x_0, x_1 \in X$, and hence as a 1-function for X. This 1-function will be denoted by $\delta_X^0 c^0$. Thus $\delta_X^0 c^0$ is the 1-function for X defined by the formula

$$(\delta_X^0 c^0)(x_0, x_1) = c^0(x_1) - c^0(x_0). \tag{2}$$

Similarly, given $c^1 \in C_F^1(X)$, the increment of c^1 over a triple of points x_0, x_1, x_2 is a 2-function for X, to be denoted by $\delta_X^1 c^1$ and given by the formula

$$(\delta_X^1 c^1)(x_0, x_1, x_2) = c^1(x_1, x_2) - c^1(x_0, x_2) + c^1(x_0, x_1). \tag{3}$$

In generalizing this process, the following notation will be convenient. Given an array of points $x_0, \ldots, x_m \in X$, the symbol $x_0, \ldots, \hat{x}_j, \ldots, x_m$ will denote the array $x_0, \ldots, x_{j-1}, x_{j+1}, \ldots, x_m$ obtained by deleting the point x_j. Given then a p-function $c^p \in C_F^p(X)$, the symbol $\delta_X^p c^p$ will designate the $(p+1)$-function defined by the formula

$$(\delta_X^p c^p)(x_0, \ldots, x_{p+1}) = \sum_{j=0}^{p+1} (-1)^j c^p(x_0, \ldots, \hat{x}_j, \ldots, x_{p+1}). \tag{4}$$

For $p=0$ and $p=1$, the general formula (4) reduces to the special formulas (2) and (3) respectively.

If $c^p \in C_F^p(X)$, then $\delta_X^p c^p$ is clearly an element of $C_F^{p+1}(X)$. Accordingly, the symbol δ_X^p may be interpreted as a mapping

$$\delta_X^p : C_F^p(X) \to C_F^{p+1}(X), \quad p \geq 0, \tag{5}$$

which assigns to the generic element c^p of $C_F^p(X)$ the element $\delta_X^p c^p$ of $C_F^{p+1}(X)$ as its image.

Lemma 1. δ_X^p is a homomorphism (see **I.3.2**).

Proof. The assertion means that on choosing any two elements $c^p, d^p \in C_F^p(X)$, one has the relation

$$\delta_X^p(c^p + d^p) = \delta_X^p c^p + \delta_X^p d^p. \tag{6}$$

Observe that the expressions appearing in (6) are $(p+1)$-functions. The equality of two functions means that they have the same value for every choice of the variable arguments. Accordingly, (6) means that on choosing any array $x_0, \ldots, x_{p+1}$ of $p+2$ points in X, one has [see (1)]

$$(\delta_X^p(c^p + d^p))(x_0, \ldots, x_{p+1}) = (\delta_X^p c^p)(x_0, \ldots, x_{p+1}) + (\delta_X^p d^p)(x_0, \ldots, x_{p+1}). \tag{7}$$

Now (7) can be verified by direct calculation [using (1) and (4)] as follows.

$$(\delta_X^p(c^p + d^p))(x_0, \ldots, x_{p+1}) = \sum_{j=0}^{p+1} (-1)^j (c^p + d^p)(x_0, \ldots, \hat{x}_j, \ldots, x_{p+1})$$
$$= \sum_{j=0}^{p+1} (-1)^j c^p(x_0, \ldots, \hat{x}_j, \ldots, x_{p+1}) + \sum_{j=0}^{p+1} (-1)^j d^p(x_0, \ldots, \hat{x}_j, \ldots x_{p+1})$$
$$= (\delta_X^p c^p)(x_0, \ldots, x_{p+1}) + (\delta_X^p d^p)(x_0, \ldots, x_{p+1}),$$

and (7) is proved.

Remark 1. The simple proof of lemma 1 has been carried out in full detail as an illustration of the steps involved in proving an identity between p-functions. The detailed discussion of similar identities in the sequel has the purpose of providing exercises for the reader who is studying this subject for the first time.

Remark 2. If p-functions relative to two sets X, Y are to be considered simultaneously, then notations like c_X^p, c_Y^p, δ_Y^p, and so forth will be used to avoid ambiguity. However, consistent use of fully descriptive notations would soon lead to prohibitive complications, and hence we shall use the simplest *ad hoc* notations consistent with clarity.

While the homomorphism δ_X^p has the effect of *raising the dimension,* another basic homomorphism to be introduced presently operates in the opposite direction. Let a point v, to be termed the *vertex,* be selected in X. A homomorphism

$$\Gamma_v^p : C_F^p(X) \to C_F^{p-1}(X), \quad p \geq 1,$$

is then defined as follows: if $c^p \in C_F^p(X)$, then $\Gamma_v^p c^p$ is the $(p-1)$-function defined by the formula

$$(\Gamma_v^p c^p)(x_0, \ldots, x_{p-1}) = (-1)^p c^p(x_0, \ldots, x_{p-1}, v), \quad p \geq 1. \tag{8}$$

For $p=1$ and $p=2$, for example, one has

$$(\Gamma_v^1 c^1)(x_0) = -c^1(x_0, v), \quad (\Gamma_v^2 c^2)(x_0, x_1) = c^2(x_0, x_1, v).$$

Since c^p is integral-valued, clearly $\Gamma_v^p c^p$ is also integral-valued. It is also immediate that if c^p, d^p are any two elements of $C_F^p(X)$, then $\Gamma_v^p(c^p + d^p) = \Gamma_v^p c^p + \Gamma_v^p d^p$. It is customary to refer to Γ_v^p as the cone homomorphism with vertex v.

Lemma 2. If $p \geq 1$ and c_1^p, c_2^p are elements of $C_F^p(X)$, then

$$c_1^p = c_2^p \tag{9}$$

if and only if

$$\Gamma_v^p c_1^p = \Gamma_v^p c_2^p \tag{10}$$

for every choice of the vertex $v \in X$.

Proof. Clearly (9) implies (10). Suppose, conversely, that (10) holds for every $v \in X$. Let the points $x_0, \ldots, x_p \in X$ be arbitrarily assigned. On choosing $v = x_p$, (8) and (10) yield

$$\begin{aligned} c_1^p(x_0, \ldots, x_p) &= (-1)^p (\Gamma_{x_p}^p c_1^p)(x_0, \ldots, x_{p-1}) \\ &= (-1)^p (\Gamma_{x_p}^p c_2^p)(x_0, \ldots, x_{p-1}) = c_2^p(x_0, \ldots, x_p), \end{aligned}$$

and (9) is verified.

Remark 3. This lemma enables us to reduce the verification of the equality of two p-functions c_1^p, c_2^p to the verification of the equality of two $(p-1)$-functions $\Gamma_v^p c_1^p, \Gamma_v^p c_2^p$. This remark will be utilized in proofs based upon induction with respect to p.

Lemma 3. If $c^p \in C_F^p(X)$, then

$$\delta_X^{p-1} \Gamma_v^p c^p + \Gamma_v^{p+1} \delta_X^p c^p = c^p, \quad p \geq 1, \tag{11}$$

for every choice of the vertex $v \in X$.

Proof. On choosing any array $x_0, \ldots, x_p$ of $p+1$ points in X, (4) and (8) yield

$$\left.\begin{aligned} (\delta_X^{p-1} \Gamma_v^p c^p)(x_0, \ldots, x_p) &= \sum_{j=0}^{p} (-1)^j (\Gamma_v^p c^p)(x_0, \ldots, \hat{x}_j, \ldots, x_p) \\ &= (-1)^p \sum_{j=0}^{p} (-1)^j c^p(x_0, \ldots, \hat{x}_j, \ldots, x_p, v), \end{aligned}\right\} \tag{12}$$

$$\left.\begin{aligned} &(\Gamma_v^{p+1} \delta_X^p c^p)(x_0, \ldots, x_p) = (-1)^{p+1} (\delta_X^p c^p)(x_0, \ldots, x_p, v) \\ &= (-1)^{p+1} \left[\sum_{j=0}^{p} (-1)^j c^p(x_0, \ldots, \hat{x}_j, \ldots, x_p, v) + (-1)^{p+1} c^p(x_0, \ldots, x_p)\right]. \end{aligned}\right\} \tag{13}$$

Adding (12) and (13), we obtain

$$(\delta_X^{p-1} \Gamma_v^p c^p + \Gamma_v^{p+1} \delta_X^p c^p)(x_0, \ldots, x_p) = c^p(x_0, \ldots, x_p).$$

Since the points $x_0, \ldots, x_p$ were arbitrarily chosen, (11) follows.

Remark 4. (11) is termed *the cone identity*. It enables us to express the combination $\Gamma\delta$ in terms of the combination $\delta\Gamma$, and *vice versa*.

Lemma 4. If $c^p \in C_F^p(X)$, then

$$\delta_X^{p+1} \delta_X^p c^p = 0, \quad p \geq 0. \tag{14}$$

Proof. The assertion means that

$$(\delta_X^{p+1} \delta_X^p c^p)(x_0, \ldots, x_{p+2}) = 0$$

for every choice of the points $x_0, \ldots, x_{p+2} \in X$. For $p = 0$, direct calculation yields

$$\begin{aligned} (\delta_X^1 \delta_X^0 c^0)(x_0, x_1, x_2) &= (\delta_X^0 c^0)(x_1, x_2) - (\delta_X^0 c^0)(x_0, x_2) + (\delta_X^0 c^0)(x_0, x_1) \\ &= \big(c^0(x_2) - c^0(x_1)\big) - \big((c^0(x_2) - c^0(x_0)\big) + \big(c^0(x_1) - c^0(x_0)\big) = 0, \end{aligned}$$

and (14) is verified for $p=0$. Proceeding by induction, assume that (14) is known to hold for a certain integer $p \geq 0$. We have to show that, as a consequence of (14),

$$\delta_X^{p+2}\, \delta_X^{p+1}\, c^{p+1} = 0 \tag{15}$$

for $c^{p+1} \in C_F^{p+1}(X)$. By lemma 2, (15) will be proved if we show that

$$\Gamma_v^{p+3}\, \delta_X^{p+2}\, \delta_X^{p+1}\, c^{p+1} = 0 \tag{16}$$

for every choice of the vertex $v \in X$. Two successive applications of the cone identity (11) yield

$$\begin{aligned} \Gamma_v^{p+3}\, \delta_X^{p+2}\, \delta_X^{p+1}\, c^{p+1} &= \delta_X^{p+1}\, c^{p+1} - \delta_X^{p+1}\, \Gamma_v^{p+2}\, \delta_X^{p+1}\, c^{p+1} \\ &= \delta_X^{p+1}\, c^{p+1} - \delta_X^{p+1}\,(c^{p+1} - \delta_X^{p}\, \Gamma_v^{p+1}\, c^{p+1}) = \delta_X^{p+1}\, \delta_X^{p}\, \Gamma_v^{p+1}\, c^{p+1} = 0, \end{aligned}$$

in view of the induction assumption (14). Thus the inductive proof of the lemma is complete.

I.5.2. The formal complex $M_F(X)$. Given a set $X \neq \emptyset$, the corresponding formal complex

$$M_F(X) = \left(C_F^p(X),\, \delta_X^p\right) \tag{1}$$

is a MAYER complex (see **§ I.4**) defined as follows. For $p \geq 0$, the groups $C_F^p(X)$ and the homomorphisms δ_X^p are set up as explained in **I.5.1**. For $p < 0$, one agrees to set $C_F^p(X) = 0$ (that is, $C_F^p(X)$ consists of a zero-element alone for $p < 0$). Accordingly, for $p < 0$ the homomorphism δ_X^p from $C_F^p(X)$ into $C_F^{p+1}(X)$ is a zero-homomorphism (see **I.3.2**). To show that these definitions yield a MAYER complex, we have to verify that

$$\delta_X^{p+1}\, \delta_X^p\, c^p = 0 \quad \text{for every } p, \tag{2}$$

where $c^p \in C_F^p(X)$. If $p < 0$, then $c^p = 0$ and (2) is obvious. If $p \geq 0$, then (2) follows by **I.5.1**, lemma 4.

The cochains, cocycles, coboundaries, and cohomology groups of the formal complex $M_F(X)$ will be termed the cochains, cocycles, coboundaries and cohomology groups respectively of X. In analogy with the notation $C_F^p(X)$ for the group of p-cochains, the symbols $Z_F^p(X)$ and $B_F^p(X)$ will designate the groups of p-cocycles and p-coboundaries of X respectively, while the cohomology groups will be denoted by $H_F^p(X)$. One sees that for $p \geq 0$ the p-cochains of X are merely the p-functions of X, while for $p < 0$ the only p-cochain of X is the zero-element of the trivial group $C_F^p(X)$.

The introduction of the trivial cochain groups $C_F^p(X) = 0$ for $p < 0$ is a matter of convenience. Agreement with the general theory of MAYER complexes is accomplished in this manner, and various basic theorems can be stated in a uniform manner. A significant exception arises, however, in connection with the cone homomorphism (see **I.5.1**). This point will be discussed presently in some detail.

Definition 1. The cone homomorphism

$$\Gamma_v^p : C_F^p(X) \to C_F^{p-1}(X), \tag{3}$$

with vertex $v \in X$, is defined as in **I.5.1** (8) for $p \geq 1$. As $C_F^{p-1}(X) = 0$ for $p \leq 0$, in accordance with **I.3.2** the homomorphism Γ_v^p is a zero-homomorphism if $p \leq 0$.

Thus definition 1 extends, in a trivial manner, the cone homomorphism considered in **I.5.1** to the case $p \leq 0$. There arises the question as to the status of the cone identity **I.5.1** (11) in this new situation. In considering this question, the following definition will be useful.

Definition 2. The unit 0-cocycle of X, to be denoted by ζ_X^0, is the 0-function which has the constant value 1 on X.

Remark 1. To justify this definition, we have to show that $\zeta_X^0 \in Z_F^0(X)$. It is obvious, however, that a 0-function c^0 is a cocycle if and only if it has a constant value on X. Thus $Z_F^0(X)$ consists of those 0-functions c^0 that reduce to constants on X. This result implies that $Z_F^0(X)$ is isomorphic to the additive group I of integers. Indeed, choose a point $\bar{x} \in X$. Then the mapping $m : Z_F^0(X) \to \mathsf{I}$, defined by $m z^0 = z^0(\bar{x})$ for $z^0 \in Z_F^0(X)$, is obviously an isomorphism from $Z_F^0(X)$ onto I.

Lemma 1. If $c^p \in C_F^p(X)$, then (see definition 2)

$$\delta_X^{p-1} \Gamma_v^p c^p + \Gamma_v^{p+1} \delta_X^p c^p = c^p \quad \text{if} \quad p \neq 0, \tag{4}$$

$$\delta_X^{-1} \Gamma_v^0 c^0 + \Gamma_v^1 \delta_X^0 c^0 = c^0 - c^0(v)\, \zeta_X^0. \tag{5}$$

Proof. If $p < 0$, then $c^p = 0$ and (4) is obvious. If $p \geq 1$, then (4) is merely a re-statement of **I.5.1** (11). Turning now to (5), observe that $\Gamma_v^0 c^0 = 0$ by definition 1, and hence (5) reduces to

$$\Gamma_v^1 \delta_X^0 c^0 = c^0 - c^0(v)\, \zeta_X^0. \tag{6}$$

On choosing an arbitrary point $x \in X$, one obtains

$$(\Gamma_v^1 \delta_X^0 c^0)(x) = -(\delta_X^0 c^0)(x, v) = c^0(x) - c^0(v) = \left(c^0 - c^0(v)\, \zeta_X^0\right)(x),$$

and (6) is verified.

Remark 2. Comparison of (4) and (5) reveals that the cone identity (4) generally fails to hold if $p = 0$. From (5) we infer the precise statement: for $p = 0$ the cone identity (4) holds for those and only those 0-cochains c^0 that have the value zero at the vertex v.

Lemma 2. The cohomology groups $H_F^p(X)$ of a set $X \neq \emptyset$ are as follows:

$$H_F^p(X) = 0 \quad \text{if} \quad p \neq 0, \tag{7}$$

$$H_F^0(X) \approx \mathsf{I}, \tag{8}$$

where I is the additive group of integers.

Proof. Since $H_F^p(X) = Z_F^p(X)/B_F^p(X)$, (7) is equivalent (see **I.3.1**, exercise 3) to the relation $Z_F^p(X) = B_F^p(X)$, which in turn is equivalent to the relation

$$Z_F^p(X) \subset B_F^p(X) \quad \text{for} \quad p \neq 0, \tag{9}$$

since $B_F^p(X)$ is a subgroup of $Z_F^p(X)$. Now if z^p is any element of $Z_F^p(X)$, then $\delta_X^p z^p = 0$, and (4) yields (see **§ I.4**)

$$z^p = \delta_X^{p-1} \Gamma_v^p z^p \in B_F^p(X).$$

As z^p was an arbitrary element of $Z_F^p(X)$, (9) follows. Turning now to (8), observe first that $B_F^0(X) = \delta_X^{-1} C_F^{-1}(X) = 0$ since $C_F^{-1}(X) = 0$. Thus (8) is equivalent (see **I.3.1**, exercise 2) to the relation $Z_F^0(X) \approx \mathsf{I}$, and this last relation has already been established in remark 1.

I.5.3. Induced cochain mappings. Given two sets $X \neq \emptyset$ and $Y \neq \emptyset$, consider a mapping (see **I.1.1**)

$$f: X \to Y. \tag{1}$$

Then f gives rise, for every integer p, to a homomorphism

$$f_F^p: C_F^p(Y) \to C_F^p(X) \tag{2}$$

in the following manner. If $p \geqq 0$ and $c_Y^p \in C_F^p(Y)$, then the image $f_F^p c_Y^p$ of c_Y^p is the p-function for X defined by the formula

$$(f_F^p c_Y^p)(x_0, \ldots, x_p) = c_Y^p(f x_0, \ldots, f x_p), \quad p \geqq 0. \tag{3}$$

If $p < 0$, then both of the groups $C_F^p(X)$, $C_F^p(Y)$ are trivial, and f_F^p is the trivial zero-homomorphism, in accordance with **I.3.2**. It is immediate that

$$f_F^p(c_Y^p + d_Y^p) = f_F^p c_Y^p + f_F^p d_Y^p$$

for any two elements c_Y^p, d_Y^p of $C_F^p(Y)$, and thus f_F^p is a homomorphism.

Lemma 1. The homomorphisms f_F^p constitute a cochain mapping from $M_F(Y)$ to $M_F(X)$.

Proof. The assertion means (see **I.4.2**) that

$$\delta_X^p f_F^p c_Y^p = f_F^{p+1} \delta_Y^p c_Y^p \tag{4}$$

for $c_Y^p \in C_F^p(Y)$. If $p < 0$, then $c_Y^p = 0$ and (4) is obvious. If $p \geqq 0$, then on choosing an arbitrary array of $p+2$ points $x_0, \ldots, x_{p+1} \in X$, direct calculation yields

$$\begin{aligned}(\delta_X^p f_F^p c_Y^p)(x_0, \ldots, x_{p+1}) &= \sum_{j=0}^{p+1} (-1)^j (f_F^p c_Y^p)(x_0, \ldots, \hat{x}_j, \ldots, x_{p+1}) \\ &= \sum_{j=0}^{p+1} (-1)^j c_F^p(f x_0, \ldots, \widehat{f x_j}, \ldots, f x_{p+1}) \\ &= (\delta_Y^p c_Y^p)(f x_0, \ldots, f x_{p+1}) = (f_F^{p+1} \delta_Y^p c_Y^p)(x_0, \ldots, x_{p+1}),\end{aligned}$$

and (4) is verified.

The cochain mapping considered in lemma 1 is termed the cochain mapping *induced* by the mapping f.

Let there be given three sets $X \neq \emptyset$, $Y \neq \emptyset$, $Z \neq \emptyset$, and mappings

$$f: X \to Y, \tag{5}$$

$$g: Y \to Z. \tag{6}$$

Then we have also the product mapping (see **I.1.1**)

$$g f: X \to Z. \tag{7}$$

The mappings (5), (6), (7) induce cochain mappings

$$f_F^p: C_F^p(Y) \to C_F^p(X), \tag{8}$$

$$g_F^p: C_F^p(Z) \to C_F^p(Y), \tag{9}$$

$$(g f)_F^p: C_F^p(Z) \to C_F^p(X). \tag{10}$$

The mappings (5) to (10) give rise to the following diagram.

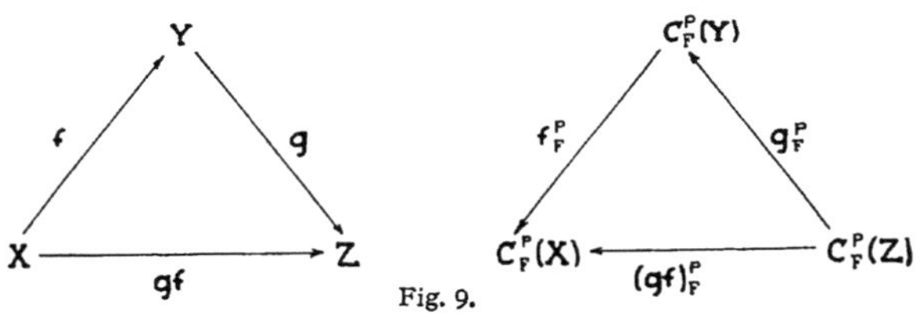

Fig. 9.

Lemma 2 (the *vector law* for induced cochain mappings). Under the conditions just stated, we have the relation

$$(g f)_F^p = f_F^p g_F^p. \tag{11}$$

Proof. Explicitly, (11) means that

$$(g f)_F^p c_Z^p = f_F^p g_F^p c_Z^p \quad \text{for} \quad c_Z^p \in C_F^p(Z). \tag{12}$$

If $p < 0$, then $c_Z^p = 0$ and (12) is obvious. If $p \geq 0$, then on choosing an arbitrary array of $p+1$ points $x_0, \ldots, x_p \in X$, direct calculation yields:

$$\begin{aligned}[(g f)_F^p c_Z^p](x_0, \ldots, x_p) &= c_Z^p(g f x_0, \ldots, g f x_p) \\ &= (g_F^p c_Z^p)(f x_0, \ldots, f x_p) = (f_F^p g_F^p c_Z^p)(x_0, \ldots, x_p),\end{aligned}$$

and (12) is proved.

Remark 1. Given a set $X \neq \emptyset$, let $i: X \to X$ denote the identity mapping in X (that is, $i x = x$ for every point $x \in X$). Then the corresponding induced homomorphism $i_F^p: C_F^p(X) \to C_F^p(X)$ is the identity homomorphism in $C_F^p(X)$. Explicitly:

$$i_F^p c^p = c^p \quad \text{for} \quad c^p \in C_F^p(X). \tag{13}$$

If $p < 0$, then $c^p = 0$ and (13) is obvious. If $p \geq 0$, then

$$(i_F^p c^p)(x_0, \ldots, x_p) = c^p(i x_0, \ldots, i x_p) = c^p(x_0, \ldots, x_p)$$

for arbitrary points $x_0, \ldots, x_p \in X$, and (13) follows.

Remark 2. Suppose that $X \neq \emptyset$ is a subset of Y. Then the identity mapping i in X is also a mapping from X into Y, and hence we have induced homomorphisms

$$i_F^p: C_F^p(Y) \to C_F^p(X). \tag{14}$$

Observe that the symbol i_F^p has altogether different meanings in (13) and (14). As we noted earlier, completely descriptive notations in this theory would lead to prohibitive complications, and accordingly we shall have to use (as in the present case) identical symbols with different meanings in different situations. Returning to (14), the following comments are instructive. Take $p \geq 0$, and consider a p-function c_Y^p for Y. On choosing an array of $p+1$ points $x_0, \ldots, x_p \in X$, the inclusion $X \subset Y$ implies that $c_Y^p(x_0, \ldots, x_p)$ is a well-defined integer. In other words, on restricting the arguments of c_Y^p to the subset X of Y, we obtain a p-function for X which we denote by $c_Y^p | X$ (representing the function c_Y^p *cut down* to the subset X). We have then the relation

$$i_F^p c_Y^p = c_Y^p | X. \tag{15}$$

Indeed, for arbitrary choice of the points $x_0, \ldots, x_p \in X$, we obtain

$$\begin{aligned}(i_F^p c_Y^p)(x_0, \ldots, x_p) &= c_Y^p(i x_0, \ldots, i x_p) \\ &= c_Y^p(x_0, \ldots, x_p) = (c_Y^p | X)(x_0, \ldots, x_p).\end{aligned}$$

Remark 3. Consider the following diagram.

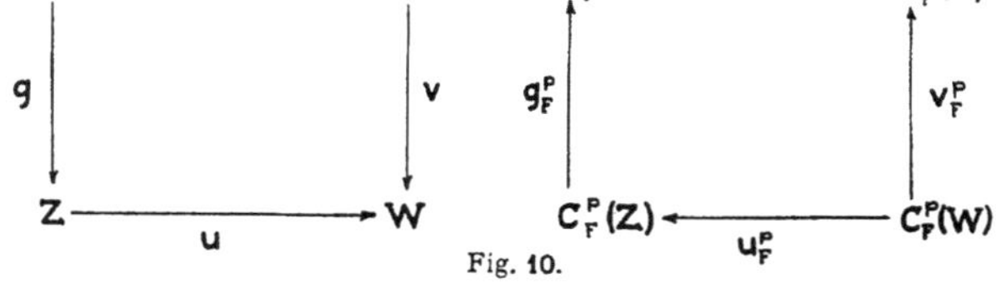

Fig. 10.

In this diagram, X, Y, Z, W denote non-empty sets, f, g, u, v denote mappings as indicated by the arrows, and $f_F^p, g_F^p, u_F^p, v_F^p$ designate the corresponding induced homomorphisms. Then vf and ug are mappings from X into W which induce homomorphisms $(vf)_F^p$ and $(ug)_F^p$ respectively from $C_F^p(W)$ into $C_F^p(X)$. Suppose now that

$$vf = ug. \tag{16}$$

Then we assert that

$$f_F^p v_F^p = g_F^p u_F^p. \tag{17}$$

Indeed, in view of (16) we conclude from lemma 2 that

$$f_F^p v_F^p = (vf)_F^p = (ug)_F^p = g_F^p u_F^p,$$

and (17) is verified.

Remark 4. Given the mapping $f: X \to Y$ as in (1), choose a point $v \in X$. Then we have (see **I.5.2**) the cone homomorphism

$$\Gamma_v^p : C_F^p(X) \to C_F^{p-1}(X).$$

Similarly, the image point $fv \in Y$ gives rise to the cone homomorphism

$$\Gamma_{fv}^p : C_F^p(Y) \to C_F^{p-1}(Y).$$

There arises the following diagram (in which the homomorphisms f_F^p and f_F^{p-1} are induced by the mapping f).

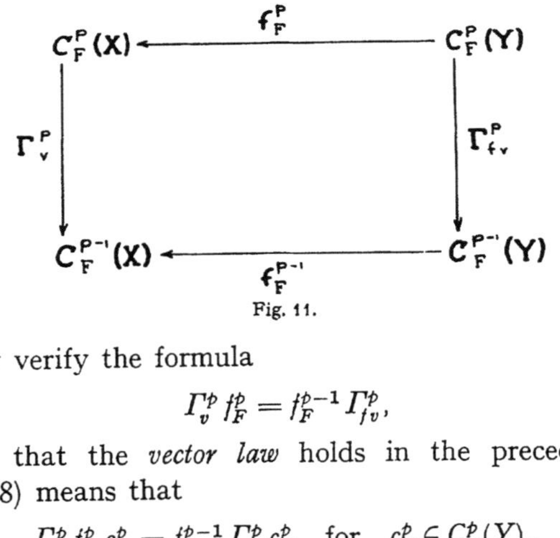

Fig. 11.

We shall now verify the formula

$$\Gamma_v^p f_F^p = f_F^{p-1} \Gamma_{fv}^p, \tag{18}$$

which states that the *vector law* holds in the preceding diagram. Explicitly, (18) means that

$$\Gamma_v^p f_F^p c_Y^p = f_F^{p-1} \Gamma_{fv}^p c_Y^p \quad \text{for} \quad c_Y^p \in C_F^p(Y). \tag{19}$$

If $p \leq 0$, then (19) reduces to $0 = 0$ (since $C_F^{p-1}(X) = 0$ for $p \leq 0$). If $p \geq 1$, then on choosing an arbitrary array of p points $x_0, \ldots, x_{p-1} \in X$, we obtain [(see **I.5.1** (8)]:

$$\begin{aligned}(\Gamma_v^p f_F^p c_Y^p)(x_0, \ldots, x_{p-1}) &= (-1)^p (f_F^p c_Y^p)(x_0, \ldots, x_{p-1}, v) \\ &= (-1)^p c_Y^p(f x_0, \ldots, f x_{p-1}, fv) = (\Gamma_{fv}^p c_Y^p)(f x_0, \ldots, f x_{p-1}) \\ &= (f_F^{p-1} \Gamma_{fv}^p c_Y^p)(x_0, \ldots, x_{p-1}),\end{aligned}$$

and (19) follows.

I.5.4. The homotopy operator D^p. Let there be given two sets $X \neq \emptyset$, $Y \neq \emptyset$, and two mappings

$$f: X \to Y, \quad g: X \to Y, \tag{1}$$

from X into Y. We have then the corresponding induced homomorphisms

$$f_F^p: C_F^p(Y) \to C_F^p(X), \quad g_F^p: C_F^p(Y) \to C_F^p(X). \tag{2}$$

In terms of f and g the *homotopy operator* D^p is defined as a homomorphism

$$D^p: C_F^p(Y) \to C_F^{p-1}(X), \tag{3}$$

in the following manner. If $p \leq 0$, then $C_F^{p-1}(X) = 0$, and D^p is the trivial zero-homomorphism, in accordance with **I.3.2**. If $p \geq 1$, then the image $D^p c_Y^p$ of an element $c_Y^p \in C_F^p(Y)$ is the $(p-1)$-function for X defined by the formula

$$(D^p c_Y^p)(x_0, \ldots, x_{p-1}) = \sum_{j=0}^{p-1} (-1)^j c_Y^p(f x_0, \ldots, f x_j, g x_j, \ldots, g x_{p-1}). \tag{4}$$

For example, for $p = 1, 2, 3$ the detailed expressions for D^p are as follows:

$$(D^1 c_Y^1)(x_0) = c_Y^1(f x_0, g x_0),$$

$$(D^2 c_Y^2)(x_0, x_1) = c_Y^2(f x_0, g x_0, g x_1) - c_Y^2(f x_0, f x_1, g x_1),$$

$$(D^3 c_Y^3)(x_0, x_1, x_2) = c_Y^3(f x_0, g x_0, g x_1, g x_2) -$$
$$- c_Y^3(f x_0, f x_1, g x_1, g x_2) + c_Y^3(f x_0, f x_1, f x_2, g x_2).$$

Clearly, if c_Y^p, d_Y^p are any two elements of $C_F^p(Y)$, then

$$D^p(c_Y^p + d_Y^p) = D^p c_Y^p + D^p d_Y^p,$$

and thus D^p is a homomorphism. Observe that the defining formula (4) is *not* symmetric in f and g.

The homotopy operator D^p is related to the cone homomorphism [see **I.5.1** (8), **I.5.2**] in the following manner. Take an arbitrary vertex $v \in X$ and an integer $p \geq 1$. Then we have the cone homomorphism

$$\Gamma_v^p: C_F^p(X) \to C_F^{p-1}(X).$$

The images fv, gv of v give rise, in Y, to the cone homomorphisms

$$\Gamma_{gv}^{p+1}: C_F^{p+1}(Y) \to C_F^p(Y), \quad \Gamma_{fv}^p: C_F^p(Y) \to C_F^{p-1}(Y).$$

We have then the following two identities (the straightforward verifications are left to the reader).

$$\Gamma_v^p D^{p+1} c_Y^{p+1} = -D^p \Gamma_{gv}^{p+1} c_Y^{p+1} - f_F^{p-1} \Gamma_{fv}^p \Gamma_{gv}^{p+1} c_Y^{p+1}, \quad p \geq 1, \tag{5}$$

where $c_Y^{p+1} \in C_F^{p+1}(Y)$. In terms of the auxiliary expression

$$\Delta^p c_Y^p = \delta_X^{p-1} D^p c_Y^p + D^{p+1} \delta_Y^p c_Y^p - g_F^p c_Y^p + f_F^p c_Y^p, \quad c_Y^p \in C_F^p(Y), \tag{6}$$

we have the identity

$$\Gamma_v^{p+1} \Delta^{p+1} c_Y^{p+1} = \Delta^p \Gamma_{gv}^{p+1} c_Y^{p+1} \quad \text{for} \quad c_Y^{p+1} \in C_F^{p+1}(Y). \tag{7}$$

The fundamental property of the homotopy operator D^p is expressed by the *homotopy identity* (see **I.4.5**)

$$\delta_X^{p-1} D^p c_Y^p + D^{p+1} \delta_Y^p c_Y^p = g_F^p c_Y^p - f_F^p c_Y^p, \quad c_Y^p \in C_F^p(Y). \tag{8}$$

The proof to be presented here is based on induction with respect to p. If $p<0$, then $c_Y^p = 0$ and (8) is trivial. If $p=0$, then (8) reduces to

$$D^1 \delta_Y^0 c_Y^0 = g_F^0 c_Y^0 - f_F^0 c_Y^0. \tag{9}$$

On choosing an arbitrary point $x_0 \in X$, direct calculation yields

$$\begin{aligned}(D^1 \delta_Y^0 c_Y^0)(x_0) &= (\delta_Y^0 c_Y^0)(f x_0, g x_0) \\ &= c_Y^0(g x_0) - c_Y^0(f x_0) = (g_F^0 c_Y^0 - f_F^0 c_Y^0)(x_0),\end{aligned}$$

and (9) follows. A similar explicit calculation, left to the reader, shows that (8) holds for $p=1$. Proceeding by induction, we assume that (8) is known to hold for a certain integer $p \geq 1$. In terms of Δ^p, defined in (6), the induction assumption appears in the form

$$\Delta^p c_Y^p = 0 \quad \text{for} \quad c_Y^p \in C_F^p(Y), \tag{10}$$

and we have to prove, on the basis of (10), that

$$\Delta^{p+1} c_Y^{p+1} = 0 \quad \text{for} \quad c_Y^{p+1} \in C_F^{p+1}(Y). \tag{11}$$

Take any element $c_Y^{p+1} \in C_F^{p+1}(Y)$ and choose a point $v \in X$. On setting

$$c_Y^p = \Gamma_{gv}^{p+1} c_Y^{p+1},$$

we obtain from (10) and (7) the relation

$$\Gamma_v^{p+1} \Delta^{p+1} c_Y^{p+1} = \Delta^p \Gamma_{gv}^{p+1} c_Y^{p+1} = \Delta^p c_Y^p = 0. \tag{12}$$

Since the vertex $v \in X$ was arbitrary, (12) implies (11) by **I.5.1**, lemma 2, and the inductive proof of (8) is complete.

I.5.5. The formal complex $M_F(X, A)$. Given a set $X \neq 0$ and a subset A of X, take an integer $p \geq 0$. Then an element c^p of the group $C_F^p(X)$ is said to vanish on A if $c^p(x_0, \ldots, x_p) = 0$ whenever $x_0, \ldots, x_p \in A$. Clearly, those elements $c^p \in C_F^p(X)$ that vanish on A form a subgroup of $C_F^p(X)$ which will be denoted by $C_F^p(X, A)$. For $p<0$, we have

$C_F^p(X)=0$, and in this case we set $C_F^p(X, A)=C_F^p(X)=0$. We assert that

$$\delta_X^p C_F^p(X, A) \subset C_F^{p+1}(X, A). \tag{1}$$

If $p<0$, then $C_F^p(X, A)=0$ and (1) is obvious. So assume that $p \geq 0$, and let c^p be any element of $C_F^p(X, A)$. On choosing an arbitrary array of points $a_0, \ldots, a_{p+1} \in A$, we obtain

$$(\delta_X^p c^p)(a_0, \ldots, a_{p+1}) = \sum_{j=0}^{p+1} (-1)^j c^p(a_0, \ldots, \hat{a}_j, \ldots, a_{p+1}) = 0,$$

since $a_0, \ldots, \hat{a}_j, \ldots, a_{p+1} \in A$ and c^p vanishes on A. Thus $\delta_X^p c^p \in C_F^{p+1}(X, A)$, and (1) follows. Accordingly (see **I.4.6**), the groups $C_F^p(X, A)$ and the homomorphisms δ_X^p constitute a subcomplex

$$M_F(X, A) = \left(C_F^p(X, A), \delta_X^p\right) \tag{2}$$

of the complex $M_F(X)$. The groups of p-cocycles and p-coboundaries of the complex (2) will be denoted by $Z_F^p(X, A)$, $B_F^p(X, A)$ and will be termed the groups of p-cocycles and p-coboundaries respectively of the pair (X, A). Similarly, the cohomology groups of the complex (2) will be denoted by $H_F^p(X, A)$ and will be termed the cohomology groups of the pair (X, A). The MAYER complex $M_F(X, A)$ will be referred to as the *formal complex* of the pair (X, A). Let us recall that the concept of a pair (X, A) implies that $A \subset X$ (see **I.1.1**).

Remark 1. Actually, we should write $\delta_X^p \mid C_F^p(X, A)$ in (2), in conformity with **I.1.1**, since δ_X^p is now operating from the subgroup $C_F^p(X, A)$ of $C_F^p(X)$. However, since this fact is clear from the context, we continue to use the symbol δ_X^p for the sake of simplicity. Let us note that clearly $Z_F^p(X, A)$ and $B_F^p(X, A)$ are subgroups of $Z_F^p(X)$ and $B_F^p(X)$ respectively. For $p=0$, it follows that $Z_F^0(X, A)=0$ if $A \neq \emptyset$. Indeed, if $z^0 \in Z_F^0(X, A)$, then z^0 is constant on X by **I.5.2**, remark 1, and since z^0 vanishes on $A \neq \emptyset$, it follows that z^0 vanishes identically on X.

Remark 2. If $A = \emptyset$, then every element c^p of $C_F^p(X)$ vacuously satisfies the condition of vanishing on A. Accordingly, we have

$$C_F^p(X, \emptyset) = C_F^p(X), \quad Z_F^p(X, \emptyset) = Z_F^p(X), \quad B_F^p(X, \emptyset) = B_F^p(X),$$
$$H_F^p(X, \emptyset) = H_F^p(X), \quad M_F(X, \emptyset) = M_F(X).$$

Lemma 1. If (X, A) is a pair of sets such that $A \neq \emptyset$, then

$$H_F^p(X, A) = 0 \quad \text{for every} \quad p. \tag{3}$$

Proof. In view of **I.3.1**, exercise 3, (3) is equivalent to the relation $Z_F^p(X, A)=B_F^p(X, A)$, which in turn is equivalent to the relation

$$Z_F^p(X, A) \subset B_F^p(X, A), \tag{4}$$

since $B_F^p(X, A) \subset Z_F^p(X, A)$. If $p=0$, then $Z_F^0(X, A)=0$ since $A \neq \emptyset$ (see remark 1) and (4) is obvious. If $p<0$, then $Z_F^p(X, A)=0$ and (4) is again trivial. Consider finally the case $p \geq 1$. Observe that $z^p \in Z_F^p(X, A)$ implies that $z^p \in Z_F^p(X)$, and hence **I.5.2** (4) yields

$$z^p = \delta_X^{p-1} \Gamma_v^p z^p. \tag{5}$$

Since $A \neq \emptyset$, we can select the vertex v as a point of A. We assert then that

$$\Gamma_v^p z^p \in C_F^{p-1}(X, A). \tag{6}$$

Indeed, on choosing an arbitrary array of points $a_0, \ldots, a_{p-1} \in A$, we obtain

$$(\Gamma_v^p z^p)(a_0, \ldots, a_{p-1}) = (-1)^p z^p(a_0, \ldots, a_{p-1}, v) = 0,$$

since $a_0, \ldots, a_{p-1}, v \in A$ and z^p vanishes on A. Thus $\Gamma_v^p z^p$ vanishes on A, and (6) is verified. From (5) and (6) it follows that $z^p \in B_F^p(X, A)$, and (4) is proved.

Remark 3. Observe that (3) fails to hold if $p=0$ and $A=\emptyset$, since then [see remark 2 and **I.5.2** (8)] $H_F^0(X, \emptyset) = H_F^0(X) \approx \mathsf{I}$.

Let there be given two sets $X \neq \emptyset$, $Y \neq \emptyset$, and a mapping

$$f: X \to Y. \tag{7}$$

We have then the induced homomorphisms (see **I.5.3**)

$$f_F^p: C_F^p(Y) \to C_F^p(X), \tag{8}$$

which satisfy the relation

$$\delta_X^p f_F^p = f_F^{p+1} \delta_Y^p. \tag{9}$$

Let A and B be subsets of X and Y respectively, such that (see **I.1.1**)

$$f: (X, A) \to (Y, B). \tag{10}$$

We assert then that

$$f_F^p C_F^p(Y, B) \subset C_F^p(X, A). \tag{11}$$

If $p<0$, then $C_F^p(Y, B)=0$ and (11) is obvious. So assume that $p \geq 0$, and consider any element $c^p \in C_F^p(Y, B)$. On choosing an arbitrary array of points $a_0, \ldots, a_p \in A$, we obtain

$$(f_F^p c^p)(a_0, \ldots, a_p) = c^p(f a_0, \ldots, f a_p) = 0,$$

since $f a_0, \ldots, f a_p \in B$ by (10) and c^p vanishes on B. Thus $f_F^p c^p$ vanishes on A. Hence $f_F^p c^p \in C_F^p(X, A)$, and (11) follows. Setting (see **I.1.1**)

$$f_{\# F}^p = f_F^p | C_F^p(Y, B), \tag{12}$$

we obtain in view of (11) homomorphisms

$$f_{\# F}^p: C_F^p(Y, B) \to C_F^p(X, A), \tag{13}$$

which by (9) satisfy the relations

$$\delta_X^p f_{\#F}^p = f_{\#F}^{p+1} \delta_Y^p. \tag{14}$$

Thus the homomorphisms $f_{\#F}^p$ constitute a cochain mapping (see **I.4.2**) from $M_F(Y, B)$ to $M_F(X, A)$ which is said to be induced by the mapping (10).

If (X, A), (Y, B), (Z, C) are three pairs of sets, and if there are given mappings $f: (X, A) \to (Y, B)$, $g: (Y, B) \to (Z, C)$, then lemma 2 in **I.5.3** yields [see (12)] the *vector law*

$$(g f)_{\#F}^p = f_{\#F}^p g_{\#F}^p. \tag{15}$$

Remark 4. If $A = \emptyset$, $B = \emptyset$ in the situation considered in (7) to (14), then clearly $f_{\#F}^p$ reduces to f_F^p. If $A = \emptyset$, $B \neq \emptyset$, then (11) reduces to $f_F^p C_F^p(Y, B) \subset C_F^p(X, \emptyset)$, an obvious relation, since $C_F^p(Y, B) \subset C_F^p(Y)$ and hence

$$f_F^p C_F^p(Y, B) \subset f_F^p C_F^p(Y) \subset C_F^p(X) = C_F^p(X, \emptyset).$$

Similarly, the reader will readily see that the vector law (15) remains meaningful and valid in the special cases when $A = \emptyset$, $B = \emptyset$, $C = \emptyset$, or $A = \emptyset$, $B = \emptyset$, $C \neq \emptyset$, or $A = \emptyset$, $B \neq \emptyset$, $C \neq \emptyset$.

Consider now two pairs of sets (X, A), (Y, B) and two mappings

$$f: (X, A) \to (Y, B), \quad g: (X, A) \to (Y, B). \tag{16}$$

The relations (16) imply that

$$f: X \to Y, \quad g: X \to Y, \tag{17}$$

$$f A \subset B, \quad g A \subset B. \tag{18}$$

Corresponding to the mappings (17) we have (see **I.5.4**) the homotopy operator

$$D^p: C_F^p(Y) \to C_F^{p-1}(X). \tag{19}$$

We assert that

$$D^p C_F^p(Y, B) \subset C_F^{p-1}(X, A). \tag{20}$$

If $p \leq 0$, then D^p is a zero-homomorphism, and (20) is obvious. So assume that $p \geq 1$, and consider any element c of $C_F^p(Y, B)$. On choosing an arbitrary array of points $a_0, \ldots, a_{p-1} \in A$, we obtain

$$(D^p c)(a_0, \ldots, a_{p-1}) = \sum_{j=0}^{p-1} (-1)^j c(f a_0, \ldots, f a_j, g a_j, \ldots, g a_{p-1}) = 0,$$

since $f a_0, \ldots, f a_j, g a_j, \ldots, g a_{p-1} \in B$ by (18) and c vanishes on B. Thus $D^p c$ vanishes on A. Hence $D^p c \in C_F^{p-1}(X, A)$, and (20) follows. On setting (see **I.1.1**)

$$D_{\#}^p = D^p \mid C_F^p(Y, B), \tag{21}$$

we obtain in view of (20) homomorphisms

$$D_{\#}^{p}: C_F^{p}(Y, B) \to C_F^{p-1}(X, A), \tag{22}$$

for which the homotopy identity in **I.5.4** yields directly [see (12)] the identity

$$\delta_X^{p-1} D_{\#}^{p} c + D_{\#}^{p+1} \delta_Y^{p} c = g_{\# F}^{p} c - f_{\# F}^{p} c, \quad c \in C_F^{p}(Y, B). \tag{23}$$

Remark 5. The reader will readily see that (20) to (23) remain trivially meaningful and valid in the special cases when $A = \emptyset$, $B = \emptyset$ or $A = \emptyset$, $B \neq \emptyset$.

I.5.6. Excision. Given a pair (X, A) and a set $U \subset A$, the pair $(X - U, A - U)$ is thought of as being derived from the pair (X, A) by the process of *excising* the set U. If i is the identity mapping in $X - U$ (that is, $i x = x$ for $x \in X - U$), then clearly

$$i: (X - U, A - U) \to (X, A)$$

in the sense of **I.1.1**. This mapping is termed an *excision mapping*. It induces (see **I.5.5**) homomorphisms

$$i_{\# F}^{p}: C_F^{p}(X, A) \to C_F^{p}(X - U, A - U), \tag{1}$$

which constitute a cochain mapping. To avoid trivialities, we assume that

$$A \neq \emptyset, \quad U \neq \emptyset, \quad X - A \neq \emptyset. \tag{2}$$

The case when simultaneously

$$A - U = \emptyset \quad \text{and} \quad p = 1 \tag{3}$$

will be termed *the exceptional case*. The following diagram will be helpful to the reader in connection with the lemmas to be discussed presently.

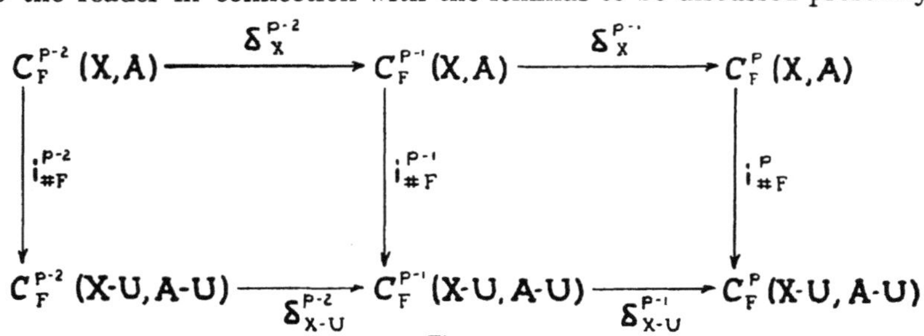

Fig. 12.

Lemma 1. Assuming (2), we have

$$i_{\# F}^{p} C_F^{p}(X, A) = C_F^{p}(X - U, A - U). \tag{4}$$

Proof. We have to show that the homomorphism (1) is *onto*. If $p < 0$, then both of the groups occurring in (1) are trivial, and (4) is obvious. So assume that $p \geqq 0$, and assign any element

$$c^p \in C_F^p(X - U, A - U). \tag{5}$$

Define a p-function d^p on X as follows:

$$d^p(x_0, \ldots, x_p) = \begin{Bmatrix} c^p(x_0, \ldots, x_p) & \text{if} \quad x_0, \ldots, x_p \in X - U, \\ 0 \text{ otherwise}. & \end{Bmatrix} \tag{6}$$

We assert that

$$d^p \in C_F^p(X, A). \tag{7}$$

To verify (7), we have to show that d^p vanishes on A. Consider any array of points $a_0, \ldots, a_p \in A$. If at least one of these points lies in U, then $d^p(a_0, \ldots, a_p) = 0$ by (6). On the other hand, if $a_0, \ldots, a_p \in A - U$, then (since $A - U \subset X - U$)

$$d^p(a_0, \ldots, a_p) = c^p(a_0, \ldots, a_p) \tag{8}$$

by (6) and

$$c^p(a_0, \ldots, a_p) = 0, \tag{9}$$

since c^p vanishes on $A - U$ by (5). From (8) and (9) it follows that $d^p(a_0, \ldots, a_p) = 0$ in this case also, and the lemma is proved since clearly $i^p_{\# F} d^p = c^p$.

Lemma 2. Assuming (2) and excluding the case when simultaneously $A - U = \emptyset$ and $p = 0$, we have

$$i^p_{\# F} Z_F^p(X, A) = Z_F^p(X - U, A - U). \tag{10}$$

Proof. Since the homomorphisms (1) constitute a cochain mapping, by **I.4.2** we have the inclusion

$$i^p_{\# F} Z_F^p(X, A) \subset Z_F^p(X - U, A - U).$$

Hence (10) will be proved if we can show that

$$i^p_{\# F} Z_F^p(X, A) \supset Z_F^p(X - U, A - U). \tag{11}$$

Since the groups involved in (11) are trivial if $p < 0$, we can assume that $p \geqq 0$. Taking first $p \geqq 1$, we assert that

$$Z_F^p(X - U, A - U) = B_F^p(X - U, A - U) \quad \text{for} \quad p \geqq 1. \tag{12}$$

Indeed, if $A - U \neq \emptyset$, then (12) follows from **I.5.5**, lemma 1. On the other hand, if $A - U = \emptyset$, then by **I.5.5**, remark 2 and **I.5.2**, lemma 2,

$$Z_F^p(X - U, A - U) = Z_F^p(X - U) = B_F^p(X - U) = B_F^p(X - U, A - U),$$

since $p \geqq 1$. Thus (12) is verified. Assign now any p-cocycle

$$z^p \in Z_F^p(X - U, A - U). \tag{13}$$

In view of (12), there exists a $(p-1)$-cochain

$$c^{p-1} \in C_F^{p-1}(X-U, A-U), \tag{14}$$

such that

$$z^p = \delta_{X-U}^{p-1} c^{p-1}. \tag{15}$$

By lemma 1 there exists a $(p-1)$-cochain

$$c_1^{p-1} \in C_F^p(X, A), \tag{16}$$

such that

$$c^{p-1} = i_{\#F}^{p-1} c_1^{p-1}. \tag{17}$$

Let us put

$$z_1^p = \delta_X^{p-1} c_1^{p-1}. \tag{18}$$

Then $z_1^p \in B_F^p(X, A) \subset Z_F^p(X, A)$ by (16), and [see (18), (17), (15)]

$$i_{\#F}^p z_1^p = i_{\#F}^p \delta_X^{p-1} c_1^{p-1} = \delta_{X-U}^{p-1} i_{\#F}^{p-1} c_1^{p-1} = \delta_{X-U}^{p-1} c^{p-1} = z^p.$$

As z^p was an arbitrary element of $Z_F^p(X-U, A-U)$, the inclusion (11) is verified for the case $p \geq 1$. Finally, if $p=0$, then $A-U \neq \emptyset$ by assumption, and hence $Z_F^0(X-U, A-U)=0$ (see **I.5.5**, remark 1). Thus (11) is trivial in this case, and the proof is complete.

Lemma 3. Let there be given a p-cocycle

$$z^p \in Z_F^p(X, A), \tag{19}$$

and a $(p-1)$-cochain

$$c^{p-1} \in C_F^{p-1}(X-U, A-U), \tag{20}$$

such that

$$i_{\#F}^p z^p = \delta_{X-U}^{p-1} c^{p-1}. \tag{21}$$

Assuming (2) and excluding the exceptional case (3), there exists then a $(p-1)$-cochain

$$d^{p-1} \in C_F^{p-1}(X, A), \tag{22}$$

such that

$$\delta_X^{p-1} d^{p-1} = z^p, \quad i_{\#F}^{p-1} d^{p-1} = c^{p-1}. \tag{23}$$

Proof. Note first that $Z_F^p(X, A) = B_F^p(X, A)$ by **I.5.5**, lemma 1, since $A \neq \emptyset$. Thus (19) implies the existence of a $(p-1)$-cochain

$$c_1^{p-1} \in C_F^{p-1}(X, A), \tag{24}$$

such that

$$\delta_X^{p-1} c_1^{p-1} = z^p. \tag{25}$$

On setting $c_2^{p-1} = i_{\#F}^{p-1} c^{p-1}$, we have

$$c_2^{p-1} = i_{\#F}^{p-1} c_1^{p-1} \in C_F^{p-1}(X-U, A-U), \tag{26}$$

and furthermore [see (25)]

$$\delta_{X-U}^{p-1} c_2^{p-1} = \delta_{X-U}^{p-1} i_{\#F}^{p-1} c_1^{p-1} = i_{\#F}^{p} \delta_X^{p-1} c_1^{p-1} = i_{\#F}^{p} z^p. \tag{27}$$

From (21) and (27) it follows that

$$\delta_{X-U}^{p-1} (c^{p-1} - c_2^{p-1}) = 0,$$

and hence

$$c^{p-1} - c_2^{p-1} \in Z_F^{p-1}(X-U, A-U). \tag{28}$$

Since the exceptional case (3) has been excluded, we cannot have simultaneously $p-1=0$ and $A-U=\emptyset$. Accordingly, we can apply lemma 2 (with p replaced by $p-1$), and we conclude thus from (28) the existence of a $(p-1)$-cocycle

$$z_1^{p-1} \in Z_F^{p-1}(X, A),$$

such that

$$i_{\#F}^{p-1} z_1^{p-1} = c^{p-1} - c_2^{p-1}. \tag{29}$$

Consider now the $(p-1)$-cochain

$$d^{p-1} = c_1^{p-1} + z_1^{p-1} \in C_F^{p-1}(X, A). \tag{30}$$

In view of (30), (26), (25), (29) we obtain

$$\delta_X^{p-1} d^{p-1} = \delta_X^{p-1} c_1^{p-1} = z^p, \quad i_{\#F}^{p-1} d^{p-1} = c_2^{p-1} + (c^{p-1} - c_2^{p-1}) = c^{p-1},$$

and (23) is proved.

§ I.6. General cohomology theory[1].

I.6.1. The cohomology groups $H^p(X, A)$. Given a pair (X, A) of topological spaces, where A is a subspace of X, we set up a corresponding Mayer complex $M(X, A)$ as follows. For each integer p, we first define a subgroup $C^p(X, A)$ of the formal cochain group $C_F^p(X, A)$ by the following agreements.

[1] *Cohomology theory* is in a certain sense dual to *homology theory*. For a comprehensive study of the algebraic apparatus and geometrical motivation of *singular homology theory* see the Alexandroff-Hopf treatise listed in the *Bibliography*. There exist several versions of cohomology theory (see the treatises by Lefschetz and Eilenberg-Steenrod, listed in the *Bibliography*, for historical comments and a comprehensive discussion). The idea of defining cohomology groups in terms of p-functions is credited to Alexander and Kolmogoroff. For a detailed discussion of a version of this approach (credited to A. D. Wallace), see E. H. Spanier, Cohomology theory for general spaces [Ann. of Math. **49**, 407—427 (1948)]. For the version used here, see T. Rado, On general cohomology theory [Proc. Amer. Math. Soc. **4**, 244—246 (1953)].

(i) For $p<0$, the group $C_F^p(X, A)$ is trivial, and hence we have no choice but to set $C^p(X, A)=0$.

(ii) For $p\geq 0$, the group $C^p(X, A)$ is defined as consisting of those elements c^p of $C_F^p(X, A)$ that *vanish locally* on X. Explicitly, this conditions means that for every point $x\in X$ there exists an open set $U=U(x, c^p)$, which depends upon both x and c^p, such that $c^p(x_0, \ldots, x_p)=0$ whenever $x_0, \ldots, x_p\in U$. Clearly, these elements c^p form a subgroup of $C_F^p(X, A)$. Explicitly, a p-function c^p, $p\geq 0$, belongs to $C^p(X, A)$ if and only if (α) c^p vanishes locally on X and (β) c^p vanishes on A in the sense that $c^p(x_0, \ldots, x_p)=0$ whenever $x_0, \ldots, x_p$ lie in A.

We observe that $c^p\in C^p(X, A)$ implies that $\delta_X^p c^p\in C^{p+1}(X, A)$. For $p<0$, the assertion is trivial, since then $c^p=0$. For $p\geq 0$, we have surely $\delta_X^p c^p\in C_F^{p+1}(X, A)$, and there remains to show that $\delta_X^p c^p$ vanishes locally on X. Now since $c^p\in C^p(X, A)$, there exists an open covering $\mathfrak{U}$ of X such that c^p vanishes on each set $U\in\mathfrak{U}$. For $x_0, \ldots, x_{p+1}\in U\in\mathfrak{U}$, in the formula

$$(\delta_X^p c^p)(x_0, \ldots, x_{p+1}) = \sum_{j=0}^{p+1} (-1)^j c^p(x_0, \ldots, \hat{x}_j, \ldots, x_{p+1})$$

every term of the summation vanishes, since $x_0, \ldots, \hat{x}_j, \ldots, x_{p+1}\in U$. Thus $\delta_X^p c^p$ vanishes on each set $U\in\mathfrak{U}$. Hence the groups $C^p(X, A)$, jointly with the homomorphisms δ_X^p, constitute a MAYER complex

$$M(X, A) = \{C^p(X, A), \delta_X^p\},$$

which is a subcomplex of the formal complex $M_F(X, A)$. Applying to this situation the general concepts relating to MAYER complexes, we have the groups $Z^p(X, A)$, $B^p(X, A)$ of the p-cocycles and p-coboundaries of the complex $M(X, A)$. The cohomology groups

$$H^p(X, A) = Z^p(X, A)/B^p(X, A)$$

will be referred to as the cohomology groups of the pair (X, A)[1].

While the formal complex $M_F(X, A)$ did not involve topological concepts, the complex $M(X, A)$ depends upon the topology of X and A through the use of locally vanishing p-cochains c^p [see condition (ii)

[1] For the case when A is closed in X (which is the only one relevant for applications in this volume) the cohomology group $H^p(X, A)$ so defined is isomorphic to the cohomology group of dimension $p-1$ occurring in alternative versions. This *shift in dimension* could be avoided, of course, by a mere change of notation. However, since we apply the general theory primarily to subsets of Euclidean spaces, there seems to be a certain advantage in operating with the dimensions directly obtained from the definition. A further aspect of the version adopted here is that it yields directly the *reduced theory*. Thus it is unnecessary to introduce separately the so-called reduced cohomology groups (see the EILENBERG-STEENROD treatise listed in the *Bibliography*).

above]. In point of notation, observe that we use the subscript F whenever the formal complex is considered.

The definition of the complex $M(X, A)$ remains meaningful if $A=\emptyset$. In this case, we write $M(X), C^p(X), \ldots, H^p(X)$ instead of $M(X, \emptyset), C^p(X, \emptyset), \ldots, H^p(X, \emptyset)$. Then $C^p(X)$ for example, for $p \geqq 0$, consists of those elements c^p of the formal cochain group $C_F^p(X)$ that vanish locally on X.

Clearly, $C^p(X, A)$ is a subgroup of $C^p(X)$, and hence $M(X, A)$ is a subcomplex of $M(X)$, since the same homomorphisms δ_X^p are used throughout.

To illustrate and to clarify the meaning of these general concepts, a series of remarks will be discussed presently.

Remark 1. For $p<0$, $C^p(X, A)=0$ by definition, and thus clearly $H^p(X, A)=0$ for $p<0$. In particular $H^p(X)=0$ for $p<0$.

Remark 2. In fact, $C^0(X, A)=0$ always. Indeed, consider any element $c^0 \in C^0(X, A)$. For any point $y \in X$, we must have an open set U containing y such that $c^0(x)=0$ whenever $x \in U$. Since $y \in U$, it follows that $c^0(y)=0$. As y was an arbitrary point of X, it follows that $c^0=0$. Thus $C^0(X, A)$ consists of a zero-element alone, and hence $C^0(X, A)=0$. Accordingly, $H^0(X, A)=0$ always. In particular, $H^0(X)=0$ always.

In view of the preceding two remarks, one sees that $H^p(X, A)=0$, $H^p(X)=0$ for $p \leqq 0$.

Remark 3. $H^p(X, X)=0$ for every p. For $p \leqq 0$, this follows from the preceding remarks. For $p \geqq 1$, an element c^p of $C^p(X, X)$ must vanish, by definition, on X, and hence $c^p=0$. Thus $C^p(X, X)$ consists of a zero-element alone, and the assertion $H^p(X, X)=0$ follows.

Remark 4. If y is any point of X, then $H^p(X, y)=H^p(X)$, for every p. Indeed, one has $M(X, y)=M(X)$. To see this, note that for $p \leqq 0$ one has $C^p(X)=0=C^p(X, y)$ by Remarks 1 and 2. For $p \geqq 1$, consider any p-cochain $c^p \in C^p(X)$. Since c^p vanishes locally on X, we have an open set U such that $y \in U$ and $c^p(x_0, \ldots, x_p)=0$ whenever $x_0, \ldots, x_p \in U$. In particular, it follows that $c^p(y, \ldots, y)=0$. Thus c^p vanishes on y (considered as a subspace of X). This means that $c^p \in C^p(X, y)$, and hence $C^p(X) \subset C^p(X, y)$. Since clearly $C^p(X, y) \subset C^p(X)$, it follows that $C^p(X)=C^p(X, y)$. As the same homomorphisms δ_X^p are used in setting up the complexes $M(X)$ and $M(X, y)$, it follows that $M(X, y)=M(X)$, and hence $H^p(X, y)=H^p(X)$.

Remark 5. If X consists of a single point x, then $H^p(X)=0$ for every p. Indeed, by the Remarks 3 and 4 we have then

$$H^p(X)=H^p(X, x)=H^p(X, X)=0.$$

Remark 6. If $c^p \in C^p(X, A)$, and $p \geq 0$, then $c^p(x, \ldots, x) = 0$ for every point $x \in X$. This follows directly from the fact that c^p vanishes locally on X.

Remark 7. Always $H^1(X, A) \approx Z^1(X, A)$. Indeed, by Remark 2 we have $C^0(X, A) = 0$, and hence $B^1(X, A) = \delta^0_X C^0(X, A) = 0$. Thus $H^1(X, A) = Z^1(X, A)/0 \approx Z^1(X, A)$ (see **I.3.1**, exercise 2). In particular, always $H^1(X) \approx Z^1(X)$.

Remark 8. Let X be a HAUSDORFF space consisting of just two distinct points a and b. Then $H^1(X) \approx \mathsf{I}$ (the additive group of integers).

Proof. In view of Remark 7, we have to show that $Z^1(X) \approx \mathsf{I}$ in this case. To accomplish this, we introduce a homomorphism $h: Z^1(X) \to \mathsf{I}$ by setting $hz = z(a, b)$, where z is a generic notation for an element of $Z^1(X)$. We first show that h is *onto*. To see this, let there be assigned any integer n. Consider the (formal) zero-cochain $c^0_F \in C^0_F(X)$ defined by $c^0_F(a) = 0$, $c^0_F(b) = n$. Then we assert that

$$z = \delta^0_X \, c^0_F \in Z^1(X). \tag{1}$$

Indeed, since $\delta\delta = 0$, surely $z \in Z^1_F(X)$. To verify (1), we have to show yet that z vanishes locally on X. In the present case, this merely means that $z(a, a) = 0$, $z(b, b) = 0$, and these relations are immediate. Indeed,

$$z(a, a) = (\delta^0_X \, c^0_F)(a, a) = c^0_F(a) - c^0_F(a) = 0,$$

$$z(b, b) = (\delta^0_X \, c^0_F)(b, b) = c^0_F(b) - c^0_F(b) = 0.$$

Once (1) is established, we conclude that

$$hz = z(a, b) = (\delta^0_X \, c^0_F)(a, b) = c^0_F(b) - c^0_F(a) = n - 0 = n.$$

Thus h is *onto*. To show that h is an *isomorphism into*, we have to verify that if $z \in Z^1(X)$ and $z(a, b) = 0$, then $z(x_0, x_1) = 0$ for every choice of $x_0, x_1 \in X$. The only choices are now (a, a), (b, b), (a, b), (b, a). Since z vanishes locally on X, we have $z(a, a) = 0$, $z(b, b) = 0$, while $z(a, b) = 0$ by assumption. Finally, since z is a 1-cocycle, we have

$$0 = (\delta^1_X z)(a, b, a) = z(b, a) - z(a, a) + z(a, b) = z(b, a) - 0 + 0.$$

Thus $z(b, a) = 0$ also. Summing up: h is an *isomorphism onto*, and consequently $H^1(X) \approx \mathsf{I}$.

Remark 9. A central objective of cohomology theory is to establish relations between geometrical properties of spaces on the one hand and algebraic properties of their cohomology groups on the other hand. In the way of illustration, we prove the following *theorem*: a topological space X is connected if and only if $H^1(X) = 0$.

It is convenient to divide the proof of this theorem into three parts.

(i) Let $z \in Z^1(X)$. Suppose A and B are subsets of X having a point x in common such that z vanishes on each of the sets A, B. For points $a \in A$, $b \in B$, one finds $z(a, b) = (\delta^1_X z)(a, b, x) = 0$. Thus z clearly vanishes on $A \cup B$.

(ii) Again consider $z \in Z^1(X)$. For a fixed point $x \in X$ let $\mathfrak{U}$ be the class of all open sets containing x on which z vanishes. This class is not empty, since z vanishes locally on X. Denote by U the union of those sets in $\mathfrak{U}$. Clearly z vanishes on U by (i), and thus U is the largest open set containing x on which z vanishes. Let y be any point in $\overline{U}$. Since z vanishes locally on X there is an open set V containing y on which z vanishes. Because V meets U (see **I.1.3**, exercise 2) it follows by (i) that z vanishes on the open set $V \cup U$. But U is the largest open set containing x on which z vanishes. Thus $y \in V \subset U$, and hence U is closed. Summarizing, with each element $z \in Z^1(X)$ and each point $x \in X$ there is associated an open and closed set U containing x on which z vanishes.

(iii) The theorem may now be proved as follows. In view of remark 7 it suffices to show that X is connected if and only if $Z^1(X) = 0$. First, suppose that X is connected. Then the only non-empty set in X which is both open and closed is the space X itself (see **I.1.3**, exercise 13). From (ii) it is thus evident that every $z \in Z^1(X)$ must vanish on X. That is, $Z^1(X) = 0$. Next, assume that X is not connected. Then $X = A \cup B$, where A and B are non-empty disjoint open (and closed) sets (see **I.1.3**, exercise 12). Define

$$c(x) = \begin{cases} 0 & \text{if} \quad x \in A, \\ 1 & \text{if} \quad x \in B, \end{cases}$$

$$z(x, y) = (\delta^0_X c)(x, y) = c(y) - c(x) \quad \text{for} \quad x, y \in X.$$

Obviously z vanishes on each of the sets A, B, and thus it vanishes locally on X. On the other hand, for $a \in A$, $b \in B$, $z(a, b) = 1$. Since $\delta^1_X z = 0$, one finds in z a non-trivial element of $Z^1(X)$. This establishes the theorem.

I.6.2. Induced cochain mappings. Let (X, A), (Y, B) be two pairs of topological spaces, where A is a subspace of X and B is a subspace of Y, and let (see **I.1.5**, definition 2) $f: (X, A) \to (Y, B)$ be a continuous mapping. By **I.5.5**, f induces then a cochain mapping

$$f^p_{\#F}: C^p_F(Y, B) \to C^p_F(X, A).$$

Clearly

$$f^p_{\#F}\, C^p(Y, B) \subset C^p(X, A). \tag{1}$$

In view of **I.4.6**, lemma 3, we infer from (1) that the homomorphisms

$$f_{\#}^{p} = f_{\#F}^{p} \,|\, C^{p}(Y, B) \tag{2}$$

constitute a cochain mapping

$$f_{\#}^{p} : C^{p}(Y, B) \to C^{p}(X, A).$$

By **I.4.2**, we obtain corresponding induced homomorphisms

$$f_{\#*}^{p} : H^{p}(Y, B) \to H^{p}(X, A)$$

for the cohomology groups of the complexes $M(Y, B)$, $M(X, A)$. If $z^{p} \in Z^{p}(Y, B)$, then the homomorphism $f_{\#*}^{p}$ is given explicitly by the formula

$$f_{\#*}^{p}\,[z^{p}]_{YB} = [f_{\#}^{p}\, z^{p}]_{XA},$$

where the square brackets, with subscripts YB, XA denote the cohomology classes [relative to the complexes $M(Y, B)$ and $M(X, A)$ respectively], that contain the indicated p-cocycles.

If (X, A), (Y, B), (Z, C) are three pairs of topological spaces, where A, B, C are subspaces of X, Y, Z respectively, and if $f:(X, A) \to (Y, B)$, $g:(Y, B) \to (Z, C)$ are continuous mappings, then gf: $(X, A) \to (Z, C)$ is also a continuous mapping. Accordingly, we have the induced homomorphisms indicated by the diagram in Fig. 13.

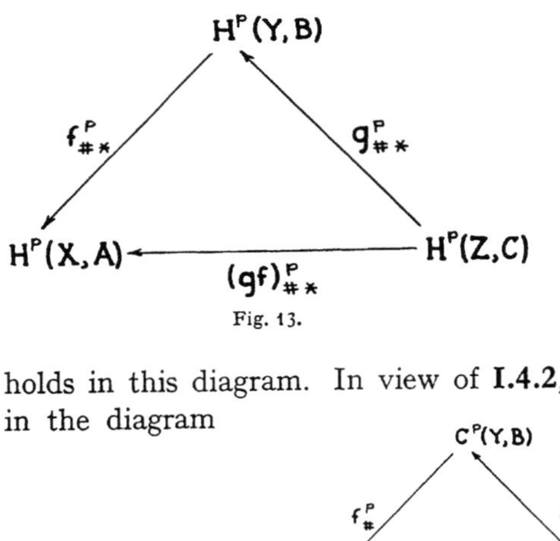

Fig. 13.

We assert that the *vector law*

$$(g\,f)_{\#*}^{p} = f_{\#*}^{p}\, g_{\#*}^{p}$$

holds in this diagram. In view of **I.4.2**, it is sufficient to verify that in the diagram

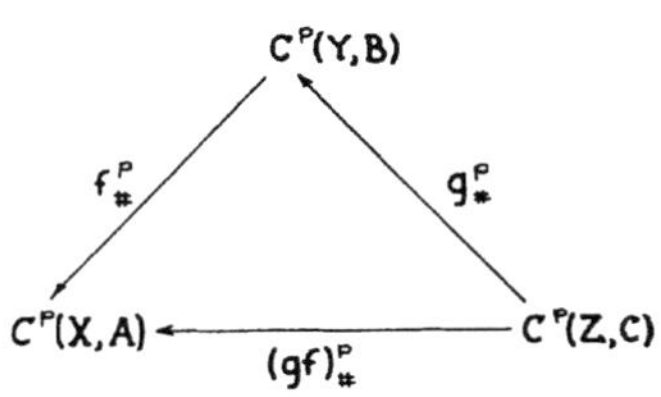

Fig. 14.

the vector law

$$(g\,f)_{\#}^{p} = f_{\#}^{p}\, g_{\#}^{p} \tag{3}$$

holds. In view of (2), the relation (3) is an immediate consequence of the vector law established in **I.5.5**.

The preceding definitions and statements remain obviously meaningful and valid if one or more of the subspaces A, B, C reduces to the empty set $\emptyset$. Whenever an empty subspace occurs, we agree to drop the subscript $\#$ to simplify notations. In particular, if $A=\emptyset$, $B=\emptyset$, $C=\emptyset$, then we have the situation indicated by the following diagrams.

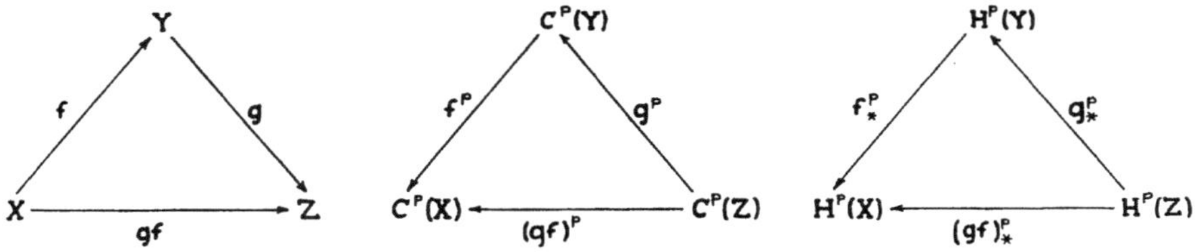

Fig. 15.

The vector law holds throughout. The reader should set up the diagrams corresponding to the cases $A=\emptyset$, $B\neq\emptyset$, $C\neq\emptyset$ and $A=\emptyset$, $B=\emptyset$, $C\neq\emptyset$.

Remark 1. An important special case arises in connection with identity maps. Let A be a subspace of the topological space X, and let $i:X\to X$ be the identity map defined by $ix=x$ for every $x\in X$. Then $i:(X,A)\to(X,A)$ is a continuous mapping and hence we have the induced homomorphisms

$$\left.\begin{aligned} i^p_{\#}&:C^p(X,A)\to C^p(X,A),\\ i^p_{\#*}&:H^p(X,A)\to H^p(X,A).\end{aligned}\right\}\tag{4}$$

In view of (2), it follows from **I.5.3**, remark 1, that $i^p_{\#}c^p=c^p$ for every $c^p\in C^p(X,A)$, and hence obviously $i^p_{\#*}h^p=h^p$ for every $h^p\in H^p(X,A)$. Thus we have the *theorem*: the identity map $i:(X,A)\to(X,A)$ induces the identity homomorphism in the cohomology groups $H^p(X,A)$.

Remark 2. As an immediate consequence, we obtain the *topological invariance* of the cohomology groups in the following precise form. Let (X,A), (Y,B) be two pairs of topological spaces. The pairs (X,A), (Y,B) are termed homeomorphic if there exists a homeomorphism $f:X\to Y$ such that $fA=B$. Then $f:(X,A)\to(Y,B)$. Given such a homeomorphism f, we assert that the induced homomorphism

$$f^p_{\#*}:H^p(Y,B)\to H^p(X,A)$$

is an *isomorphism onto*, for every p. To prove this, let

$$g:(Y,B)\to(X,A)$$

be the inverse of f (since f is a homeomorphism, $g=f^{-1}$ is single-valued and continuous, and $gB=A$). Then $gf=i$ is the identity map in X, and $fg=j$ is the identity map in Y. Using the vector law and remark 1, we infer that

$$\begin{aligned} f^p_{\#*}\,g^p_{\#*}\,u&=u \quad\text{for}\quad u\in H^p(X,A),\\ g^p_{\#*}\,f^p_{\#*}\,v&=v \quad\text{for}\quad v\in H^p(Y,B).\end{aligned}$$

The first relation shows that $f^p_{\#*}$ is a homomorphism *onto* $H^p(X, A)$. The second relation shows that $f^p_{\#*}v = 0$ implies that $v = 0$, and thus $f^p_{\#*}$ is also an *isomorphism into.*

As a matter of fact, it is obvious from the basic definitions that even the cochain groups $C^p(X, A)$ are topologically invariant in a similar sense.

Remark 3. Given a continuous mapping

$$f: (X, A) \to (Y, B), \tag{5}$$

suppose that

$$fX \subset B. \tag{6}$$

Then we have also the relation

$$f: (X, A) \to (B, B), \tag{7}$$

as well as the inclusion mapping

$$i: (B, B) \to (Y, B). \tag{8}$$

The mappings (5), (7), (8) yield the following diagram.

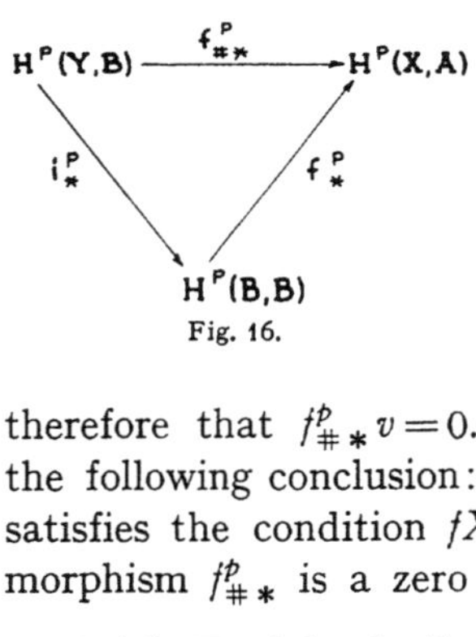

Fig. 16.

The homomorphisms $f^p_{\#*}$, f^p_*, i^p_* are induced by the mappings (5), (7), (8) respectively. If v is any element of $H^p(Y, B)$, then by the vector law

$$f^p_{\#*}v = f^p_* i^p_* v. \tag{9}$$

On the other hand, $i^p_* v = 0$ since $H^p(B, B)$ is trivial by **I.6.1**, remark 3. From (9) it follows therefore that $f^p_{\#*}v = 0$. Since $v \in H^p(Y, B)$ was arbitrary, we obtain the following conclusion: if the continuous mapping $f: (X, A) \to (Y, B)$ satisfies the condition $fX \subset B$, then the corresponding induced homomorphism $f^p_{\#*}$ is a zero homomorphism [that is, $f^p_{\#*}H^p(Y, B) = 0$].

I.6.3. Excision[1]. Let A and U be subspaces of the topological space X, such that $U \subset A$. If $i: X - U \to X - U$ is the identity mapping on $X - U$, then clearly

$$i: (X - U, A - U) \to (X, A). \tag{1}$$

Since the pair $(X - U, A - U)$ is obtained by excising the set U simultaneously from X and A, the mapping (1) is termed, in the

[1] The discussion of *excision* would be considerably simpler if we were operating with the *unreduced theory*. However, from the point of view of actual applications, the advantage of the unreduced theory is perhaps more apparent than real, due to the necessity of introducing separately the reduced cohomology groups (see the EILENBERG-STEENROD treatise listed in the *Bibliography*).

present context, an *excision mapping* (see **I.5.6**). The corresponding homomorphism

$$i^p_{\#*}: H^p(X, A) \to H^p(X - U, A - U) \tag{2}$$

is termed the *excision homomorphism*. A definitive study of the excision homomorphism will be made in **I.6.14**. However, some special lemmas to be used in that connection will be established presently, to provide some immediate applications of the information contained in **I.6.1**, **I.6.2**. The condition $U \subset A$ will be replaced, in these special lemmas, by the stronger requirement

$$\bar{U} \subset \operatorname{int} A. \tag{3}$$

By **I.1.3**, exercise 8, condition (3) is equivalent to

$$X = (\operatorname{int} A) \cup (\operatorname{int} \complement U). \tag{4}$$

To avoid trivialities, we assume that

$$A \neq \emptyset, \quad U \neq \emptyset, \quad X - A \neq \emptyset. \tag{5}$$

The case when simultaneously

$$A - U = \emptyset \quad \text{and} \quad p = 1 \tag{6}$$

will play a special role, and will be termed the *exceptional case* (see **I.5.6**).

Lemma 1. Under the assumptions (3) and (5), the excision homomorphism $i^p_{\#*}$ is *onto.*

Proof. In view of the basic definitions, the assertion is a direct consequence of the relation

$$i^p_{\#} Z^p(X, A) = Z^p(X - U, A - U), \tag{7}$$

which will be verified presently. For $p \leq 0$, the groups involved in (7) are trivial (see **I.6.1**, remarks 1 and 2) and hence we can assume that $p \geq 1$. Assign (see **I.6.1**) any p-cocycle

$$z \in Z^p(X - U, A - U) \subset Z^p_F(X - U, A - U). \tag{8}$$

By **I.5.6**, lemma 2, we have then a p-cocycle

$$z_1 \in Z^p_F(X, A), \tag{9}$$

such that

$$i^p_{\# F} z_1 = z, \tag{10}$$

and clearly (7) is proved if we can show that z_1 vanishes locally on X. So let x be a point of X. If $x \in \operatorname{int} A$, then on choosing $V = \operatorname{int} A$, we have $x \in V$ and $z_1(x_0, \ldots, x_p) = 0$ if $x_0, \ldots, x_p \in V$, since $V \subset A$ and z_1

vanishes on A by (9). If x does not lie in int A, then

$$x \in \operatorname{int} \complement U \tag{11}$$

by (4). By (10) we have

$$z_1(x_0, \ldots, x_p) = z(x_0, \ldots, x_p) \quad \text{for} \quad x_0, \ldots, x_p \in \complement U. \tag{12}$$

By (8), z vanishes locally on $\complement U$, and hence there exists an open set O of X such that $x \in O$ and

$$z(x_0, \ldots, x_p) = 0 \quad \text{if} \quad x_0, \ldots, x_p \in O \cap \complement U. \tag{13}$$

On choosing $V = O \cap \operatorname{int} \complement U$, we obtain an open set of X which contains x [see (11)]. Furthermore, if $x_0, \ldots, x_p \in V$, then by (12) and (13) we see that $z_1(x_0, \ldots, x_p) = 0$. Thus z_1 vanishes locally on X.

Lemma 2. Assuming (3) and (5), and excluding the exceptional case (6), the excision homomorphism $i^p_{\#*}$ is an *isomorphism into.*

Proof. Since the groups involved in (2) are trivial for $p \leq 0$ (see **I.6.1**, remarks 1 and 2) we can assume that $p \geq 1$. In view of the basic definitions, the assertion to be proved is equivalent to the following: if we are given a cocycle

$$z \in Z^p(X, A), \tag{14}$$

such that

$$i^p_{\#} z \in B^p(X - U, A - U), \tag{15}$$

then we have also

$$z \in B^p(X, A). \tag{16}$$

Now (15) means that there exists a cochain c such that

$$c \in C^{p-1}(X - U, A - U), \quad \delta^{p-1}_{X-U} c = i^p_{\#} z. \tag{17}$$

By **I.5.6**, lemma 3, there follows the existence of a cochain c_1 such that

$$c_1 \in C^{p-1}_F(X, A), \tag{18}$$

$$i^{p-1}_{\# F} c_1 = c, \tag{19}$$

$$\delta^{p-1}_X c_1 = z. \tag{20}$$

Thus (16) will be established if we can show that $c_1 \in C^{p-1}(X, A)$, and in view of (18) there remains to show that c_1 vanishes locally on X. Now c_1 vanishes on A by (18), and $c_1(x_0, \ldots, x_{p-1}) = c(x_0, \ldots, x_{p-1})$ for $x_0, \ldots, x_{p-1} \in \complement U$ by (19). Also, c vanishes locally on $\complement U$ by (17). From these facts, jointly with (4), it follows that c_1 vanishes locally on X, by the same argument as the one used in the proof of lemma 1.

Combining the preceding two lemmas, we obtain directly the following statement.

Lemma 3. If $\overline{U} \subset \operatorname{int} A$ and $A - U \neq \emptyset$, then the excision homomorphism $i^p_{\#*}$ is an *isomorphism onto* for every p.

The case when $\overline{U} \subset \operatorname{int} A$ and simultaneously $A - U = \emptyset$ gives rise to the following comments. In this case, we have $A = U$ and hence

$$\overline{A} = \overline{U} \subset \operatorname{int} A \subset A.$$

It follows that $\overline{A} = A = \operatorname{int} A$. Since $\overline{A}$ is closed and $\operatorname{int} A$ is open, it follows that A is both open and closed. Hence, on setting $B = \complement A$, the set B is also both open and closed, and $B \neq \emptyset$, $A \neq \emptyset$ by (5). Thus we have now the following situation:

$$X = A \cup B, \quad A \neq \emptyset, \quad B \neq \emptyset, \quad A \cap B = \emptyset, \tag{21}$$

$$A \text{ open}, \quad B \text{ open}. \tag{22}$$

Accordingly (see **I.1.3**, definition 11) the topological space X is disconnected and the sets A, B constitute a *partition* of X. The identity mapping (1) becomes now

$$i: B \to (X, A), \tag{23}$$

while the excision homomorphism reduces to

$$i^p_{\#*}: H^p(X, A) \to H^p(B). \tag{24}$$

For this case, the lemmas 1 and 2 yield the following information.

Lemma 4. The excision homomorphism (24) is *onto* for every p, and for $p \neq 1$ it is an *isomorphism onto.*

Remark 1. Let us observe that for $p = 1$ the second statement in lemma 4 breaks down. Indeed, let X be a Hausdorff space consisting of two distinct points a and b. Then we can identify the sets A and B with the points a and b respectively, obtaining for $p = 1$ the homomorphism

$$i^1_{\#*}: H^1(X, a) \to H^1(b).$$

By **I.6.1**, remarks 4, 8, 5, we have $H^1(X, a) = H^1(X) \approx \mathsf{I}$, $H^1(b) = 0$, and hence in this case the excision homomorphism is certainly not an isomorphism onto.

Remark 2. Let us add that, in the situation considered in remark 1, we have $H^p(X) = 0$ for $p \neq 1$. Indeed, using lemma 4, jointly with the remarks 4 and 5 in **I.6.1**, we conclude that for $p \neq 1$ we have

$$H^p(X) = H^p(X, a) \approx H^p(b) = 0.$$

Combining this result with remark 8 in **I.6.1**, we obtain the following *theorem*: If X is a Hausdorff space consisting of precisely two points, then $H^p(X) = 0$ for $p \neq 1$ and $H^1(X) \approx \mathsf{I}$, the additive group of integers.

I.6.4. Exactness. Let X be a topological space, and let A, B be closed subsets of X such that

$$X \supset A \supset B. \tag{1}$$

Then A and B, as subspaces of X, give rise to the three pairs (X, A), (X, B), (A, B). In view of (1) we have then the inclusion mappings

$$i\colon (X, B) \to (X, A), \tag{2}$$

$$j\colon (A, B) \to (X, B), \tag{3}$$

which in turn give rise to cochain mappings that we shall denote by

$$i^p\colon C^p(X, A) \to C^p(X, B), \tag{4}$$

$$j^p\colon C^p(X, B) \to C^p(A, B). \tag{5}$$

Thus, for each p, we have the diagram

$$C^p(X, A) \xrightarrow{i^p} C^p(X, B) \longrightarrow C^p(A, B). \tag{6}$$

We shall verify presently that the complexes $M(X, A)$, $M(X, B)$, $M(A, B)$, jointly with the cochain mappings indicated in (6), constitute a *standard triple*, in the sense of **I.4.3**. Since the groups occurring in (6) are trivial for $p \leq 0$ by **I.6.1**, remarks 1 and 2, we have to discuss (6) only for $p \geq 1$.

(a) The homomorphism j^p in (6) is *onto*. Indeed, assign $c \in C^p(A, B)$. Define a p-cochain d on X by the agreement

$$d(x_0, \ldots, x_p) = \begin{cases} c(x_0, \ldots, x_p) & \text{if} \quad x_0, \ldots, x_p \in A, \\ 0 \text{ otherwise}. \end{cases} \tag{7}$$

Since c vanishes on B, clearly d also vanishes on B. Now observe that c vanishes locally on A. Hence, there exists an open covering $\mathfrak{U}_A$ of A such that c vanishes on each set $U_A \in \mathfrak{U}_A$. Since A is a subspace of X, each set $U_A \in \mathfrak{U}_A$ can be written in the form $U_A = U_X \cap A$, where U_X is open in X. Noting that $\complement A$ is open, we define an open covering $\mathfrak{V}_X$ of X as comprised of $\complement A$ and of all the sets U_X just referred to. We assert that d vanishes on each set $V_X \in \mathfrak{V}_X$. Indeed, if $V_X = \complement A$ and $x_0, \ldots, x_p \in V_X$, then $d(x_0, \ldots, x_p) = 0$ by (7). If V_X is one of the sets U_X, then $V_X \cap A \in \mathfrak{U}_A$. If $x_0, \ldots, x_p \in V_X$, then either $x_0, \ldots, x_p \in A$, and then $x_0, \ldots, x_p \in V_X \cap A \in \mathfrak{U}_A$, and hence $d(x_0, \ldots, x_p) = c(x_0, \ldots, x_p) = 0$, since c vanishes on each set of $\mathfrak{U}_A$; or else, the points $x_0, \ldots, x_p$ do not all lie in A, and then $d(x_0, \ldots, x_p) = 0$ by (7). Thus d vanishes on B and vanishes locally on X, and hence $d \in C^p(X, B)$.

Finally, for $x_0, \ldots, x_p \in A$, we have by (7)

$$(j^p d)(x_0, \ldots, x_p) = d(j x_0, \ldots, j x_p) = d(x_0, \ldots, x_p) = c(x_0, \ldots, x_p).$$

Thus $j^p d = c$, proving that j^p is *onto*.

(b) Since $B \subset A$, it is clear that $C^p(X, A)$ is a subgroup of $C^p(X, B)$. For $c \in C^p(X, A)$ and $x_0, \ldots, x_p \in X$, we have $(i^p c)(x_0, \ldots, x_p) = c(i x_0, \ldots, i x_p) = c(x_0, \ldots, x_p)$, and thus $i^p c = c$. Hence $i^p c = 0$ implies that $c = 0$, proving that i^p is an *isomorphism into*.

(c) If $c \in C^p(X, A)$ and $x_0, \ldots, x_p \in A$, then

$$(j^p i^p c)(x_0, \ldots, x_p) = c(i j x_0, \ldots, i j x_p) = c(x_0, \ldots, x_p) = 0,$$

since c vanishes on A. Thus $j^p i^p c = 0$, proving that $i^p C^p(X, A)$ is contained in the nucleus of j^p.

(d) Let $c \in C^p(X, B)$ be such that $j^p c = 0$. For $x_0, \ldots, x_p \in A$ we have then

$$0 = (j^p c)(x_0, \ldots, x_p) = c(j x_0, \ldots, j x_p) = c(x_0, \ldots, x_p).$$

Thus c vanishes on A, and [since $c \in C^p(X, B)$] it vanishes locally on X. Thus $c \in C^p(X, A)$. But then $c = i^p c$, as we observed under (b), proving that the nucleus of j^p is comprised in $i^p C^p(X, A)$. In view of (c) it follows that $i^p C^p(X, A)$ coincides with the nucleus of j^p. Thus we have exactness in (6).

The facts established in (a) to (d) mean that the diagram (6) represents a *standard triple*. Accordingly, by the scheme described in **I.4.3**, we obtain homomorphisms

$$\delta_*^{p-1} : H^{p-1}(A, B) \to H^p(X, A), \tag{8}$$

which jointly with the induced homomorphisms

$$i_*^p : H^p(X, A) \to H^p(X, B), \quad j_*^p : H^p(X, B) \to H^p(A, B),$$

yield the cohomology sequence

$$\cdots \xrightarrow{j_*^{p-1}} H^{p-1}(A, B) \xrightarrow{\delta_*^{p-1}} H^p(X, A) \xrightarrow{i_*^p} H^p(X, B) \xrightarrow{j_*^p} H^p(A, B) \xrightarrow{\delta_*^p} \cdots, \tag{9}$$

corresponding to the complexes $M(X, A)$, $M(X, B)$, $M(A, B)$. The sequence (9) will be termed the *cohomology sequence of the triple* (X, A, B). By **I.4.3** this cohomology sequence possesses the fundamental property of *exactness*.

Remark 1. In establishing this fact, we assumed that $X \supset A \supset B$ and that A and B are closed subsets of X. The assumption that A is closed was used in showing that j^p in (6) is *onto*. On the other hand, the assumption that B is closed was not used at all. The generalization implied by this observation is irrelevant for our purposes.

Remark 2. It is instructive to visualize the explicit description of the homomorphism δ_*^{p-1} occurring in the cohomology sequence (9). For $p \leq 1$, the group $H^{p-1}(A, B)$ is trivial by **I.6.1**, remark 2, and hence it is sufficient to consider the case $p \geq 2$. The generic element of $H^{p-1}(A, B)$ is of the form $[z]_{AB}$, where $z \in Z^{p-1}(A, B)$, and $[z]_{AB}$ indicates the cohomology class containing z, relative to the complex $M(A, B)$. Define a $(p-1)$-cochain d on X by the agreement

$$d(x_0, \ldots, x_{p-1}) = \left\{ \begin{array}{ll} z(x_0, \ldots, x_{p-1}) & \text{if} \quad x_0, \ldots, x_{p-1} \in A, \\ 0 \text{ otherwise}. & \end{array} \right\} \tag{10}$$

By the argument employed under (a) above (in showing that j^p is *onto*), one sees that $d \in C^{p-1}(X, B)$. Let us denote this cochain d by $e_A^X z$, where the symbol e_A^X is interpreted as the operator effecting the *natural extension* from A to X, described explicitly by (10). One verifies readily that on setting

$$z_1 = \delta_X^{p-1} e_A^X z,$$

one has the relation $z_1 \in Z^p(X, A)$. Inspection of the general scheme in **I.4.3** reveals that the homomorphism δ_*^{p-1} in (9) is now described explicitly by the formula

$$\delta_*^{p-1} [z]_{AB} = [\delta_X^{p-1} e_A^X z]_{XA},$$

where $[\ ]_{XA}$ means the cohomology class relative to the complex $M(X, A)$.

Remark 3. An important special case arises if $B = \emptyset$. The cohomology sequence (9) reduces then to

$$\cdots \xrightarrow{j_*^{p-1}} H^{p-1}(A) \xrightarrow{\delta_*^{p-1}} H^p(X, A) \xrightarrow{i_*^p} H^p(X) \xrightarrow{j_*^p} H^p(A) \xrightarrow{\delta_*^p} \cdots, \tag{11}$$

and is referred to as the *cohomology sequence of the pair* (X, A). Thus the sequence (11) is exact. Let us recall that A is assumed to be a closed subset of the topological space X.

I.6.5. Examples. In a topological space S, let A and B be non-empty, closed sets such that $A \cap B = \emptyset$. Then the following holds.

(a) If $H^{p-1}(A) = 0$ and $H^p(A \cup B) = 0$ for a certain integer p, then $H^p(B) = 0$ for the same integer p.

(b) If $H^p(A) = 0$ and $H^p(B) = 0$ for a certain integer $p \neq 1$, then $H^p(A \cup B) = 0$ for the same integer p.

Proof of (a). On setting $X = A \cup B$, the subspace X is a closed subset of S, and hence A and B are closed relative to X also. Since $A \cap B = \emptyset$, it follows that A and B are also open relative to X. By **I.6.3**, lemma 4, the excision homomorphism

$$i^p_{\#*} : H^p(A \cup B, A) \to H^p(B) \tag{1}$$

is therefore *onto*. In the cohomology sequence of the pair $(A \cup B, A)$, the groups $H^{p-1}(A)$, $H^p(A \cup B, A)$, $H^p(A \cup B)$ are consecutive terms. The first and the third of these groups are trivial by assumption. Hence, by **I.3.3**, lemma 3, we have $H^p(A \cup B, A) = 0$. Since the homomorphism (1) is onto, it follows that $H^p(B) = 0$.

Proof of (b). Since $p \neq 1$, the homomorphism (1) is now an isomorphism into, by **I.6.3**, lemma 4. Since $H^p(B) = 0$ by assumption, it follows that

$$H^p(A \cup B, A) = 0. \tag{2}$$

In the cohomology sequence of the pair $(A \cup B, A)$, the groups $H^p(A \cup B, A)$, $H^p(A \cup B)$, $H^p(A)$ are consecutive terms. The first group is trivial by (2), while the third group is trivial by assumption. By **I.3.3**, lemma 3 it follows that $H^p(A \cup B) = 0$.

Remark. Let S be a HAUSDORFF space consisting of just two points a and b. We can then choose $A = a$, $B = b$, and for $p = 1$ we have $H^1(a) = 0$, $H^1(b) = 0$, $H^1(a \cup b) \neq 0$ by **I.6.1**, remarks 5 and 8. Thus the exclusion of the case $p = 1$ is really necessary in statement (b) above.

I.6.6. The vector law for cohomology sequences. Let (X, A, B) and $(\tilde{X}, \tilde{A}, \tilde{B})$ be two triples of topological spaces, such that $X \supset A \supset B$, $\tilde{X} \supset \tilde{A} \supset \tilde{B}$, and A, B are closed subsets of X, and $\tilde{A}$, $\tilde{B}$ are closed subsets of $\tilde{X}$. As in **I.6.4**, A and B are thought of as subspaces of X, and similarly $\tilde{A}$ and $\tilde{B}$ are thought of as subspaces of $\tilde{X}$. Now let $f: X \to \tilde{X}$ be a continuous mapping such that

$$f A \subset \tilde{A}, \quad f B \subset \tilde{B}. \tag{1}$$

We shall refer to these assumptions by writing

$$f : (X, A, B) \to (\tilde{X}, \tilde{A}, \tilde{B}).$$

Clearly, (1) implies the relations $f : (X, A) \to (\tilde{X}, \tilde{A})$, $f : (X, B) \to (\tilde{X}, \tilde{B})$, $f|A : (A, B) \to (\tilde{A}, \tilde{B})$. The corresponding cochain mappings will be denoted by

$$f^p : C^p(\tilde{X}, \tilde{A}) \to C^p(X, A), \quad f^p_1 : C^p(\tilde{X}, \tilde{B}) \to C^p(X, B),$$
$$(f|A)^p : C^p(\tilde{A}, \tilde{B}) \to C^p(A, B).$$

In accordance with **I.6.4**, we have the additional cochain mappings

$$i^p : C^p(X, A) \to C^p(X, B), \quad j^p : C^p(X, B) \to C^p(A, B).$$

The analogous cochain mappings relative to the triple $(\tilde{X}, \tilde{A}, \tilde{B})$ will be denoted by

$$\tilde{i}^p: C^p(\tilde{X}, \tilde{A}) \to C^p(\tilde{X}, \tilde{B}), \quad \tilde{j}^p: C^p(\tilde{X}, \tilde{B}) \to C^p(\tilde{A}, \tilde{B}).$$

The corresponding induced homomorphisms for the cohomology groups will be denoted by $f_*^p, i_*^p, \tilde{i}_*^p, \ldots$. Finally, $\tilde{\delta}_*^{p-1}$ will denote the homomorphism relative to the triple $(\tilde{X}, \tilde{A}, \tilde{B})$ that is associated with this triple in the same manner as the homomorphism δ_*^{p-1} was associated in **I.6.4** with the triple (X, A, B). The cohomology sequences of the triples (X, A, B) and $(\tilde{X}, \tilde{A}, \tilde{B})$ are then connected in the manner indicated by the following diagram.

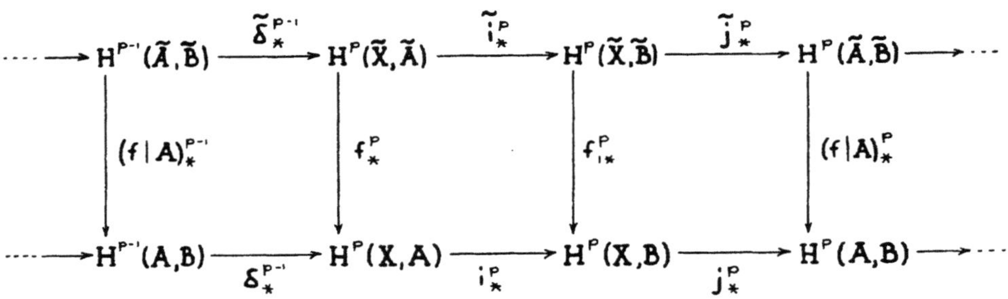

Fig. 17.

The *vector law for cohomology sequences* states that the vector law holds in each box of this diagram. To establish this statement, in view of **I.4.4**, theorem 2 it is sufficient to verify that in the diagram

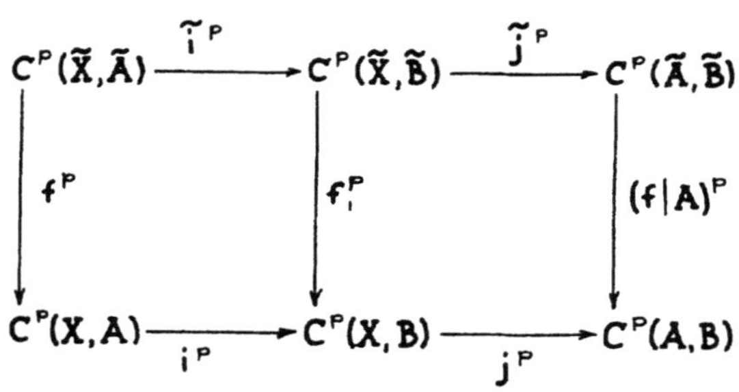

Fig. 18.

the vector law holds in each box. But this is a direct consequence of the vector law in **I.6.2**, since $i, j, \tilde{i}, \tilde{j}$ are inclusion mappings and hence clearly $\tilde{i}f = fi$, $\tilde{j}(f|A) = fj$.

Important special cases of the general vector law for cohomology sequences are considered in the following remarks.

Remark 1. If $B=0$ and $\widetilde{B}=\emptyset$, then we have the case of two pairs (X, A) and $(\widetilde{X}, \widetilde{A})$ related by a continuous mapping $f:(X, A)\to(\widetilde{X}, \widetilde{A})$. The general diagram reduces now to the form

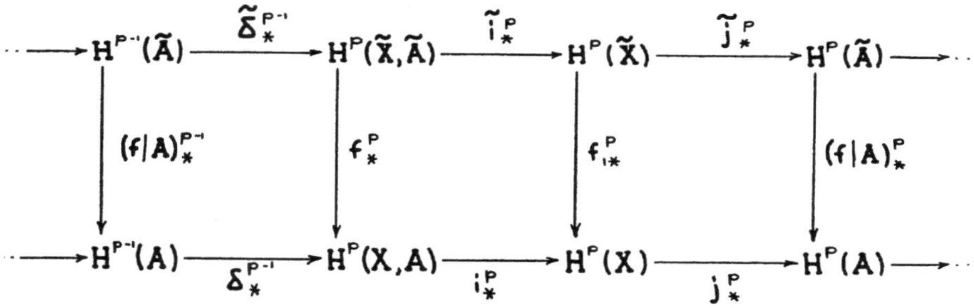

Fig. 19.

The vector law holds in each box.

Remark 2. If X is a subspace of $\widetilde{X}$, then we can take $\widetilde{A}=A$, $\widetilde{B}=B$. Let us choose f as the inclusion mapping $i:X\to\widetilde{X}$ (thus $ix=x$ for $x\in X$). Clearly $i:(X, A, B)\to(\widetilde{X}, A, B)$. The general diagram contains now the partial diagram

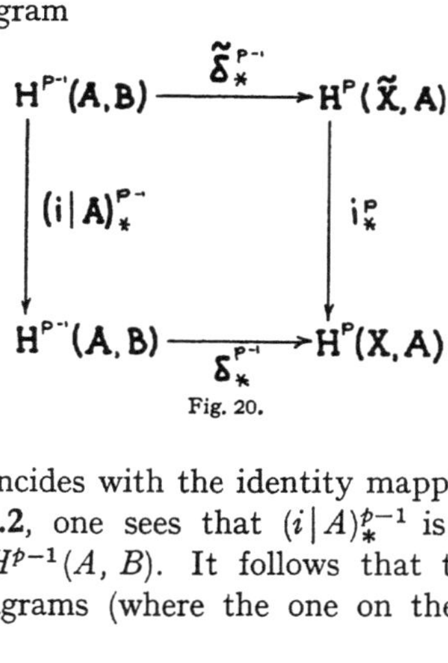

Fig. 20.

Now clearly $i|A$ coincides with the identity mapping $j:A\to A$. Hence, by remark 1 in **I.6.2**, one sees that $(i|A)_*^{p-1}$ is merely the identity homomorphism in $H^{p-1}(A, B)$. It follows that the vector law holds in the following diagrams (where the one on the right concerns the case $B=\emptyset$).

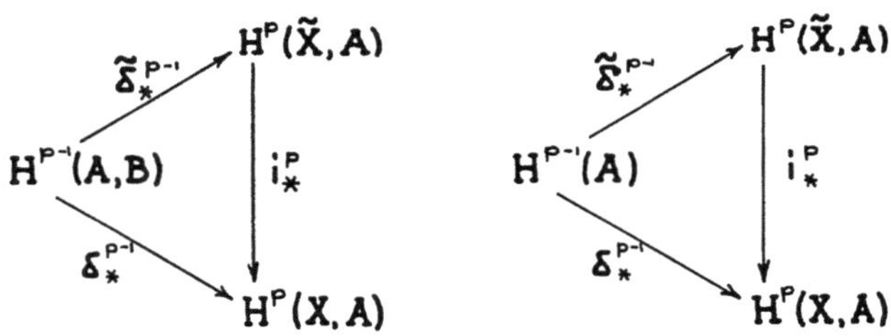

Fig. 21.

I.6.7. Extension and reduction[1]. In a fixed fully normal topological space S, we shall consider pairs of closed subspaces (X, A). Let (X_1, A_1), (X_2, A_2) be two pairs such that $(X_2, A_2) \subset (X_1, A_1)$ (see **I.1.1** for notations). If $i: X_2 \to X_2$ is the identity map on X_2, then clearly $i: (X_2, A_2) \to (X_1, A_1)$, and thus we have the corresponding induced homomorphisms

$$i^p_* : H^p(X_1, A_1) \to H^p(X_2, A_2).$$

Definition 1. If an element h_2 of $H^p(X_2, A_2)$ is the image of some element h_1 of $H^p(X_1, A_1)$ under i^p_*, then h_2 is termed *extendable* into $H^p(X_1, A_1)$.

Definition 2. If an element h_1 of $H^p(X_1, A_1)$ is carried into the zero element of $H^p(X_2, A_2)$ by i^p_*, then h_1 is said to *reduce to zero* in $H^p(X_2, A_2)$.

The following two remarks are direct consequences of the vector laws for induced homomorphisms.

Remark 1. If $(X_3, A_3) \subset (X_2, A_2) \subset (X_1, A_1)$, and an element h_3 of $H^p(X_3, A_3)$ is extendable into $H^p(X_1, A_1)$, then h_3 is also extendable into $H^p(X_2, A_2)$.

Remark 2. If $(X_3, A_3) \subset (X_2, A_2) \subset (X_1, A_1)$, and an element h_1 of $H^p(X_1, A_1)$ reduces to zero in $H^p(X_2, A_2)$, then it also reduces to zero in $H^p(X_3, A_3)$.

Now let there be given in S a closed pair (X_0, A_0) and a family F of closed pairs (X, A) such that $\mathsf{F} \Rightarrow (X_0, A_0)$, in the sense of **I.1.3**, definition 25. Then the following statements hold.

Extension Theorem. Given an element h_0 of $H^p(X_0, A_0)$, there exists a pair $(X, A) \in \mathsf{F}$ such that h_0 is extendable into $H^p(X, A)$.

Reduction Theorem. Given a pair $(X, A) \in \mathsf{F}$ and an element h of $H^p(X, A)$ such that h reduces to zero in $H^p(X_0, A_0)$, then there exists a pair $(X_1, A_1) \in \mathsf{F}$ such that $(X_1, A_1) \subset (X, A)$ and h reduces to zero in $H^p(X_1, A_1)$.

For clarity, we break down the proof into several steps.

---

[1] The *extension theorem* and the *reduction theorem* are due to A. D. WALLACE, The map excision theorem, Duke Math. Journal, vol. 19, 1952, pp. 177–182. We have stated here these theorems in a form convenient for our purposes. These two theorems summarize, in a sense, those aspects of the so-called *continuity theorem* (see the EILENBERG-STEENROD treatise in the *Bibliography*) which are independent of the assumption of compactness. From the point of view of our needs in this volume, the chief virtue of the extension theorem and of the reduction theorem is that they represent an entirely adequate substitute for the conceptually more involved general continuity theorem. The detailed discussion in **I.6.7** to **I.6.16** is meant to provide opportunities for the reader to further familiarize himself with the use of diagrams. For this same reason, we adopted (in **I.6.15**) a method of proof for the homotopy theorem (in terms of the reduction theorem) due to J. W. KEESEE, On the homotopy axiom [Ann. of Math. **54**, 247–249 (1951)].

I.6.8. Lemma. Let Y be a closed subset of a fully normal topological space S, and let z be an element of $Z^p(S, Y)$. Then there exists an open set O in S and a cocycle $z_1 \in Z^p(S, Y)$ such that $O \supset Y$, z_1 vanishes on $\bar{O}$, and $z_1 - z \in B^p(S, Y)$.

Proof. In view of remarks 1 and 2 in **I.6.1**, the assertion is trivial for $p \leq 0$, and hence we assume that $p \geq 1$. The given p-cocycle z vanishes on some open covering $\mathfrak{U}$ of S. Since S is fully normal, by the *Modification Theorem* (see **I.1.5**) there exists an open set $O \supset Y$, a star refinement $\mathfrak{V}$ of $\mathfrak{U}$, and a (generally discontinuous) mapping $f: S \to S$ such that

$$f\bar{O} \subset Y, \tag{1}$$

$$fV \subset St_{\mathfrak{V}} V \quad \text{for} \quad V \in \mathfrak{V}. \tag{2}$$

Clearly, (1) implies that $fY \subset Y$, and thus

$$f: (S, Y) \to (S, Y). \tag{3}$$

Accordingly, f induces a cochain mapping

$$f^p: C_F^p(S, Y) \to C_F^p(S, Y). \tag{4}$$

Observe that the subscript F is necessary since f is not known to be continuous (see **I.5.1**). We set

$$z_1 = f^p z, \tag{5}$$

and we proceed to verify that O and z_1 satisfy the requirements of the lemma. First, take points $x_0, \ldots, x_p \in \bar{O}$. Then $z_1(x_0, \ldots, x_p) = z(fx_0, \ldots, fx_p) = 0$, because $fx_0, \ldots, fx_p \in Y$ by (1) and z vanishes on Y. Observe next that in view of (3), $z_1 \in Z_F^p(S, Y)$ (see **I.5.5**). To see that

$$z_1 \in Z^p(S, Y), \tag{6}$$

there remains to show that z_1 vanishes locally on S. In fact, it is immediate that z_1 vanishes on each $V \in \mathfrak{V}$. Indeed, if $x_0, \ldots, x_p \in V$, then by (5) we have $z_1(x_0, \ldots, x_p) = z(fx_0, \ldots, fx_p) = 0$, since $fx_0, \ldots, fx_p \in fV \subset St_{\mathfrak{V}} V$ by (2), $St_{\mathfrak{V}} V$ is a subset of some $U \in \mathfrak{U}$ by assumption, and z vanishes on each $U \in \mathfrak{U}$. Thus (6) is established. Now let $i: S \to S$ be the identity map on S. Then

$$i: (S, Y) \to (S, Y). \tag{7}$$

Let $D^p: C_F^p(S, Y) \to C_F^{p-1}(S, Y)$ be the homotopy operator corresponding to the pair of mappings f and i (see **I.5.5**). Let us set

$$c = D^p z. \tag{8}$$

Then, by **I.5.5**, we know that $c \in C_F^{p-1}(S, Y)$. We verify that actually

$$c \in C^{p-1}(S, Y), \tag{9}$$

by showing that c vanishes locally on S. In fact, it is immediate that c vanishes on each set $V \in \mathfrak{V}$. Indeed, take $x_0, \ldots, x_{p-1} \in V$. Then [see **I.5.5** and **I.5.4** (4)]

$$c(x_0, \ldots, x_{p-1}) = \sum_{j=0}^{p-1} (-1)^j z(i\,x_0, \ldots, i\,x_j, f\,x_j, \ldots, f\,x_{p-1}). \tag{10}$$

Now $f\,x_j, \ldots, f\,x_{p-1} \in f\,V \subset St_{\mathfrak{V}}\,V$ by (2), and $i\,x_k = x_k \in V \subset St_{\mathfrak{V}}\,V$. Thus $i\,x_0, \ldots, i\,x_j, f\,x_j, \ldots, f\,x_{p-1} \in St_{\mathfrak{V}}\,V$, and $St_{\mathfrak{V}}\,V$ is a subset of some $U \in \mathfrak{U}$ by assumption. Since z vanishes on each $U \in \mathfrak{U}$, it follows that each individual term of the summation in (10) vanishes, and hence $c(x_0, \ldots, x_{p-1}) = 0$, proving (9). Finally, by **I.5.5** (23) we have

$$\delta_S^{p-1}\, c = f^p z - i^p z = z_1 - z.$$

In view of (9), this shows that $z_1 - z \in B^p(S, Y)$, and the lemma is proved.

I.6.9. Lemma. Let Y be a closed subset of the fully normal topological space S, and let an element h of $H^p(S, Y)$ be assigned. Then there exists an open set O containing Y such that h reduces to zero in $H^p(\overline{O}, Y)$.

Proof. The given h is of the form $h = [z]$, where $z \in Z^p(S, Y)$ and $[z]$ denotes the cohomology class containing z. We apply the lemma in **I.6.8** to this p-cocycle z, obtaining O and z_1 as described there. Consider the inclusion mapping $j: (\overline{O}, Y) \to (S, Y)$. Since $z_1 - z \in B^p(S, Y)$, we have $h = [z] = [z_1]$, and hence $j_*^p h = j_*^p [z_1]$. Thus the lemma is proved if we can show that $j^p z_1 \in B^p(\overline{O}, Y)$. Now for $x_0, \ldots, x_p \in \overline{O}$ we have $(j^p z_1)(x_0, \ldots, x_p) = z_1(x_0, \ldots, x_p) = 0$, since z_1 vanishes on $\overline{O}$. Thus actually $j^p z_1 = 0$, and hence surely $j^p z_1 \in B^p(\overline{O}, Y)$.

I.6.10. Lemma. Let Y be a closed subset of the fully normal topological space S, and let O be an open set containing Y. Let an element h of $H^p(\overline{O}, Y)$ be assigned. Then there exists an open set U such that $Y \subset U \subset O$ and h reduces to zero in $H^p(\overline{U}, Y)$.

Proof. Setting $S' = \overline{O}$, we observe that S' is again fully normal (see **I.1.3**, exercise 10), and we apply the lemma in **I.6.9** to (S', Y). According to that lemma, we have a set O' which is open relative to S', satisfies $Y \subset O' \subset S'$, and is such that h reduces to zero in $H^p(\overline{O'}, Y)$. Observe now that since O' is open relative to $S' = \overline{O}$, we have $O' = \overline{O} \cap O_1$, where O_1 is open in S. On choosing $U = O \cap O_1$, clearly $Y \subset U \subset O'$, U is open in S, and $Y \subset \overline{U} \subset \overline{O'}$. Thus $(\overline{U}, Y) \subset (\overline{O'}, Y)$. Hence, by remark 2 in **I.6.7**, h reduces to zero in $H^p(\overline{U}, Y)$ also, and the lemma is proved.

Remark. Since $S' = \overline{O}$ is closed in S, there was no need to distinguish between closure relative to S and closure relative to S' in the preceeding argument.

I.6.11. The extension lemma. Let (X, A) be a closed pair in a fully normal space S, and let an element h of $H^p(X, A)$ be assigned. Then there exist open sets O and U such that $X \subset O$, $A \subset U$, $\overline{U} \subset O$, and h is extendable into $H^p(\overline{O}, \overline{U})$.

Proof. The diagram in Fig. 22 is meant to assist the reader in visualizing the situation.

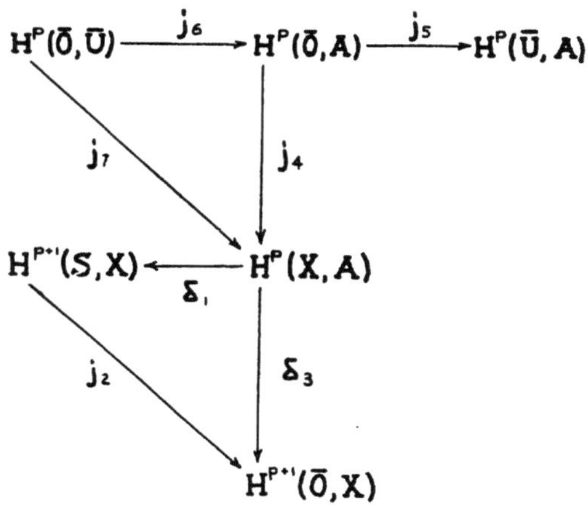

Fig. 22.

Starting with the assigned element $h \in H^p(X, A)$, denote by δ_1 the homomorphism $H^p(X, A) \to H^{p+1}(S, X)$ occurring in the cohomology sequence of the triple (S, X, A), and set $h_1 = \delta_1 h$. By the lemma in **I.6.9**, applied in dimension $p+1$, we have an open set $O \supset X$ such that $j_2 h_1 = 0$, where j_2 is induced by the inclusion mapping $(\overline{O}, X) \to (S, X)$. On denoting by δ_3 the homomorphism $H^p(X, A) \to H^{p+1}(\overline{O}, X)$ ocurring in the cohomology sequence of the triple $(\overline{O}, X, A)$, we have $\delta_3 h = j_2 \delta_1 h = j_2 h_1 = 0$, by the vector law for cohomology sequences. By the exactness of the cohomology sequence of the triple $(\overline{O}, X, A)$, the relation $\delta_3 h = 0$ implies the existence of an element $h_4 \in H^p(\overline{O}, A)$ such that $j_4 h_4 = h$, where j_4 is induced by the inclusion mapping $(X, A) \to (\overline{O}, A)$. By the lemma in **I.6.10**, there exists an open set U such that $A \subset U \subset O$ and h_4 reduces to zero in $H^p(\overline{U}, A)$. This means that $j_5 h_4 = 0$, where j_5 is induced by the inclusion mapping $(\overline{U}, A) \to (\overline{O}, A)$. By the exactness of the cohomology sequence of the triple $(\overline{O}, \overline{U}, A)$, there follows the existence of an element $h_6 \in H^p(\overline{O}, \overline{U})$ such that $j_6 h_6 = h_4$, where j_6 is induced by the inclusion mapping $(\overline{O}, A) \to (\overline{O}, \overline{U})$. On denoting by j_7 the homomorphism induced by the inclusion mapping $(X, A) \to (\overline{O}, \overline{U})$, the vector law yields $j_7 h_6 = j_4 j_6 h_6 = j_4 h_4 = h$. Thus h is extendable into $H^p(\overline{O}, \overline{U})$, and the extension lemma is proved.

Remark. For the special case when $A = \emptyset$, the extension lemma reduces to the following statement: If X is a closed subset of the fully normal space S, and if an element h of $H^p(X)$ is assigned, then there exists an open set $O \supset X$ such that h is extendable into $H^p(\overline{O})$.

Inspection of the preceding diagram reveals that in this special case the proof is complete at the moment when the homomorphism j_4 is brought into play.

I.6.12. The reduction lemma. In the fully normal space S, let (X, A) be a closed pair, and let O and U be open sets such that $X \subset O$, $A \subset U$, $U \subset O$. Finally, let h be an element of $H^p(\overline{O}, \overline{U})$ such that h reduces to zero in $H^p(X, A)$. Then there exist open sets O_1, U_1 such that $X \subset O_1 \subset O$, $A \subset U_1 \subset U$, $U_1 \subset O_1$, and h reduces to zero in $H^p(\overline{O}_1, \overline{U}_1)$.

The march of the proof is indicated by the following diagram.

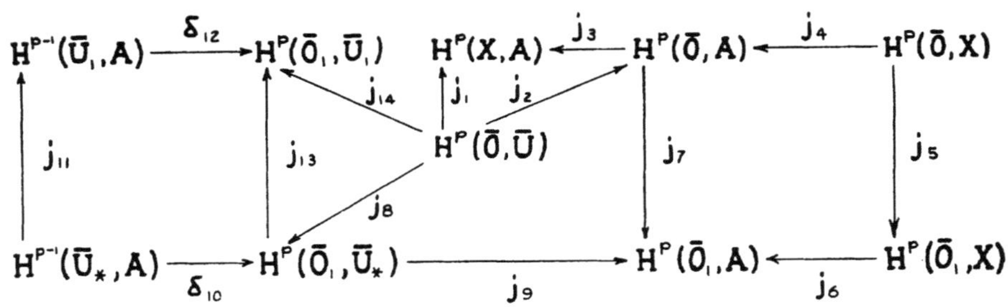

Fig. 23.

The proof is made as follows. By assumption, $j_1 h = 0$, where j_1 is induced by the inclusion mapping $(X, A) \to (\overline{O}, \overline{U})$. Let j_2 and j_3 be induced by the inclusion mappings $(\overline{O}, A) \to (\overline{O}, \overline{U})$ and $(X, A) \to (\overline{O}, A)$ respectively, and set $h_1 = j_2 h$. By the vector law it follows that $j_3 h_1 = j_3 j_2 h = j_1 h = 0$. By the exactness of the cohomology sequence of the triple $(\overline{O}, X, A)$, there follows the existence of an element $h_2 \in H^p(\overline{O}, X)$ such that $j_4 h_2 = h_1$, where j_4 is induced by the inclusion mapping $(\overline{O}, A) \to (\overline{O}, X)$. By the lemma in **I.6.10** there exists an open set O_1 such that $X \subset O_1 \subset O$, and $j_5 h_2 = 0$, where j_5 is induced by the inclusion mapping $(\overline{O}_1, X) \to (\overline{O}, X)$. Let j_6 and j_7 be induced by the inclusion mappings $(\overline{O}_1, A) \to (\overline{O}_1, X)$ and $(\overline{O}_1, A) \to (\overline{O}, A)$ respectively. By the vector law we have then $j_7 h_1 = j_7 j_4 h_2 = j_6 j_5 h_2 = j_6 0 = 0$. Now introduce the set $U_* = U \cap O_1$. Then U_* is open and $A \subset U_* \subset O_1$. Accordingly, we have homomorphisms j_8 and j_9 induced by the inclusion mappings $(\overline{O}_1, \overline{U}_*) \to (\overline{O}, \overline{U})$ and $(\overline{O}_1, A) \to (\overline{O}_1, \overline{U}_*)$ respectively. Set $h_3 = j_8 h$. By the vector law it follows that $j_9 h_3 = j_9 j_8 h = j_7 j_2 h = j_7 h_1 = 0$. By the exactness of the cohomology sequence of the triple $(\overline{O}_1, \overline{U}_*, A)$ there follows the existence of an element $h_4 \in H^{p-1}(\overline{U}_*, A)$ such that $\delta_{10} h_4 = h_3$, where δ_{10} is the homomorphism $H^{p-1}(\overline{U}_*, A) \to H^p(\overline{O}_1, \overline{U}_*)$ occurring in the cohomology sequence of the triple $(\overline{O}_1, \overline{U}_*, A)$. By the lemma in **I.6.10**, applied in the dimension $p-1$, there exists an open set U_1 such that $A \subset U_1 \subset U_*$ and $j_{11} h_4 = 0$, where j_{11} is induced by the inclusion mapping $(\overline{U}_1, A) \to (\overline{U}_*, A)$. Let now j_{13} and j_{14} be induced by the inclusion mappings $(\overline{O}_1, \overline{U}_1) \to (\overline{O}_1, \overline{U}_*)$ and $(\overline{O}_1, \overline{U}_1) \to (\overline{O}, \overline{U})$ respectively. Finally, let δ_{12} denote the homomorphism $H^{p-1}(\overline{U}_1, A) \to H^p(\overline{O}_1, \overline{U}_1)$ occurring in the cohomology sequence of the triple $(\overline{O}_1, \overline{U}_1, A)$. We have then $j_{13}h_3 = j_{13}\delta_{10}h_4 = \delta_{12}j_{11}h_4 = \delta_{12}0 = 0$ by the vector law for cohomology sequences, and finally $j_{14}h = j_{13}j_8h = j_{13}h_3 = 0$ by the vector law for induced homomorphisms. The relation $j_{14}\,h = 0$ means that h reduces to zero in $H^p(\overline{O}_1, \overline{U}_1)$, proving the reduction lemma.

Remark. In the special case when $U = \emptyset$ and $A = \emptyset$, the reduction lemma yields the following statement: If X is closed and O is open and $X \subset O$, and if h is an element of $H^p(\overline{O})$ such that h reduces to zero in $H^p(X)$, then there exists an open set O_1 such that $X \subset O_1 \subset O$ and h reduces to zero in $H^p(\overline{O}_1)$. For this special case, the homomorphisms j_2 and j_3 are not needed in the preceding diagram, and the proof is completed at the moment when the homomorphism j_7 has been reached.

I.6.13. Proof of the extension and reduction theorems. Returning to **I.6.7**, consider first the extension theorem. Given $h_0 \in H^p(X_0, A_0)$, by the extension lemma in **I.6.11** there exist open sets O, U such that $X_0 \subset O$, $A_0 \subset U$, $U \subset O$, and h_0 is extendable into $H^p(\overline{O}, \overline{U})$. Since $\mathsf{F} \Rightarrow (X_0, A_0)$, there exists a pair $(X, A) \in \mathsf{F}$ such that $X \subset O$, $A \subset U$. Then we have the relations $(X_0, A_0) \subset (X, A) \subset (\overline{O}, \overline{U})$. By remark 1 in **I.6.7** it follows that h_0 is extendable into $H^p(X, A)$, and the extension theorem is proved.

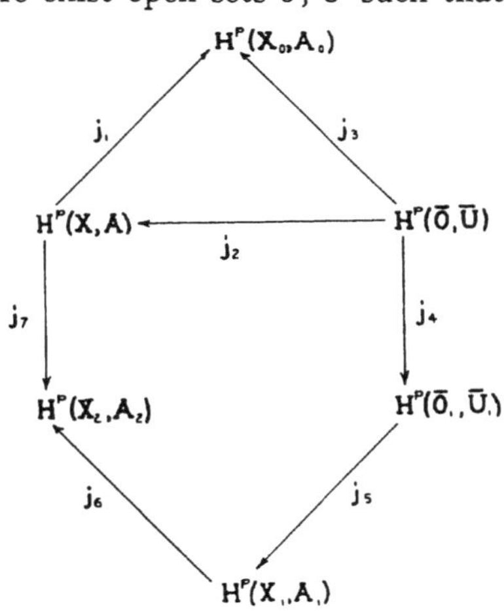

Fig. 24.

Next, consider the situation assumed in the statement of the reduction theorem (see **I.6.7**). The diagram in Fig. 24 illustrates the sequence of steps in the proof.

By assumption, we are given an element $h \in H^p(X, A)$ such that $j_1h = 0$, where j_1 is induced by the inclusion mapping $(X_0, A_0) \to (X, A)$. By the extension lemma in **I.6.11**, there exist open sets O, U such that $X \subset O$, $A \subset U$, $U \subset O$ and there exists an element $h_1 \in H^p(\overline{O}, \overline{U})$ such that $j_2h_1 = h$, where j_2 is induced by the inclusion mapping $(X, A) \to (\overline{O}, \overline{U})$. Let j_3 be induced by the inclusion mapping $(X_0, A_0) \to (\overline{O}, \overline{U})$. Observe that this mapping is available since $(X_0, A_0) \subset (X, A) \subset (\overline{O}, \overline{U})$. The vector law yields now $j_3h_1 = j_1j_2h_1 = j_1h = 0$. By the reduction lemma in **I.6.12**, there follows the existence of open sets O_1, U_1 such that $X_0 \subset O_1 \subset O$, $A_0 \subset U_1 \subset U$, $U_1 \subset O_1$, and $j_4h_1 = 0$, where j_4 is induced by the inclusion mapping $(\bar{O}_1, \bar{U}_1) \to (\bar{O}, \bar{U})$. Now since $\mathsf{F} \Rightarrow (X_0, A_0)$, there exists a pair $(X_1, A_1) \in \mathsf{F}$ such that $(X_1, A_1) \subset (\bar{O}_1, \bar{U}_1)$ and finally there exists a pair $(X_2, A_2) \in \mathsf{F}$ such that $(X_2, A_2) \subset (X, A)$, $(X_2, A_2) \subset (X_1, A_1)$. The inclusion mappings $(X_1, A_1) \to (\bar{O}_1, \bar{U}_1)$, $(X_2, A_2) \to (X_1, A_1)$ and $(X_2, A_2) \to (X, A)$ induce homomorphisms that are designated by j_5, j_6, j_7 respectively in the diagram. By the vector law we conclude now that $j_7 h = j_7 j_2 h_1 = j_6 j_5 j_4 h_1 = j_6 j_5 0 = 0$. Thus h reduces to zero in $H^p(X_2, A_2)$, where $(X_2, A_2) \in \mathsf{F}$ and $(X_2, A_2) \subset (X, A)$. This completes the proof of the reduction theorem.

I.6.14. Strong excision. Let A and U be two subsets of the fully normal space X, such that

$$U \subset A \subset X, \quad A - U \neq \emptyset. \tag{1}$$

The inclusion mapping $(X - U, A - U) \to (X, A)$ induces then (see **I.6.3** for terminology) the *excision homomorphism* which we now denote by

$$e^p : H^p(X, A) \to H^p(X - U, A - U). \tag{2}$$

In **I.6.3** we proved that e^p is an isomorphism onto, provided that $\bar{U} \subset \operatorname{int} A$. We proceed to show that if

$$A \text{ closed}, \quad U \text{ open in } X, \tag{3}$$

then the condition $\bar{U} \subset \operatorname{int} A$ can be dropped. Thus we have the

Strong excision theorem. If X is fully normal, and the conditions (1) and (3) are satisfied, then the excision homomorphism (2) is an isomorphism onto, for every p.

Proof. Let F be the family of all the pairs $(X - V, A - V)$ where V denotes an arbitrary (perhaps empty) open set such that $\bar{V} \subset U$. By **I.1.3**, exercise 26 we have then the relations

$$\mathsf{F} \Rightarrow (X - U, A - U), \quad (X, A) \in \mathsf{F}. \tag{4}$$

Let $(X - V, A - V) \in \mathsf{F}$. The inclusion mappings $(X - U, A - U) \to (X - V, A - V)$, $(X - V, A - V) \to (X, A)$ give rise to induced homomorphisms designated by j_2^p, j_1^p respectively in the diagram in Fig. 25 [where e^p is the excision homomorphism (2)].

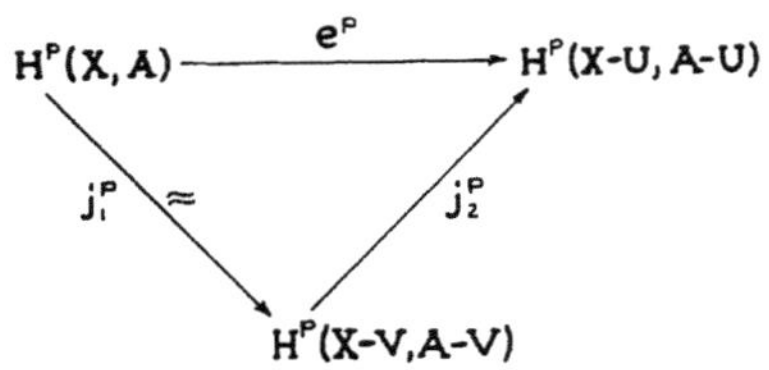

Fig. 25.

Since $\bar{V} \subset U$ and $U \subset A$, clearly $\bar{V} \subset \operatorname{int} A$, and $A - V \neq \emptyset$ since $A - U \neq \emptyset$. Thus j_1^p is an isomorphism onto by **I.6.3**, lemma 3. The symbol $\approx$ in the diagram refers to this fact. Now assign any element $h \in H^p(X - U, A - U)$. In view of (4) and the extension theorem in **I.6.7**, we can choose V in such a way that $h = j_2^p h_1$, $h_1 \in H^p(X-V, A-V)$. Next, since j_1^p is *onto*, we have $h_1 = j_1^p h_2$, $h_2 \in H^p(X, A)$. By the vector law it follows that $h = j_2^p j_1^p h_2 = e^p h_2$. Thus e^p is *onto*. Next, let $h \in H^p(X, A)$ and $e^p h = 0$. In view of (4), it follows from the reduction theorem that we can choose V in such a manner that $j_1^p h = 0$. Since j_1^p is an isomorphism, it follows that $h = 0$. Thus e^p is also an *isomorphism into*, and the proof is complete.

Remark 1. The case when $A - U = \emptyset$ has been considered in **I.6.3.**

Remark 2. In a fully normal space X, let X_1, X_2, Y be closed subsets such that

$$X = X_1 \cup X_2, \quad Y = X_1 \cap X_2 \neq \emptyset. \tag{4*}$$

Then the inclusion mapping

$$(X_1, Y) \to (X, X_2) \tag{5}$$

is an excision mapping. Indeed, set $U = \complement X_1$. Then U is open, and clearly $U \subset X_2$ by (4*). Furthermore $X - U = X_1$, and $X_2 - U = X_2 \cap \complement U = X_2 \cap X_1 = Y$. Thus (5) coincides with the excision mapping obtained by excising the open set $\complement X_1$. Observe that since $X_2 - U = Y \neq \emptyset$, the theorem derived in this section applies. Hence, on denoting by

$$e_2^p : H^p(X, X_2) \to H^p(X_1, Y), \tag{6}$$

the excision homomorphism induced by (5), we see that e_2^p is an isomorphism onto for every p. Since the assumptions (4*) are symmetric with respect to X_1, X_2, it follows in the same manner that the excision homomorphism

$$e_1^p : H^p(X, X_1) \to H^p(X_2, Y), \tag{7}$$

induced by the inclusion mapping $(X_2, Y) \to (X, X_1)$, is an isomorphism onto for every p.

I.6.15. The homotopy theorem. This basic theorem is actually valid in general topological spaces, but we restrict ourselves to the case of compact HAUSDORFF spaces which is adequate for our purposes. In this special case, several elegant proofs are available. The one to be presented here has been selected because it constitutes an instructive application of the *reduction lemma.*

The special homotopy theorem. Let A be a compact subspace of the compact HAUSDORFF space X. Denoting by I the closed unit interval $0 \leq u \leq 1$, consider the compact HAUSDORFF pair $(X \times I, A \times I)$. The *first homotopy diagram* in **I.1.1** gives rise to the following diagram of induced homomorphisms.

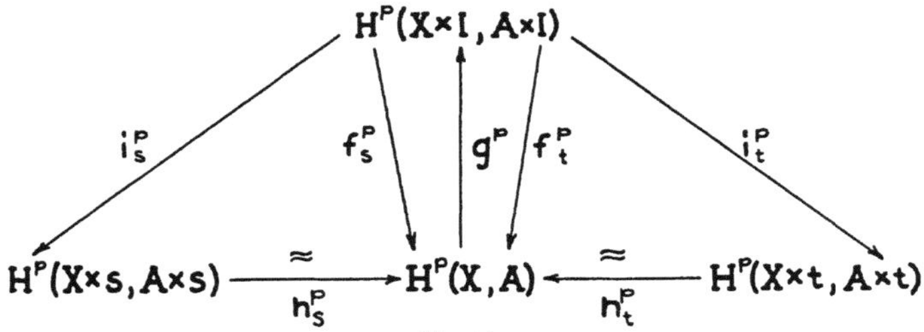

Fig. 26.

In this diagram, the symbol g^p, for example, denotes the homomorphism induced by the mapping $g:(X\times I, A\times I)\to(X, A)$ explained in **I.1.1**, with similar interpretations for the other homomorphisms appearing in the diagram. Since the mappings h_s, h_t are homeomorphisms, the homomorphisms h_s^p, h_t^p are isomorphisms onto by **I.6.2**, remark 2. The symbols $\approx$ in the diagram refer to this fact. The *special homotopy theorem*, to be proved presently, states that for any two numbers $s, t\in I$ we have

$$f_s^p = f_t^p. \tag{1}$$

The proof is based on the following series of observations.

(i) By **I.1.1**, we have the relation $f_s = i_s h_s$. By the vector law it follows that $f_s^p = h_s^p i_s^p$. Since h_s^p is an isomorphism, we infer by **I.3.2**, exercise 2 that f_s^p and i_s^p have the same nucleus, to be denoted by $\mathfrak{N}_s$. Similarly, f_t^p and i_t^p have the same nucleus $\mathfrak{N}_t$.

(ii) Set $\mathfrak{J}^* = g^p H^p(X, A)$. Then (see **I.3.1**)

$$H^p(X\times I, A\times I) = \mathfrak{J}^* \dotplus \mathfrak{N}_t. \tag{2}$$

Indeed, by **I.1.1** we have $g f_t = i$, the inclusion mapping $(X, A)\to(X, A)$. Hence, by the vector law and remark 1 in **I.6.2** it follows that

$$f_t^p g^p = 1, \tag{3}$$

where 1 denotes the identity homomorphism in $H^p(X, A)$. Thus (2) is a direct consequence of **I.3.2**, exercise 5.

(iii) For $e\in\mathfrak{N}_t$, there exists an $\varepsilon=\varepsilon(e, t)>0$, such that $e\in\mathfrak{N}_s$ for $|s-t|<\varepsilon$. Indeed, the assumption means, by (i), that $i_t^p e=0$. By the reduction lemma in **I.6.12** there follows the existence of open sets O, U in $X\times I$ such that $O\supset X\times t$, $U\supset A\times t$, $U\subset O$, and e reduces to zero in $H^p(\bar{O},\bar{U})$. By **I.1.3**, exercise 40, we have then an $\varepsilon>0$ such that $X\times s\subset\bar{O}$, $A\times s\subset\bar{U}$ for $|s-t|<\varepsilon$. If s is so restricted, then by remark 2 in **I.6.7** the element e reduces to zero in $H^p(X\times s, A\times s)$ also. In other words, $i_s^p e=0$ for $|s-t|<\varepsilon$, proving our assertion.

(iv) If $e \in \mathfrak{J}^*$, then $f_t^p e = f_s^p e$ for any two numbers $s, t \in I$. Indeed, the assumption means that $e = g^p a$, $a \in H^p(X, A)$. But then, by (3), it follows that $f_t^p e = f_t^p g^p a = a$. Thus $f_t^p e$ is independent of t for $e \in \mathfrak{J}^*$.

(v) Given $a \in H^p(X \times I, A \times I)$, $b \in H^p(X, A)$, denote by $E(a, b)$ the set of those numbers $u \in I$ for which $f_u^p a = b$. Then the set $E(a, b)$, which may be empty of course, is both open and closed (in I).

Indeed, in case $E(a, b)$ is non-empty, take any $t \in E(a, b)$. By (ii), we have

$$a = a' + a'', \quad a' \in \mathfrak{J}^*, \quad a'' \in \mathfrak{N}_t. \tag{4}$$

By (iii) we have then

$$a'' \in \mathfrak{N}_s \quad \text{for} \quad |s - t| < \varepsilon(a'', t). \tag{5}$$

Consider now any $s \in I$ such that $|s - t| < \varepsilon(a'', t)$. Then, by (4), (5) and (iv), we obtain successively $f_s^p a = f_s^p a' = f_t^p a' = f_t^p a = b$. Thus $s \in E(a, b)$ for $|s - t| < \varepsilon(a'', t)$, proving that $E(a, b)$ is open. Now clearly

$$I - E(a, b) = \cup E(a, c), \quad c \in H^p(X, A), \quad c \neq b.$$

By what we just proved, each set $E(a, c)$ is open in I, hence $I - E(a, b)$ is open as a union of open sets. Thus $E(a, b)$ is also closed, proving our assertion.

(vi) Since I is connected, it follows (see **I.1.3**, exercise 13) that either $E(a, b) = I$ or else $E(a, b)$ is empty. Now take any element $a \in H^p(X \times I, A \times I)$, and set $b = f_0^p(a)$. Then $E(a, b)$ is non-empty, and hence $E(a, b) = I$. In other words, $f_t^p a = b = f_0^p a$ for every $t \in I$, proving the special homotopy theorem.

The homotopy theorem. Let (X, A) and (Y, B) be two pairs of compact HAUSDORFF spaces, and let the continuous mappings $m_0: (X, A) \to (Y, B)$, $m_1: (X, A) \to (Y, B)$ be homotopic. Denote by m_0^p, m_1^p the corresponding induced homomorphisms from $H^p(Y, B)$ into $H^p(X, A)$. Then $m_0^p = m_1^p$.

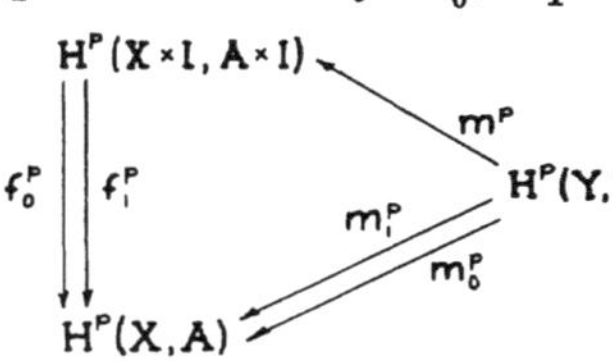

Fig. 27.

Proof. The induced homomorphisms, corresponding to the transformations in the *second homotopy diagram* in **I.1.1**, give rise to the diagram in Fig. 27.

Since $m_0 = m f_0$, by the vector law we have $m_0^p = f_0^p m^p$, and similarly $m_1^p = f_1^p m^p$. Now $f_0^p = f_1^p$ by the special homotopy theorem, and hence $m_0^p = m_1^p$.

Remark. In the preceding discussion, the only fact about I that we actually used was that I is compact and connected. This observation yields a generalized homotopy theorem which the reader will readily visualize.

I.6.16. Examples of trivial cohomology groups. In many applications, it is a matter of importance to know that certain cohomology groups involved are trivial. In this section, some theorems are presented that have a bearing upon this subject. For convenience, *all the spaces involved are assumed to be compact* HAUSDORFF *spaces.*

Lemma 1. If A is a retract of X, then the homomorphism $j^p: H^p(X) \to H^p(A)$, induced by the inclusion mapping $j: A \to X$, is *onto.*

Proof. By assumption, we have a continuous mapping $f: X \to A$ such that $fx = x$ for $x \in A$. Then f induces a homomorphism $f^p: H^p(A) \to H^p(X)$. Now $fj = i$, the identity mapping $A \to A$. By the vector law and remark 1 in **I.6.2** it follows that $j^p f^p = i^p = 1$, the identity homomorphism in $H^p(A)$. Now assign any element $u \in H^p(A)$ and set $v = f^p u$. Then $j^p v = j^p f^p u = i^p u = u$, proving the lemma.

Theorem 1. If $H^p(X) = 0$ for a certain p, and A is a retract of X, then $H^p(A) = 0$ for the same p.

This is a direct consequence of lemma 1.

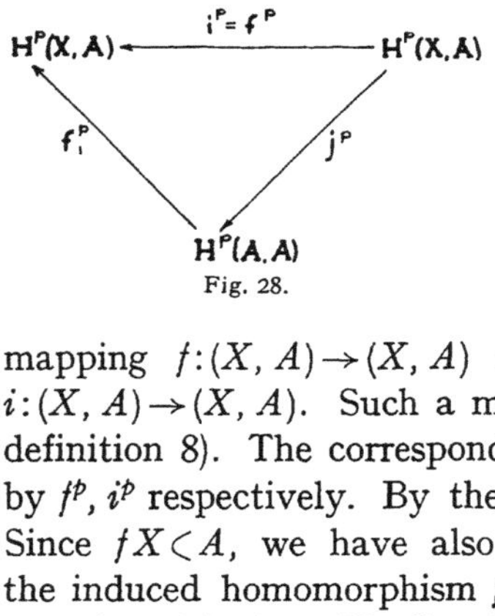

Fig. 28.

Theorem 2. If A is a deformation retract of X, then $H^p(X, A) = 0$ for every p.

Proof. Consider the diagram in Fig. 28.

In this diagram, f denotes a retraction from X onto A such that the mapping $f: (X, A) \to (X, A)$ is homotopic to the identity mapping $i: (X, A) \to (X, A)$. Such a mapping f exists by assumption (see **I.1.5**, definition 8). The corresponding induced homomorphisms are denoted by f^p, i^p respectively. By the homotopy theorem we have then $f^p = i^p$. Since $fX \subset A$, we have also the relation $f: (X, A) \to (A, A)$, yielding the induced homomorphism f_1^p. Finally, j^p is induced by the inclusion mapping $(A, A) \to (X, A)$. Clearly $f = jf$, since $fX \subset A$, and hence $f^p = f_1^p j^p$ by the vector law. Now take any element $e \in H^p(X, A)$. Then, on the one hand, $f^p e = i^p e = e$ by remark 1 in **I.6.2**, and on the other hand $f^p e = f_1^p j^p e = f_1 0 = 0$, since $H^p(A, A) = 0$ by remark 3 in **I.6.1**. Thus $e = 0$, proving the theorem.

Corollary to Theorem 2. If A is a deformation retract of X, then the homomorphism $j^p: H^p(X) \to H^p(A)$, induced by the inclusion mapping $j: A \to X$, is an isomorphism onto, for every p.

Proof. The groups $H^p(X, A)$, $H^p(X)$, $H^p(A)$, $H^{p+1}(X, A)$ are consecutive terms in the cohomology sequence of the pair (X, A). The first group and the fourth group are trivial by theorem 2, and thus the corollary follows directly from **I.3.3**, lemma 4.

Theorem 3. Let X_1, X_2, Y be subspaces of X such that $X_1 \cup X_2 = X$, $X_1 \cap X_2 = Y \neq \emptyset$. Suppose that, for some integer p,

$$H^p(X_1) = 0, \quad H^p(X_2) = 0, \quad H^{p-1}(Y) = 0. \tag{1}$$

Then $H^p(X) = 0$, for the same p.

Proof. By **I.6.14**, remark 2, we have

$$H^p(X, X_2) \approx H^p(X_1, Y). \tag{2}$$

Now the groups $H^{p-1}(Y), H^p(X_1, Y), H^p(X_1)$ are consecutive terms in the cohomology sequence of the pair (X_1, Y). The first group and the third group are trivial by (1). Hence, by **I.3.3**, lemma 3,

$$H^p(X_1, Y) = 0. \tag{3}$$

Next, the groups $H^p(X, X_2), H^p(X), H^p(X_2)$ are consecutive terms in the cohomology sequence of the pair (X, X_2). The first group is trivial by (2) and (3), while the third group is trivial by (1). Hence $H^p(X) = 0$ by **I.3.3**, lemma 3.

Theorem 4. Let X_1, X_2, Y be subspaces of X such that $X_1 \cup X_2 = X$, $X_1 \cap X_2 = Y \neq \emptyset$. Suppose that, for some integer p,

$$H^p(X) = 0, \quad H^p(Y) = 0, \quad H^{p-1}(X_2) = 0. \tag{4}$$

Then $H^p(X_1) = 0$, for the same integer p.

Proof. By **I.6.14**, remark 2, we have again (2). The groups $H^{p-1}(X_2)$, $H^p(X, X_2)$, $H^p(X)$ are consecutive terms in the cohomology sequence of the pair (X, X_2). The first and the third groups are trivial by (4). Hence, by **I.3.3**, lemma 3 it follows that

$$H^p(X, X_2) = 0. \tag{5}$$

Next, the groups $H^p(X_1, Y), H^p(X_1), H^p(Y)$ are consecutive terms in the cohomology sequence of the pair (X_1, Y). The first group is trivial by (5) and (2), while the third group is trivial by (4). Hence, by **I.3.3**, lemma 3, $H^p(X_1) = 0$.

Theorem 5. Let X be a subspace of S. Suppose that, for some integer p, the following holds: for every open set $O \supset X$ there exists a subspace Y such that $X \subset Y \subset O$ and $H^p(Y) = 0$. Then $H^p(X) = 0$, for the same p (recall that all the spaces occurring in the present section are assumed to be compact HAUSDORFF spaces).

Proof. Select an element $e \in H^p(X)$. By the extension lemma in **I.6.11** (see the *remark* there) we have then an open set $O \supset X$ such that e

is extendable into $H^p(\overline{O})$. By assumption, we have next a subspace Y such that $X \subset Y \subset O$ and $H^p(Y)=0$. The homomorphisms induced by the inclusion mappings $X \to Y$, $Y \to \overline{O}$, $X \to \overline{O}$ give rise to the following diagram.

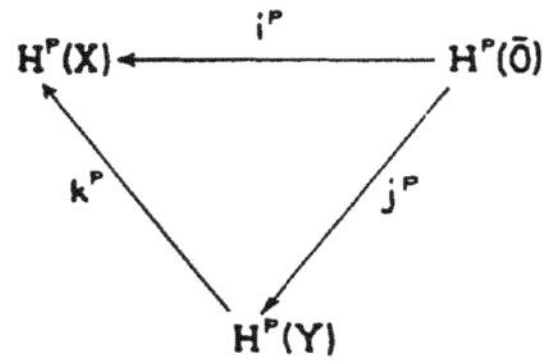

Fig. 29.

Since e is extendable into $H^p(\overline{O})$, we have $e = i^p e'$, $e' \in H^p(\overline{O})$. Then $j^p e' = 0$ since $H^p(Y)=0$, and hence $e = i^p e' = k^p j^p e' = k^p 0 = 0$. Since e was an arbitrary element of $H^p(X)$, it follows that $H^p(X)=0$.

1.6.17. A structure theorem. It will be a matter of importance in the sequel to decompose additively certain cohomology groups, as well as homomorphisms connecting such groups. We proceed to derive the basic facts needed for this purpose. Let X be a compact HAUSDORFF space and let Y be a closed (and hence compact) subset of X. We assume that

$$Y \neq \emptyset, \quad Y \neq X. \tag{1}$$

Then $O = \complement Y$ is a non-empty open set. Let O be decomposed, in any manner, into a union of pair-wise disjoint open sets $O_1, \ldots, O_n, \ldots$, where the sequence $\{O_n\}$ may be finite or infinite. Set

$$Y_n = \complement O_n, \quad Y_n^* = Y \cup O_n. \tag{2}$$

Then (see **I.1.3**, exercise 41) Y_n and Y_n^* are non-empty, closed (and hence compact) sets, and $Y_n \supset Y$. The inclusion mapping $(X, Y) \to (X, Y_n)$ induces a homomorphism

$$i_n^p : H^p(X, Y_n) \to H^p(X, Y). \tag{3}$$

On setting

$$I(i_n^p) = i_n^p H^p(X, Y_n), \tag{4}$$

we can formulate the basic *structure theorem* as follows.

I. i_n^p is an isomorphism into, and hence $I(i_n^p) \approx H^p(X, Y_n)$.

II. $H^p(X, Y)$ is the weak direct sum or the direct sum of the subgroups $I(i_n^p)$, according as the sequence $\{O_n\}$ is infinite or finite (see **I.3.1**).

Proof. We fix the integer p, and will not display it in designating the various homomorphisms to be used. Thus i_n^p will be written simply as i_n. The proof is based now on the following diagram.

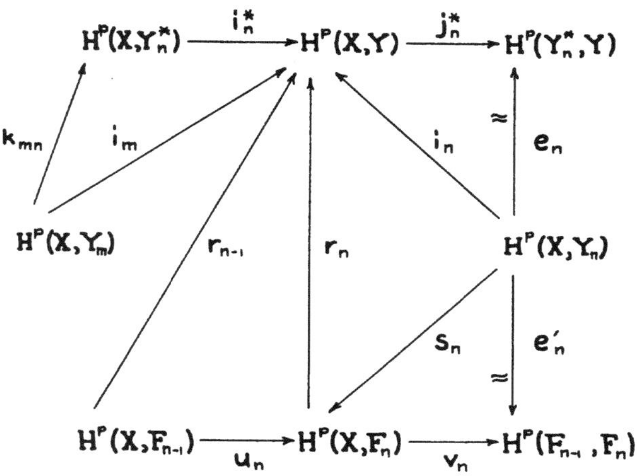

Fig. 30.

All the homomorphisms involved are induced by inclusion mappings. i_n is the i_n^p in (3), and i_m is the analogous homomorphism corresponding to an integer $m \neq n$. By **I.1.3**, exercise 41 and (2) we have

$$Y_n^* \supset Y, \quad Y_m \supset Y_n^* \quad \text{for} \quad m \neq n. \tag{5}$$

One has thus the inclusion mappings $(Y_n^*, Y) \to (X, Y)$, $(X, Y) \to (X, Y_n^*)$, $(Y_n^*, Y) \to (X, Y_n)$, and $(X, Y_n^*) \to (X, Y_m)$ for $m \neq n$, yielding the induced homomorphisms j_n^*, i_n^*, e_n, k_{nm} respectively. Next, on setting

$$F_n = \cap Y_k, \quad k = 1, \ldots, n, \tag{6}$$

it is clear that

$$F_1 \supset F_2 \supset \cdots \supset Y \neq \emptyset, \quad \cap F_n = Y, \quad F_n \subset Y_n. \tag{7}$$

By **I.1.3**, exercise 41 it follows that

$$(X, F_n) \Rightarrow (X, Y). \tag{8}$$

By (7) one has the inclusion mappings $(X, F_n) \to (X, Y_n)$, $(X, Y) \to (X, F_n)$, yielding the induced homomorphisms s_n, r_n. Finally, for the case when $n \geqq 2$, one has also the inclusion mappings $(X, Y) \to (X, F_{n-1})$, $(F_{n-1}, F_n) \to (X, F_n)$, $(X, F_n) \to (X, F_{n-1})$, $(F_{n-1}, F_n) \to (X, Y_n)$, yielding the induced homomorphisms r_{n-1}, v_n, u_n, e'_n respectively. Clearly

$$Y_n \cup Y_n^* = X, \quad Y_n \cap Y_n^* = Y \neq \emptyset,$$
$$Y_n \cup F_{n-1} = X, \quad Y_n \cap F_{n-1} = F_n \neq \emptyset.$$

Accordingly (see **I.6.14**, remark 2) e_n and e'_n are excision homomorphisms and hence isomorphisms onto, as indicated by the symbols $\approx$. The proof of the structure theorem is now made by the following series of observations.

(i) By the vector law, $j_n^* i_n = e_n$, an isomorphism. Hence clearly i_n is an isomorphism into.

(ii) If $b_m \in H^p(X, Y_m)$, then

$$j_n^* i_m b_m = \begin{cases} 0 & \text{if } m \neq n, \\ e_n b_n & \text{if } m = n. \end{cases}$$

The second assertion is obvious since $j_n^* i_n = e_n$. On the other hand, if $m \neq n$, then $j_n^* i_m b_m = j_n^* i_n^* k_{nm} b_m = 0$, because of the exactness of the cohomology sequence of the triple (X, Y_n^*, Y).

(iii) The subgroups $I(i_n) = i_n H^p(X, Y_n)$ of $H^p(X, Y)$ form an *independent system* in the following sense: if $a_m \in I(i_m)$, $m = 1, \ldots, N$, and

$$\sum_{m=1}^{N} a_m = 0, \tag{9}$$

then $a_1 = \cdots = a_N = 0$. Indeed, by assumption

$$a_m = i_m b_m, \quad b_m \in H^p(X, Y_m), \qquad m = 1, \ldots, N, \tag{10}$$

and hence (9) yields

$$\sum_{m=1}^{N} i_m b_m = 0. \tag{11}$$

Take any integer n such that $1 \leq n \leq N$. Then (11) yields

$$\sum_{m=1}^{N} j_n^* i_m b_m = 0.$$

By (ii) this formula reduces to $e_n b_n = 0$. Since e_n is an isomorphism, it follows that $b_n = 0$ and hence $a_n = 0$, $1 \leq n \leq N$, proving the assertion (iii).

(iv) If $a \in H^p(X, Y)$ is such that

$$a \in I(r_n) = r_n H^p(X, F_n), \tag{12}$$

then a can be written in the form

$$a = \sum_{k=1}^{n} i_k b_k, \quad b_k \in H^p(X, Y_k). \tag{13}$$

Denoting this assertion by A_n, we make its proof by induction. For $n = 1$, we have $F_1 = Y_1$, $r_1 = i_1$, and thus A_1 is obvious. So let $n \geq 2$, and assume that A_{n-1} is known to be true. Observe that since

e'_n is an isomorphism onto, the relation $e'_n = v_n s_n$ (which holds by the vector law) can be re-written in the form

$$((e'_n)^{-1} v_n)\, s_n = 1,$$

the identity homomorphism in $H^p(X, Y_n)$. By **I.3.2**, exercise 5, it follows that any element c_n of $H^p(X, F_n)$ can be written in the form

$$c_n = s_n b_n + c'_n, \quad b_n \in H^p(X, Y_n), \quad c'_n \in N((e'_n)^{-1} v_n).$$

Since e'_n is an isomorphism, it follows that $c'_n \in N(v_n)$, the nucleus of v_n. But $N(v_n) = u_n H^p(X, F_{n-1})$ by the exactness of the cohomology sequence of the triple (X, F_{n-1}, F_n). Hence c'_n can be written in the form $c'_n = u_n c_{n-1}$, $c_{n-1} \in H^p(X, F_{n-1})$. Thus we obtain the representation

$$c_n = u_n c_{n-1} + s_n b_n. \tag{14}$$

Now if a satisfies (12), then $a = r_n c_n$, $c_n \in H^p(X, F_n)$. By (14), we obtain

$$a = r_n u_n c_{n-1} + r_n s_n b_n = r_{n-1} c_{n-1} + i_n b_n.$$

By the induction assumption A_{n-1}, the term $r_{n-1} c_{n-1}$ can be written in the form

$$\sum_{k=1}^{n-1} i_k b_k, \quad b_k \in H^p(X, Y_k),$$

and thus A_n follows, completing the induction.

(v) The *structure theorem* is now proved as follows. In view of (i) and (iii), there remains to show that any assigned element $a \in H^p(X, Y)$ can be written as a (finite) sum of the form

$$a = \sum_{k=1}^{n} i_k b_k, \quad b_k \in H^p(X, Y_k). \tag{15}$$

To verify this, observe that by (8) and the extension theorem in **I.6.7** there exists an integer n (depending upon a) such that $a \in I(r_n)$, and hence for this n there exists a representation of the form (15) by (iv).

Remark 1. If the sequence $\{O_n\}$ is finite, and contains say N terms, then for $n = N$ one has $F_N = Y$ and $r_N = 1$ [the identity homomorphism in $H^p(X, Y)$]. Accordingly, $I(r_N) = H^p(X, Y)$, and thus by (iv) every element a of $H^p(X, Y)$ can be written in the form (15) with $n = N$.

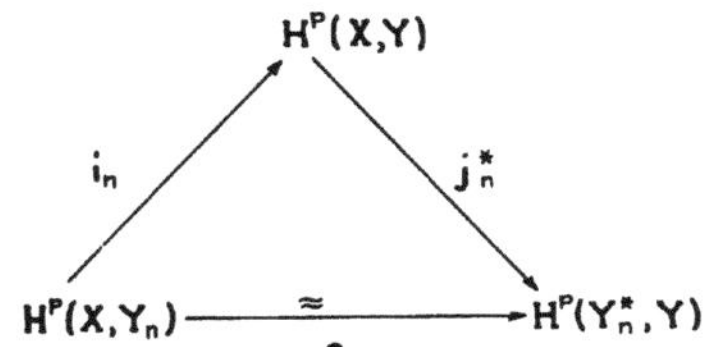

Fig. 31.

Remark 2. Consider Fig. 31 (which is a portion of Fig. 30).

Since e_n is an isomorphism onto, e_n^{-1} is also a (single-valued) isomorphism. We assert that for every element $a \in H^p(X, Y)$ we have the relation

$$a = \sum{}' i_n e_n^{-1} j_n^* a, \tag{16}$$

where the prime attached to the summation symbol means that the summation is finite (that is, all but a finite number of the terms are equal to zero). To prove (16), observe that by the structure theorem we have a representation of the form

$$a = \sum_{k=1}^{N} i_k b_k, \quad b_k \in H^p(X, Y_k). \tag{17}$$

Take any positive integer n. From (17) we get

$$i_n e_n^{-1} j_n^* a = \sum_{k=1}^{N} i_n e_n^{-1} j_n^* i_k b_k. \tag{18}$$

By (ii) above, it follows that

$$j_n^* i_k b_k = \begin{cases} 0 & \text{if } k \neq n, \\ e_n b_n & \text{if } k = n. \end{cases}$$

Thus (18) reduces to

$$i_n e_n^{-1} j_n^* a = \left\{ \begin{array}{ll} i_n b_n & \text{if } 1 \leq n \leq N, \\ 0 & \text{if } n > N. \end{array} \right\} \tag{19}$$

Thus all the terms in (16) vanish for $n > N$, and hence that summation is finite. From (17) and (19) we infer now that

$$\sum{}' i_n e_n^{-1} j_n^* a = \sum_{n=1}^{N} i_n b_n = a.$$

Remark 3. By **I.1.3**, exercise 41, we have the relations $\overline{O}_n \subset Y_n^*$, $fr\, O_n \subset Y \subset Y_n$, yielding the inclusion mapping $(\overline{O}_n, fr\, O_n) \to (Y_n^*, Y)$. Accordingly, we can complete the last diagram as indicated in Fig. 32.

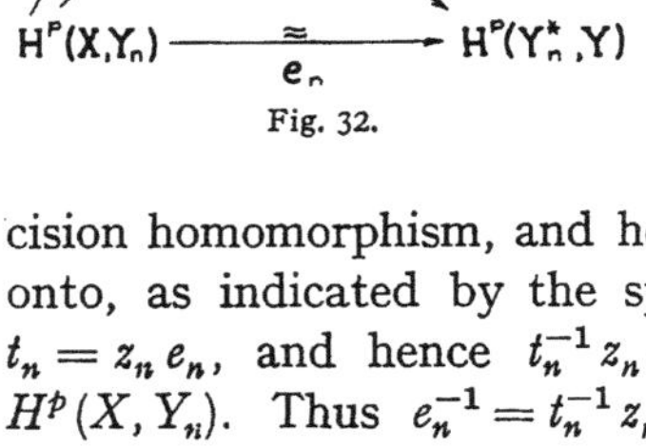

Fig. 32.

All the homomorphisms are induced by inclusion mappings. Let us now add the assumption

$$fr\, O_n \neq \emptyset \text{ for all } n. \tag{20}$$

By **I.1.3**, exercise 41 we have $X = Y_n \cup \overline{O}_n$, $fr\, O_n = Y_n \cap \overline{O}_n$. Thus (see **I.6.14**, remark 2) it follows that t_n is an excision homomorphism, and hence in view of (20) it is an isomorphism onto, as indicated by the symbol $\approx$. By the vector law we have $t_n = z_n e_n$, and hence $t_n^{-1} z_n e_n = 1$, the identity homomorphism in $H^p(X, Y_n)$. Thus $e_n^{-1} = t_n^{-1} z_n$. We conclude that

$$i_n e_n^{-1} j_n^* = i_n t_n^{-1} z_n j_n^* = i_n t_n^{-1} w_n. \tag{21}$$

From (16) and (21) we obtain, for $a \in H^p(X,Y)$, the representation

$$a = \sum{}' i_n t_n^{-1} w_n a, \tag{22}$$

which will be of importance later on. Recall that the summation is finite (all but a finite number of the terms are equal to zero), a fact referred to by the prime attached to the summation symbol.

Remark 4. Since t_n is an isomorphism onto, in view of part I of the structure theorem it follows that $I(i_n) \approx H^p(\overline{O}_n, fr\,O_n)$. Part II of the structure theorems yields then the following fundamental result.

Theorem. $H^p(X,Y)$ is the weak direct sum (or direct sum if the sequence $\{O_n\}$ is finite) of a set of its subgroups which are isomorphic to the groups $H^p(\overline{O}_n, fr\,O_n)$.

This theorem describes the structure of $H^p(X,Y)$ in terms of the *holes* O_n.

Remark 5. Take a second pair (X^*, Y^*) of compact HAUSDORFF spaces and consider a continuous mapping

$$f: (X, Y) \to (X^*, Y^*). \tag{23}$$

Also, take a closed subset Z of X such that $Z \subset Y$. Then we have also

$$f: (X, Z) \to (X^*, Y^*). \tag{24}$$

Furthermore, since $fr\,O_n \subset Y$, we have the relation

$$f|\overline{O}_n: (\overline{O}_n, fr\,O_n) \to (X^*, Y^*). \tag{25}$$

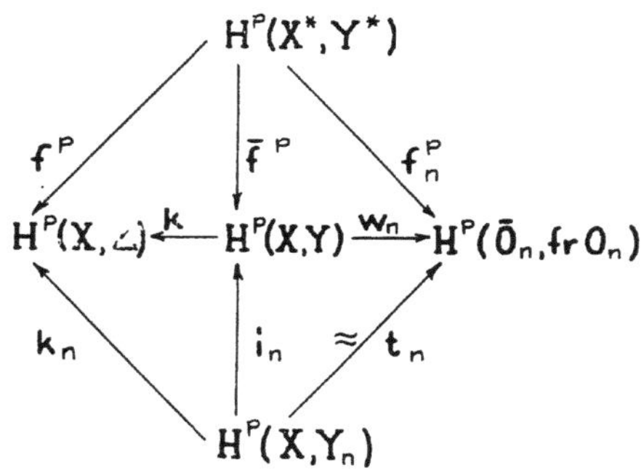

Fig. 33.

Consider now the diagram in Fig. 33.

The homomorphisms $\bar{f}^p$, f^p, f_n^p are induced by the mappings (23), (24), (25) respectively. The homomorphisms w_n, t_n, i_n are taken from the diagram used in remark 3. Finally, the homomorphisms k and k_n are induced by the inclusion mappings $(X, Z) \to (X, Y)$, $(X, Z) \to (X, Y_n)$ respectively. Now take an element $a^* \in H^p(X^*, Y^*)$. Using (22), we calculate

$$\begin{aligned} {}^p a^* &= k\bar{f}^p a^* = k \sum{}' i_n t_n^{-1} w_n \bar{f}^p a^* = k \sum{}' i_n t_n^{-1} f_n^p a^* \\ &= \sum{}' k\, i_n t_n^{-1} f_n^p a^* = \sum{}' k_n t_n^{-1} f_n^p a^*. \end{aligned}$$

Symbolically, this result may be written in the form

$$f^p = \sum{}' k_n t_n^{-1} f_n^p. \tag{26}$$

This formula will play an important part later on. It yields a decomposition of the homomorphism f^p in terms of the *holes* O_n corresponding to the subspace $Y > Z$.

Remark 6. Due to the fact that the case $p=1$ required special treatment in connection with excision homomorphisms (see **I.6.3**), we excluded in the preceding discussion the cases when $Y=\emptyset$ or $fr\,O_n=\emptyset$. However, if $p\neq 1$, then the excision homomorphisms involved are isomorphisms onto even if $Y=\emptyset$ or $fr\,O_n=\emptyset$, by the lemmas 1 and 2 in **I.6.3**. It follows that for $p\neq 1$ the structure theorem and the various other results derived from it remain valid even if $Y=\emptyset$ or $fr\,O_n=\emptyset$, as inspection of the proofs reveals.

§ I.7. Cohomology groups in Euclidean spaces[1].

I.7.1. Cells and spheres. The cohomology groups of the n-cell E^n and of the n-sphere S^n (see **I.2.2**) are given by the following theorems for all integers $n \geq 0$.

Theorem 1. $H^p(E^n)=0$ for all p.

Theorem 2. $H^p(S^n)=0$ for $p\neq n+1$, and $H^{n+1}(S^n)\approx \mathsf{I}$.

Proof. Since E^0 is a single point, we have $H^p(E^0)=0$ by **I.6.1**, remark 5. For $n\geq 1$, the origin $P=(0,\ldots,0)$ is a deformation retract of E^n (see **I.2.2**, exercise 13). Thus $H^p(E^n, P)=0$ by theorem 2 in **I.6.16**, and hence $H^p(E^n)=H^p(E^n, P)=0$ by **I.6.1**, remark 4, proving theorem 1.

Theorem 2 will be proved by induction on n. For clarity, denote by T_n that theorem stated for a specific n. Since S^0 consists of precisely two points, T_0 is a direct consequence of remark 2 in **I.6.3**. So take $n\geq 1$ and assume that T_{n-1} has been already established. From the *basic diagram* in R^{n+1} (see **I.2.2**) select first the pair $(E^n_-, \hat{S}^{n-1})$, and choose an integer q. The groups

$$H^q(E^n_-),\quad H^q(\hat{S}^{n-1}),\quad H^{q+1}(E^n_-, \hat{S}^{n-1}),\quad H^{q+1}(E^n_-)$$

are consecutive terms in the cohomology sequence of the pair $(E^n_-, \hat{S}^{n-1})$. Since E^n_- is homeomorphic to E^n, the first and the last groups are trivial by theorem 1. By **I.3.3**, lemma 4, it follows that

$$H^q(\hat{S}^{n-1}) \approx H^{q+1}(E^n_-, \hat{S}^{n-1}). \tag{1}$$

[1] For historical comments and further details, see the various treatises in Algebraic Topology listed in the *Bibliography*. If this volume is used in conjunction with other treatises, then the *shift in dimension* referred to on p. 64 in footnote 1 should be kept in mind. Many of the results derived in § I.7 are treated in the literature as applications of the ALEXANDER Duality Theorem. The inductive proofs given here provide instructive exercises for the reader in making effective use of elementary properties of Euclidean spaces.

In **I.2.2**, exercise 1 we noted the relation $S^n = E^n_+ \cup E^n_-$, $\hat{S}^{n-1} = E^n_+ \cap E^n_- \neq \emptyset$. The excision theorem (see **I.6.14**, remark 2) yields therefore

$$H^{q+1}(S^n, E^n_+) \approx H^{q+1}(E^n_-, \hat{S}^{n-1}). \tag{2}$$

Observe now that the groups

$$H^q(E^n_+), \quad H^{q+1}(S^n, E^n_+), \quad H^{q+1}(S^n), \quad H^{q+1}(E^n_+)$$

are consecutive terms in the cohomology sequence of the pair (S^n, E^n_+). Since E^n_+ is homeomorphic to E^n, the first and the last groups are trivial by theorem 1. Hence, by **I.3.3**, lemma 4,

$$H^{q+1}(S^n, E^n_+) \approx H^{q+1}(S^n). \tag{3}$$

As $\hat{S}^{n-1}$ is homeomorphic to S^{n-1}, we have

$$H^q(S^{n-1}) \approx H^q(\hat{S}^{n-1}). \tag{4}$$

From (1) to (4) we infer finally that

$$H^q(S^{n-1}) \approx H^{q+1}(S^n). \tag{5}$$

By the induction assumption T_{n-1}, the group on the left in (5) is trivial or isomorphic to I according as $q \neq n$ or $q = n$, and thus (5) yields directly T_n, completing the induction.

Remark. For the case $q = n$ of (5) it is a matter of interest to have an explicit isomorphism from $H^n(S^{n-1})$ onto $H^{n+1}(S^n)$. Consider the following diagram.

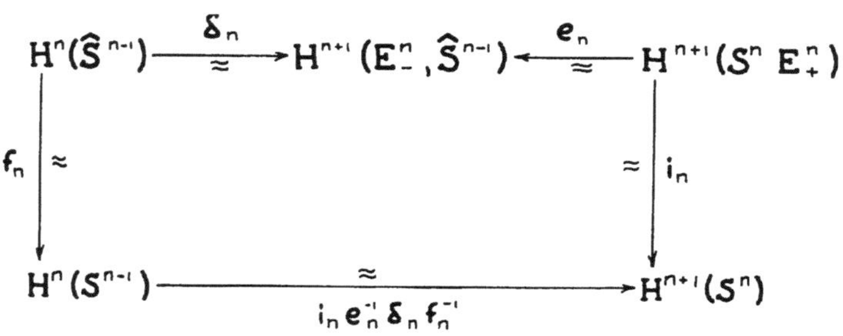

Fig. 34.

The homomorphisms δ_n, e_n, i_n are those that yielded the relations (1), (2), (3). As regards f_n, it is induced by the mapping $f: S^{n-1} \to \hat{S}^{n-1}$, given by $f(x^1, \ldots, x^n) = (x^1, \ldots, x^n, 0)$. Since f is clearly a homeomorphism, f_n is also an isomorphism onto, as indicated by the symbol $\approx$, in view of **I.6.2**, remark 2. The diagram yields the desired explicit isomorphism in the form $i_n e_n^{-1} \delta_n f_n^{-1}$.

I.7.2. A general property of compact sets in R^n and S^n. For any topological space X, we have $H^p(X)=0$ for $p \leqq 0$ by **I.6.1**, remarks 1 and 2. For compact sets in R^n and S^n, the following two theorems yield an upper limitation for the dimension of non-trivial cohomology groups.

Theorem 1. If F is a non-empty, compact subset of R^n, then $H^p(F)=0$ for $p>n$.

Theorem 2. If F is a non-empty, compact subset of S^n such that $S^n-F \neq \emptyset$, then $H^p(F)=0$ for $p>n$.

Since the proof will be made by induction on n, it is convenient to denote these theorems by $T(R^n)$ and $T(S^n)$ respectively. The following auxiliary theorem, to be denoted by $T(R^n, m, s)$, will also be proved and utilized in the course of the proof. Let $n \geqq 1$, $m \geqq 1$, $s \geqq 1$ be integers, and let $F(n, m, s)$ be a generic notation for a set in R^n that is the union of precisely s different cubes of the subdivision $\Delta(n, m)$ of R^n (see **I.2.3**, definition 1). Then $T(R^n, m, s)$ is the theorem that $H^p(F(n, m, s))=0$ for $p>n$.

The proof of $T(R^n)$, $T(S^n)$, $T(R^n, m, s)$ is made by simultaneous induction, and is based on the following series of observations.

(i) $T(R^0)$ and $T(S^0)$ are both true. Indeed, R^0 consists of a single point, and thus $T(R^0)$ is a direct consequence of remark 5 in **I.6.1**. As regards $T(S^0)$, observe that a non-empty, proper subset of S^0 consists of a single point, and thus $T(S^0)$ is also a direct consequence of remark 5 in **I.6.1**.

(ii) $T(R^n, m, 1)$ is true. Indeed, a set $F(n, m, 1)$ is merely a cube of the subdivision $\Delta(n, m)$, and hence it is homeomorphic to the n-cell E^n. By theorem 1 in **I.7.1** and remark 2 in **I.6.2** it follows that $H^p(F(n, m, 1))=0$, and indeed not only for $p>n$ but for every p.

(iii) For $n \geqq 1$, $T(R^n)$ and $T(S^n)$ mutually imply each other. Indeed, in view of the availability of the stereographic projection (see **I.2.2**) it is evident that a non-empty compact set $F \subset R^n$ is homeomorphic to a non-empty, compact set $F^* \subset S^n$ such that $S^n-F^* \neq \emptyset$, and *vice versa*. Since then $H^p(F) \approx H^p(F^*)$ for every p by remark 2 in **I.6.2**, the equivalence of $T(R^n)$ and $T(S^n)$ for $n \geqq 1$ is obvious.

(iv) If $n \geqq 1$, and $T(R^{n-1})$ is already established, then $T(R^n, m, s)$ follows for all integers $m \geqq 1$, $s \geqq 1$. To see this, observe first that $T(S^{n-1})$ is also available [by (i) if $n=1$ and by (iii) if $n>1$]. Fix now the integer $m \geqq 1$. Noting that $T(R^n, m, 1)$ holds by (ii), we proceed by induction on s. So assume that $s \geqq 2$, and that $T(R^n, m, s-1)$ has already been established. Take a set

$$X = F(n, m, s) = \bigcup_{j=1}^{s} q_j,$$

where $q_1, \ldots, q_s$ are different cubes of the subdivision $\Delta(n, m)$ of R^n. Set

$$X_1 = \bigcup_{j=1}^{s-1} q_j, \quad X_2 = q_s, \quad Y = X_1 \cap X_2.$$

Then X_1 is a set $F(n, m, s-1)$, and hence

$$H^p(X_1) = 0 \quad \text{for} \quad p > n, \tag{1}$$

by the induction assumption $T(R^n, m, s-1)$. Also, since $X_2 = q_s$ is homeomorphic to the n-cell E^n, we have

$$H^p(X_2) = 0 \quad \text{for} \quad p > n, \tag{2}$$

and indeed for every p, by theorem 1 in **I.7.1**. By **I.2.3**, exercise 3, the notations may be so chosen that Y is a proper subset (perhaps empty) of $fr\, q_s$. Since $fr\, q_s$ is homeomorphic to S^{n-1} (see **I.2.2**, exercise 12), and since $T(S^{n-1})$ is available as noted above, we have the relation

$$H^{p-1}(Y) = 0 \quad \text{for} \quad Y \neq \emptyset, \quad p-1 > n-1. \tag{3}$$

Assume first that $Y \neq \emptyset$. Noting that

$$X = F(n, m, s) = X_1 \cup X_2, \quad Y = X_1 \cap X_2, \tag{4}$$

the relations (1) to (4) yield, by theorem 3 in **I.6.16**,

$$H^p(F(n, m, s)) = 0 \quad \text{for} \quad p > n. \tag{5}$$

On the other hand, if $Y = \emptyset$, then (5) follows already from (1), (2) and (4) by **I.6.5**(b), since $p > n \geqq 1$ and hence $p \neq 1$. Thus the inductive proof is complete.

(v) If $T(R^n, m, s)$ is known to hold for some $n \geqq 1$ and for all integers $m \geqq 1$, $s \geqq 1$, then $T(R^n)$ follows. To see this, take a non-empty compact set $F \subset R^n$, and let O be any open set such that $O \supset F$. Consider the subdivision $\Delta(n, m)$ of R^n, and denote by F^* the union of all those cubes $q \in \Delta(n, m)$ that intersect F. Then F^* is a set $F(n, m, s)$, and hence by $T(R^n, m, s)$ we have

$$H^p(F^*) = 0 \quad \text{for} \quad p > n. \tag{6}$$

For m sufficiently large, we have however (see **I.2.3**, exercise 5) the inclusions $F \subset F^* \subset O$. Since the open set $O \supset F$ was arbitrary, (6) implies by theorem 5 in **I.6.16** that $H^p(F) = 0$ for $p > n$. Thus the inductive proof of $T(R^n)$ is complete. Hence, in view of (iii), $T(S^n)$ is also established.

Remark 1. The condition $S^n - F \neq \emptyset$ in theorem 2 is necessary. Indeed, if $F = S^n$, then $H^{n+1}(F) = H^{n+1}(S^n) \neq 0$ by theorem 2 in **I.7.1**.

Remark 2. As an immediate corollary to theorem 1, we obtain the statement that if X is a topological space which is homeomorphic to R^n and Y is a non-empty compact subset of X, then $H^p(Y)=0$ for $p>n$. Indeed, the assumption concerning X implies that Y is homeomorphic to a non-empty compact subset F of R^n. Hence by remark 2 in **I.6.2** we conclude that, for $p>n$, $H^p(Y)\approx H^p(F)=0$. A similar argument, starting with theorem 2, yields the following corollary: if X is a topological space which is homeomorphic to S^n and Y is a non-empty, compact, proper subset of X, then $H^p(Y)=0$ for $p>n$.

I.7.3. A lemma. Let $n\geq 2$, $m\geq 1$ be arbitrary integers which will be kept fixed in this section. Consider the subdivision $\Delta^*(n, m)$ of the unit cube Q^n in R^n (see **I.2.3**, definition 2). The letter q is used as a generic notation for a cube of $\Delta^*(n, m)$, and q^* refers to the unique cube q which contains the point $(0, 0, \ldots, 0)$. Then clearly

$$(0, 0, \ldots, 0)\in fr\, q^*, \qquad (0, 0, \ldots, 0)\notin q \quad \text{if} \quad q\neq q^*. \tag{1}$$

Let N be the number of cubes q and s be an integer such that

$$0\leq s< N. \tag{2}$$

The symbol K_s is used to refer to a collection of s cubes q with the following properties (see **I.2.3**, definition 7).

(i) K_s is strongly connected.

(ii) $q^*\in K_s$ if $s\geq 1$.

The symbol K_s^* is used to refer to the collection complementary to K_s (that is, K_s^* contains those cubes q that are not contained in K_s). The union of the cubes of K_s, K_s^* will be denoted by $|K_s|$, $|K_s^*|$ respectively. Observe that $|K_s^*|\neq\emptyset$ in view of (2).

Lemma $L(s)$. $H^n(|K_s^*|)=0$.

Proof. As the notation $L(s)$ indicates, we proceed by induction on s. For $s=0$, the lemma is true since $|K_1^*|=Q^n$, and $H^n(Q^n)=0$ by **I.7.1** and remark 2 in **I.6.2**, in view of the fact that Q^n is homeomorphic to the n-cell E^n (see **I.2.2**, exercise 12). Now that $L(0)$ is established, take $s\geq 1$ and assume that $L(s-1)$ has been proved. We arrange the s cubes of K_s in the following manner into a sequence

$$q_1, \ldots, q_s. \tag{3}$$

(a) If $s=1$, then K_1 consists [in view of (ii)] of the single cube q^*, and (3) is of course the one-term sequence where $q_1=q^*$.

(b) If $s\geq 2$, then by **I.2.3**, exercise 8 we can choose (3) in such a manner that the cubes $q_1, \ldots, q_{s-1}$ form a strongly connected collection containing q^* and the cubes q_{s-1} and q_s are strongly adjacent. Observe that these properties of (3) imply that $q_s\neq q^*$.

Introduce the notations

$$X_1 = |K_s^*|, \quad X_2 = q_s, \quad X = X_1 \cup X_2, \quad Y = X_1 \cap X_2. \tag{4}$$

Since X_2 is then homeomorphic to the n-cell E^n, we have

$$H^{n-1}(X_2) = 0. \tag{5}$$

Next we verify that

$$H^n(X) = 0. \tag{6}$$

If $s=1$, then $X = Q^n$, and (6) follows since Q^n is homeomorphic to E^n. If $s \geq 2$, then in view of (b) it follows that X is a set of the form $|K_{s-1}^*|$, where K_{s-1}^* is the complementary collection to the collection K_{s-1} consisting of $q_1, \ldots, q_{s-1}$. Thus (6) holds by the induction assumption $L(s-1)$.

Assume now first that $Y = \emptyset$. Then by **I.6.5** (a) it follows from (5) and (6) that $H^n(X_1) = 0$, proving $L(s)$ in this case. Assume next that

$$Y \neq \emptyset. \tag{7}$$

Since $q_s \notin K_s^*$ and $Y \subset X_1 = |K_s^*|$, clearly

$$Y \subset fr\, q_s. \tag{8}$$

We proceed to verify that

$$fr\, q_s - Y \neq \emptyset. \tag{9}$$

If $s=1$, observe that in this case $q_s = q_1 = q^*$, and hence by (1)

$$(0, 0, \ldots, 0) \in fr\, q^* - |K_1^*| \subset fr\, q^* - Y,$$

showing that (9) holds. If $s \geq 2$, then by (b) the cubes q_{s-1} and q_s are strongly adjacent, and hence (see **I.2.3**, exercise 6) there exists a point x such that

$$x \in fr\, q_s, \quad x \notin q \quad \text{if} \quad q \neq q_s,\, q_{s-1}. \tag{10}$$

Since all the cubes $q \in K_s^*$ satisfy the condition $q \neq q_s, q_{s-1}$, we infer from (4) and (10) that

$$x \in fr\, q_s - X_1 \subset fr\, q_s - Y,$$

and thus (9) holds again. In view of (7), (8), (9) it follows that Y is a non-empty, compact, proper subset of $fr q_s$. Since $fr q_s$ is homeomorphic to the $(n-1)$-sphere S^{n-1} (see **I.2.2**, exercise 12) we conclude by remark 2 in **I.7.2** that

$$H^n(Y) = 0. \tag{11}$$

Finally, by theorem 4 in **I.6.16** the relations (7), (5), (6), (11) imply that $H^n(X_1) = H^n(|K_s^*|) = 0$, and the inductive proof of the lemma is complete.

I.7.4. Applications. *Theorem 1.* Let F be a non-empty compact subset of the interior of the unit cube Q^n, $n \geq 2$, such that $Q^n - F$ is connected. Then $H^n(F) = 0$.

Proof. Let there be assigned any set U^* such that $F \subset U^* \subset Q^n$ and U^* is open relative to Q^n. Setting $U = U^* \cap \operatorname{int} Q^n$, we have $F \subset U \subset U^*$. Clearly, the point $(0, 0, \ldots, 0)$ does not lie in U. Since $Q^n - F$ is connected, we have (see **I.1.3**, exercise 35) an open set O such that

(i) $F \subset O \subset U$,

(ii) $Q^n - O$ is connected.

Since $(0, 0, \ldots, 0) \notin U$, it follows that

(iii) $(0, 0, \ldots, 0) \in Q^n - O$.

In view of (i), F and $Q^n - O$ are non-empty, compact, disjoint subsets of Q^n. Consider now the subdivision $\Delta^*(n, m)$ of Q^n (see **I.2.3**). Denote again by q a generic cube of this subdivision, and by q^* the unique cube q which contains the point $(0, 0, \ldots, 0)$. Let K_s be the collection of those cubes q which intersect $Q^n - O$, where s is the number of the cubes of K_s, and let K_s^* be the complementary collection (thus K_s^* consists of those cubes q which do not intersect $Q^n - O$). For m large enough, by **I.2.3**, exercise 4 we have then

$$q \cap F = \emptyset \quad \text{if} \quad q \in K_s. \tag{1}$$

Let m be so chosen. In view of (iii), clearly $q^* \in K_s$. In view of (ii), it follows by **I.2.3**, exercise 7 that K_s is strongly connected. Since $F \neq \emptyset$, (1) implies that K_s^* is non-empty. Thus K_s and K_s^* satisfy the assumptions of the lemma in **I.7.3**, and hence

$$H^n(|K_s^*|) = 0. \tag{2}$$

In view of (1), (i), and the definition of K_s^*, it follows that

$$F \subset |K_s^*| \subset O \subset U \subset U^*. \tag{3}$$

Since the (relative to Q^n) open set $U^* \supset F$ was arbitrary, the relations (2) and (3) imply, by theorem 5 in **I.6.16**, that $H^n(F) = 0$.

Theorem 2. If F is a non-empty compact subset of R^n, $n \geq 2$, such that $R^n - F$ is connected, then $H^n(F) = 0$.

Proof. By **I.2.2**, exercise 11, F is homeomorphic to a non-empty, compact subset F^* of the interior of the unit cube Q^n such that $Q^n - F^*$ is connected. In view of theorem 1 and **I.6.2**, remark 2, it follows that $H^n(F) \approx H^n(F^*) = 0$.

Theorem 3. If F is a non-empty, compact subset of S^n, $n \geq 1$, such that $S^n - F$ is connected, then $H^n(F) = 0$.

Proof. If $S^n - F = \emptyset$, then $H^n(F) = H^n(S^n) = 0$ by **I.7.1**. If $S^n - F \neq \emptyset$ and $n \geq 2$, then F is homeomorphic (see **I.2.2**, exercise 10) to a non-empty compact subset F^* of R^n such that $R^n - F^*$ is connected. By theorem 2 and **I.6.2**, remark 2 it follows that $H^n(F) \approx H^n(F^*) = 0$. Finally, if $n = 1$ and $S^1 - F \neq \emptyset$, then either F is a single point and $H^1(F) = 0$ by **I.6.1**, remark 5, or else F is homeomorphic (see **I.2.2**, exercise 14) to the 1-cell E^1, and $H^1(F) \approx H^1(E^1) = 0$ by **I.7.1** and **I.6.2**, remark 2.

Theorem 4. Let X and $Y \subset X$ be non-empty, compact subsets of the n-sphere S^n, $n \geq 1$, such that $X - Y$ is connected and non-empty. Then $H^{n+1}(X, Y) \approx \mathsf{I}$ or $H^{n+1}(X, Y) = 0$, according as $X - Y$ is or is not open, or equivalently, according as Y does or does not contain $fr\, X$.

Proof. Since $X - Y$ is a non-empty connected subset of $\complement Y$, it is contained in a certain component O_1 of $\complement Y$. Let O_2 be the union of the components of $\complement Y$ other than O_1 (thus O_2 is empty if $\complement Y$ is connected). Since S^n is locally connected, O_1 and O_2 are disjoint open sets, and O_1 is connected. Thus we are dealing with the situation studied in **I.1.3**, exercise 19. Accordingly, on setting $X^* = X \cup O_2$, $Y^* = Y \cup O_2$, the sets X^*, Y^* are closed (and hence compact since S^n is compact). We first verify that

$$H^{n+1}(X^*, Y^*) \approx H^{n+1}(X, Y). \tag{4}$$

Of course, (4) is obvious if $O_2 = \emptyset$, since then $X^* = X$, $Y^* = Y$. In any case, by **I.1.3**, exercise 19 we have the relations

$$X \cup Y^* = X^*, \quad X \cap Y^* = Y \neq \emptyset,$$

and (4) follows from the excision theorem as stated in **I.6.14**, remark 2. Next (see **I.1.3**, exercise 19), $S^n - Y^* = O_1 \neq \emptyset$. Thus Y^* is a non-empty, compact subset of S^n such that $S^n - Y^*$ is connected. By theorem 3 it follows that

$$H^n(Y^*) = 0. \tag{5}$$

Furthermore, by **I.7.2**, theorem 2 we have

$$H^{n+1}(Y^*) = 0. \tag{6}$$

Observe now that the groups $H^n(Y^*)$, $H^{n+1}(X^*, Y^*)$, $H^{n+1}(X^*)$, $H^{n+1}(Y^*)$ are consecutive terms in the cohomology sequence of the pair (X^*, Y^*). The first and the last group are trivial by (5) and (6). Hence, by **I.3.3**, lemma 4,

$$H^{n+1}(X^*, Y^*) \approx H^{n+1}(X^*). \tag{7}$$

By **I.7.1**, theorem 2 and **I.7.2**, theorem 2 we have $H^{n+1}(X^*) \approx \mathsf{I}$ or $H^{n+1}(X^*) = 0$ according as $S^n - X^* = \emptyset$ or $S^n - X^* \neq \emptyset$, and hence (see **I.1.3**, exercise 19) according as $X - Y$ is or is not open, or equivalently, according as Y does or does not contain frX. In view of the isomorphisms (7) and (4), the proof is complete.

Theorem 5. Let X and $Y \subset X$ be non-empty, compact subsets of R^n, $n \geq 1$, such that $X - Y$ is connected and non-empty. Then $H^{n+1}(X, Y) \approx \mathsf{I}$ or $H^{n+1}(X, Y) = 0$, according as $X - Y$ is or is not open, or equivalently, according as Y does or does not contain frX.

Proof. Let $f: R^n \to S^n - P$ be the inverse of the stereographic projection (see **I.2.2**). Thus f is a homeomorphism from R^n onto $S^n - P$ (the n-sphere with the north pole P deleted). The pair (X, Y) is carried by f into a pair (X^*, Y^*) satisfying the assumptions of theorem 4. Observe now that $X - Y$ is open in R^n if and only if $f(X - Y) = X^* - Y^*$ is open in $S^n - P$ (since f is a homeomorphism). On the other hand, since $S^n - P$ is open in S^n, clearly $X^* - Y^*$ is open in $S^n - P$ if and only if it is open in S^n. Thus $X - Y$ is open in R^n if and only if $X^* - Y^*$ is open in S^n. Since $H^{n+1}(X^*, Y^*) \approx H^{n+1}(X, Y)$ by **I.6.2**, remark 2, and since by **I.1.3**, exercise 19 the set $X - Y$ is open if and only if $frX \subset Y$, theorem 5 appears as a corollary of theorem 4.

Theorem 6. Let Y be a non-empty, compact, proper subset of S^n, $n \geq 1$. Denote by $q(\complement Y)$ the number of components of $\complement Y$. Then (see **I.3.1**)

$$H^{n+1}(S^n, Y) \approx \mathsf{I}^{q(\complement Y)}. \tag{8}$$

Proof. Since S^n is locally connected and separable, the components of $\complement Y$ constitute a finite or countably infinite sequence of non-empty, pair-wise disjoint domains $\Gamma_1, \ldots, \Gamma_k, \ldots$ (see **I.1.3**, exercise 28). The set $Y_k = \complement \Gamma_k$ is a non-empty, compact, proper subset of S^n, and $\complement Y_k = \Gamma_k$ is connected. By **I.7.4**, theorem 3 and **I.7.2**, theorem 2 it follows that

$$H^n(Y_k) = 0, \quad H^{n+1}(Y_k) = 0. \tag{9}$$

The groups $H^n(Y_k)$, $H^{n+1}(S^n, Y_k)$, $H^{n+1}(S^n)$, $H^{n+1}(Y_k)$ are consecutive terms of the cohomology sequence of the pair (S^n, Y_k). The first and the last group are trivial by (9), and hence by **I.3.3**, lemma 4 and **I.7.1**,

$$H^{n+1}(S^n, Y_k) \approx H^{n+1}(S^n) \approx \mathsf{I}. \tag{10}$$

By the *structure theorem* in **I.6.17**, the group $H^{n+1}(S^n, Y)$ is the weak direct sum or the direct sum of a sequence of its subgroups which are isomorphic, term for term, to the groups of the sequence $H^{n+1}(S^n, Y_k)$, $k = 1, 2, \ldots$, and thus (8) follows in view of (10).

Theorem 7. Let Y be a non-empty, compact, proper subset of S^n, $n \geq 1$, and denote again by $q(\complement Y)$ the number of components of $\complement Y$. Then

$$H^n(Y) \approx \mathsf{I}^{q(\complement Y) - 1}. \tag{11}$$

Proof. Let Γ be a component of $\complement Y$, and let O be the union of the other components of $\complement Y$ (thus O is empty if $\complement Y$ is connected). As noted in the proof of theorem 6, the components of $\complement Y$ are open sets. It follows that the sets $X = Y \cup O$, $Y^* = Y \cup \Gamma$ are closed (and hence compact). Indeed, clearly

$$\complement X = \Gamma, \quad \complement Y^* = O \tag{12}$$

are open. (12) shows that X is a non-empty, compact, proper subset of S^n such $\complement X$ is connected. Hence, by **I.7.2**, theorem 2 and **I.7.4**, theorem 3,

$$H^n(X) = 0, \quad H^{n+1}(X) = 0. \tag{13}$$

The groups $H^n(X)$, $H^n(Y)$, $H^{n+1}(X, Y)$, $H^{n+1}(X)$ are consecutive terms in the cohomology sequence of the pair (X, Y). The first and the last group are trivial by (13). Hence, by **I.3.3**, lemma 4,

$$H^n(Y) \approx H^{n+1}(X, Y). \tag{14}$$

We verify next that

$$H^{n+1}(X, Y) \approx H^{n+1}(S^n, Y^*). \tag{15}$$

Clearly $S^n - \Gamma = X$, $Y^* - \Gamma = Y$, and thus (15) follows from the excision theorem in **I.6.14**. Obviously, the number of components of $\complement Y^*$ is one less than the number of components of $\complement Y$. By theorem 6 it follows that

$$H^{n+1}(S^n, Y^*) \approx \mathsf{I}^{q(\complement Y)-1}, \tag{16}$$

and (11) is obtained finally by means of the isomorphisms (14) to (16).

Theorem 8. Let Y be a non-empty, compact subset of R^n, $n \geq 1$, and denote by $k(\complement Y)$ the number of *bounded* components of $\complement Y$. Then

$$H^n(Y) \approx \mathsf{I}^{k(\complement Y)}.$$

Proof. By **I.2.2**, exercise 10, the set Y is homeomorphic to a non-empty, compact, proper subset Y^* of S^n such that $k(\complement Y) = q(\complement Y^*) - 1$, where $q(\complement Y^*)$ denotes the number of the components of $\complement Y^*$. By **I.6.2**, remark 2 and the preceding theorem 7 we conclude that

$$H^n(Y) \approx H^n(Y^*) \approx \mathsf{I}^{q(\complement Y^*)-1} = \mathsf{I}^{k(\complement Y)}.$$

Theorem 9. If the non-empty, compact subsets Y_1, Y_2 of R^n, $n \geq 1$, are homeomorphic, then the number of the components of $\complement Y_1$ is equal to the number of the components of $\complement Y_2$.

Proof. Let again $k(\complement Y_i)$, $i = 1, 2$, denote the number of the *bounded* components of $\complement Y_i$. Since (see **I.2.1**, exercise 6) the total number of the components of $\complement Y_i$ is $k(\complement Y_i) + 2$ if $n = 1$ and $k(\complement Y_i) + 1$ if

$n \geqq 2$, it is sufficient to show that

$$k(\mathsf{C}Y_1) = k(\mathsf{C}Y_2). \tag{17}$$

Now we have, by **I.6.2**, remark 2,

$$H^n(Y_1) \approx H^n(Y_2). \tag{18}$$

In view of theorem 8 it follows that

$$\mathsf{I}^{k(\mathsf{C}Y_1)} \approx \mathsf{I}^{k(\mathsf{C}Y_2)}. \tag{19}$$

By **I.3.1**, exercises 5 and 6, the isomorphism (19) implies (17).

Theorem 10 (the JORDAN-BROUWER *theorem).* If the subset Y of R^n, $n \geqq 2$, is homeomorphic to the $(n-1)$-sphere S^{n-1}, then the following holds.

(i) $\mathsf{C}Y$ has precisely two components D_1, D_2.

(ii) One of these components is bounded, while the other one is unbounded. Both of these components are domains.

(iii) $fr\,D_1 = fr\,D_2 = Y$.

Proof. Since Y is homeomorphic to S^{n-1}, $n \geqq 2$, Y is a non-empty, compact set which does not reduce to a single point, and precisely one component of $\mathsf{C}Y$ is unbounded (see **I.2.1**, exercise 6). Furthermore, by **I.2.1**, exercise 4, each component of $\mathsf{C}Y$ is open and connected, and hence it is a domain. By **I.1.3**, exercise 33, for each component D of $\mathsf{C}Y$ we have the inclusion $fr\,D \subset Y$. Thus the theorem will be verified if we show yet that (i) is true and that

$$Y \subset fr\,D_i, \quad i = 1, 2. \tag{20}$$

Now since Y is homeomorphic to S^{n-1}, by **I.7.1** and **I.6.2**, remark 2 it follows that $H^n(Y) \approx H^n(S^{n-1}) \approx \mathsf{I}$. By theorem 8 it follows that the number of bounded components of $\mathsf{C}Y$ is equal to one. Since $n \geqq 2$, we conclude by **I.2.1**, exercise 6 that the total number of components of $\mathsf{C}Y$ is equal to two, and (i) is proved. Now let x_0 be any point of Y, and consider the spherical neighborhood

$$\Delta_r(x_0) = \{x \mid \|x - x_0\| < r\},$$

where $r > 0$ is a real number. We show first that

$$\Delta_r(x_0) \cap D_i \neq \emptyset, \quad i = 1, 2. \tag{21}$$

Suppose that (21) fails to hold for some $r_0 > 0$. Then it also fails to hold for any r such that $0 < r < r_0$, since $\Delta_r(x_0)$ decreases as r decreases. Choose now $r < r_0$ in such a manner that $Y - \Delta_r(x_0) \neq \emptyset$ (this is possible since Y does not reduce to a single point). Consider the set

$$F = \mathsf{C}(D_1 \cup D_2 \cup \Delta_r(x_0)).$$

This defining formula can be re-written in the form

$$F = \complement(D_1 \cup D_2) \cap \complement \Delta_r(x_0) = Y \cap \complement \Delta_r(x_0) = Y - \Delta_r(x_0).$$

It follows that F is a non-empty, compact subset of Y. Furthermore, F is a proper subset of Y, since it does not contain the point $x_0 \in Y$. Since Y is homeomorphic to S^{n-1}, we conclude that F is homeomorphic to a non-empty, compact, proper subset of S^{n-1}, and hence by **I.7.2**, theorem 2, we have $H^n(F) = 0$. By theorem 8 it follows that $\complement F$ is connected. Now

$$\complement F = D_1 \cup D_2 \cup \Delta_r(x_0).$$

Since (21) fails to hold by assumption, $\Delta_r(x_0)$ does not intersect both D_1 and D_2. As $D_1 \cap D_2 = \emptyset$, it follows from **I.1.3**, exercise 15, that $\complement F$ is disconnected, a contradiction. Thus (21) is established. Since (21) holds for every $r > 0$, by **I.2.1**, exercise 3 we conclude that

$$x_0 \in \overline{D}_i, \quad i = 1, 2. \tag{22}$$

On the other hand, $x_0 \in Y = \complement(D_1 \cup D_2) = \complement D_1 \cap \complement D_2$, and thus

$$x_0 \in \complement D_i, \quad i = 1, 2. \tag{23}$$

Since D_i is open, (22) and (23) imply that

$$x_0 \in \overline{D}_i \cap \complement D_i = fr\, D_i, \quad i = 1, 2.$$

As x_0 was an arbitrary point of Y, (20) follows, and the proof is complete.

Theorem 11. Let Y be a non-empty set in R^n, $n \geq 1$, and let x_0 be a point of Y. Then $x_0 \in \operatorname{int} Y$ if and only if there exists a pair of subsets A, B of Y with the following properties.

(i) $\emptyset \neq B \subset A \subset Y$.
(ii) A and B are compact.
(iii) $A - B$ is non-empty and connected.
(iv) $x_0 \in A - B$.
(v) $H^{n+1}(A, B) \approx \mathsf{I}$.

Proof. Suppose first that $x_0 \in \operatorname{int} Y$. Then for $r > 0$ sufficiently small the spherical neighborhood $\Delta = \Delta_r(x_0)$ satisfies the inclusions $x_0 \in \Delta$, $\overline{\Delta} \subset \operatorname{int} Y$. Define the sets A, B as follows: $A = \overline{\Delta}$, $B = fr\,\overline{\Delta}$. Then (i), (ii), (iii), (iv) hold (see **I.2.1**, exercise 1). As $A - B = \Delta$ is open, we conclude from **I.7.4**, theorem 5 that (v) also holds.

Suppose now that there exists, for the point $x_0 \in Y$, a pair of sets A, B satisfying the conditions (i) to (v). From **I.7.4**, theorem 5 we conclude that $A - B$ is open, and hence $A - B \subset \operatorname{int} Y$ (see **I.1.3**, definition 4). Thus $x_0 \in A - B \subset \operatorname{int} Y$, and the theorem is proved.

Theorem 12 (invariance of the interior and of the frontier). If the subsets Y_1, Y_2 of R^n, $n \geq 1$, are homeomorphic and if $T: Y_1 \to Y_2$ is a topological mapping from Y_1 onto Y_2, then $T \operatorname{int} Y_1 = \operatorname{int} Y_2$. Furthermore, if Y_1 (and hence also Y_2) is compact, then $T\,fr\,Y_1 = fr\,Y_2$.

Proof. Take a point $x_1 \in \operatorname{int} Y_1$. By theorem 11, we have then a pair of compact, non-empty subsets A_1, B_1 of Y_1 such that $B_1 \subset A_1$, $A_1 - B_1$ is non-empty and connected, $x_1 \in A_1 - B_1$, and $H^{n+1}(A_1, B_1) \approx I$. Put $x_2 = T x_1$, $A_2 = T A_1$, $B_2 = T B_1$. Since T is a topological mapping, by **I.1.5**, exercise 8 it follows that A_2, B_2 are compact, non-empty subsets of Y_2, $B_2 \subset A_2$, $A_2 - B_2$ is non-empty and connected, $x_2 \in A_2 - B_2$, and (by **I.6.2**, remark 2)

$$H^{n+1}(A_2, B_2) \approx H^{n+1}(A_1, B_1) \approx I .$$

By theorem 11 we conclude that $x_2 \in \operatorname{int} Y_2$. Since x_1 was an arbitrary point of $\operatorname{int} Y_1$, it follows that $T \operatorname{int} Y_1 \subset \operatorname{int} Y_2$. The same argument applied to T^{-1} yields the complementary inclusion $T \operatorname{int} Y_1 \supset \operatorname{int} Y_2$. If Y_1, Y_2 are compact, then $fr\,Y_i = Y_i - \operatorname{int} Y_i$, $i = 1, 2$, and the relation $T\,fr\,Y_1 = fr\,Y_2$ appears as a direct consequence of the relation $T \operatorname{int} Y_1 = \operatorname{int} Y_2$.

Part II. Topological study of continuous transformations in R^n.

§ II.1. Orientation in R^n [1].

II.1.1. Bounded domains in R^n. The integer $n \geq 1$ will be arbitrary but fixed until further notice. A bounded domain D in R^n is a non-empty, bounded, connected, open set. A pair of non-empty compact sets X, Y in R^n will be said to form a *frame* for D if $X \supset Y$ and $X - Y = D$. The symbol $[X, Y, D]$ will be used to refer to this situation. For brevity, we shall speak simply of the frame $[X, Y, D]$. Thus the use of this term is merely an abbreviation for the following set of statements.

(i) X and Y are non-empty compact sets in R^n.

(ii) $X \supset Y$.

(iii) $X - Y = D$ is non-empty, open, and connected.

By **I.7.4**, theorem 5 we have the relation

$$H^{n+1}(X, Y) \approx I \tag{1}$$

for every frame $[X, Y, D]$ in R^n.

[1] The relationship between the abstract concept of orientation and the intuitive idea of orientation should become clearer to the reader after he had completed the study of Parts V and VI which deal with the case of the line and of the plane.

For any frame $[X, Y, D]$, the following holds by **I.1.3**, exercise 31.

(iv) $fr D$ and $\overline{D}$ are non-empty and compact.

(v) $X = \overline{D} \cup Y$, $fr D = \overline{D} \cap Y$.

(vi) $\overline{D} \supset fr D$, $\overline{D} - fr D = D$.

In view of (iv) and (vi), we have the special frame $[\overline{D}, fr D, D]$. From (v) it follows that

(vii) $(\overline{D}, fr D) \subset (X, Y)$ for any frame $[X, Y, D]$.

Accordingly, for a given D, the frame $[\overline{D}, fr D, D]$ is the smallest available frame [in the sense of the inclusion stated in (vii)]. This smallest frame $[\overline{D}, fr D, D]$ will be termed the *reduced frame* for D. From (iv) and (v) we obtain, by the strong excision theorem as stated in **I.6.14**, remark 2, the *excision isomorphism*

$$e: H^{n+1}(X, Y) \xrightarrow{\approx} H^{n+1}(\overline{D}, fr D), \tag{2}$$

where the symbol $\approx$ indicates that e is an *isomorphism onto*. Applying (1) to the reduced frame, there follows the relation

$$H^{n+1}(\overline{D}, fr D) \approx \mathsf{I}. \tag{3}$$

Consider now a pair of bounded domains D_1, D_2 such that

$$D_1 \subset D_2. \tag{4}$$

On setting

$$F_{12} = \overline{D}_2 - D_1, \tag{5}$$

we have (see **I.1.3**, exercise 7) the following relations.

$$\overline{D}_2 \supset F_{12} \supset fr D_2 \neq \emptyset, \quad \overline{D}_2 - F_{12} = D_1, \tag{6}$$

$$\overline{D}_2 = \overline{D}_1 \cup F_{12}, \quad fr D_1 = \overline{D}_1 \cap F_{12} \neq \emptyset. \tag{7}$$

From (5) and (7) we see that F_{12} is a non-empty compact set. In view of (6), we obtain a homomorphism

$$i_{12}: H^{n+1}(\overline{D}_2, F_{12}) \to H^{n+1}(\overline{D}_2, fr D_2), \tag{8}$$

induced by the inclusion mapping $(\overline{D}_2, fr D_2) \to (\overline{D}_2, F_{12})$. Furthermore, the relations (7) yield, by the strong excision theorem in **I.6.14**, remark 2, the *excision isomorphism*

$$e_{12}: H^{n+1}(\overline{D}_2, F_{12}) \xrightarrow{\approx} H^{n+1}(\overline{D}_1, fr D_1). \tag{9}$$

The relations (6) also yield the frame $[\overline{D}_2, F_{12}, D_1]$. Hence [see (1)]

$$H^{n+1}(\overline{D}_2, F_{12}) \approx \mathsf{I}. \tag{10}$$

Lemma 1. The homomorphism i_{12} occurring in (8) is an isomorphism onto.

Proof. Assume first that $n=1$. Then D_1, D_2 are merely bounded open intervals in R^1 (the real number-line), and clearly $fr\,D_2$ is a deformation retract of F_{12}. Accordingly, $H^1(F_{12}, fr\,D_2)=0$, $H^2(F_{12}, fr\,D_2)=0$ by **I.6.16**, theorem 2. The cohomology sequence of the triple $(\bar{D}_2, F_{12}, fr\,D_2)$ contains the section

$$H^1(F_{12}, fr\,D_2) \to H^2(\bar{D}_2, F_{12}) \xrightarrow{i_{12}} H^2(\bar{D}_2, fr\,D_2) \to H^2(F_{12}, fr\,D_2).$$

As we just observed, the first and the last group in this section are trivial. Hence (see **I.3.3**, lemma 4) i_{12} is an isomorphism onto, and the lemma is proved for the case $n=1$. Assume now that $n \geqq 2$. Choose a point $x^0 \in D_1$, and consider the open spherical neighborhood

$$D_0: \|x - x^0\| < r.$$

Since D_1 is open, we have the inclusion $\bar{D}_0 \subset D_1$ and hence also $\bar{D}_0 \subset D_2$, if $r>0$ is chosen sufficiently small. Furthermore (see **I.2.1**, exercise 1), D_0 is a bounded domain, and $fr\,D_0$ is connected. As a consequence (see **I.1.3**, exercise 34) the set $D_2 - D_0$ is non-empty, connected, and *not* open. We set now $F_{02} = \bar{D}_2 - D_0$, and assert that

$$H^{n+1}(F_{02}, fr\,D_2) = 0. \tag{11}$$

To see this, observe that F_{02}, $fr\,D_2$ are non-empty compact sets such that $F_{02} \supset fr\,D_2$ [by (6) applied to the domains D_0, D_2]. Also,

$$\begin{aligned} F_{02} - fr\,D_2 &= (\bar{D}_2 \cap \complement D_0) \cap \complement fr\,D_2 = (\bar{D}_2 \cap \complement D_0) \cap (\complement \bar{D}_2 \cup D_2) \\ &= \bar{D}_2 \cap D_2 \cap \complement D_0 = D_2 \cap \complement D_0 = D_2 - D_0. \end{aligned}$$

Thus $F_{02} - fr\,D_2$ is non-empty, connected, and *not* open. Accordingly, (11) follows by **I.7.4**, theorem 5. Next, by (10) applied to the domains D_0, D_2,

$$H^{n+1}(\bar{D}_2, F_{02}) \approx \mathsf{I}. \tag{12}$$

Consider now the following diagram, where all the homomorphisms are induced by inclusion mappings (which are available by virtue of inclusions already established).

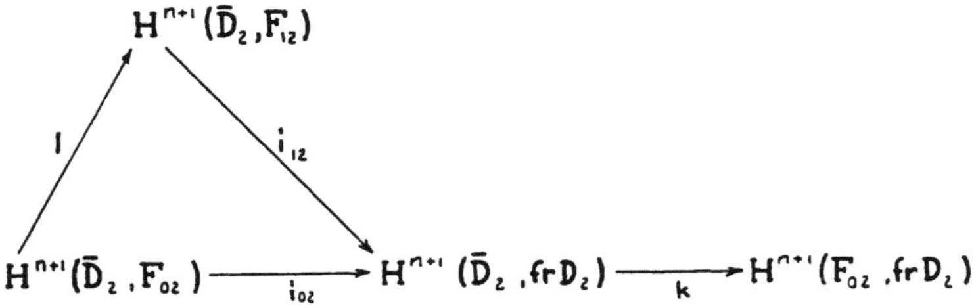

Fig. 35.

The portion of this diagram containing i_{02} and k is a three-term section of the cohomology sequence of the triple $(\overline{D}_2, F_{02}, fr\,D_2)$. The third group of this section is trivial by (11). Hence, by **I.3.3**, lemma 1, i_{02} is *onto*. Since $i_{02} = i_{12}l$ by the vector law, it follows (see **I.3.2**, exercise 3) that i_{12} is also *onto*. But the groups joined by i_{12} are both isomorphic to I [see (10) and (3)]. By **I.3.2**, exercise 4, the fact that i_{12} is *onto* implies therefore that i_{12} is an isomorphism onto, and the lemma is proved.

Our next lemma is concerned with two frames $[X_1, Y_1, D_1]$, $[X_2, Y_2, D_2]$ satisfying the condition

$$(X_2, Y_2) \subset (X_1, Y_1). \tag{13}$$

The inclusion mapping $(X_2, Y_2) \to (X_1, Y_1)$ induces then a homomorphism

$$j_{12}: H^{n+1}(X_1, Y_1) \to H^{n+1}(X_2, Y_2). \tag{14}$$

In view of **I.1.3**, exercise 42, we have either $D_1 \subset D_2$ or else $D_1 \cap D_2 = \emptyset$ under the present circumstances.

Lemma 2. The homomorphism j_{12} occurring in (14) is an isomorphism onto or a zero homomorphism (see **I.3.2**) according as $D_1 \subset D_2$ or $D_1 \cap D_2 = \emptyset$.

Proof. For the case $D_1 \subset D_2$ we use the following diagram.

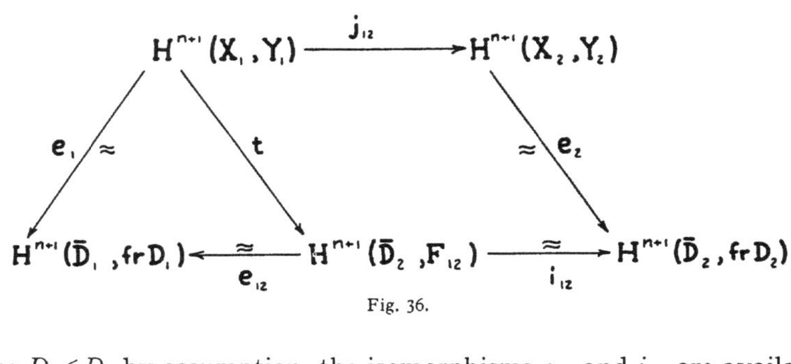

Fig. 36.

Since $D_1 \subset D_2$ by assumption, the isomorphisms e_{12} and i_{12} are available, in the sense of (9), (8) and lemma 1. The excision isomorphisms e_1 and e_2 are used on the basis of (2). Finally, t is obtained as follows. We have obviously $\overline{D}_2 \subset X_1$, since $\overline{D}_2 \subset X_2 \subset X_1$. Also,

$$F_{12} = \overline{D}_2 - D_1 \subset X_1 \cap \complement D_1 \subset X_1 \cap (\complement X_1 \cup Y_1) = Y_1,$$

since $X_1 \supset Y_1$. Thus the inclusion mapping $(\overline{D}_2, F_{12}) \to (X_1, Y_1)$ is available, and t is the homomorphism induced by this mapping. In summary, all the homomorphisms in the diagram are induced by inclusion mappings. The vector law yields the relations $e_2 j_{12} = i_{12} t$

and $e_{12}t = e_1$, and hence

$$\dot{j}_{12} = e_2^{-1} i_{12} e_{12}^{-1} e_1. \tag{15}$$

Since each factor on the right is an isomorphism onto, it follows that $\dot{j}_{12}$ is also an isomorphism onto.

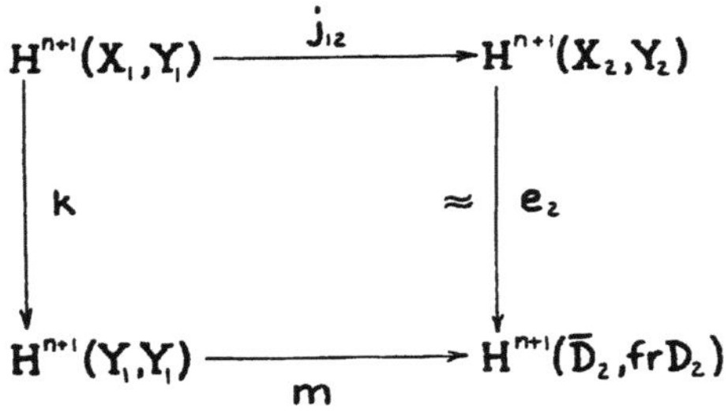

Fig. 37.

For the case when $D_1 \cap D_2 = \emptyset$, we use the diagram in Fig. 37.

The excision isomorphism e_2 is used again in the sense of (2), while k is induced by the inclusion mapping $(Y_1, Y_1) \to (X_1, Y_1)$. Finally, m is induced by the inclusion mapping $(\bar{D}_2, fr D_2) \to (Y_1, Y_1)$, which is available in view of the inclusion

$$(\bar{D}_2, fr\, D_2) \subset (Y_1, Y_1). \tag{16}$$

To verify (16), note first that $\bar{D}_2 \subset X_2 \subset X_1$, and also $\bar{D}_2 \subset \complement D_1 = \complement X_1 \cup Y_1$, since now $D_1 \cap D_2 = \emptyset$. Thus clearly $\bar{D}_2 \subset Y_1$, and hence *a fortiori* $fr D_2 \subset \bar{D}_2 \subset Y_1$, establishing (16). Take now any element $u \in H^{n+1}(X_1, Y_1)$. The vector law yields

$$e_2 \dot{j}_{12} u = m k u = 0, \tag{17}$$

since $ku \in H^{n+1}(Y_1, Y_1) = 0$ (see **I.6.1**, remark 3). As e_2 is an isomorphism onto, (17) implies that $\dot{j}_{12} u = 0$ for every $u \in H^{n+1}(X_1, Y_1)$ and the proof of lemma 2 is complete.

II.1.2. Standard isomorphisms. Given in R^n two bounded domains D_2 and $D_1 \subset D_2$, we introduce the *auxiliary isomorphism*

$$\left.\begin{array}{l} s_{12}: H^{n+1}(\bar{D}_1, fr\, D_1) \xrightarrow{\approx} \\ \quad H^{n+1}(\bar{D}_2, fr\, D_2) \end{array}\right\} \tag{1}$$

on the basis of the diagram in Fig. 38.

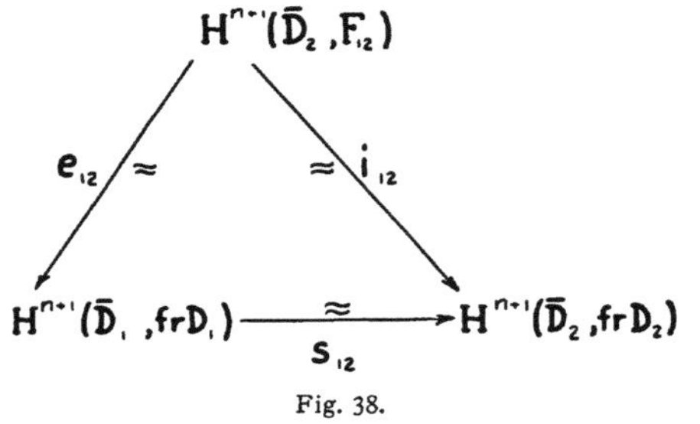

Fig. 38.

The isomorphisms e_{12}, i_{12} are used in the sense of (9), (8), and lemma 1 in **II.1.1**, and s_{12} is defined by

$$s_{12} = i_{12} e_{12}^{-1}. \tag{2}$$

Since both of the factors on the right are isomorphisms onto, s_{12} is also an isomorphism onto.

Consider now three bounded domains D_1, D_2, D_3 such that $D_1 \subset D_2 \subset D_3$. There arise, in the sense of the preceding definition, three auxiliary isomorphisms s_{12}, s_{23}, s_{13}. We proceed to verify the *vector law*

$$s_{13} = s_{23}\, s_{12}. \tag{3}$$

The proof is obtained from the following diagram.

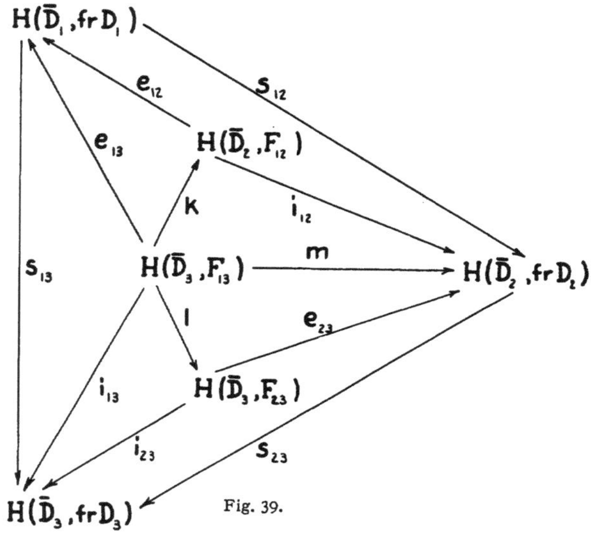

Fig. 39.

To save space, H^{n+1} is written merely as H in this diagram. The sets F_{13}, F_{23} and the isomorphisms $e_{13}, i_{13}, e_{23}, i_{23}$ are associated with the pairs of domains $D_1 \subset D_3$ and $D_2 \subset D_3$ in the same manner as F_{12}, e_{12}, i_{12} were associated with the pair of domains $D_1 \subset D_2$. Accordingly

$$F_{13} = \bar{D}_3 - D_1, \quad F_{23} = \bar{D}_3 - D_2, \quad s_{13} = i_{13}\, e_{13}^{-1}, \quad s_{23} = i_{23}\, e_{23}^{-1}.$$

Finally, the homomorphisms k, l, m are induced by the inclusion mappings

$$(\bar{D}_2, F_{12}) \to (\bar{D}_3, F_{13}), \quad (\bar{D}_3, F_{23}) \to (\bar{D}_3, F_{13}), \quad (\bar{D}_2, fr\, D_2) \to (\bar{D}_3, F_{13}),$$

which are available in view of the obvious inclusions

$$F_{12} \subset F_{13}, \quad F_{23} \subset F_{13}, \quad fr\, D_2 \subset F_{12} \subset F_{13}.$$

Thus all the homomorphisms in the diagram are induced by inclusion mappings. By the vector law we conclude now that

$$\begin{aligned} s_{23}\, s_{12} &= (i_{23}\, e_{23}^{-1})\,(i_{12}\, e_{12}^{-1}) = i_{23}\, e_{23}^{-1}\, i_{12}\, k\, e_{13}^{-1} \\ &= i_{23}\, e_{23}^{-1}\, m\, e_{13}^{-1} = i_{23}\, l\, e_{13}^{-1} = i_{13}\, e_{13}^{-1} = s_{13}, \end{aligned}$$

and (3) is proved.

Consider now the special case of (1) when $D_1 = D_2$. Then clearly $F_{12} = \bar{D}_2 - D_1 = \bar{D}_1 - D_1 = fr\,D_1$. Accordingly, by remark 1 in **I.6.2**, it follows that e_{12} and i_{12} reduce in this case to the identity homomorphism in the group $H^{n+1}(\bar{D}_1, fr\,D_1)$, and hence the same is true for s_{12}. We express this fact by the formula

$$s_{11} = 1. \tag{4}$$

Let now D_1, D_2 be two bounded domains in arbitrary relative position (thus neither $D_1 < D_2$ nor $D_2 < D_1$ is required). We proceed to define a *standard isomorphism*

$$\sigma_{12}: H^{n+1}(\bar{D}_1, fr\,D_1) \xrightarrow{\approx} H^{n+1}(\bar{D}_2, fr\,D_2) \tag{5}$$

in the following manner. Take any bounded domain D_3 such that $D_3 > D_1, D_3 > D_2$. Then σ_{12} is given, in terms of the auxiliary isomorphisms s_{13} and s_{23}, by means of the agreement

$$\sigma_{12} = s_{23}^{-1} s_{13}. \tag{6}$$

To justify this definition, we have to show that on taking another bounded domain D_4 such that $D_4 > D_1, D_4 > D_2$, we have the relation

$$s_{23}^{-1} s_{13} = s_{24}^{-1} s_{14}. \tag{7}$$

To prove (7), take a bounded domain D_5 such that $D_5 > D_3, D_5 > D_4$. By the vector law (3), applied with the appropriate subscripts, we obtain

$$s_{25}^{-1} s_{15} = (s_{35} s_{23})^{-1} (s_{35} s_{13}) = s_{23}^{-1} s_{35}^{-1} s_{35} s_{13} = s_{23}^{-1} s_{13},$$

and similarly

$$s_{25}^{-1} s_{15} = (s_{45} s_{24})^{-1} (s_{45} s_{14}) = s_{24}^{-1} s_{14},$$

and (7) follows. Thus σ_{12}, as defined by (6), is independent of the choice of D_3.

Lemma 1. If $D_1 = D_2$, then σ_{12} reduces to the identity homomorphism in the group $H^{n+1}(\bar{D}_1, fr\,D_1)$. Symbolically $\sigma_{11} = 1$. This is a direct consequence of the defining formula (6).

Lemma 2. $\sigma_{12} = \sigma_{21}^{-1}$.

Proof. By the definition (6), we have $\sigma_{12} = s_{23}^{-1} s_{13}$, $\sigma_{21} = s_{13}^{-1} s_{23}$, and the assertion follows.

Lemma 3 (the *vector law*). If D_1, D_2, D_3 are any three bounded domains, then $\sigma_{13} = \sigma_{23} \sigma_{12}$.

Proof. Take a bounded domain $D_4 > D_1 \cup D_2 \cup D_3$. Then $\sigma_{12} = s_{24}^{-1} s_{14}$, $\sigma_{23} = s_{34}^{-1} s_{24}$, $\sigma_{13} = s_{34}^{-1} s_{14}$, and the assertion follows.

Consider now two frames $[X_1, Y_1, D_1]$, $[X_2, Y_2, D_2]$ in arbitrary relative position. We proceed to define a *standard isomorphism*

$$\tau_{12}: H^{n+1}(X_1, Y_1) \xrightarrow{\approx} H^{n+1}(X_2, Y_2) \tag{8}$$

by means of the formula

$$\tau_{12} = e_2^{-1} \sigma_{12} e_1. \tag{9}$$

In this definition, e_1 and e_2 are the *excision isomorphisms*

$$e_1: H^{n+1}(X_1, Y_1) \xrightarrow{\approx} H^{n+1}(\bar{D}_1, fr\, D_1),$$
$$e_2: H^{n+1}(X_2, Y_2) \xrightarrow{\approx} H^{n+1}(\bar{D}_2, fr\, D_2),$$

explained in connection with (2) in **II.1.1**, while σ_{12} is the standard isomorphism

$$\sigma_{12}: H^{n+1}(\bar{D}_1, fr\, D_1) \xrightarrow{\approx} H^{n+1}(\bar{D}_2, fr\, D_2).$$

Lemma 4. If both of the frames $[X_1, Y_1, D_1]$ and $[X_2, Y_2, D_2]$ are reduced, then $\tau_{12} = \sigma_{12}$.

Proof. The assumption means that $X_i = \bar{D}_i$, $Y_i = fr D_i$, $i = 1, 2$. By **I.6.2**, remark 1, e_i in (9) reduces then to the identity homomorphism in $H^{n+1}(\bar{D}_i, fr\, D_i)$, $i = 1, 2$, and the assertion follows from the definition (9).

Lemma 5. $\tau_{12} = \tau_{21}^{-1}$.

This is a direct consequence of the definition, in view of lemma 2.

Lemma 6 (the vector law). If $[X_i, Y_i, D_i]$, $i = 1, 2, 3$, are any three frames, then $\tau_{13} = \tau_{23} \tau_{12}$.

Proof. In terms of the excision isomorphisms

$$e_i: H^{n+1}(X_i, Y_i) \xrightarrow{\approx} H^{n+1}(\bar{D}_i, fr\, D_i)$$

we have by definition the relations $\tau_{13} = e_3^{-1} \sigma_{13} e_1$, $\tau_{23} = e_3^{-1} \sigma_{23} e_2$, $\tau_{12} = e_2^{-1} \sigma_{12} e_1$, and the assertion follows in view of lemma 3.

Lemma 7. Let $[X_1, Y_1, D_1]$, $[X_2, Y_2, D_2]$ be two frames such that $(X_1, Y_1) \supset (X_2, Y_2)$. Then the homomorphism

$$j_{12}: H^{n+1}(X_1, Y_1) \to H^{n+1}(X_2, Y_2),$$

induced by the inclusion mapping $(X_2, Y_2) \to (X_1, Y_1)$, is related to the standard isomorphism τ_{12} [given by (9)] as follows: $j_{12} = \tau_{12}$ if $D_1 \subset D_2$, and j_{12} is a zero-homomorphism if $D_1 \cap D_2 = \emptyset$.

Proof. Let us recall that the assumption $(X_1, Y_1) \supset (X_2, Y_2)$ implies that either $D_1 \subset D_2$ or else $D_1 \cap D_2 = \emptyset$. In the second case, j_{12} is a zero-homomorphism by **II.1.1**, lemma 2. On the other hand, if $D_1 \subset D_2$, then by (15) in **II.1.1** we have $j_{12} = e_2^{-1} i_{12} e_{12}^{-1} e_1$. Since $D_1 \subset D_2$, we can choose $D_3 = D_2$ in the formula (6) for σ_{12}, obtaining in view of (4) the relation $\sigma_{12} = s_{22}^{-1} s_{12} = s_{12}$. Finally, $s_{12} = i_{12} e_{12}^{-1}$ by the definition (2). Hence by (9)

$$j_{12} = e_2^{-1} \sigma_{12} e_1 = \tau_{12}.$$

Lemma 8. If $[X_1, Y_1, D_1] = [X_2, Y_2, D_2]$, then the standard isomorphism τ_{12} reduces to the identity homomorphism in the group $H^{n+1}(X_1, Y_1)$. Symbolically: $\tau_{11} = 1$.

This follows directly from lemma 7, in view of remark 1 in **I.6.2** (or else more directly from the definition of τ_{12}).

II.1.3. Simultaneous orientation of frames in $\boldsymbol{R}^n$. For any frame $[X, Y, D]$ in R^n, the group $H^{n+1}(X, Y)$ is infinite cyclic by **II.1.1** (1). Accordingly, this group contains precisely two elements g', g'' each of which generates the group, and these two elements satisfy the relation $g'' = -g'$. *Orienting the frame* $[X, Y, D]$ means, for our purposes, the selection of one of g', g'' as the *preferred generator* for the group. We use the following process to orient simultaneously all the frames in R^n. Consider the $(n-1)$-sphere S^{n-1} and the n-cell E^n in R^n (see **I.2.2**). The cohomology sequence of the pair (E^n, S^{n-1}) contains the section

$$H^n(E^n) \to H^n(S^{n-1}) \xrightarrow[\delta_*^{n-1}]{} H^{n+1}(E^n, S^{n-1}) \to H^{n+1}(E^n),$$

where the undesignated homomorphisms are induced by inclusion mappings (see **I.6.4**). The first and the last group are trivial by **I.7.1**. Accordingly, by **I.3.3**, lemma 4, δ_*^{n-1} is an isomorphism onto. In symbols:

$$\delta_*^{n-1} : H^n(S^{n-1}) \xrightarrow{\approx} H^{n+1}(E^n, S^{n-1}). \tag{1}$$

By **I.7.1**, the group $H^n(S^{n-1})$ is infinite cyclic. Let $\mathfrak{g}_{n-1}$ denote one of the two elements of this group which can be used to generate it. Once chosen, this generator $\mathfrak{g}_{n-1}$ will be kept fixed, and will be termed the *preferred generator* for S^{n-1}. Observe now that $fr E^n = S^{n-1}$, and int E^n is non-empty, open, and connected (see **I.2.2**, exercise 5). Thus we have the frame $[E^n, S^{n-1}, \text{int } E^n]$. Accordingly, the group $H^{n+1}(E^n, S^{n-1})$ is infinite cyclic [a fact which is clear from (1) also], and for this group the isomorphism (1) yields a *preferred generator* in the form $\delta_*^{n-1} \mathfrak{g}_{n-1}$.

Consider now any frame $[X, Y, D]$ in R^n. The standard isomorphism τ (see **II.1.2**) from $H^{n+1}(E^n, S^{n-1})$ onto $H^{n+1}(X, Y)$ yields the *preferred generator* $\tau \delta_*^{n-1} \mathfrak{g}_{n-1}$ for the (infinite cyclic) group $H^{n+1}(X, Y)$. This preferred generator will be denoted, if such explicit notation is needed in some situation, by $\mathfrak{g}[X, Y, D]$. In view of lemma 8 in **II.1.2**, this definition leads to no discrepancy for the special frame $[E^n, S^{n-1}, \text{int } E^n]$.

Lemma 1. If $[X_1, Y_1, D_1]$, $[X_2, Y_2, D_2]$ are two frames in R^n, then

$$\tau_{12}\,\mathfrak{g}[X_1, Y_1, D_1] = \mathfrak{g}[X_2, Y_2, D_2],$$

where τ_{12} is the standard isomorphism defined by (8), (9) in **II.1.2.**

Proof. Denote by $\tau_i, i=1, 2,$ the standard isomorphism from $H^{n+1}(E^n, S^{n-1})$ onto $H^{n+1}(X_i, Y_i)$, in the sense of **II.1.2** (8), (9). By definition

$$\mathfrak{g}[X_i, Y_i, D_i] = \tau_i\,\delta_*^{n-1}\,\mathfrak{g}_{n-1}.$$

On the other hand, the *vector law* yields (see **II.1.2**, lemma 6) $\tau_{12}\tau_1 = \tau_2$. We conclude that

$$\tau_{12}\,\mathfrak{g}[X_1, Y_1, D_1] = \tau_{12}\,\tau_1\,\delta_*^{n-1}\,\mathfrak{g}_{n-1} = \tau_2\,\delta_*^{n-1}\,\mathfrak{g}_{n-1} = \mathfrak{g}[X_2, Y_2, D_2].$$

Lemma 2. Let $[X_1, Y_1, D_1]$, $[X_2, Y_2, D_2]$ be two frames in R^n such that $(X_1, Y_1) \supset (X_2, Y_2)$. On denoting by

$$j_{12}: H^{n+1}(X_1, Y_1) \to H^{n+1}(X_2, Y_2)$$

the homomorphism induced by the inclusion mapping $(X_2, Y_2) \to (X_1, Y_1)$, one has

$$j_{12}\,\mathfrak{g}[X_1, Y_1, D_1] = \mathfrak{g}[X_2, Y_2, D_2] \quad \text{or} \quad j_{12}\,\mathfrak{g}[X_1, Y_1, D_1] = 0,$$

according as $D_1 \subset D_2$ or $D_1 \cap D_2 = \emptyset$.

This is a direct consequence of the preceding lemma 1 and of **II.1.2**, lemma 7.

II.1.4. Positive orientation of frames in R^n. In the preceding discussion, the choice of the preferred generator $\mathfrak{g}_{n-1}$ for the group $H^n(S^{n-1})$ was quite arbitrary. For certain purposes, it is convenient to make this choice a definite one (corresponding to the choice of the positive or counter-clockwise orientation in the Euclidean plane R^2, for example). Starting with the case $n=1$, we have to consider first the matter of making a definite selection for a generator for the infinite cyclic group $H^1(S^0)$. Now S^0 consists of the two points -1 and $+1$ on the real number-line R^1. By **I.6.1**, remark 8, the group $H^1(S^0)$ is isomorphic to (and indeed may be considered as identical with) the group Z^1 of the 1-cocycles z^1 of S^0. Furthermore, as it has been proved there, we have at our disposal the explicit isomorphism $h: Z^1 \xrightarrow{\approx} \mathsf{I}$, given by the formula $hz^1 = z^1(-1, 1)$.

Thus the (unique) 1-cocycle z^1 for which $z^1(-1, 1) = 1$ corresponds to the generator $+1$ of I, and hence this particular 1-cocycle is a generator of $Z^1 = H^1(S^0)$. We select this generator as the preferred generator $\mathfrak{g}_0$ for $H^1(S^0)$. Thus (see **I.6.1**, remark 8) $\mathfrak{g}_0$ is the 1-cocycle of S^0 which is fully determined by the following table of its values:

$$\mathfrak{g}_0(-1, 1) = 1, \quad \mathfrak{g}_0(1, -1) = -1, \quad \mathfrak{g}_0(-1, -1) = 0, \quad \mathfrak{g}_0(1, 1) = 0.$$

There follows a natural selection of a preferred generator $\mathfrak{g}_{n-1}$ for $H^n(S^{n-1})$, for all values of $n \geq 2$, based on the explicit isomorphism

$$H^n(S^{n-1}) \xrightarrow{\approx} H^{n+1}(S^n),$$

discussed in the *remark* to **I.7.1.** Denoting that explicit isomorphism by h_{n-1} for brevity, we define successively

$$\mathfrak{g}_1 = h_0\, \mathfrak{g}_0, \ldots, \mathfrak{g}_n = h_{n-1}\, \mathfrak{g}_{n-1}, \ldots,$$

as preferred generators for $H^2(S^1), \ldots, H^{n+1}(S^n), \ldots$. Once $\mathfrak{g}_0, \mathfrak{g}_1, \ldots$ have been selected in this manner, the process described in **II.1.3** yields, for each $n \geq 1$, preferred generators $\mathfrak{g}[X, Y, D]$ for the frames in R^n. For convenience, the orientation so constructed will be referred to as the *positive orientation* of the frames in R^n.

§ II.2. The topological index[1].

II.2.1. Preliminaries. We shall operate in Euclidean n-space R^n, where the integer $n \geq 1$ is arbitrary but fixed. Let

$$T: U \to R^n \tag{1}$$

be a continuous mapping from a non-empty subset U of R^n into R^n. To increase clarity of notations, x will be used as a generic notation for a point of R^n, while u will be used to refer to a generic point of U (thus a point denoted by x may or may not lie in U). Let us recall (see **I.1.1**) that the concept of a mapping T has been so formulated that it includes the set from which T operates. Accordingly, if we have occasion to consider the mapping T in (1) as operating from a subset $U_* \subset U$, then we shall write $T | U_*$ to refer to T cut down to U_*. At times, when results from general cohomology theory will be applied, U will play the role of *space*, and to avoid ambiguity we agree that terms like open set, closed set, closure will always be used relative to R^n. If we wish to consider, for example, frontier relative to U, then some special notation like fr^* will be employed to secure clarity.

In the present § **II.2** we shall be concerned mainly with the following situation. Given the continuous mapping T as in (1), a bounded domain D will be considered such that

$$\bar{D} \subset U. \tag{2}$$

[1] While it is now customary to treat the topological index in terms of cohomology groups, a reader desiring to achieve a better understanding of the geometrical motivation should study the excellent presentation of the degree of a mapping *(Abbildungsgrad)* in terms of singular homology groups in the ALEXANDROFF-HOPF treatise listed in the *Bibliography*. In Part VI we shall discuss the description of the topological index in terms of the *variation of the argument* for the case of the plane.

The following definitions will be used.

Definition 1. Given T and D as in (1), (2), a point x is termed (T, D)-admissible if $x \notin T\,fr D$.

Definition 2. Given T and D as in (1), (2), a frame $[A, B, \Delta]$ in R^n (see **II.1.1**) is termed (T, D)-admissible if the relation

$$T|\bar{D}: (\bar{D}, fr\, D) \to (A, B) \tag{3}$$

holds. Thus the frame is (T, D)-admissible if and only if $T\bar{D} \subset A$, $T\,fr D \subset B$.

Definition 3. Given T and D as in (1), (2) and a point x, a frame $[A, B, \Delta]$ in R^n is termed (x, T, D)-admissible if it is (T, D)-admissible and $x \in \Delta$.

Definition 4. Given a point x and a real number r such that $0 < r < 1$, the *spherical frame* $[A_r(x), B_r(x), \Delta_r(x)]$ with center x is defined as follows:

$$A_r(x) = \{x' \mid \|x' - x\| \leqq 1/r\},$$
$$B_r(x) = \{x' \mid r \leqq \|x' - x\| \leqq 1/r\},$$
$$\Delta_r(x) = \{x' \mid \|x' - x\| < r\}.$$

Clearly (see **I.2.1**, exercise 2), the *spherical frame* is a *frame* in the sense of **II.1.1**. The domain $\Delta_r(x)$ is the open spherical neighborhood of radius r for x (see **I.1.4**, definition 2).

Lemma 1. If x is (T, D)-admissible, then there exist (x, T, D)-admissible frames. In particular, if $r > 0$ is sufficiently small, then the spherical frame $[A_r(x), B_r(x), \Delta_r(x)]$ is (x, T, D)-admissible.

Proof. Since D in (2) is bounded, by **I.2.1**, exercise 4, the sets $\bar{D}$ and $fr D$ are compact and non-empty (see **I.1.3**, exercise 18). Hence (see **I.1.5**, exercise 8) the sets $T\bar{D}$ and $T\,fr D$ are also compact. By assumption, $x \notin T\,fr D$. Since $\complement T\,fr D$ is open, for r small enough we have the inclusion $T\,fr D \subset \complement \Delta_r(x)$. Also, since $T\bar{D}$ is compact and hence bounded, for $r > 0$ sufficiently small we have $T\bar{D} \subset A_r(x)$. On choosing r sufficiently small for both of these purposes, we shall have $T\bar{D} \subset A_r(x)$, $T\,fr D \subset B_r(x)$, and the lemma is proved.

Lemma 2. A point x is (T, D)-admissible if and only if $\bar{D} \cap T^{-1}x \subset D$.

Proof. Since $\bar{D} = D \cup fr D$ and $D \cap fr D = \emptyset$, one has

$$\bar{D} \cap T^{-1}x = (D \cap T^{-1}x) \cup (fr D \cap T^{-1}x), \quad (D \cap T^{-1}x) \cap (fr D \cap T^{-1}x) = \emptyset.$$

Now x is (T, D)-admissible if and only if $fr D \cap T^{-1}x = \emptyset$, so the lemma is obvious.

Definition 5. Given T and D as in (1), (2), let $\{D_j\}$ be a (finite or infinite) sequence of pair-wise disjoint domains in D. Then the sequence $\{D_j\}$ is termed (x, T, D)-complete if $\bar{D} \cap T^{-1}x \subset \cup D_j$.

Remark 1. If $\{D_j\}$ is (x, T, D)-complete, then $\bar{D} \cap T^{-1}x \subset \bigcup D_j \subset D$, and hence x is (T, D)-admissible by lemma 2.

Remark 2. If $x \notin T\bar{D}$, then $\bar{D} \cap T^{-1}x = \emptyset$, and hence *any* sequence $\{D_j\}$ of pair-wise disjoint domains in D is (x, T, D)-complete.

Remark 3. If x is (T, D)-admissible, then in view of lemma 2 the sequence consisting of D alone is (x, T, D)-complete.

Definition 6. Given T and D as in (1), (2), a domain $D_* \subset D$ is (x, T)-admissible if $x \notin T\,fr\,D_*$. Equivalently, the domain $D_* \subset D$ is (x, T)-admissible if x is (T, D_*)-admissible.

Lemma 3. Let the sequence $\{D_j\}$ be (x, T, D)-complete, in the sense of definition 5. Set $\bar{D} - \bigcup D_j = Y$. Then the following holds.

(i) $\emptyset \neq fr\,D_j \subset Y$.

(ii) Each domain D_j is (x, T)-admissible.

(iii) $fr\,D_j$ coincides with the frontier of D_j relative to $\bar{D}$.

(iv) The number of those domains D_j for which $D_j \cap T^{-1}x \neq \emptyset$ is finite.

Proof. Set $O = \bigcup D_j$. By **I.1.3**, exercise 4, it follows that $fr\,D_j \subset \complement O$. Also, $fr\,D_j \subset \bar{D}_j \subset \bar{D}$. Thus

$$fr\,D_j \subset \bar{D} \cap \complement O = \bar{D} - O = Y,$$

and (i) is proved (note that $fr\,D_j \neq \emptyset$ by **I.1.3**, exercise 18). Observe now that $\bar{D} \cap T^{-1}x \subset \bigcup D_j$ by assumption, and hence $Y \cap T^{-1}x = \emptyset$. Thus $x \notin TY$, and in view of (i) (which we already proved) it follows that $x \notin T\,fr\,D_j$, proving (ii). To verify (iii), note that since $\bar{D}$ is closed in R^n, the terms *closed set* and *closure of a set* have the same meaning relative to $\bar{D}$ as relative to R^n. Accordingly, if $fr^* D_j$ is the frontier of D_j relative to $\bar{D}$, then

$$fr^* D_j = \bar{D}_j \cap (\bar{D} - D_j) = \bar{D}_j \cap \bar{D} \cap \complement D_j = \bar{D}_j \cap \complement D_j = fr\,D_j,$$

where we used the fact that $\bar{D}_j \cap \bar{D} = \bar{D}_j$ as a consequence of the inclusion $\bar{D}_j \subset \bar{D}$.

Finally, to prove (iv), observe that the set

$$\bar{D} \cap T^{-1}x = (T|\bar{D})^{-1}x$$

is a closed subset of $\bar{D}$ (see **I.1.5**, exercise 4), and hence it is compact since $\bar{D}$ is compact (see **I.1.3**, exercise 22, **I.2.1**, exercise 4). By assumption, the domains D_j cover $\bar{D} \cap T^{-1}x$. Since each D_j is open and $\bar{D} \cap T^{-1}x$ is compact, it follows that $\bar{D} \cap T^{-1}x$ is covered by a *finite* subsequence of the sequence $\{D_j\}$, and (iv) is proved.

Lemma 4. Let x be (T, D)-admissible and let $\{D_j\}$, $j = 1, \ldots, m$, be a *finite* sequence of pair-wise disjoint (x, T)-admissible domains in D such that $\{D_j\}$ is *not* (x, T, D)-complete. Then $\{D_j\}$ can be extended to a sequence $D_1, \ldots, D_m, D_{m+1}, \ldots$ of pair-wise disjoint domains in D such that the extended sequence is (x, T, D)-complete.

Proof. Introduce the auxiliary set

$$F = fr\, D \cup \bar{D}_1 \cup \cdots \cup \bar{D}_m.$$

Then F, as a finite union of compact sets, is compact and contains $fr D$. Hence, by **I.1.3**, exercise 5, the set $O = \bar{D} - F$ is open. Since $\bar{D} = F \cup O$, we obtain

$$\bar{D} \cap T^{-1} x = (F \cap T^{-1} x) \cup (O \cap T^{-1} x). \tag{4}$$

By assumption, $fr D \cap T^{-1} x = \emptyset$, $fr D_j \cap T^{-1} x = \emptyset$, and hence

$$F \cap T^{-1} x = (\cup D_j) \cap T^{-1} x. \tag{5}$$

From (4) and (5) we infer that

$$\bar{D} \cap T^{-1} x = \big((\cup D_j) \cup O\big) \cap T^{-1} x,$$

and hence

$$\bar{D} \cap T^{-1} x \subset (\cup D_j) \cup O. \tag{6}$$

Since the sequence $\{D_j\}$ is *not* (x, T, D)-complete, (6) shows that $O \neq \emptyset$. Furthermore, since $\cup D_j \subset F$ and $O \cap F = \emptyset$, we have

$$D_j \cap O = \emptyset, \quad j = 1, \ldots, m. \tag{7}$$

Now let $D_{m+1}, \ldots, D_{m+k}, \ldots$ be the sequence of the components of $O \neq \emptyset$. Note that each D_{m+k} is a domain and that the set of the components of O is countable (see **I.1.3**, exercise 28). By (6) and (7) it follows that

$$D_1, \ldots, D_m, D_{m+1}, \ldots$$

is a sequence of pair-wise disjoint domains in D whose union contains $\bar{D} \cap T^{-1} x$, and the lemma is proved.

Remark 4. In view of part (iv) of lemma 3, only a finite number of the new domains D_{m+k} intersect $T^{-1} x$. Hence only a finite number of these new domains are needed to obtain an (x, T, D)-complete sequence. Accordingly, lemma 4 can be strengthened to the statement that the initial finite sequence $\{D_j\}$ can be extended to a *finite* (x, T, D)-complete sequence of pair-wise disjoint domains in D.

II.2.2. Definition of the topological index. Given T and D as in **II.2.1** (1), (2), take a point x which is (T, D)-admissible, and select a frame $[A, B, \Delta]$ which is (x, T, D)-admissible (see **II.2.1**, lemma 1). Then the mapping

$$T|\bar{D} : (\bar{D}, fr\, D) \to (A, B) \tag{1}$$

induces a homomorphism

$$(T|\bar{D})_*: H^{n+1}(A, B) \to H^{n+1}(\bar{D}, fr\, D). \tag{2}$$

The image of the *preferred generator* $\mathfrak{g}[A, B, \Delta]$ of the group $H^{n+1}(A, B)$ is then an element of the group $H^{n+1}(\bar{D}, fr\, D)$, and hence it is a certain integral multiple of the *preferred generator* $\mathfrak{g}[\bar{D}, fr\, D, D]$ (see **II.1.3**). Thus we have a relation of the form

$$(T|\bar{D})_*\, \mathfrak{g}[A, B, \Delta] = \mu\, \mathfrak{g}[\bar{D}, fr\, D, D], \tag{3}$$

where μ is an integer which is uniquely determined by T, D, and the frame $[A, B, \Delta]$. Let us recall that this frame must be (x, T, D)-admissible.

Lemma. The integer μ is independent of the choice of the (x, T, D)-admissible frame $[A, B, \Delta]$.

Proof. An (x, T, D)-admissible frame $[A, B, \Delta]$ having been selected, consider the spherical frame $[A_r(x), B_r(x), \Delta_r(x)]$. By lemma 1 in **II.2.1**, this frame is (x, T, D)-admissible if $r>0$ is sufficiently small. If r is so chosen, then the mapping

$$T|\bar{D}: (\bar{D}, fr\, D) \to (A_r(x), B_r(x)) \tag{4}$$

induces a homomorphism

$$(T|\bar{D})^r_*: H^{n+1}(A_r(x), B_r(x)) \to H^{n+1}(\bar{D}, fr\, D), \tag{5}$$

which in turn gives rise to the relation [analogous to (3)]

$$(T|\bar{D})^r_*\, \mathfrak{g}[A_r(x), B_r(x), \Delta_r(x)] = \mu_r\, \mathfrak{g}[\bar{D}, fr\, D, D]. \tag{6}$$

Observe now that since $[A, B, \Delta]$ is (x, T, D)-admissible, we have the inclusion $x \in \Delta$. Hence, if $r>0$ is sufficiently small, we have the inclusion $\Delta_r(x) \subset \Delta$. Furthermore, since A is compact and hence bounded, for r sufficiently small we have the inclusion $A_r(x) \supset A$, and consequently also

$$B_r(x) = A_r(x) - \Delta_r(x) \supset A - \Delta = B.$$

Summing up: for $r>0$ sufficiently small, we have the inclusions

$$A_r(x) \supset A, \quad B_r(x) \supset B, \quad \Delta_r(x) \subset \Delta. \tag{7}$$

The identity mapping $(A, B) \to (A_r(x), B_r(x))$ induces then a homomorphism

$$j: H^{n+1}(A_r(x), B_r(x)) \to H^{n+1}(A, B),$$

for which lemma 2 in **II.1.3** yields the relation

$$j\, \mathfrak{g}[A_r(x), B_r(x), \Delta_r(x)] = \mathfrak{g}[A, B, \Delta]. \tag{8}$$

The vector law yields $(T|\overline{D})^r_* = (T|\overline{D})_* j$. In view of (6), (8) and (3) we conclude that

$$\mu_r\, \mathfrak{g}[\overline{D}, fr\, D, D] = \mu\, \mathfrak{g}[\overline{D}, fr\, D, D]. \tag{9}$$

Since $\mathfrak{g}[\overline{D}, fr\, D, D]$ is a generator of the infinite cyclic group $H^{n+1}(\overline{D}, fr\, D)$, (9) implies that

$$\mu = \mu_r. \tag{10}$$

Now consider any two (x, T, D)-admissible frames $[A^k, B^k, \Delta^k]$, $k=1, 2$, and denote by μ^1, μ^2 the corresponding integers μ. If $r>0$ is sufficiently small in relation to both of these frames, then (10) holds for either frame, and hence $\mu^1 = \mu_r = \mu^2$, proving the lemma.

In view of this lemma, the integer μ determined by (1), (2), (3) depends only upon T, D, and the point x [which must be (T, D)-admissible]. Accordingly, we are justified in denoting this integer by $\mu(x, T, D)$.

Definition. $\mu(x, T, D)$ is termed the *topological index* of the point x, with respect to the mapping T and the domain D.

Remark. According to this definition, $\mu(x, T, D)$ may be thought of as an integral-valued function of three variables x, T, D which are subject to the following conditions: (i) T is a continuous mapping from a set $U \subset R^n$ into R^n. (ii) D is a bounded domain such that $\overline{D} \subset U$. (iii) $x \notin T\, fr D$.

II.2.3. Properties of the topological index. Given T and D as in **II.2.1** (1), (2), consider a (T, D)-admissible point x_0 and an (x_0, T, D)-admissible frame $[A, B, \Delta]$. Then $T\, fr D \subset \complement \Delta$, and hence $\Delta \cap T\, fr D = \emptyset$. It follows that every point $x \in \Delta$ is (T, D)-admissible and that $[A, B, \Delta]$ is (x, T, D)-admissible for $x \in \Delta$. Thus we can use this same frame $[A, B, \Delta]$ in evaluating $\mu(x, T, D)$ for all the points $x \in \Delta$. In view of the defining formulas (1), (2), (3) in **II.2.2** it follows that

$$\mu(x, T, D) = \mu(x_0, T, D) \quad \text{for} \quad x \in \Delta. \tag{1}$$

Consider now a component Γ of the (open) set $\complement T\, fr D$, and take any point $x_0 \in \Gamma$. Then x_0 is (T, D)-admissible, and by (1) there exists an open set Δ such that $x_0 \in \Delta$ and $\mu(x, T, D)$ is constant on Δ. Since Γ is also open, it follows that for every point $x_0 \in \Gamma$ there exists an open set $O = \Gamma \cap \Delta$ such that $x_0 \in O$ and $\mu(x, T, D)$ is constant on O. Since Γ is connected and $\mu(x, T, D)$ is integral-valued, by **I.1.3**, exercise 37 we obtain the following statement.

Theorem 1. $\mu(x, T, D)$ is defined and has a constant value on each component of the set $\complement T\, fr D$.

Remark 1. Let E be any connected subset of $\complement T\,fr D$. Then E is contained in a certain component Γ of $\complement T\,fr D$, and by theorem 1 we conclude that $\mu(x, T, D)$ is defined and has a constant value on each connected subset of $\complement T\,fr D$.

Remark 2. Suppose $x \notin T\bar{D}$. Then $x \notin T\,fr D$, and so $\mu(x, T, D)$ is defined. Since $T\bar{D}$ is compact, for $r>0$ sufficiently small the spherical frame $[A_r(x), B_r(x), \Delta_r(x)]$ satisfies now not merely the inclusions (see **II.2.1**, lemma 1) $T\bar{D} \subset A_r(x)$, $T\,fr D \subset B_r(x)$, but also the inclusion $T\bar{D} \subset B_r(x)$. By **I.6.2**, remark 3 it follows that the induced homomorphism $(T|\bar{D})_*$ in **II.2.2** (2) is a zero-homomorphism, and thus **II.2.2** (3) yields $\mu(x, T, D)=0$. Summarizing: if $x \notin T\bar{D}$, then x is (T, D)-admissible and $\mu(x, T, D)=0$.

Theorem 2. If x is (T, D)-admissible and $\mu(x, T, D) \neq 0$, then $x \in TD$.

Proof. Observe first that by remark 1 it follows that $x \in T\bar{D}$. But since $x \notin T\,fr D$ by assumption, one must have $x \in TD$.

Remark 3. Consider again a component Γ of $\complement T\,fr D$, and suppose that Γ is not completely covered by TD. Then there exists a point $x_0 \in \Gamma - TD$. Clearly x_0 is (T, D)-admissible, and $\mu(x_0, T, D)=0$ by theorem 2. Since $\mu(x, T, D)$ is constant on Γ by theorem 1, it follows that $\mu(x, T, D) \equiv 0$ on Γ. This discussion is applicable, in particular, to the unbounded component (see **I.2.1**, exercise 6) of $\complement T\,fr D$. Indeed, TD is a subset of the compact set $T\bar{D}$, and hence TD is a bounded set. Accordingly, the unbounded component of $\complement T\,fr D$ is certainly not completely covered by TD, and hence $\mu(x, T, D) \equiv 0$ on this particular component. If $n=1$, then $\complement T\,fr D$ has two unbounded components (see **I.2.1**, exercise 6), and the preceding comments apply to both of them.

Theorem 3. If the sequence $\{D_j\}$ is (x, T, D)-complete (see **II.2.1**, definition 5), then

$$\mu(x, T, D) = \sum_j{}' \mu(x, T, D_j), \tag{1*}$$

where the prime attached to the summation symbol indicates that the summation is actually finite (that is, only a finite number of terms can be different from zero).

Proof. In view of remark 1 of **II.2.1** the point x is (T, D)-admissible and $\mu(x, T, D)$ is defined. By lemma 3 in **II.2.1**, each domain D_j is (x, T)-admissible and hence $\mu(x, T, D_j)$ is defined. To simplify notations, put $\mu = \mu(x, T, D)$, $\mu_j = \mu(x, T, D_j)$, and introduce the auxiliary sets

$$Y = \bar{D} - \cup D_j, \quad Y_j = \bar{D} - D_j.$$

By assumption $\overline{D} \cap T^{-1}x \subset \bigcup D_j$, so clearly $x \notin TY$. Since TY is compact, it follows that for $r>0$ sufficiently small the spherical neighborhood $\Delta_r(x)$ satisfies the inclusion

$$\Delta_r(x) \subset \complement TY.$$

In view of the compactness of $T\overline{D}$ we conclude that for $r>0$ sufficiently small the spherical frame $[A_r(x), B_r(x), \Delta_r(x)]$ satisfies the following inclusions: $T\overline{D} \subset A_r(x)$, $TY \subset B_r(x)$. Since $fr\, D_j \subset Y$ by lemma 3 in **II.2.1**, there follows the inclusion

$$T\, fr\, D_j \subset B_r(x), \quad j = 1, 2, \ldots .$$

Furthermore, since $\overline{D}_j \subset \overline{D}$, we also have the inclusion

$$T\, \overline{D}_j \subset A_r(x), \quad j = 1, 2, \ldots .$$

These inclusions mean that for $r>0$ sufficiently small the spherical frame $[A_r(x), B_r(x), \Delta_r(x)]$ is (x, T, D_j)-admissible for every j. From lemma 1 in **II.2.1** we also know that this frame is (x, T, D)-admissible if $r>0$ is sufficiently small. Choose $r>0$ sufficiently small for all these purposes, and write $[A, B, \Delta]$ for $[A_r(x), B_r(x), \Delta_r(x)]$ to simplify notations. For the frame $[A, B, \Delta]$ we have then the relations

$$T|\overline{D}: (\overline{D}, fr\, D) \to (A, B), \quad T|\overline{D}_j: (\overline{D}_j, fr\, D_j) \to (A, B),$$

and the corresponding induced homomorphisms

$$(T|\overline{D})_*: H^{n+1}(A, B) \to H^{n+1}(\overline{D}, fr\, D),$$

$$(T|\overline{D}_j)_*: H^{n+1}(A, B) \to H^{n+1}(\overline{D}_j, fr\, D_j).$$

By the definition of the topological index we obtain

$$(T|\overline{D})_*\, \mathfrak{g}[A, B, \Delta] = \mu\, \mathfrak{g}[\overline{D}, fr\, D, D], \tag{2}$$

$$(T|\overline{D}_j)_*\, \mathfrak{g}[A, B, \Delta] = \mu_j\, \mathfrak{g}[\overline{D}_j, fr\, D_j, D_j]. \tag{3}$$

To this situation we apply the result in **I.6.17**, remark 5. The sets now denoted by $\overline{D}, fr D, A, B, Y_j, D_j$ correspond to the sets denoted there by X, Z, X^*, Y^*, Y_n, O_n respectively. The inclusions already noted in this section yield readily the verification of the fact that the assumptions made in **I.6.17**, remark 5 are all satisfied. While the term *frontier* is used at present relative to R^n, we noted in **II.2.1**, lemma 3 that the frontier of D_j relative to $\overline{D}$ is the same as its frontier relative to R^n, and thus no discrepancy arises on this account. The needed portion of the diagram of **I.6.17**, remark 5 appears in the present context in the following form.

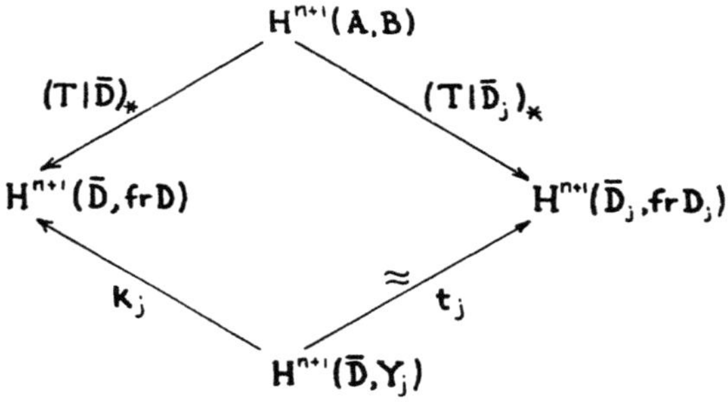

Fig. 40.

By formula (26) in **I.6.17**, we have the relation

$$(T|\bar{D})_* \mathfrak{g}[A, B, \Delta] = \sum{}' k_j t_j^{-1} (T|\bar{D}_j)_* \mathfrak{g}[A, B, \Delta], \tag{4}$$

where the notation $\sum'$ indicates that the summation is actually finite. In view of (3) we have

$$k_j t_j^{-1} (T|\bar{D}_j)_* \mathfrak{g}[A, B, \Delta] = \mu_j k_j t_j^{-1} \mathfrak{g}[\bar{D}_j, fr\, D_j, D_j]. \tag{5}$$

Observe that t_j is induced by the inclusion mapping $(\bar{D}_j, fr\, D_j) \to (\bar{D}, Y_j)$. Since $\bar{D}_j - fr\, D_j = D_j = \bar{D} - Y_j$, lemma 2 in **II.1.3** yields

$$t_j \mathfrak{g}[\bar{D}, Y_j, D_j] = \mathfrak{g}[\bar{D}_j, fr\, D_j, D_j].$$

Since t_j is an isomorphism onto, it follows that

$$t_j^{-1} \mathfrak{g}[\bar{D}_j, fr\, D_j, D_j] = \mathfrak{g}[\bar{D}, Y_j, D_j]. \tag{6}$$

Observe next that k_j is induced by the inclusion mapping $(\bar{D}, fr\, D) \to (\bar{D}, Y_j)$. Since $\bar{D} - Y_j = D_j \subset D = \bar{D} - fr\, D$, from lemma 2 in **II.1.3** it follows that

$$k_j \mathfrak{g}[\bar{D}, Y_j, D_j] = \mathfrak{g}[\bar{D}, fr\, D, D]. \tag{7}$$

From (2), (4), (5), (6), (7) we obtain

$$\mu \mathfrak{g}[\bar{D}, fr\, D, D] = (\sum{}' \mu_j) \mathfrak{g}[\bar{D}, fr\, D, D].$$

Since $\mathfrak{g}[\bar{D}, fr\, D, D]$ is a generator of the infinite cyclic group $H^{n+1}(\bar{D}, fr\, D)$, it follows that $\mu = \sum' \mu_j$, and (1*) is proved.

Remark 4. Consider a (T, D)-admissible point x. Then $\bar{D} \cap T^{-1} x \subset D$. Let now D_* be a domain such that $\bar{D} \cap T^{-1} x \subset D_* \subset D$. Then the one-term sequence consisting of D_* alone is (x, T, D)-complete, and theorem 3 yields therefore $\mu(x, T, D) = \mu(x, T, D_*)$.

Definition. Given T and D as in **II.2.1** (1), (2) and a (T, D)-admissible point x, let $D_1, \ldots, D_m$ be a *finite* sequence of pair-wise disjoint (x, T)-admissible domains in D. The difference

$$d = \mu(x, T, D) - \sum_{j=1}^{m} \mu(x, T, D_j) \tag{8}$$

is termed the *defect* of the sequence $\{D_j\}$, relative to x, T, D.

Theorem 4. If the defect d (in the sense of the preceding definition) is different from zero, then there exists in D a domain D_* with the following properties. (i) D_* is (x, T)-admissible. (ii) $D_* \cap D_j = \emptyset$, $j = 1, \ldots, m$. (iii) $\operatorname{sgn} \mu(x, T, D_*) = \operatorname{sgn} d$.

Proof. Since $d \neq 0$, we infer from theorem 3 that the sequence $D_1, \ldots, D_m$ is *not* (x, T, D)-complete. By lemma 4 in **II.2.1** there follows the existence of an extension $D_1, \ldots, D_m, D_{m+1}, \ldots$ to an (x, T, D)-complete sequence. Applying theorem 3 to this extended sequence, we obtain the formula

$$\mu(x, T, D) = \sum_{j=1}^{m} \mu(x, T, D_j) + \sum_k{}' \mu(x, T, D_{m+k}). \tag{9}$$

From (8) and (9) it follows that

$$d = \sum_k{}' \mu(x, T, D_{m+k}). \tag{10}$$

Since $d \neq 0$, we must have at least one integer k such that

$$\operatorname{sgn} \mu(x, T, D_{m+k}) = \operatorname{sgn} d.$$

If k satisfies this condition, then clearly the domain $D_* = D_{m+k}$ possesses the required properties (i), (ii), (iii).

Remark 5. If $D_\#$ is an (x, T)-admissible domain in D such that

$$\mu(x, T, D_\#) \neq \mu(x, T, D),$$

then $D_\#$ by itself constitutes a one-term sequence satisfying the assumptions of the preceding theorem. Accordingly, there follows the existence of an (x, T)-admissible domain $D_* \subset D$ such that $D_* \cap D_\# = \emptyset$ and

$$\operatorname{sgn} \mu(x, T, D_*) = \operatorname{sgn}[\mu(x, T, D) - \mu(x, T, D_\#)].$$

Remark 6. Consider two continuous mappings

$$T: U \to R^n, \qquad \tilde{T}: \tilde{U} \to R^n, \tag{11}$$

where $U, \tilde{U} \subset R^n$, and a bounded domain D such that

$$\bar{D} \subset U \cap \tilde{U}. \tag{12}$$

Suppose that a frame $[A, B, \Delta]$ is both (T, D)-admissible and $(\tilde{T}, D)$-admissible. Then the mappings

$$T|\overline{D}, \tilde{T}|\overline{D} : (\overline{D}, fr\, D) \to (A, B) \tag{13}$$

induce homomorphisms

$$(T|\overline{D})_*, (\tilde{T}|\overline{D})_* : H^{n+1}(A, B) \to H^{n+1}(\overline{D}, fr\, D). \tag{14}$$

If $x, \tilde{x}$ are any two points in Δ, then by the definition of the topological index

$$(T|\overline{D})_* \mathfrak{g}[A, B, \Delta] = \mu(x, T, D)\, \mathfrak{g}[\overline{D}, fr\, D, D], \tag{15}$$

$$(\tilde{T}|\overline{D})_* \mathfrak{g}[A, B, \Delta] = \mu(\tilde{x}, \tilde{T}, D)\, \mathfrak{g}[\overline{D}, fr\, D, D]. \tag{16}$$

Now suppose that the mappings (13) are homotopic (see **I.6.15**). Then $(T|\overline{D})_* = (\tilde{T}|\overline{D})_*$, and from (15) and (16) it follows that $\mu(x, T, D) = \mu(\tilde{x}, \tilde{T}, D)$ for any two points $x, \tilde{x} \in \Delta$. In particular, $\mu(x, T, D) = \mu(x, \tilde{T}, D)$ for $x \in \Delta$.

Theorem 5. Given $T, \tilde{T}$ and D as in (11) and (12), let Δ be a bounded domain in R^n such that the following holds: for every point $u \in fr D$ there exists a convex set $K(u) \subset R^n$ such that

$$T u, \tilde{T} u \in K(u), \quad K(u) \cap \Delta = \emptyset.$$

Then $\mu(x, T, D) = \mu(\tilde{x}, \tilde{T}, D)$ for any two points $x, \tilde{x} \in \Delta$.

Proof. For $r > 0$ large enough, the set $A = \{x \mid \|x\| \leq r\}$ satisfies the conditions

$$A \supset T\overline{D} \cup \tilde{T}\overline{D} \cup \Delta,$$

since the sets $T\overline{D}, \tilde{T}\overline{D}, \Delta$ are bounded. Let r be so chosen, and put $B = A - \Delta$, obtaining a frame $[A, B, \Delta]$ which is both (T, D)-admissible and $(\tilde{T}, D)$-admissible. For this choice of $[A, B, \Delta]$, we assert that the mappings (13) are homotopic (note that if this fact is established, then the present theorem follows by remark 6). For $0 \leq t \leq 1$, consider the continuous mapping

$$T_t : \overline{D} \to R^n,$$

defined for $u \in \overline{D}$ by the formula (see **I.2.1**)

$$T_t u = t\, T u + (1 - t)\, \tilde{T} u.$$

Since $T u, \tilde{T} u \in A$ and A is convex, clearly $T_t u \in A$, and hence

$$T_t \overline{D} \subset A \quad \text{for} \quad 0 \leq t \leq 1. \tag{17}$$

For $u \in fr D$, we have by assumption $T u, \tilde{T} u \in K(u) \subset \complement \Delta$, and hence

$$T u, \tilde{T} u \in A \cap K(u) \subset A \cap \complement \Delta = B. \tag{18}$$

Noting that $A \cap K(u)$, as the intersection of convex sets, is itself convex, we conclude from (18) that $T_t u \in B$ if $u \in fr D$. Hence

$$T_t\, fr\, D \subset B \quad \text{for} \quad 0 \leq t \leq 1. \tag{19}$$

By (17) and (19) we have the relation

$$T_t : (\bar{D}, fr\, D) \to (A, B) \quad \text{for} \quad 0 \leq t \leq 1. \tag{20}$$

Clearly $T_t u$ is continuous on $\bar{D} \times I$, where I is the unit interval $0 \leq t \leq 1$, and $T_0 = \tilde{T} | \bar{D}$, $T_1 = T | \bar{D}$. Thus (20) implies that the mappings (13) are homotopic, and the proof is complete.

Theorem 6. Given $T, \tilde{T}, D$ as in (11) and (12), let $x_0, \tilde{x}_0$ be two points such that x_0 is (T, D)-admissible and (see **I.1.5**, definition 5 and **I.1.4**, definition 7),

$$\|x - \tilde{x}_0\| + \varrho(T, \tilde{T}, fr\, D) < e(x_0, T\, fr\, D). \tag{21}$$

Then $\mu(x_0, T, D) = \mu(\tilde{x}_0, \tilde{T}, D)$.

Proof. In view of (21) we can choose r to satisfy the inequalities

$$\|x_0 - \tilde{x}_0\| < r < e(x_0, T\, fr\, D) - \varrho(T, \tilde{T}, fr\, D). \tag{22}$$

To simplify notations, put

$$\varrho = \varrho(T, \tilde{T}, fr\, D).$$

For $u \in fr D$, introduce the set

$$K(u) = \{x \mid \|T u - x\| \leq \varrho\}. \tag{23}$$

In view of the definition of ϱ we have then $T u, \tilde{T} u \in K(u)$. For the spherical neighborhood $\Delta_r(x_0)$, where r satisfies (22), we have the relation

$$K(u) \cap \Delta_r(x_0) = \emptyset. \tag{24}$$

Indeed, if one assumes that a point x lies in both $K(u)$ and $\Delta_r(x_0)$, then in view of (22) it would follow that

$$e(x_0, T\, fr\, D) \leq \|T u - x_0\| \leq \|T u - x\| + \|x - x_0\| \leq \varrho + r < e(x_0, T\, fr\, D),$$

a contradiction. Thus (24) holds. Since $K(u)$ is convex by **I.2.2**, exercise 7, the assumptions of theorem 5 are satisfied, with $\Delta = \Delta_r(x_0)$. In view of (22), clearly $x_0, \tilde{x}_0 \in \Delta_r(x_0)$, and hence $\mu(x_0, T, D) = \mu(\tilde{x}_0, \tilde{T}, D)$ by theorem 5.

Remark 7. If $T | fr D = \tilde{T} | fr D$, then $\varrho(T, \tilde{T}, fr D) = 0$, and on choosing $\tilde{x}_0 = x_0 \notin T\, fr D$ the preceding theorem yields the relation $\mu(x_0, T, D) = \mu(x_0, \tilde{T}, D)$. This result shows that the topological index $\mu(x, T, D)$ depends only upon the behavior of T on $fr D$. Explicitly: if the continuous mappings $T, \tilde{T}$ are both defined on $\bar{D}$, and if $T|fr D = \tilde{T}|fr D$, then $\mu(x, T, D)$ and $\mu(x, \tilde{T}, D)$ are defined on the same set $\complement T fr D = \complement \tilde{T} fr D$ and $\mu(x, T, D) \equiv \mu(x, \tilde{T}, D)$ on this set.

Remark 8. On choosing $\tilde{x}_0 = x_0 \notin T fr D$ in theorem 6, we obtain the following statement. If the continuous mappings $T, \tilde{T}$ are both defined on $\bar{D}$, then

$$\mu(x_0, \tilde{T}, D) = \mu(x_0, T, D) \quad \text{if} \quad e(x_0, T\, fr\, D) > \varrho(T, \tilde{T}, fr\, D).$$

II.2.4. Applications to homeomorphisms. Given T and D as in **II.2.1** (1), (2), assume now that $T|\bar{D}$ is *one-to-one* (that is, $u_1, u_2 \in \bar{D}$ and $T u_1 = T u_2$ imply that $u_1 = u_2$). Since D is bounded by assumption, the set $\bar{D}$ is compact (see **I.2.1**, exercise 4). By **I.1.5**, exercise 8 it follows that the image set

$$T\bar{D} = X \tag{1}$$

is also compact, and by **I.1.5**, exercise 7 it follows further that

$$T|\bar{D}: \bar{D} \to X \tag{2}$$

is a homeomorphism from $\bar{D}$ onto X. Put

$$TD = \Delta, \tag{3}$$

$$T\, fr\, D = Y. \tag{4}$$

Now since the mapping (2) is a homeomorphism and $D \subset \bar{D}$, the mapping

$$T|D: D \to \Delta$$

is a homeomorphism from D onto Δ. As D is open, we have the relation $D = \text{int}\, D$. By **I.7.4**, theorem 12 it follows that

$$\Delta = TD = T\, \text{int}\, D = \text{int}\, \Delta. \tag{5}$$

Since $\text{int}\, \Delta$ is open, we conclude that Δ itself is open (relative to R^n). As D is connected, $\Delta = TD$ is also connected (see **I.1.5**, exercise 8). Finally, $\Delta = TD \subset T\bar{D}$ implies (since $T\bar{D}$ is compact and hence bounded) that Δ is bounded. Thus $\Delta = TD$ is a bounded domain. Since $\bar{D}$ is compact, by **I.1.5**, exercise 9 we infer that

$$\bar{\Delta} = \overline{TD} = T\bar{D} = X. \tag{6}$$

Since D and Δ are bounded domains, and $T|\bar{D}$ is one-to-one, we conclude now [see (1), (5), (6)] that

$$T\, fr D = T(\bar{D} - D) = T\bar{D} - TD = \bar{\Delta} - \Delta = fr\, \Delta. \tag{7}$$

In view of (1), (6), (7), we conclude that

$$T|\bar{D}: (\bar{D}, fr\, D) \to (\bar{\Delta}, fr\, \Delta) \tag{8}$$

is a homeomorphism from the pair $(\overline{D}, fr D)$ onto the pair $(\overline{\Delta}, fr\,\Delta)$. Accordingly, the induced homomorphism

$$(T|\overline{D})_*: H^{n+1}(\overline{\Delta}, fr\,\Delta) \xrightarrow{\approx} H^{n+1}(\overline{D}, fr\,D) \tag{9}$$

is an isomorphism onto (see **I.6.2**, remark 2), a fact indicated by the symbol $\approx$ in (9). It follows that

$$(T|\overline{D})_*\, \mathfrak{g}[\overline{\Delta}, fr\,\Delta, \Delta] = \iota(T, D)\, \mathfrak{g}[\overline{D}, fr\,D, D], \tag{10}$$

where $\iota(T, D)$ is an integer equal to 1 or to -1. In view of the defining formula (10), this integer is determined by T and D and thus the notation $\iota(T, D)$ is justified.

Definition. $\iota(T, D)$ is termed the index of T relative to D, and T is termed even or odd in D according as $\iota(T, D) = 1$ or $\iota(T, D) = -1$.

Remark 1. For the symbol $\iota(T, D)$ to be meaningful in the sense of this definition, T must be a continuous mapping from a subset U of R^n into R^n, D must be a bounded domain such that the inclusion $\overline{D} \subset U$ holds, and $T|\overline{D}$ must be one-to-one. In view of (3), (6), (7) the defining equation (10) may be re-written in the form

$$(T|\overline{D})_*\, \mathfrak{g}[T\overline{D}, T\, fr\,D, T\,D] = \iota(T, D)\, \mathfrak{g}[\overline{D}, fr\,D, D]. \tag{11}$$

Observe also that $\iota(T, D)$ is determined as soon as T is known on $\overline{D}$, and thus

$$\iota(T, D) = \iota(T|\overline{D}, D).$$

Remark 2. Since $\Delta = TD$ is a bounded domain, $[\overline{\Delta}, fr\Delta, \Delta]$ is actually a frame, and (8) implies that this frame is (T, D)-admissible. By **II.2.3** it follows that

$$(T|\overline{D})_*\, \mathfrak{g}[\overline{\Delta}, fr\,\Delta, \Delta] = \mu(x, T, D)\, \mathfrak{g}[\overline{D}, fr\,D, D] \quad \text{for} \quad x \in \Delta. \tag{12}$$

Comparison of (10) and (12) yields the following statement.

Theorem 1. $\iota(T, D) = \mu(x, T, D)$ for $x \in \Delta = TD$.

Remark 3. Since $\mu(x, T, D)$ is undefined for $x \in T\,fr D$ and vanishes if $x \notin T\overline{D}$ (see **II.2.3**, theorem 2), we obtain the following picture under the present circumstances. If $\mu(x, T, D)$ is defined for a point x, then either $x \in TD$ or $x \in \complement T\overline{D}$. On TD, $\mu(x, T, D)$ is defined and has the constant value $\iota(T, D)$, and thus on TD it has the constant value 1 or the constant value -1. On $\complement T\overline{D}$, $\mu(x, T, D)$ is defined and has the constant value zero.

Theorem 2. If $T|\overline{D}$ is the identity mapping on $\overline{D}$ (that is, if $Tu = u$ for $u \in \overline{D}$), then $\iota(T, D) = 1$.

Proof. In this case, $T|\overline{D}$ is certainly one-to-one on $\overline{D}$, and hence $\iota(T, D)$ is defined. Also, $T\overline{D} = \overline{D}$, $T\,fr D = fr D$, $TD = D$. Accordingly, in the defining formula (11) the homomorphism $(T|\overline{D})_*$ reduces now to the identity homomorphism by **I.6.2**, remark 1, and thus (11) reduces to

$$\mathfrak{g}[\overline{D}, fr\,D, D] = \iota(T, D)\,\mathfrak{g}[\overline{D}, fr\,D, D].$$

Since $\mathfrak{g}[\overline{D}, fr\,D, D]$ is a generator of the infinite cyclic group $H^{n+1}(\overline{D}, fr\,D)$, it follows that $\iota(T, D) = 1$.

Theorem 3. If $\iota(T, D)$ is defined (see remark 1) and D_* is any domain in D, then $\iota(T, D_*)$ is also defined and $\iota(T, D_*) = \iota(T, D)$.

Proof. The assumptions imply that T is defined, continuous, and one-to-one on $\overline{D}$, and that D is bounded. The inclusion $D_* \subset D$ implies that D_* is also bounded and $\overline{D}_* \subset \overline{D}$, and thus T is defined, continuous, and one-to-one on $\overline{D}_*$. Hence $\iota(T, D_*)$ is defined. Take a point $u_* \in D_*$ and put $x_* = T u_*$. Then $\overline{D} \cap T^{-1} x_* = u_* \subset D_* \subset D$, since $T|\overline{D}$ is one-to-one. By remark 4 in **II.2.3** and theorem 1 in the present section we conclude that

$$\iota(T, D_*) = \mu(x_*, T, D_*) = \mu(x_*, T, D) = \iota(T, D).$$

Theorem 4. Let U_1, U_2 be subsets of R^n, and let

$$T_1: U_1 \to R^n, \quad T_2: U_2 \to R^n$$

be continuous mappings. Let D_1, D_2, D_3 be bounded domains in R^n, such that (i) $\overline{D}_1 \subset U_1$, $\overline{D}_2 \subset U_2$, (ii) the relations

$$T_1|\overline{D}_1: (\overline{D}_1, fr\,D_1) \to (\overline{D}_2, fr\,D_2), \tag{13}$$

$$T_2|\overline{D}_2: (\overline{D}_2, fr\,D_2) \to (\overline{D}_3, fr\,D_3) \tag{14}$$

hold, and (iii) the mappings (13), (14) are homeomorphisms from the pairs $(\overline{D}_1, fr\,D_1)$, $(\overline{D}_2, fr\,D_2)$ onto the pairs $(\overline{D}_2, fr\,D_2)$, $(\overline{D}_3, fr\,D_3)$ respectively. Then $\iota(T_1, D_1)$, $\iota(T_2, D_2)$, $\iota(T_2 T_1, D_1)$ are defined, and

$$\iota(T_2 T_1, D_1) = \iota(T_1, D_1)\,\iota(T_2, D_2). \tag{15}$$

Proof. In view of remark 1, it is immediate that the three indices ι occurring in (15) are defined. Observe now that the mappings (13), (14) yield

$$T_2 T_1|\overline{D}_1: (\overline{D}_1, fr\,D_1) \to (\overline{D}_3, fr\,D_3).$$

From the defining formula (10), in combination with the vector law for induced homomorphisms, we conclude that

$$\begin{aligned}\iota(T_2 T_1, D_1)\,\mathfrak{g}[\overline{D}_1, fr\,D_1, D_1] &= (T_2 T_1|\overline{D}_1)_*\,\mathfrak{g}[\overline{D}_3, fr\,D_3, D_3]\\ &= (T_1|\overline{D}_1)_*(T_2|\overline{D}_2)_*\,\mathfrak{g}[\overline{D}_3, fr\,D_3, D_3] = \iota(T_2, D_2)(T_1|\overline{D}_1)_*\,\mathfrak{g}[\overline{D}_2, fr\,D_2, D_2]\\ &= \iota(T_2, D_2)\,\iota(T_1, D_1)\,\mathfrak{g}[\overline{D}_1, fr\,D_1, D_1],\end{aligned}$$

and (15) follows by **I.3.1**, exercise 4.

Theorem 5. Let $\boldsymbol{D}$ be a (not necessarily bounded) domain in R^n, and let

$$T: \boldsymbol{D} \to R^n \tag{16}$$

be a one-to-one continuous mapping. Then the following holds.

(i) $T\boldsymbol{D}$ is a domain.

(ii) T is a homeomorphism from $\boldsymbol{D}$ onto $T\boldsymbol{D}$.

(iii) If D_1, D_2 are any two bounded domains in $\boldsymbol{D}$ such that $\overline{D}_1, \overline{D}_2 \subset \boldsymbol{D}$, then $\iota(T, D_1)$, $\iota(T, D_2)$ are both defined, and

$$\iota(T, D_1) = \iota(T, D_2). \tag{17}$$

Proof. In view of remark 1 (applied with $U = \boldsymbol{D}$), $\iota(T, D_1)$ and $\iota(T, D_2)$ are both defined. Denote now by Ω^+ (Ω^-) the class of all those bounded domains D in $\boldsymbol{D}$ for which $\overline{D} \subset \boldsymbol{D}$ and $\iota(T, D) = 1$ (-1). Put

$$G^+ = \bigcup D, \quad D \in \Omega^+; \quad G^- = \bigcup D, \quad D \in \Omega^-.$$

Then G^+, G^- are open (perhaps empty) sets in $\boldsymbol{D}$. We first verify that

$$\boldsymbol{D} = G^+ \cup G^-. \tag{18}$$

Indeed, consider any point $u_0 \in \boldsymbol{D}$. Then there exists a bounded domain $D(u_0)$ such that $u_0 \in D(u_0)$ and $\overline{D(u_0)} \subset \boldsymbol{D}$ (for example, the spherical neighborhood $\Delta_r(u_0)$ may be used if r is sufficiently small). Then $\iota\big(T, D(u_0)\big)$ is defined, and clearly $u_0 \in G^+$ or $u_0 \in G^-$ according as $\iota\big(T, D(u_0)\big)$ equals 1 or -1. Thus (18) holds. Next we verify that

$$G^+ \cap G^- = \emptyset. \tag{19}$$

If this assertion is denied, then we should have two bounded domains D', D'' in $\boldsymbol{D}$ such that $\overline{D}', \overline{D}'' \subset \boldsymbol{D}$ and

$$\iota(T, D') = 1, \quad \iota(T, D'') = -1, \quad D' \cap D'' \neq \emptyset. \tag{20}$$

Then $D = D' \cup D''$ would be a bounded domain in $\boldsymbol{D}$ such that $\overline{D} \subset \boldsymbol{D}$, and theorem 3 would yield

$$\iota(T, D') = \iota(T, D) = \iota(T, D''),$$

in contradiction with (20), and (19) is proved. Since $\boldsymbol{D}$ is connected, (18) and (19) imply that one of the sets G^+, G^- must be empty. Accordingly, one of the classes Ω^+, Ω^- must be empty, and part (iii) of the theorem follows.

Consider now any open set $O \subset \boldsymbol{D}$, and take any point $x_0 \in TO$. Then $x_0 = Tu_0$, $u_0 \in O$. Since O is open, there exists a bounded domain D_0 such that $u_0 \in D_0, \overline{D}_0 \subset O$ [for example, the spherical neighborhood $\Delta_r(u_0)$ will serve if r is sufficiently small]. Then, since T is one-to-one,

TD_0 is a domain (by the discussion at the beginning of the present section), and $x_0 = Tu_0 \in TD_0 \subset TO$. Thus every point x_0 of TO is contained in an open set (namely, in TD_0) which is contained in TO. Hence TO is open. Thus the mapping (16) is continuous, one-to-one, and carries open sets into open sets, and hence part (ii) of the theorem follows by **I.1.5**, exercise 6. Observe that the term *open set* is used here relative to R^n. However, if a set G is open relative to R^n and is contained in a subset S of R^n, then G is also open relative to S, and hence there arises no discrepancy. Finally, part (i) of the theorem follows from part (ii) [see (5)].

We turn presently to a study of *one-parameter families of homeomorphisms*

$$T_t : \bar{D} \to R^n, \tag{21}$$

defined as follows. D is a bounded domain in R^n, and for each $t \in I$, where I is the unit interval $0 \leq t \leq 1$, T_t is continuous and one-to-one on $\bar{D}$. Finally, we require that $T_t u$, interpreted as a mapping from $\bar{D} \times I$ into R^n, be continuous on $\bar{D} \times I$.

Theorem 6. Given a one-parameter family of homeomorphisms as in (21), the index $\iota(T_t, D)$ is defined for each $t \in I$ and is independent of t. In particular, $\iota(T_0, D) = \iota(T_1, D)$.

Proof. In view of remark 1, applied with $U = \bar{D}$, $\iota(T_t, D)$ is defined for $t \in I$. Observe now that $\iota(T_t, D)$ may be considered as an integral-valued function of t on I. Since I is connected, in view of **I.1.3**, exercise 37, it is sufficient to show that $\iota(T_t, D)$ is locally constant on I. To simplify notations, put $\Delta(t) = T_t D$. Choose a point $u_* \in D$ and set $T_t u_* = x(t)$. Now take any $t_0 \in I$. Then (see the discussion at the beginning of the present section), $x(t_0) \in \Delta(t_0)$, and thus $x(t_0) \notin fr\,\Delta(t_0)$. Since

$$fr\,\Delta(t_0) = T_{t_0} fr\,D,$$

it follows that

$$e(x(t_0), T_{t_0} fr\,D) > 0.$$

As $T_t u$ is continuous on $\bar{D} \times I$, there follows the existence of an $\varepsilon > 0$ such that

$$\|x(t) - x(t_0)\| + \varrho(T_t, T_{t_0}, fr\,D) < e(x(t_0), T_{t_0} fr\,D), \tag{22}$$

for $|t - t_0| < \varepsilon$, $t \in I$. Let t be so restricted. By theorem 6 in **II.2.3**, (22) implies that

$$\mu(x(t), T_t, D) = \mu(x(t_0), T_{t_0}, D). \tag{23}$$

By theorem 1 we have

$$\mu(x(t_0), T_{t_0}, D) = \iota(T_{t_0}, D) = \pm 1. \tag{24}$$

From (23) and (24) it follows that

$$\mu(x(t), T_t, D) \neq 0. \tag{25}$$

By remark 3, we infer from (25) that $\iota(T_t, D)$ is defined and

$$\iota(T_t, D) = \mu(x(t), T_t, D). \tag{26}$$

From (26), (23), (24) we see that

$$\iota(T_t, D) = \iota(T_{t_0}, D) \quad \text{for} \quad |t - t_0| < \varepsilon,\ t \in I.$$

Thus $\iota(T_t, D)$ is locally constant on I, and the proof is complete.

II.2.5. Linear transformations in R^n. Returning to the mapping $T: U \to R^n$ in **II.2.1** (1), let us note that for certain purposes it is convenient to describe T in the form

$$T: x^i = f_i(u^1, \ldots, u^n), \quad i = 1, \ldots, n, \tag{1}$$

by giving the coordinates $x^1, \ldots, x^n$ of the image point x as functions of the coordinates $u^1, \ldots, u^n$ of the point $u \in U$. We shall study presently the important special case when $U = R^n$ and the functions f_i in (1) are linear functions of $u^1, \ldots, u^n$. The mapping T is then termed a *linear transformation* from R^n into R^n and will be denoted, generically, by $\mathfrak{L}$. Thus a linear transformation is determined by formulas of the type

$$\mathfrak{L}: x^i = b^i + \sum_{k=1}^{n} a_{ik} u^k, \quad i = 1, \ldots, n, \tag{2}$$

where b^i, a_{ik} are real numbers. Clearly $\mathfrak{L}$ is a continuous mapping from R^n into R^n.

In the following discussion of linear transformations, the integer $n \geq 1$ is arbitrary, but fixed until further notice.

The matrix (a_{ik}) will be termed the matrix $\mathfrak{M}(\mathfrak{L})$ of $\mathfrak{L}$. The determinant $\det(a_{ik})$ of this matrix will be termed the determinant of $\mathfrak{L}$ and will be denoted by $\det \mathfrak{L}$.

Definition 1. A linear transformation $\mathfrak{L}$ is termed non-singular or singular according as $\det \mathfrak{L} \neq 0$ or $\det \mathfrak{L} = 0$.

Remark 1. The non-singular linear transformations $\mathfrak{L}$ fall naturally into two categories, according as $\det \mathfrak{L} > 0$ or $\det \mathfrak{L} < 0$. The first category contains the identity mapping in R^n, given by $x^i = u^i$, $i = 1, \ldots, n$. An important linear transformation in the second category is that given by $x^1 = -u^1$, $x^i = u^i$ for $i \neq 1$. For convenient reference, these two particular transformations will be denoted by $\mathfrak{L}^+$ and $\mathfrak{L}^-$ respectively. Thus the defining formulas are as follows.

$$\mathfrak{L}^+: x^i = u^i, \quad i = 1, \ldots, n, \tag{3}$$

$$\mathfrak{L}^-: x^1 = -u^1, \quad x^i = u^i \quad \text{for} \quad i \neq 1. \tag{4}$$

Clearly

$$\det \mathfrak{L}^+ = 1, \quad \det \mathfrak{L}^- = -1. \tag{5}$$

The following two lemmas are direct consequences of standard theorems on systems of linear equations.

Lemma 1. If $\mathfrak{L}$ is non-singular, then $\mathfrak{L}$ is a homeomorphism from R^n onto R^n, and $\mathfrak{L}^{-1}$ is linear and non-singular.

Lemma 2. If $\mathfrak{L}_1$, $\mathfrak{L}_2$ are linear transformations, then their product (see **I.1.1**) $\mathfrak{L}_2\mathfrak{L}_1$ is a linear transformation and $\det \mathfrak{L}_2\mathfrak{L}_1 = \det \mathfrak{L}_2 \det \mathfrak{L}_1$. In particular, if $\mathfrak{L}_1$, $\mathfrak{L}_2$ are both non-singular, then $\mathfrak{L}_2\mathfrak{L}_1$ is also non-singular.

Definition 2. Let I denote the unit interval $0 \leqq t \leqq 1$, and let $b^i(t)$, $a_{ik}(t)$, $i, k = 1, \ldots, n$, be real-valued continuous functions on I such that $\det a_{ik}(t) \neq 0$ for all $t \in I$. For each $t \in I$ we have then the non-singular linear transformation

$$\mathfrak{L}_t : x^i = b^i(t) + \sum_{k=1}^{n} a_{ik}(t)\, u^k, \quad i = 1, \ldots, n, \quad t \in I. \tag{6}$$

These linear transformations will be said to constitute a one-parameter family of non-singular linear transformations.

Remark 2. Since $\det \mathfrak{L}_t$ is a continuous function of t which is different from zero on I, it follows that either $\det \mathfrak{L}_t > 0$ for all $t \in I$ or $\det \mathfrak{L}_t < 0$ for all $t \in I$.

Definition 3. Two non-singular linear transformations $\mathfrak{L}'$, $\mathfrak{L}''$ will be termed equivalent, in symbols $\mathfrak{L}' \sim \mathfrak{L}''$, if and only if there exists a one-parameter family $\mathfrak{L}_t$ of non-singular linear transformations such that $\mathfrak{L}_0 = \mathfrak{L}'$, $\mathfrak{L}_1 = \mathfrak{L}''$. For brevity, such a family $\mathfrak{L}_t$ will be termed an admissible family connecting $\mathfrak{L}'$ to $\mathfrak{L}''$.

Remark 3. In view of remark 2, the relation $\mathfrak{L}' \sim \mathfrak{L}''$ implies that $\operatorname{sgn} \det \mathfrak{L}' = \operatorname{sgn} \det \mathfrak{L}''$. Thus this condition is necessary for the relation $\mathfrak{L}' \sim \mathfrak{L}''$ to hold. It follows that the transformations $\mathfrak{L}^+$, $\mathfrak{L}^-$ [see (3), (4)] are not equivalent in the sense of definition 3.

Lemma 3. The binary relation $\mathfrak{L}' \sim \mathfrak{L}''$ is reflexive, symmetric, and transitive.

This is an obvious consequence of the definitions involved.

Lemma 4. Given a non-singular linear transformation $\mathfrak{L}$ as in (2), let $\mathfrak{L}_*$ be given by

$$\mathfrak{L}_* : x^i = \sum_{k=1}^{n} a_{ik}\, u^k, \quad i = 1, \ldots, n. \tag{7}$$

That is, $\mathfrak{L}_*$ is obtained from $\mathfrak{L}$ by dropping the constant terms b^i. Then $\mathfrak{L}_*$ is non-singular and $\mathfrak{L}_* \sim \mathfrak{L}$.

Proof. Since $\det \mathfrak{L}_* = \det \mathfrak{L} \neq 0$, $\mathfrak{L}_*$ is non-singular. Now define, for $0 \leq t \leq 1$,

$$\mathfrak{L}_t : x^i = (1-t)\, b^i + \sum_{k=1}^{n} a_{ik} u^k, \quad i = 1, \ldots, n.$$

Clearly, $\mathfrak{L}_t$ is an admissible family connecting $\mathfrak{L}$ to $\mathfrak{L}_*$, and the lemma is proved.

Remark 4. The next few lemmas all assert that certain pairs of linear transformations $\mathfrak{L}$, $\mathfrak{L}_*$ are equivalent. The pattern of the proof will be the same as in lemma 4: it will be clear that $\mathfrak{L}_*$ is non-singular, and there will remain to exhibit an admissible family $\mathfrak{L}_t$ connecting $\mathfrak{L}$ to $\mathfrak{L}_*$. Accordingly, we shall merely indicate in each case the connecting admissible family $\mathfrak{L}_t$, leaving trivial verifications to the reader.

Lemma 5. Given a non-singular linear transformation of the form

$$\mathfrak{L} : x^i = \sum_{k=1}^{n} a_{ik} u^k, \quad i = 1, \ldots, n, \tag{8}$$

let $\mathfrak{L}_*$ be the linear transformation obtained by multiplying the elements of the r-th column (row) of the matrix (a_{ik}) by the same real number $\lambda > 0$. Then $\mathfrak{L} \sim \mathfrak{L}_*$.

Proof. Use (for the column case) the connecting family

$$\mathfrak{L}_t : x^i = \sum_{k=1}^{n} a_{ik}(t)\, u^k, \quad i = 1, \ldots, n, \quad 0 \leq t \leq 1,$$

where

$$a_{ik}(t) \equiv a_{ik} \quad \text{for} \quad k \neq r, \quad i = 1, \ldots, n,$$
$$a_{ir}(t) = (1 - t + \lambda t)\, a_{ir}, \quad i = 1, \ldots, n.$$

Lemma 6. Given a non-singular linear transformation $\mathfrak{L}$ as in (8), denote by $\mathfrak{L}_*$ the linear transformation obtained by multiplying the elements of the r-th row (column) of the matrix (a_{ik}) by an arbitrary real number λ and adding to the corresponding elements of the s-th row (column), where $r \neq s$. Then $\mathfrak{L} \sim \mathfrak{L}_*$.

Proof. Use (in the row case) the connecting family

$$\mathfrak{L}_t : x^i = \sum_{k=1}^{n} a_{ik}(t)\, u^k, \quad i = 1, \ldots, n, \quad 0 \leq t \leq 1,$$

where

$$a_{ik}(t) \equiv a_{ik} \quad \text{for} \quad i \neq s, \quad k = 1, \ldots, n,$$
$$a_{sk}(t) = a_{sk} + t \lambda a_{rk}, \qquad k = 1, \ldots, n.$$

Definition 4. Let $\varepsilon_1, \ldots, \varepsilon_n$ be integers such that $|\varepsilon_i| = 1$, $i = 1, \ldots, n$ (thus each ε_i is equal to either 1 or -1). Then the linear transformation

$$\mathfrak{L} : x^i = \varepsilon_i u^i, \quad i = 1, \ldots, n, \tag{9}$$

is non-singular, since $\det \mathfrak{L} = \varepsilon_1 \ldots \varepsilon_n = \pm 1$. The transformation $\mathfrak{L}$ given by (9) will be denoted by $[\varepsilon_1, \ldots, \varepsilon_n]$.

Lemma 7. Suppose that $n \geqq 2$, and let j be any integer such that $1 \leqq j \leqq n-1$. Then

$$[\ldots, \varepsilon_j, \varepsilon_{j+1}, \ldots] \sim [\ldots, -\varepsilon_j, -\varepsilon_{j+1}, \ldots].$$

Proof. Use the connecting family $\mathfrak{L}_t$ defined as follows:

$$x^i = \varepsilon_i u^i \quad \text{for} \quad i \neq j, j+1,$$
$$x^j = \varepsilon_j (\cos \pi t) u^j - \varepsilon_j (\sin \pi t) u^{j+1},$$
$$x^{j+1} = \varepsilon_{j+1} (\sin \pi t) u^j + \varepsilon_{j+1} (\cos \pi t) u^{j+1}.$$

Theorem 1. Given a non-singular linear transformation $\mathfrak{L}$ as in (2), one has $\mathfrak{L} \sim \mathfrak{L}^+$ or $\mathfrak{L} \sim \mathfrak{L}^-$ [see (3), (4)] according as $\det \mathfrak{L} > 0$ or $\det \mathfrak{L} < 0$.

Proof. The idea of the proof is to exhibit a finite sequence of non-singular linear transformations $\mathfrak{L}_1, \ldots, \mathfrak{L}_m$, such that $\mathfrak{L} \sim \mathfrak{L}_1, \mathfrak{L}_1 \sim \mathfrak{L}_2, \ldots, \mathfrak{L}_{m-1} \sim \mathfrak{L}_m$, and $\mathfrak{L}_m$ coincides either with $\mathfrak{L}^+$ or $\mathfrak{L}^-$. By lemma 3 it follows then that $\mathfrak{L} \sim \mathfrak{L}_m$, and hence $\mathfrak{L} \sim \mathfrak{L}^+$ or $\mathfrak{L} \sim \mathfrak{L}^-$. By remark 3 we must have in the first case $\det \mathfrak{L} > 0$ and in the second case $\det \mathfrak{L} < 0$ [see (5)], and the theorem will be proved. We proceed now to exhibit the sequence $\mathfrak{L}_1, \mathfrak{L}_2, \ldots$. Starting with $\mathfrak{L}$ given as in (2), we first drop the constant terms b^i, obtaining by lemma 4 a non-singular linear transformation $\mathfrak{L}_1 \sim \mathfrak{L}$, where $\mathfrak{L}_1$ is of the form

$$x^i = \sum_{k=1}^{n} a_{ik} u^k, \quad i = 1, \ldots, n.$$

Observe now that the matrix (a_{ik}) can be brought into the *diagonal form* by repeated application of the process of multiplying all the elements of a column (row) by a real number and adding to the corresponding elements of another column (row). In view of lemma 6, it follows that we obtain in this manner a sequence $\mathfrak{L}_1, \ldots, \mathfrak{L}_r$ of non-singular linear transformations such that $\mathfrak{L}_1 \sim \cdots \sim \mathfrak{L}_r$, and $\mathfrak{L}_r$ is of the form

$$x_i = c_i u^i, \quad c_i \neq 0, \quad i = 1, \ldots, n.$$

Repeated application of lemma 5 (using successively $\lambda = 1/|c_1|, \ldots, \lambda = 1/|c_n|$), yields a non-singular linear transformation $\mathfrak{L}_{r+1} \sim \mathfrak{L}_r$ of the form $[\varepsilon_1, \ldots, \varepsilon_n]$ (see definition 4). If $n = 1$, then the proof is complete, since $[\varepsilon_1] = \mathfrak{L}^+$ or $[\varepsilon_1] = \mathfrak{L}^-$ according as $\varepsilon_1 = 1$ or $\varepsilon_1 = -1$. So assume that $n \geqq 2$. We treat then

$$\mathfrak{L}_{r+1} = [\varepsilon_1, \ldots, \varepsilon_n]$$

as follows. Suppose that for some index j we have $\varepsilon_j = 1$, $\varepsilon_{j+1} = -1$. By lemma 7 we have then (since now $-\varepsilon_j = \varepsilon_{j+1}$, $-\varepsilon_{j+1} = \varepsilon_j$)

$$[\ldots, \varepsilon_j, \varepsilon_{j+1}, \ldots] \sim [\ldots, \varepsilon_{j+1}, \varepsilon_j, \ldots].$$

In other words, if in the sequence $\varepsilon_1, \ldots, \varepsilon_n$ a term 1 is followed by a term -1, then the process of exchanging these two consecutive terms yields an equivalent transformation. Repeated application of this remark yields a linear transformation

$$\mathfrak{L}_{r+2} = [\varepsilon_1', \ldots, \varepsilon_n'],$$

such that $\mathfrak{L}_{r+2} \sim \mathfrak{L}_{r+1}$ and either $\varepsilon_1' = \cdots = \varepsilon_n' = 1$ (in which case $\mathfrak{L}_{r+2} = \mathfrak{L}^+$ and the proof is complete), or else we have an integer j such that $\varepsilon_1' = \cdots = \varepsilon_j' = -1$, and (in case $j < n$) $\varepsilon_{j+1}' = \cdots = \varepsilon_n' = 1$. If $j = 1$, then clearly $\mathfrak{L}_{r+2} = \mathfrak{L}^-$, and the proof is complete. If $j \geq 2$, then $\varepsilon_{j-1}' = \varepsilon_j' = -1$, and by lemma 7 we have

$$[\ldots, \varepsilon_{j-1}', \varepsilon_j', \ldots] \sim [\ldots, -\varepsilon_{j-1}', -\varepsilon_j', \ldots].$$

Since now $-\varepsilon_{j-1}' = -\varepsilon_j' = 1$, the number of terms equal to -1 has been reduced by two. Repeated application of this remark yields a linear transformation

$$\mathfrak{L}_{r+3} = [\varepsilon_1'', \ldots, \varepsilon_n''],$$

such that $\mathfrak{L}_{r+3} \sim \mathfrak{L}_{r+2}$, and either $\varepsilon_1'' = \cdots = \varepsilon_n'' = 1$, or else $\varepsilon_1'' = -1$, $\varepsilon_2'' = \cdots = \varepsilon_n'' = 1$. In the first case $\mathfrak{L}_{r+3} = \mathfrak{L}^+$, in the second case $\mathfrak{L}_{r+3} = \mathfrak{L}^-$, and the proof is complete.

Theorem 2. Two non-singular linear transformations $\mathfrak{L}', \mathfrak{L}''$ are equivalent if and only if

$$\operatorname{sgn} \det \mathfrak{L}' = \operatorname{sgn} \det \mathfrak{L}''. \tag{10}$$

Proof. The necessity of the condition (10) follows by remark 3 To prove the sufficiency, assume that (10) holds. If $\det \mathfrak{L}', \det \mathfrak{L}''$ are both positive (negative), then by theorem 1 we have $\mathfrak{L}', \mathfrak{L}'' \sim \mathfrak{L}^+ (\mathfrak{L}^-)$, and hence $\mathfrak{L}' \sim \mathfrak{L}''$ in view of the transitivity of the relation $\sim$ (see lemma 3).

Consider now a non-singular linear transformation $\mathfrak{L}$. Then $\mathfrak{L}$ is a homeomorphism from R^n onto R^n (see lemma 1) and hence theorem 5 in **II.2.4** applies (with $D = R^n$, $T = \mathfrak{L}$). We conclude that if D, D' are any two bounded domains in R^n, then $\iota(L, D) = \iota(L, D')$. This shows that the index $\iota(\mathfrak{L}, D)$ is independent of the choice of the bounded domain $D \subset R^n$, and hence this index depends only upon $\mathfrak{L}$. Thus the following definition is justified.

Definition 5. The *index* $\iota(\mathfrak{L})$ of a non-singular linear transformation $\mathfrak{L}$ is defined by the formula

$$\iota(\mathfrak{L}) = \iota(\mathfrak{L}, D), \tag{11}$$

where D is any bounded domain in R^n.

Lemma 8. If $\mathfrak{L}'$, $\mathfrak{L}''$ are non-singular linear transformations such that $\mathfrak{L}' \sim \mathfrak{L}''$, then $\iota(\mathfrak{L}') = \iota(\mathfrak{L}'')$.

Proof. Choose a bounded domain $D \subset R^n$. By definition

$$\iota(\mathfrak{L}') = \iota(\mathfrak{L}', D), \quad \iota(\mathfrak{L}'') = \iota(\mathfrak{L}'', D). \tag{12}$$

By assumption (see definition 3) there exists an admissible family $\mathfrak{L}_t$ connecting $\mathfrak{L}'$ to $\mathfrak{L}''$. On setting $T_t = \mathfrak{L}_t | \overline{D}$, clearly T_t is a one-parameter family of homeomorphisms [see **II.2.4** (21)] such that

$$T_0 = \mathfrak{L}' | \overline{D}, \quad T_1 = \mathfrak{L}'' | \overline{D}.$$

Hence, by theorem 6 and remark 1 in **II.2.4**,

$$\iota(\mathfrak{L}', D) = \iota(\mathfrak{L}'', D). \tag{13}$$

From (12) and (13) it follows that $\iota(\mathfrak{L}') = \iota(\mathfrak{L}'')$.

Theorem 3. If $\mathfrak{L}'$, $\mathfrak{L}''$ are non-singular linear transformations such that $\operatorname{sgn} \det \mathfrak{L}' = \operatorname{sgn} \det \mathfrak{L}''$, then $\iota(\mathfrak{L}') = \iota(\mathfrak{L}'')$.

Proof. By theorem 2 the assumption implies that $\mathfrak{L}' \sim \mathfrak{L}''$, and hence $\iota(\mathfrak{L}') = \iota(\mathfrak{L}'')$ by lemma 8.

In view of theorem 1 it is clear that $\iota(\mathfrak{L})$ can be explicitly calculated for any non-singular linear transformation $\mathfrak{L}$ provided that one knows the index ι for the particular transformations $\mathfrak{L}^+$, $\mathfrak{L}^-$ [see (3), (4)]. We proceed to derive the lemmas to be used in determining $\iota(\mathfrak{L}^+)$, $\iota(\mathfrak{L}^-)$.

Consider the $(n-1)$-sphere S^{n-1} in R^n and the linear transformation $\mathfrak{L}^-$ [see (4)]. Put

$$T^n = \mathfrak{L}^- | S^{n-1}. \tag{14}$$

Displaying the integer n in the notation T^n is now necessary since we shall employ induction on n later. Clearly T^n is a homeomorphism from S^{n-1} onto S^{n-1}. Accordingly, the induced homomorphism

$$T^n_*: H^n(S^{n-1}) \xrightarrow{\approx} H^n(S_{n-1}) \tag{15}$$

is an isomorphism onto (see **I.6.2**, remark 2), as indicated by the symbol $\approx$.

Lemma 9. The isomorphism T^n_* is *odd* (see **I.3.2**).

Proof. Consider first the case $n=1$. Recall that S^0 consists of the two real numbers 1 and -1, and that (by **I.6.1**, remark 7) $H^1(S^0)$ can be identified with the group $Z^1(S^0)$ of the 1-cocycles of S^0. Let z be any element of $Z^1(S^0)$. We first note that

$$z(1, -1) = -z(-1, 1). \tag{16}$$

Indeed, since z is a 1-cocycle, one has

$$0 = (\delta^1 z)(-1, 1, -1) = z(1, -1) - z(-1, -1) + z(-1, 1).$$

The term $z(-1,-1)$ vanishes by **I.6.1**, remark 6, and (16) follows. We noted in **I.6.1** that we have at our disposal the isomorphism

$$h: Z^1(S^0) \xrightarrow{\approx} \mathrm{I},$$

defined by the formula $hz = z(-1, 1)$. By (14), (4), (16) and **I.6.2** it follows that

$$h\, T^1_* z = (T^1_* z)(-1, 1) = z(1, -1) = -z(-1, 1) = -hz.$$

Since h is an isomorphism, we conclude that $T^1_* z = -z$. Thus T^1_* is odd.

Proceeding by induction, we assume that for a certain integer $n \geqq 1$ we already know that T^n_* is odd, and we show that T^{n+1}_* is also odd by using the following diagram.

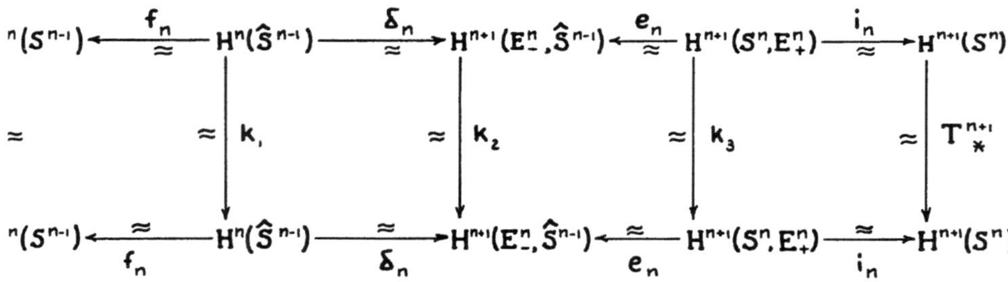

Fig. 41.

The isomorphisms f_n, δ_n, e_n, i_n, as well as the cohomology groups involved are taken from the diagram used at the end of section **I.7.1**. The isomorphism T^{n+1}_* corresponds to T^n_* in the dimension $n+1$, while T^n_* itself is defined in (15). The isomorphisms k_1, k_2, k_3 are induced by the homeomorphisms

$$T^{n+1}|\hat{S}^{n-1}: \hat{S}^{n-1} \to \hat{S}^{n-1},$$
$$T^{n+1}|E^n_-: (E^n_-, \hat{S}^{n-1}) \to (E^n_-, \hat{S}^{n-1}),$$
$$T^{n+1}: (S^n, E^n_+) \to (S^n, E^n_+)$$

respectively. By **I.6.2**, remark 2 it follows that k_1, k_2, k_3 are isomorphisms onto. Finally, the vector law holds in all four boxes of the diagram (by **I.6.6** in the second box from the left, and by **I.6.2** in the other three boxes). Now since T^n_* is odd by the induction assumption, it follows by proceeding from the left to the right in the preceding diagram that k_1, k_2, k_3, T^{n+1}_* are odd, and the inductive proof of the lemma is complete.

In preparation for our next lemma, we consider the mapping

$$\tilde{T}^n = \mathfrak{L}^-|E^n: (E^n, S^{n-1}) \to (E^n, S^{n-1}), \tag{17}$$

where E^n is the n-cell in R^n. Clearly, $\tilde{T}^n$ is a topological mapping from the pair (E^n, S^{n-1}) onto itself. Hence, by remark 2 in **I.6.2**, the induced homomorphism

$$\tilde{T}^n_*: H^{n+1}(E^n, S^{n-1}) \xrightarrow{\approx} H^{n+1}(E^n, S^{n-1}) \tag{18}$$

is an isomorphism onto, as indicated by the symbol $\approx$.

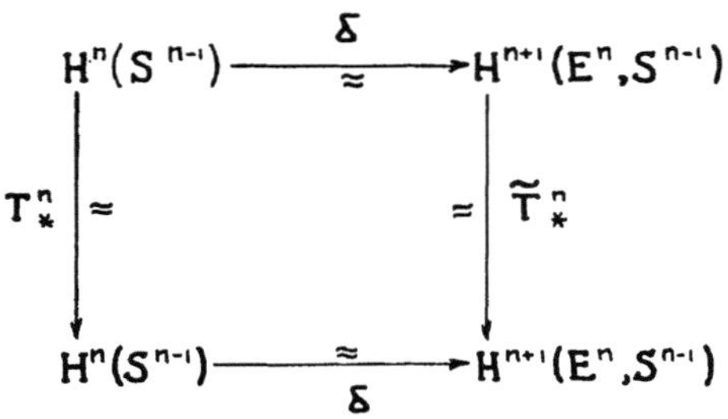

Fig. 42.

Lemma 10. The isomorphism $\tilde{T}^n_*$ is *odd* (see **I.3.2**).

Proof. Consider the diagram in Fig. 42.

The isomorphisms T^n_*, $\tilde{T}^n_*$ are those defined in (15) and (18) respectively. The homomorphism δ is taken from the cohomology sequence (see **I.6.4**) of the pair (E^n, S^{n-1}). In the four-term section

$$H^n(E^n), \quad H^n(S^{n-1}), \quad H^{n+1}(E^n, S^{n-1}), \quad H^{n+1}(E^n)$$

of that cohomology sequence the first and the fourth group are trivial by **I.7.1**, and hence we conclude by **I.3.3**, lemma 4 that the homomorphism δ from the second to the third group is an isomorphism onto. Observe finally that the vector law holds in the diagram by **I.6.6**. Now since T^n_* is odd by lemma 9, we conclude that $\tilde{T}^n_*$ is also odd, and the lemma is proved.

Lemma 11. $\mu(x, \mathfrak{L}^-, \operatorname{int} E^n) = -1$ for $x \in \operatorname{int} E^n$ [see (4) and **II.2.2**].

Proof. Note first that $[E^n, S^{n-1}, \operatorname{int} E^n]$ is a *frame* in the sense of **II.1.1**. We observed above that $\mathfrak{L}^- | E^n = \tilde{T}^n$ is a topological mapping from the pair (E^n, S^{n-1}) onto itself. Clearly, $\mathfrak{L}^- \operatorname{int} E^n = \operatorname{int} E^n$. Thus if $x \in \operatorname{int} E^n$, then x is $(\mathfrak{L}^-, \operatorname{int} E^n)$-admissible and the frame $[E^n, S^{n-1}, \operatorname{int} E^n]$ is $(x, \mathfrak{L}^-, \operatorname{int} E^n)$-admissible (see **II.2.1**). By the definition of the topological index (see **II.2.2**) we conclude that

$$\mu(x, \mathfrak{L}^-, \operatorname{int} E^n)\, \mathfrak{g}[E^n, S^{n-1}, \operatorname{int} E^n] = \tilde{T}^n_*\, \mathfrak{g}[E^n, S^{n-1}, \operatorname{int} E^n]. \tag{19}$$

Since $\tilde{T}^n_*$ is odd by lemma 10, we have

$$\tilde{T}^n_*\, \mathfrak{g}[E^n, S^{n-1}, \operatorname{int} E^n] = -\,\mathfrak{g}[E^n, S^{n-1}, \operatorname{int} E^n]. \tag{20}$$

As the $\mathfrak{g}$ occurring in (19) and (20) is a generator of the infinite cyclic group $H^{n+1}(E^n, S^{n-1})$, it follows from (19) and (20) that $\mu(x, \mathfrak{L}^-, \operatorname{int} E^n) = -1$, and the lemma is proved.

Lemma 12. $\iota(\mathfrak{L}^-) = -1$.

Proof. In view of definition 5 we have

$$\iota(\mathfrak{L}^-) = \iota(\mathfrak{L}^-, \operatorname{int} E^n). \tag{21}$$

Theorem 1 in **II.2.4** yields

$$\iota(\mathfrak{L}^-, \operatorname{int} E^n) = \mu(x, \mathfrak{L}^-, \operatorname{int} E^n) \quad \text{for} \quad x \in \operatorname{int} E^n. \tag{22}$$

Finally, by lemma 11,

$$\mu(x, \mathfrak{L}^-, \operatorname{int} E^n) = -1, \quad x \in \operatorname{int} E^n. \tag{23}$$

From (21) to (23) one sees that $\iota(\mathfrak{L}^-) = -1$.

Lemma 13. $\iota(\mathfrak{L}^+) = 1$.

Proof. If D is any bounded domain in R^n, then $\mathfrak{L}^+|\overline{D}$ is the identity mapping on $\overline{D}$. By definition 5 and **II.2.4**, theorem 2 we conclude that

$$\iota(\mathfrak{L}^+) = \iota(\mathfrak{L}^+, D) = 1.$$

Theorem 4. If $\mathfrak{L}$ is a non-singular linear transformation in R^n, then the following holds.

(i) $\iota(\mathfrak{L}) = \operatorname{sgn} \det \mathfrak{L}$.

(ii) If D is any bounded domain, then $\mu(x, \mathfrak{L}, D) = \operatorname{sgn} \det \mathfrak{L}$ for $x \in \mathfrak{L}D$ and $= 0$ for $x \in \mathrm{C}\mathfrak{L}\overline{D}$.

Proof. If $\det \mathfrak{L} > 0$, then $\iota(\mathfrak{L}) = \iota(\mathfrak{L}^+) = 1$ by theorem 3 and lemma 13. Similarly, if $\det \mathfrak{L} < 0$, then $\iota(\mathfrak{L}) = \iota(\mathfrak{L}^-) = -1$ by theorem 3 and lemma 12. Thus (i) is proved. Finally, (ii) follows from (i) by definition 5 and **II.2.4**, remark 3.

Remark 5. In view of the *definition* in **II.2.4** and definition 5 in **II.2.5**, part (i) of the preceding theorem may be re-stated as follows: a non-singular linear transformation $\mathfrak{L}$ is even or odd according as $\det \mathfrak{L} > 0$ or $\det \mathfrak{L} < 0$.

§ II.3. Multiplicity functions and index functions[1].

II.3.1. Preliminaries. We shall consider a continuous mapping

$$T: U \to R^n, \tag{1}$$

using the notations agreed upon in **II.2.1**. Let $\boldsymbol{D}$ be a bounded domain such that

$$\boldsymbol{D} \subset U. \tag{2}$$

[1] The concepts and results discussed in § **II.3** were first developed for the case $n = 2$ (for historical comments and a comprehensive discussion, see the treatise on *Length and Area* listed in the *Bibliography*). For the extension to a general $n \geq 2$ see H. FEDERER, Essential multiplicity and Lebesgue area [Proc. Nat. Acad. Sci. U.S.A. **34**, 611–616 (1948)], and T. RADO and P. V. REICHELDERFER, On n-dimensional concepts of bounded variation, absolute continuity and generalized Jacobian [Proc. Nat. Acad. Sci. U.S.A. **35**, 678–681 (1949)].

We also assume that the mapping

$$T|\boldsymbol{D}: \boldsymbol{D} \to R^n \tag{3}$$

is *bounded*; that is, there exists a (finite) positive constant λ such that

$$\|T u\| < \lambda \quad \text{for} \quad u \in \boldsymbol{D}. \tag{4}$$

Observe that we do *not* require now the inclusion $\overline{\boldsymbol{D}} \subset U$. As a consequence, T may not be defined on $fr\,\boldsymbol{D}$ (the special case when $U = \boldsymbol{D}$ illustrates the generality of the situation to be studied).

A point $x \in R^n$ may or may not lie in $T\boldsymbol{D}$. Our main objective at present is to assign a *multiplicity number* to each point $x \in R^n$ which should indicate, in a useful and convenient manner, the number of times that x is covered under the mapping (3). If $x \notin T\boldsymbol{D}$, then clearly the multiplicity number should be equal to zero. On the other hand, if $x \in T\boldsymbol{D}$, then various plausible approaches may be considered in defining a multiplicity number. The number of distinct points in the set $\boldsymbol{D} \cap T^{-1}x$ would be the most obvious choice, but the literature on this subject reveals that this choice is inadequate beyond the most elementary cases. Experience has shown that a detailed study of the components of the set $\boldsymbol{D} \cap T^{-1}x$ is necessary to gain useful information concerning multiplicity numbers. In the present section we discuss background material needed in the subsequent study of multiplicity numbers. It will be assumed throughout that T and $\boldsymbol{D}$ are given as in (1) to (4).

Lemma 1. Let G be an open set in R^n, O an open set in $\boldsymbol{D}$, and suppose that $O \cap T^{-1}G \neq \emptyset$. Let D be a component of $O \cap T^{-1}G$ such that $\overline{D} \subset O$. Then the following holds.

(i) D is a bounded domain.

(ii) $TD \subset G$.

(iii) $T\overline{D} \subset \overline{G}$.

(iv) $T\,fr\,D \subset fr\,G \subset \complement G$.

(v) D is a component of $\boldsymbol{D} \cap T^{-1}G$.

Proof. Since $D \subset O \subset \boldsymbol{D}$ and $\boldsymbol{D}$ is bounded, D is also bounded. Since G is open, the set $O \cap T^{-1}G$ is open relative to O and hence also relative to R^n, because O is open in R^n. As D is a component of an open set in R^n, it follows that D is connected and open (see **I.2.1**, exercise 4). Thus (i) is proved. The inclusion $D \subset T^{-1}G$ yields $TD \subset G$, and (ii) is proved. By **I.1.5**, exercise 9 we conclude (since $\overline{D} \subset \boldsymbol{D}$) that $T\overline{D} = \overline{TD} \subset \overline{G}$, and (iii) is proved. Since the components of the open set $O \cap T^{-1}G$ constitute a family of disjoint open sets, from **I.1.3**, exercise 33 we infer that

$$fr\,D \subset fr(O \cap T^{-1}G) \subset \complement(O \cap T^{-1}G) = \complement O \cup \complement T^{-1}G. \tag{5}$$

By assumption, we have $\overline{D} \subset O$, and hence

$$fr\, D \subset \overline{D} \subset O. \tag{6}$$

From (5) and (6) we conclude that $fr D \subset \complement T^{-1} G$, and hence

$$T\, fr\, D \subset \complement G. \tag{7}$$

By (6) and (iii) (which we already proved) we have

$$T\, fr\, D \subset T \overline{D} \subset \overline{G}. \tag{8}$$

From (7) and (8) we obtain (since G is open)

$$T\, fr\, D \subset \complement G \cap \overline{G} = fr\, G \subset \complement G,$$

and (iv) is proved. To verify (v), note that

$$D \subset O \cap T^{-1} G \subset \boldsymbol{D} \cap T^{-1} G.$$

Thus D is a connected subset of $\boldsymbol{D} \cap T^{-1} G$. Accordingly, there exists a component D_* of $\boldsymbol{D} \cap T^{-1} G$ such that $D \subset D_*$, and (v) will be proved if we can show that $D = D_*$. If this relation is denied, then D should be a proper subset of the connected set D_*, and hence (see **I.1.3**, exercise 17) one would have

$$D_* \cap fr\, D \neq \emptyset. \tag{9}$$

As $D_* \subset T^{-1} G$, in view of (iv) (already proved) one obtains

$$T(D_* \cap fr\, D) \subset T D_* \cap T\, fr\, D \subset G \cap \complement G = \emptyset,$$

and hence $D_* \cap fr D = \emptyset$, in contradiction with (9). Thus the assumption that $D \neq D_*$ leads to a contradiction, and the proof is complete.

Lemma 2. Let G be an open set in R^n such that $\boldsymbol{D} \cap T^{-1} G \neq \emptyset$, and let D be a component of $\boldsymbol{D} \cap T^{-1} G$ such that $\overline{D} \subset \boldsymbol{D}$. Then every point $x \in G$ is (T, D)-admissible, or equivalently, D is (x, T)-admissible for every point $x \in G$ (see **II.2.1**).

Proof. By lemma 1 (applied with $O = \boldsymbol{D}$), we have $T fr D \subset \complement G$. Hence, if $x \in G$, then $x \notin T fr D$, and the lemma is proved.

Lemma 3. Let D be a domain such that $\overline{D} \subset \boldsymbol{D}$ and let x be a point such that D is (x, T)-admissible (that is, $x \notin T fr D$). Finally, let G be an open set such that $x \in G \subset \complement T fr D$. Then those components of the set $\boldsymbol{D} \cap T^{-1} G$ that lie in D constitute a sequence of domains which is (x, T, D)-complete (see **II.2.1**, definition 5).

Proof. The lemma is trivial if $\overline{D} \cap T^{-1} x = \emptyset$ (see **II.2.1**, remark 2). So assume that $\overline{D} \cap T^{-1} x \neq \emptyset$. Then $\boldsymbol{D} \cap T^{-1} G \neq \emptyset$, since $\boldsymbol{D} \supset \overline{D}$, and hence

$$\boldsymbol{D} \cap T^{-1} G \supset \overline{D} \cap T^{-1} x \neq \emptyset.$$

The components of the open set $\boldsymbol{D} \cap T^{-1}G$ constitute a sequence $\{D_j\}$ of disjoint domains (see **I.1.3**, exercise 28). We assert that

$$D_j \cap fr\,D = \emptyset \quad \text{for every } j. \tag{10}$$

Indeed, since $D_j < T^{-1}G$ and $G < \complement T\,fr\,D$, we have

$$T(D_j \cap fr\,D) < T\,D_j \cap T\,fr\,D < G \cap T\,fr\,D = \emptyset,$$

and (10) follows. We next assert that

$$D_j \cap \overline{D} \neq \emptyset \quad \text{implies} \quad D_j < D. \tag{11}$$

Indeed, if $D_j \cap \overline{D} \neq \emptyset$, then (10) shows that $D_j \cap D \neq \emptyset$. If the relation $D_j < D$ is denied, then D_j should intersect both D and $\complement D$, and hence by **I.1.3**, exercise 17, one would have $D_j \cap fr\,D \neq \emptyset$, in contradiction with (10). Thus (11) holds. Turning to the proof of the lemma, we start with the relations

$$\overline{D} \cap T^{-1}x < \overline{D} \cap \boldsymbol{D} \cap T^{-1}G = \overline{D} \cap (\cup D_j) = \cup(\overline{D} \cap D_j), \tag{12}$$

where the used the relations $\overline{D} < \boldsymbol{D}$, $\boldsymbol{D} \cap T^{-1}G = \cup D_j$. Now in view of (11) a term $\overline{D} \cap D_j$ in (12) is either empty or else $D_j < D$ and then $\overline{D} \cap D_j = D_j$. Thus (12) reduces to

$$\overline{D} \cap T^{-1}x < \cup D_j, \quad D_j < D.$$

In other words, the union of those components D_j that lie in D contains the set $\overline{D} \cap T^{-1}x$, which means precisely that these components constitute an (x, T, D)-complete sequence.

Lemma 4. Let x be a point and let G, G^* be open sets in R^n such that $x \in G$, $\overline{G} < G^*$. Let D^* be a component of $\boldsymbol{D} \cap T^{-1}G^*$ such that $\overline{D^*} < \boldsymbol{D}$. Finally, let $\{D_k\}$ be the sequence (perhaps empty) of those components of $\boldsymbol{D} \cap T^{-1}G$ that lie in D^*. Then the following holds.

(i) D^* is (x, T)-admissible (see **II.2.1**).

(ii) $\overline{D}_k < D^*$ for every k.

(iii) The sequence $\{D_k\}$ is (x, T, D^*)-complete (see **II.2.1**, definition 5).

Proof. (i) is a direct consequence of lemma 2 (applied to G^*). As regards (ii), note that $D_k < D^*$ implies that $\overline{D}_k < \overline{D^*}$, and hence (ii) is proved if we can show that

$$\overline{D}_k \cap fr\,D^* = \emptyset. \tag{13}$$

To verify this, we start with the relation

$$T(\overline{D}_k \cap fr\,D^*) < T\overline{D}_k \cap T\,fr\,D^*. \tag{14}$$

By lemma 1 we have the inclusions $T\overline{D}_k \subset \overline{G}$, $T\,fr\,D^* \subset \complement G^*$. Hence (14) yields

$$T(\overline{D}_k \cap fr\,D^*) \subset \overline{G} \cap \complement G^* = \emptyset, \tag{15}$$

since $\overline{G} \subset G^*$ by assumption. Clearly (15) implies (13). Observe now that (by lemma 1)

$$T\,fr\,D^* \subset fr\,G^* \subset \complement G^*,$$

and hence

$$x \in G \subset G^* \subset \complement T\,fr\,D^*.$$

Thus x, G, D^* satisfy the assumptions of lemma 3, and hence (iii) follows directly from that lemma.

Remark 1. Take a point $x \in T\boldsymbol{D}$. Then $\boldsymbol{D} \cap T^{-1}x$ is a non-empty bounded set which is generally not closed (even though it is closed relative to $\boldsymbol{D}$ by **I.1.5**, exercise 4). Accordingly, if C is a component of $\boldsymbol{D} \cap T^{-1}x$, then C is a connected bounded set which may or may not be closed. If C happens to be closed, then it is also compact (see **I.2.1**, exercise 4) and hence it is a continuum (see **I.1.3**, definition 15) which may reduce to a single point.

Definition 1. A *maximal model continuum* (abbreviated to m.m.c.) for $(x, T, \boldsymbol{D})$ is a component of $\boldsymbol{D} \cap T^{-1}x$ which is a continuum.

Remark 2. If C is a component of $\boldsymbol{D} \cap T^{-1}x$, then C is an m.m.c. for $(x, T, \boldsymbol{D})$ if and only if $\overline{C} \subset \boldsymbol{D}$. Indeed, if C is a continuum then $\overline{C} = C \subset \boldsymbol{D}$. Conversely, suppose that $\overline{C} \subset \boldsymbol{D}$. Then (see **I.1.5**, exercise 9)

$$T\overline{C} = \overline{TC} \subset \overline{TT^{-1}x} = x,$$

and thus

$$C \subset \overline{C} \subset \boldsymbol{D} \cap T^{-1}x. \tag{16}$$

Now since $\overline{C}$ is connected (see **I.1.3**, exercise 16) and C is a component of $\boldsymbol{D} \cap T^{-1}x$, (16) implies (see **I.1.3**, exercise 11) that $C = \overline{C}$. Thus C is a continuum, and hence an m.m.c. for $(x, T, \boldsymbol{D})$.

Lemma 5. Given a point x, let D be a domain such that $\overline{D} \subset \boldsymbol{D}$ and D is (x, T)-admissible, and finally let Γ be a component of $\boldsymbol{D} \cap T^{-1}x$. Then either $\Gamma \subset D$ or $\Gamma \subset \complement\overline{D}$.

Proof. Since $T\Gamma = x \notin T\,fr\,D$, clearly

$$\Gamma \cap fr\,D = \emptyset. \tag{17}$$

Since Γ is connected, by **I.1.3**, exercise 17 we infer from (17) that Γ cannot intersect both D and $\complement D$, and hence either $\Gamma \subset D$ or else $\Gamma \subset \complement D$. There remains to show that

$$\Gamma \subset \complement D \quad \text{implies} \quad \Gamma \subset \complement\overline{D}. \tag{18}$$

Now since $\bar{D} = D \cup fr D$, we have

$$\complement \bar{D} = \complement D \cap \complement fr D. \tag{19}$$

By (17) we have $\Gamma < \complement fr D$. Hence, if $\Gamma < \complement D$, then [see (19)]

$$\Gamma < \complement D \cap \complement fr D = \complement \bar{D},$$

proving (18).

Definition 2. A *determining sequence* $\{D_j\}$ for $(x, T, \boldsymbol{D})$ is a sequence of (non-empty) domains D_j with the following properties.

(i) $\boldsymbol{D} > \bar{D}_1$, $D_j > \bar{D}_{j+1}$ for every j.

(ii) Each D_j is (x, T)-admissible (see **II.2.1**).

(iii) $\delta T \bar{D}_j \to 0$ for $j \to \infty$ (see **I.1.4**, definition 4).

(iv) $x \in T \bar{D}_j$ for every j.

Lemma 6. If $\{D_j\}$ is a determining sequence for $(x, T, \boldsymbol{D})$, then the set

$$C = \cap \bar{D}_j \tag{20}$$

is an m.m.c. for $(x, T, \boldsymbol{D})$, and $C < D_j$ for every j.

Proof. Since D_j is a bounded domain, by **I.1.3**, exercise 16 and **I.2.1**, exercise 4 it follows that $\bar{D}_j$ is a continuum. By (i) in definition 2, the sequence $\{\bar{D}_j\}$ is a nested sequence of continua, and hence the set C in (20) is a continuum by **I.1.3**, exercise 30. If $r > 0$ is arbitrarily assigned, then (iii) and (iv) in definition 2 imply that $T\bar{D}_j$ lies in the spherical neighborhood $\Delta_r(x)$ (see **II.2.1**) for j sufficiently large. In view of (20) it follows that $TC < \Delta_r(x)$ for every $r > 0$, and this clearly implies that $TC = x$. Hence

$$C < \boldsymbol{D} \cap T^{-1} x. \tag{21}$$

By (20) we have $C < \bar{D}_{j+1}$. Hence, by (i) in definition 2,

$$C < D_j \quad \text{for every } j. \tag{22}$$

There remains to show that C is a component of $\boldsymbol{D} \cap T^{-1} x$. Now since C is a connected subset of $\boldsymbol{D} \cap T^{-1} x$ [see (21)], by **I.1.3**, exercise 11 there exists a component Γ of $\boldsymbol{D} \cap T^{-1} x$ such that $C < \Gamma$, and the proof will be complete if we can show that $C = \Gamma$. If this relation is denied, then there should exist a point $u \in \Gamma - C$. In view of (20) there follows the existence of an integer j such that $u \in \complement \bar{D}_j$. Thus the component Γ of $\boldsymbol{D} \cap T^{-1} x$ would intersect both D_j [in view of (22)] and $\complement \bar{D}_j$. Since $\bar{D}_j < \boldsymbol{D}$ and D_j is (x, T)-admissible [see (ii) in definition 2), this conclusion contradicts lemma 5, and the proof is complete.

Remark 3. In view of lemma 6 there arises the question whether every m.m.c. C for $(x, T, \boldsymbol{D})$ can be represented in the form (20) in terms of a properly chosen determining sequence $\{D_j\}$ for $(x, T, \boldsymbol{D})$.

This is indeed the case, as we shall see presently. For the purpose of subsequent applications, we shall exhibit an especially convenient special determining sequence for each m.m.c. The following preliminary comments will facilitate the discussion. Let C be an m.m.c. for $(x, T, \boldsymbol{D})$. Consider the spherical neighborhood $\Delta_r(x)$ (see **II.2.1**, definition 4). Then $\boldsymbol{D} \cap T^{-1}\Delta_r(x)$ is an open set containing C, and hence this set has a unique component D_r containing C (since C is connected). Then D_r is a domain and

$$C \subset D_r \subset \boldsymbol{D}, \quad T D_r \subset \Delta_r(x). \tag{23}$$

If $0 < s < r$, then clearly

$$\boldsymbol{D} \cap T^{-1}\Delta_s(x) \subset \boldsymbol{D} \cap T^{-1}\Delta_r(x).$$

Thus D_s is a connected subset, containing C, of $\boldsymbol{D} \cap T^{-1}\Delta_r(x)$, and hence D_s is contained in D_r, the unique component of $\boldsymbol{D} \cap T^{-1}\Delta_r(x)$ containing C. So we have

$$C \subset D_s \subset D_r, \quad C \subset \overline{D}_s \subset \overline{D}_r \quad \text{for} \quad 0 < s < r. \tag{24}$$

Observe that the inclusion $\overline{D}_r \subset \boldsymbol{D}$ may not hold.

Lemma 7. Given an m.m.c. C for $(x, T, \boldsymbol{D})$ and an open set O such that $C \subset O$, $\overline{O} \subset \boldsymbol{D}$, there exists a number $\eta > 0$ such that

$$\overline{D}_r \subset O \quad \text{for} \quad 0 < r < \eta, \tag{25}$$

where D_r has the meaning explained in remark 3.

Proof. Put

$$O_r^* = O \cap T^{-1}\Delta_r(x).$$

Then O_r^* is an open set containing C, and since C is connected there exists a unique component D_r^* of O_r^* containing C. We have then

$$C \subset D_r^*, \quad \overline{D_r^*} \subset \overline{O} \subset \boldsymbol{D}. \tag{26}$$

Clearly

$$C \subset D_s^* \subset D_r^*, \quad C \subset \overline{D_s^*} \subset \overline{D_r^*} \quad \text{for} \quad 0 < s < r. \tag{27}$$

Furthermore, since $D_r^* \subset T^{-1}\Delta_r(x)$, it follows (see **I.1.5**, exercise 9) that

$$T\overline{D_r^*} = \overline{T D_r^*} \subset \overline{\Delta_r(x)}. \tag{28}$$

Consider now any sequence $r_1 > \cdots > r_j > \cdots > 0$, $r_j \to 0$. In view of (27), the sets $\overline{D_{r_j}^*}$ constitute a nested sequence of continua containing C. Hence (see **I.1.3**, exercise 30) the set

$$C^* = \cap\, \overline{D_{r_j}^*} \tag{29}$$

is a continuum containing C. From (28) we infer that

$$T C^* \subset \overline{T D_{r_j}^*} \subset \overline{\Delta_{r_j}(x)}.$$

Since $r_j \to 0$, it follows that $TC^* = x$, and hence $C^* \subset \boldsymbol{D} \cap T^{-1}x$. Now C^*, which is connected, contains the component C of $\boldsymbol{D} \cap T^{-1}x$. By **I.1.3**, exercise 11, we conclude that $C = C^*$, and hence (29) yields

$$C = \bigcap \overline{D^*_{r_j}}. \tag{30}$$

As $C \subset O$ and O is open, (30) implies (see **I.1.3**, exercise 24) that

$$\overline{D^*_{r_j}} \subset O \quad \text{for } j \text{ large}. \tag{31}$$

Choose an integer j such that (31) holds, and set

$$\eta = r_j. \tag{32}$$

For $0 < r < \eta$ we have then, in view of (27) and (31),

$$\overline{D^*_r} \subset \overline{D^*_\eta} \subset O. \tag{33}$$

By lemma 1 [applied with $G = \Delta_r(x)$] it follows from (33) that D^*_r is also a component of $\boldsymbol{D} \cap T^{-1}\Delta_r(x)$, and hence D^*_r must coincide with D_r, the unique component of $\boldsymbol{D} \cap T^{-1}\Delta_r(x)$ containing C [observe that D^*_r contains C by (26)]. Thus (33) yields $\overline{D}_r \subset O$ for $0 < r < \eta$, and the proof is complete.

Lemma 8. Let C be an m.m.c. for $(x, T, \boldsymbol{D})$, and let O be an open set such that $C \subset O$, $\overline{O} \subset \boldsymbol{D}$. Then there exists a sequence

$$r_1 > \cdots > r_j > \cdots > 0, \quad r_j \to 0, \tag{34}$$

such that the following holds.

(i) On denoting by D_{r_j} the (unique) component of $\boldsymbol{D} \cap T^{-1}\Delta_{r_j}(x)$ that contains C (see remark 3), the sequence $\{D_{r_j}\}$ is a determining sequence for $(x, T, \boldsymbol{D})$.

(ii) $C = \bigcap \overline{D}_{r_j}$.

(iii) $\overline{D}_{r_j} \subset O$ for every j.

Proof. By lemma 7, we have a number $\eta > 0$ such that (25) holds. Choose any sequence $\{r_j\}$ satisfying (34) and also the additional condition $r_1 < \eta$. Then

$$r_j < \eta \quad \text{for every } j. \tag{35}$$

Since clearly

$$\overline{\Delta_{r_{j+1}}(x)} \subset \Delta_{r_j}(x),$$

we infer from lemma 4, part (ii) that

$$\overline{D}_{r_{j+1}} \subset D_{r_j}. \tag{36}$$

From part (i) of the same lemma it follows that D_{r_j} is (x, T)-admissible. By lemma 1 we have

$$T\overline{D}_{r_j} \subset \overline{\Delta_{r_j}(x)},$$

and hence $\delta T \overline{D}_{r_j} \to 0$ by (34). Since $C \subset D_{r_j}$, clearly

$$x = TC \subset TD_{r_j} \subset T\overline{D}_{r_j}.$$

This completes the proof of part (i) of the lemma. By lemma 6 it follows that the set $C^* = \bigcap \overline{D}_{r_j}$ is an m.m.c. for $(x, T, \boldsymbol{D})$. Now since $C \subset D_{r_j}$ for every j, we have the inclusion $C \subset C^*$. As C and C^* are components of the same set $\boldsymbol{D} \cap T^{-1} x$, it follows that $C = C^* = \bigcap \overline{D}_{r_j}$, and part (ii) of the lemma is proved. Finally, part (iii) follows directly from (35) and (25).

Lemma 9. Let u_0 be a point in $\boldsymbol{D}$ and $\{D_m\}$ a sequence of domains with the following properties.

(i) $u_0 \in D_m, \overline{D}_m \subset \boldsymbol{D}$ for every m.

(ii) Each D_m is $(T u_0, T)$-admissible (see **II.2.1**).

(iii) $\delta T \overline{D}_m < 1/m$ for every m.

Then the sequence $\{D_m\}$ contains a subsequence $\{D_{m_j}\}$, $j = 1, 2, \ldots$, which is a determining sequence for $(T u_0, T, \boldsymbol{D})$.

Proof. If $\{D_{m_j}\}$ is any subsequence, then $T u_0 \in T \overline{D}_{m_j}$ by (i) and $\delta T \overline{D}_{m_j} \to 0$ by (iii). Accordingly (see definition 2) we have to show only that the subsequence can be made to satisfy the condition

$$D_{m_j} \supset \overline{D}_{m_{j+1}}. \tag{37}$$

To see this, take any positive integer m. We assert that

$$D_m \supset \overline{D}_M \quad \text{for } M \text{ large}. \tag{38}$$

Indeed, take any $M > m$. Then $u_0 \in D_m \cap \overline{D}_M$ and hence $D_m \cap \overline{D}_M \neq \emptyset$. Since $\overline{D}_M$ is connected, it follows [see **I.1.3**, exercise (17)] that either (38) holds or else

$$\overline{D}_M \cap \mathit{fr}\, D_m \neq \emptyset. \tag{39}$$

In view of (ii) we have $T u_0 \notin T\, \mathit{fr} D_m$, and hence (see **I.1.4**, exercise 3)

$$e(T u_0, T\, \mathit{fr}\, D_m) > 0.$$

If (39) holds, then we have a point $u_m \in \overline{D}_M \cap \mathit{fr} D_m$. Then $T u_m \in T\, \mathit{fr} D_m$ and hence

$$\|T u_0 - T u_m\| \geq e(T u_0, T\, \mathit{fr}\, D_m).$$

Also, $T u_m \in T \overline{D}_M$, and hence $\|T u_0 - T u_m\| < 1/M$, since $T u_0 \in T \overline{D}_M$ and $\delta T \overline{D}_M < 1/M$. It follows that (39) implies the inequality

$$e(T u_0, T\, \mathit{fr}\, D_m) < \frac{1}{M}.$$

Accordingly, (38) must hold if M is large enough to satisfy the inequality

$$\frac{1}{M} \leq e(T u_0, T\, \mathit{fr}\, D_m).$$

The existence of a subsequence $\{D_{m_j}\}$ satisfying (37) is now immediate. Indeed, on choosing $D_{m_1}=D_1$, (38) yields the existence of an integer $m_2>m_1$ such that $D_{m_1}\supset \bar{D}_{m_2}$, and it is obvious that successive application of (38) yields a sequence of subscripts $m_1<m_2<m_3<\cdots$ such that (37) holds.

II.3.2. The multiplicity functions $\boldsymbol{K}, \boldsymbol{K}^+, \boldsymbol{K}^-$. Given T and $\boldsymbol{D}$ as in **II.3.1** and a point $x\in R^n$, an *indicator domain* D for $(x, T, \boldsymbol{D})$ is defined by the following conditions. (i) $\bar{D}\subset \boldsymbol{D}$. (ii) D is (x, T)-admissible (that is, $x\notin T\,fr\,D$). (iii) $\mu(x, T, D)\neq 0$ (see **II.2.2**). If $\mu(x, T, D)>0\,(<0)$, then D is termed a *positive (negative) indicator domain* for $(x, T, \boldsymbol{D})$. A finite (perhaps empty) collection of pair-wise disjoint indicator domains for $(x, T, \boldsymbol{D})$ will be called an *indicator system* for $(x, T, \boldsymbol{D})$, and will be denoted generically by $\mathfrak{S}(x, T, \boldsymbol{D})$. If all the domains comprised in an indicator system $\mathfrak{S}(x, T, \boldsymbol{D})$ are positive (negative) indicator domains, then $\mathfrak{S}$ will be termed a positive (negative) indicator system, and the notation $\mathfrak{S}^+(x, T, \boldsymbol{D})$ $(\mathfrak{S}^-(x, T, \boldsymbol{D}))$ will be used to indicate this fact. The symbol $M[\mathfrak{S}(x, T, \boldsymbol{D})]$ is defined by the formula

$$M[\mathfrak{S}(x, T, \boldsymbol{D})]=\sum|\mu(x, T, D)|, \quad D\in\mathfrak{S}(x, T, \boldsymbol{D}), \tag{1}$$

with the understanding that

$$M[\mathfrak{S}(x, T, \boldsymbol{D})]=0 \quad \text{if} \quad \mathfrak{S}(x, T, \boldsymbol{D}) \quad \text{is empty}. \tag{2}$$

The symbol $N[\mathfrak{S}(x, T, \boldsymbol{D})]$ is defined by

$$N[\mathfrak{S}(x, T, \boldsymbol{D})]=\text{number of domains } D\in\mathfrak{S}(x, T, \boldsymbol{D}), \tag{3}$$

with the understanding that

$$N[\mathfrak{S}(x, T, \boldsymbol{D})]=0 \quad \text{if} \quad \mathfrak{S}(x, T, \boldsymbol{D}) \quad \text{is empty}. \tag{4}$$

Since every system $\mathfrak{S}^+(x, T, \boldsymbol{D})$ is also a system $\mathfrak{S}(x, T, \boldsymbol{D})$, the preceding definitions yield

$$M[\mathfrak{S}^+(x, T, \boldsymbol{D})]=\sum\mu(x, T, D), \quad D\in\mathfrak{S}^+(x, T, \boldsymbol{D}), \tag{5}$$

and similarly

$$M[\mathfrak{S}^-(x, T, \boldsymbol{D})]=-\sum\mu(x, T, D), \quad D\in\mathfrak{S}^-(x, T, \boldsymbol{D}), \tag{6}$$

with the understanding that $M=0$ if the system $\mathfrak{S}^+(\mathfrak{S}^-)$ involved is empty.

If D is an indicator domain for $(x, T, \boldsymbol{D})$, then $\mu(x, T, D)$ is an integer different from zero, and thus clearly

$$M[\mathfrak{S}(x, T, \boldsymbol{D})]\geq N[\mathfrak{S}(x, T, \boldsymbol{D})], \tag{7}$$

$$M[\mathfrak{S}^+(x, T, \boldsymbol{D})]\geq N[\mathfrak{S}^+(x, T, \boldsymbol{D})], \tag{8}$$

$$M[\mathfrak{S}^-(x, T, \boldsymbol{D})]\geq N[\mathfrak{S}^-(x, T, \boldsymbol{D})]. \tag{9}$$

Of course, for empty indicator systems these inequalities are obvious, since then $M=N=0$.

Remark 1. Let E be a finite, non-empty set of real numbers. Then l.u.b. E is an element of E. Indeed, E contains a largest element R, and (see **I.1.2**) we have l.u.b. $E=R$.

Consider now the set of all the numbers $M[\mathfrak{S}(x, T, \boldsymbol{D})]$, as defined by (1) and (2), corresponding to all possible systems $\mathfrak{S}(x, T, \boldsymbol{D})$. This set is non-empty. Indeed, we have at least one system $\mathfrak{S}(x, T, \boldsymbol{D})$, namely the empty system, the corresponding M being equal to zero by (2). We define

$$K(x, T, \boldsymbol{D}) = \underset{\mathfrak{S}}{\text{l.u.b.}}\, M[\mathfrak{S}(x, T, \boldsymbol{D})], \tag{10}$$

and similarly

$$K^{+}(x, T, \boldsymbol{D}) = \underset{\mathfrak{S}^{+}}{\text{l.u.b.}}\, M[\mathfrak{S}^{+}(x, T, \boldsymbol{D})], \tag{11}$$

$$K^{-}(x, T, \boldsymbol{D}) = \underset{\mathfrak{S}^{-}}{\text{l.u.b.}}\, M[\mathfrak{S}^{-}(x, T, \boldsymbol{D})]. \tag{12}$$

The letter $\mathfrak{S}$ under the symbol l.u.b. in (10) is meant to indicate that the least upper bound is taken for the set of all the numbers M corresponding to all possible systems $\mathfrak{S}(x, T, \boldsymbol{D})$. The formulas (11), (12) are to be interpreted in a similar manner in terms of $\mathfrak{S}^{+}$, $\mathfrak{S}^{-}$ respectively.

Remark 2. Observing that $M>0$ for a non-empty system $\mathfrak{S}(\mathfrak{S}^{+}, \mathfrak{S}^{-})$, one sees that $K(x, T, \boldsymbol{D})=0$ if and only if there exists no indicator domain for $(x, T, \boldsymbol{D})$, with similar statements holding for $K^{+}(x, T, \boldsymbol{D})$, $K^{-}(x, T, \boldsymbol{D})$.

Consider now a point x such that $K(x, T, \boldsymbol{D})>0$. Then there exists some indicator domain D for $(x, T, \boldsymbol{D})$. Since $\mu(x, T, D)\neq 0$ for such a domain D, by **II.2.3**, theorem 2 we conclude that $x\in TD\subset T\boldsymbol{D}$. Hence: if $x\notin T\boldsymbol{D}$, then $K(x, T, \boldsymbol{D})=0$, and thus obviously also $K^{+}(x, T, \boldsymbol{D})=0$, $K^{-}(x, T, \boldsymbol{D})=0$.

Remark 3. If $K(x, T, \boldsymbol{D})<\infty$ for a point x, then the numbers M corresponding to all possible systems $\mathfrak{S}(x, T, \boldsymbol{D})$ satisfy the inequalities $0\leqq M\leqq K(x, T, \boldsymbol{D})<\infty$. Since every M is an integer, it follows that the set of these numbers M is now finite, and hence from remark 1 we conclude that there exists a system $\mathfrak{S}(x, T, \boldsymbol{D})$ such that

$$K(x, T, \boldsymbol{D}) = M[\mathfrak{S}(x, T, \boldsymbol{D})].$$

In a similar manner one sees that if $K^{+}(x, T, \boldsymbol{D})<\infty$, then there exists a system $\mathfrak{S}^{+}(x, T, \boldsymbol{D})$ such that

$$K^{+}(x, T, \boldsymbol{D}) = M[\mathfrak{S}^{+}(x, T, \boldsymbol{D})].$$

An analogous statement holds if $K^{-}(x, T, \boldsymbol{D})<\infty$.

Remark 4. If $K(x, T, \boldsymbol{D}) < \infty$, then from (7) and (10) we conclude that

$$N[\mathfrak{S}(x, T, \boldsymbol{D})] \leq K(x, T, \boldsymbol{D}) < \infty$$

for every system $\mathfrak{S}(x, T, \boldsymbol{D})$. Thus in this case the number of domains in any indicator system $\mathfrak{S}(x, T, \boldsymbol{D})$ remains under a fixed finite bound. Similar statements hold if $K^+(x, T, \boldsymbol{D}) < \infty$ or $K^-(x, T, \boldsymbol{D}) < \infty$, for systems $\mathfrak{S}^+$, $\mathfrak{S}^-$ respectively.

Remark 5. Fix a point $x \in R^n$, and consider a system $\mathfrak{S}^+(x, T, \boldsymbol{D})$. Then this system is also a system $\mathfrak{S}(x, T, \boldsymbol{D})$, and hence in view of (10)

$$M[\mathfrak{S}^+(x, T, \boldsymbol{D})] \leq K(x, T, \boldsymbol{D}).$$

By (11) it follows that

$$K^+(x, T, \boldsymbol{D}) \leq K(x, T, \boldsymbol{D}). \tag{13}$$

Similarly,

$$K^-(x, T, \boldsymbol{D}) \leq K(x, T, \boldsymbol{D}). \tag{14}$$

Consider now any system $\mathfrak{S}(x, T, \boldsymbol{D})$. The positive and the negative indicator domains contained in $\mathfrak{S}(x, T, \boldsymbol{D})$ constitute then systems $\mathfrak{S}^+(x, T, \boldsymbol{D})$, $\mathfrak{S}^-(x, T, \boldsymbol{D})$ respectively, one or both of which may be empty. Clearly, for these three systems we have

$$M[\mathfrak{S}(x, T, \boldsymbol{D})] = M[\mathfrak{S}^+(x, T, \boldsymbol{D})] + M[\mathfrak{S}^-(x, T, \boldsymbol{D})],$$

and hence [in view of (11) and (12)]

$$M[\mathfrak{S}(x, T, \boldsymbol{D})] \leq K^+(x, T, \boldsymbol{D}) + K^-(x, T, \boldsymbol{D}).$$

By (10) we conclude that

$$K(x, T, \boldsymbol{D}) \leq K^+(x, T, \boldsymbol{D}) + K^-(x, T, \boldsymbol{D}). \tag{15}$$

Remark 6. Take a point $x \in R^n$ such that

$$K(x, T, \boldsymbol{D}) < \infty. \tag{16}$$

Consider the open spherical neighborhood $\Delta_r(x)$ (see **I.1.4**, definition 2). Since $\Delta_r(x)$ is open, the set

$$O_r = \boldsymbol{D} \cap T^{-1} \Delta_r(x) \tag{17}$$

is also open. Let Ω_r be the class (perhaps empty) of those components of O_r which are indicator domains for $(x, T, \boldsymbol{D})$. If $D^1, \ldots, D^m$ is any finite set of elements of Ω_r, then these elements constitute an indicator system $\mathfrak{S}(x, T, \boldsymbol{D})$, and hence $m \leq K(x, T, \boldsymbol{D}) < \infty$ by remark 4. It follows that the class Ω_r is finite and contains at most $K(x, T, \boldsymbol{D})$ elements. Accordingly, the elements of Ω_r form an indicator system for $(x, T, \boldsymbol{D})$ which will be denoted by $\mathfrak{S}_r(x, T, \boldsymbol{D})$. The positive and the negative indicator domains contained in $\mathfrak{S}_r(x, T, \boldsymbol{D})$ form systems to be denoted by $\mathfrak{S}_r^+(x, T, \boldsymbol{D})$, $\mathfrak{S}_r^-(x, T, \boldsymbol{D})$ respectively. Clearly, for any $r > 0$,

$$M[\mathfrak{S}_r(x, T, \boldsymbol{D})] = M[\mathfrak{S}_r^+(x, T, \boldsymbol{D})] + M[\mathfrak{S}_r^-(x, T, \boldsymbol{D})]. \tag{18}$$

Since $\mathfrak{S}_r(x, T, \boldsymbol{D})$ is merely a special indicator system for $(x, T, \boldsymbol{D})$, we have

$$M[\mathfrak{S}_r(x, T, \boldsymbol{D})] \leq K(x, T, \boldsymbol{D}), \tag{19}$$

and similarly

$$M[\mathfrak{S}_r^+(x, T, \boldsymbol{D})] \leq K^+(x, T, \boldsymbol{D}), \tag{20}$$

$$M[\mathfrak{S}_r^-(x, T, \boldsymbol{D})] \leq K^-(x, T, \boldsymbol{D}). \tag{21}$$

Some or all of the systems $\mathfrak{S}_r(x, T, \boldsymbol{D})$, $\mathfrak{S}_r^+(x, T, \boldsymbol{D})$, $\mathfrak{S}_r^-(x, T, \boldsymbol{D})$ may be empty.

Lemma 1. Let x be a point such that $K(x, T, \boldsymbol{D}) < \infty$. Then there exists a number $\eta = \eta(x, T, \boldsymbol{D})$ such that (see remark 6)

$$K(x, T, \boldsymbol{D}) = M[\mathfrak{S}_r(x, T, \boldsymbol{D})], \tag{22}$$

$$K^+(x, T, \boldsymbol{D}) = M[\mathfrak{S}_r^+(x, T, \boldsymbol{D})], \tag{23}$$

$$K^-(x, T, \boldsymbol{D}) = M[\mathfrak{S}_r^-(x, T, \boldsymbol{D})], \tag{24}$$

for $0 < r < \eta$.

Proof. Since the argument is entirely analogous for each one of (22), (23), (24), we indicate merely the proof of (23) in the way of illustration. The assumption $K(x, T, \boldsymbol{D}) < \infty$ implies, by (13), that

$$K^+(x, T, \boldsymbol{D}) < \infty. \tag{25}$$

If $K^+(x, T, \boldsymbol{D}) = 0$, then by (20) it is clear that $M[\mathfrak{S}_r^+(x, T, \boldsymbol{D})] = 0$ for every $r > 0$, and thus (23) is obvious. So assume that

$$K^+(x, T, \boldsymbol{D}) > 0. \tag{26}$$

In view of remark 3, the relations (25), (26) imply the existence of a system $\mathfrak{S}^+(x, T, \boldsymbol{D})$ such that

$$M[\mathfrak{S}^+(x, T, \boldsymbol{D})] = K^+(x, T, \boldsymbol{D}). \tag{27}$$

Let $\{D_k\}$, $k = 1, \ldots, m$ be the (pair-wise disjoint) positive indicator domains for $(x, T, \boldsymbol{D})$ which constitute $\mathfrak{S}^+(x, T, \boldsymbol{D})$. In view of (5), the relation (27) may then be re-written in the form

$$K^+(x, T, \boldsymbol{D}) = \sum_k \mu(x, T, D_k). \tag{28}$$

Introduce now the auxiliary set

$$F = T\, fr\, D_1 \cup \cdots \cup T\, fr\, D_m. \tag{29}$$

Then F is compact and $x \notin F$, since the domains D_k are (x, T)-admissible. Accordingly (see **I.1.4**, exercise 3)

$$e(x, F) > 0. \tag{30}$$

Choose now any real number r such that

$$0 < r < e(x, F). \tag{31}$$

We proceed to verify that (23) holds if r is so chosen. Observe first that if (31) is satisfied, then

$$x \in \Delta_r(x) \subset \complement\, T\, fr\, D_k, \quad k = 1, \ldots, m.$$

Accordingly, we can apply **II.3.1**, lemma 3 with $G = \Delta_r(x)$, $D = D_k$. On denoting by Ω_r^k the class of those components of $\boldsymbol{D} \cap T^{-1} \Delta_r(x)$ that lie in D_k, by that lemma the domains of Ω_r^k constitute an (x, T, D_k)-complete system. By **II.2.3**, theorem 3 we conclude that

$$\mu(x, T, D_k) = \sum \mu(x, T, d^k), \quad d^k \in \Omega_r^k.$$

Summation with respect to k yields, in view of (28),

$$K^+(x, T, \boldsymbol{D}) = \sum_k \left(\sum \mu(x, T, d^k)\right), \quad d^k \in \Omega_r^k. \tag{32}$$

Let us use the symbol $\sum^+$ to indicate summation restricted to only those of the terms involved that are positive. Then (32) yields

$$K^+(x, T, \boldsymbol{D}) \leq \sum_k \left(\sum{}^+ \mu(x, T, d^k)\right), \quad d^k \in \Omega_r^k. \tag{33}$$

Observe now that those domains $d^k \in \Omega_r^k$, $k = 1, \ldots, m$, for which $\mu(x, T, d^k) > 0$ are included amongst the domains that constitute the system $\mathfrak{S}_r^+(x, T, \boldsymbol{D})$ (see remark 6). Hence the double summation in (33) cannot exceed $M[\mathfrak{S}_r^+(x, T, \boldsymbol{D})]$, and we conclude that

$$K^+(x, T, \boldsymbol{D}) \leq M[\mathfrak{S}_r^+(x, T, \boldsymbol{D})].$$

Jointly with (20), this inequality yields (23), under the assumption (31). Thus (23) is shown to hold for $\eta = e(x, F)$. The proof of (22) and (24) (with a similar choice for η) is made in an entirely analogous manner.

Remark 7. Actually, in establishing (23), we used only the condition (25). This remark yields the following statement: if x is a point such that $K^+(x, T, \boldsymbol{D}) < \infty$, then (23) holds if r is sufficiently small. Similarly it follows that if $K^-(x, T, \boldsymbol{D}) < \infty$, then (24) holds for r sufficiently small.

Theorem 1. $K(x, T, \boldsymbol{D}) = K^+(x, T, \boldsymbol{D}) + K^-(x, T, \boldsymbol{D})$.

Proof. Assume first that $K(x, T, \boldsymbol{D}) = \infty$. By (15) we conclude that one at least of $K^+(x, T, \boldsymbol{D})$, $K^-(x, T, \boldsymbol{D})$ must be equal to ∞, and thus the theorem is obvious in this case. Assume next that $K(x, T, \boldsymbol{D}) < \infty$. Using lemma 1, choose $r > 0$ so that (22), (23), (24) hold. The theorem follows then directly from (18).

In defining the multiplicity function $K(x, T, \boldsymbol{D})$, we assumed that T was defined and continuous on some set $U \supset \boldsymbol{D}$. Accordingly, if D is any domain in $\boldsymbol{D}$, then the definition of K applies and we obtain the multiplicity function $K(x, T, D)$. Thus, for fixed x and T, $K(x, T, D)$ may be thought of as a function of the variable domain $D \subset \boldsymbol{D}$. Similar comments apply to $K^+(x, T, D)$, $K^-(x, T, D)$.

Theorem 2. Given T and $\boldsymbol{D}$ as in **II.3.1**, let D be a domain in $\boldsymbol{D}$, and let $\{D_k\}$ be a (finite or infinite) sequence of pair-wise disjoint domains in D. Then

$$\sum_k K(x, T, D_k) \leq K(x, T, D), \tag{34}$$

$$\sum_k K^+(x, T, D_k) \leq K^+(x, T, D), \tag{35}$$

$$\sum_k K^-(x, T, D_k) \leq K^-(x, T, D). \tag{36}$$

Proof. The argument being entirely analogous for K, K^+, K^-, we indicate the proof merely for (35) to illustrate the method. Noting that (35) is trivial if $K^+(x, T, D) = \infty$, we can assume that

$$K^+(x, T, D) < \infty \tag{37}$$

for the point x under consideration. Observe next that if every partial sum of a series of non-negative terms is less than or equal to a fixed finite number, then the series is convergent and its sum is less than or equal to the same number. Accordingly it is sufficient to prove (35) for the case of a finite sequence $D_1, \ldots, D_m$ of pair-wise disjoint domains in D. Let then $\{D_{k_j}\}$, $j = 1, \ldots, m_k$ be a finite sequence of pair-wise disjoint domains in D_k such that each D_{k_j} is (x, T)-admissible and

$$\overline{D}_{k_j} \subset D_k, \quad \mu(x, T, D_{k_j}) > 0, \quad j = 1, \ldots, m_k.$$

Then $\{D_{k_j}\}$ is a system $\mathfrak{S}^+(x, T, D_k)$ and also a system $\mathfrak{S}^+(x, T, D)$. Accordingly [see (11)]

$$M[\mathfrak{S}^+(x, T, D_k)] \leq K^+(x, T, D) < \infty.$$

Since this holds for every system $\mathfrak{S}^+(x, T, D_k)$, by (11) (applied to D_k) it follows that

$$K^+(x, T, D_k) \leq K^+(x, T, D) < \infty.$$

Thus $K^+(x, T, D_k)$ is finite. By remark 3 (applied to D_k) we have therefore a system $\mathfrak{S}^+(x, T, D_k)$ such that

$$M[\mathfrak{S}^+(x, T, D_k)] = K^+(x, T, D_k). \tag{38}$$

Choose such a system $\mathfrak{S}^+(x, T, D_k)$ for each $k = 1, \ldots, m$. Since $D_1, \ldots, D_m$ are pair-wise disjoint domains in D, clearly the totality of the domains constituting the systems $\mathfrak{S}^+(x, T, D_k)$, $k = 1, \ldots, m$, is a system $\mathfrak{S}^+(x, T, D)$ such that

$$M[\mathfrak{S}^+(x, T, D)] = \sum_k M[\mathfrak{S}^+(x, T, D_k)]. \tag{39}$$

By (11), applied to D, we have

$$M[\mathfrak{S}^+(x, T, D)] \leq K^+(x, T, D). \tag{40}$$

From (38), (39), (40) we conclude that

$$\sum_k K^+(x, T, D_k) \leq K^+(x, T, D),$$

and the proof of (35) is complete. (34) and (36) are proved in a similar manner.

Remark 8. For fixed T and $\boldsymbol{D}$, given as in **II.3.1**, the multiplicity functions $K(x, T, \boldsymbol{D})$, $K^+(x, T, \boldsymbol{D})$, $K^-(x, T, \boldsymbol{D})$ are functions of the point $x \in R^n$. In our next theorem, we shall show that these multiplicity functions are lower semi-continuous with respect to x (see **I.1.3**, definition 28). In the sequel we shall use the following fact. Suppose that $f(x)$ is a function in R^n which assumes only values that are integers, except that at certain points $f(x)$ may have the value $+\infty$. If $f(x)$ is lower semi-continuous at a point x_0 where $f(x_0) < \infty$, then there exists an open set O containing x_0 such that $f(x) \geq f(x_0)$ for $x \in O$. This is an immediate consequence of the definition of lower semicontinuity at a point.

Theorem 3. The multiplicity functions $K(x, T, \boldsymbol{D})$, $K^+(x, T, \boldsymbol{D})$, $K^-(x, T, \boldsymbol{D})$ are lower semi-continuous functions of x (for fixed T, $\boldsymbol{D}$).

Proof. Again, the proof is entirely similar in all three cases, and this time we indicate the method of proof for the case of $K(x, T, \boldsymbol{D})$. Take a point x_0. If $K(x_0, T, \boldsymbol{D}) = 0$, then K is clearly lower semi-continuous at x_0, since always $K \geq 0$. So we can assume that $K(x_0, T, \boldsymbol{D}) > 0$. Let then any positive real number $\lambda < K(x_0, T, \boldsymbol{D})$ be assigned. In view of (10) there exists then a system $\mathfrak{S}(x_0, T, \boldsymbol{D})$ such that

$$M[\mathfrak{S}(x_0, T, \boldsymbol{D})] > \lambda. \tag{41}$$

Let $D_1, \ldots, D_m$ be the indicator domains for $(x_0, T, \boldsymbol{D})$ that constitute $\mathfrak{S}(x_0, T, \boldsymbol{D})$. Then (41) means that

$$\sum_{k=1}^{m} |\mu(x_0, T, D_k)| > \lambda. \tag{42}$$

By the definition of an indicator domain, $x_0 \in \complement T\, fr\, D_k$. Since $\complement T\, fr\, D_k$ is open, the set

$$O = \bigcap_{k=1}^{m} \complement T\, fr\, D_k$$

is an open set containing x_0. Hence O has a (unique) component $\varDelta$ containing x_0, and by **I.2.1**, exercise 4, $\varDelta$ is a domain. The lower semi-continuity of $K(x, T, \boldsymbol{D})$ at x_0 will be established if we can show that

$$K(x, T, \boldsymbol{D}) > \lambda \quad \text{for} \quad x \in \varDelta. \tag{43}$$

Now $\varDelta$ is a connected subset of $\complement T \mathit{fr} D_k$, $k = 1, \ldots, m$, and hence (by **II.2.3**, remark 1)

$$\mu(x, T, D_k) = \mu(x_0, T, D_k) \neq 0 \quad \text{for} \quad x \in \varDelta, \tag{44}$$

since D_k is an indicator domain for $(x_0, T, \boldsymbol{D})$. From (44) we see that D_k is also an indicator domain for $(x, T, \boldsymbol{D})$ if $x \in \varDelta$, and hence the system $\mathfrak{S}(x_0, T, \boldsymbol{D})$ is also a system $\mathfrak{S}(x, T, \boldsymbol{D})$ if $x \in \varDelta$. From (41), (42), (44) we infer that

$$\lambda < M[\mathfrak{S}(x, T, \boldsymbol{D})] \quad \text{if} \quad x \in \varDelta. \tag{45}$$

Since by (10)

$$M[\mathfrak{S}(x, T, \boldsymbol{D})] \leq K(x, T, \boldsymbol{D}),$$

(45) implies (43), and the proof is complete.

Remark 9. Given T and $\boldsymbol{D}$ as in **II.3.1**, let $T_j\colon U_j \to R^n$ be a sequence of continuous mappings (where $U_j \subset R^n$), and let $\boldsymbol{D}_j \subset U_j$ be a corresponding sequence of bounded domains. Suppose that the following condition holds: for every compact set $F \subset \boldsymbol{D}$ and every $\varepsilon > 0$ there exists an integer $j_0 = j_0(F, \varepsilon)$ such that (see **I.1.5**, definition 5)

$$F \subset \boldsymbol{D}_j, \quad \varrho(T, T_j, F) < \varepsilon \quad \text{for} \quad j > j_0. \tag{46}$$

Definition. Under these conditions, we shall say that the mappings $T_j|\boldsymbol{D}_j$ converge uniformly to $T|\boldsymbol{D}$ on compact subsets of $\boldsymbol{D}$.

Theorem 4. Suppose that the mappings $T_j|\boldsymbol{D}_j$ converge uniformly to $T|\boldsymbol{D}$ on compact sets in $\boldsymbol{D}$, in the sense of the preceding definition. Take a point x and any sequence $x_j \to x$. Then

$$K(x, T, \boldsymbol{D}) \leq \liminf K(x_j, T_j, \boldsymbol{D}_j), \tag{47}$$

$$K^+(x, T, \boldsymbol{D}) \leq \liminf K^+(x_j, T_j, \boldsymbol{D}_j), \tag{48}$$

$$K^-(x, T, \boldsymbol{D}) \leq \liminf K^-(x_j, T_j, \boldsymbol{D}_j). \tag{49}$$

Proof. The argument being entirely similar in all three cases, we indicate the proof only for (48). Since (48) is obvious if $K^+(x, T, \boldsymbol{D}) = 0$, we assume that $K^+(x, T, \boldsymbol{D}) > 0$. Consider any system $\mathfrak{S}^+(x, T, \boldsymbol{D})$, and let $D_1, \ldots, D_m$ be the positive indicator domains that constitute

$\mathfrak{S}^+(x, T, \boldsymbol{D})$. Then [see (5)]

$$M[\mathfrak{S}^+(x, T, \boldsymbol{D})] = \sum_{k=1}^{m} \mu(x, T, D_k), \tag{50}$$

$$\mu(x, T, D_k) > 0, \quad k = 1, \ldots, m, \tag{51}$$

$$fr\, D_k \subset \overline{D}_k \subset \boldsymbol{D}, \quad k = 1, \ldots, m, \tag{52}$$

$$x \in \bigcap_{k=1}^{m} \complement\, T\, fr\, D_k. \tag{53}$$

Choose r_0 to satisfy (see **I.1.4**, definition 7)

$$0 < r_0 < \tfrac{1}{2} e(x, T\, fr\, D_k), \quad k = 1, \ldots, m. \tag{54}$$

In view of (53), such a choice of r_0 is possible by **I.1.4**, exercise 3. Since $x_j \to x$, we shall have

$$\|x_j - x\| < r_0 \quad \text{for} \quad j > j_0', \tag{55}$$

for a properly chosen integer j_0'. Consider now the set

$$F = \bigcup_{k=1}^{m} \overline{D}_k. \tag{56}$$

In view of (52), F is then a compact set in $\boldsymbol{D}$. Hence we have an integer j_0 such that

$$F \subset \boldsymbol{D}_j, \quad \varrho(T, T_j, F) < r_0 \quad \text{for} \quad j > j_0. \tag{57}$$

Take $j > j_0' + j_0$. For each $k = 1, \ldots, m$, we have then $\overline{D}_k \subset \boldsymbol{D}_j$, and (since $fr\, D_k \subset \overline{D}_k \subset F$)

$$\varrho(T, T_j, fr\, D_k) < r_0. \tag{58}$$

From (54), (55), (58) it follows that for $k = 1, \ldots, m$,

$$\|x_j - x\| + \varrho(T, T_j, fr\, D_k) < e(x, T\, fr\, D_k) \quad \text{for} \quad j > j_0' + j_0. \tag{59}$$

By **II.2.3**, theorem 6, we conclude from (59) that x_j is (T_j, D_k)-admissible and

$$\mu(x_j, T_j, D_k) = \mu(x, T, D_k) \quad \text{for} \quad j > j_0' + j_0. \tag{60}$$

In view of (51), (56), (57), the relation (60) implies that the domains $D_1, \ldots, D_m$ constitute a system $\mathfrak{S}^+(x_j, T_j, \boldsymbol{D}_j)$ such that [see (50)]

$$M[\mathfrak{S}^+(x, T, \boldsymbol{D})] = M[\mathfrak{S}^+(x_j, T_j, \boldsymbol{D}_j)] \quad \text{for} \quad j > j_0' + j_0.$$

By (11), applied to T_j and $\boldsymbol{D}_j$, it follows that

$$M[\mathfrak{S}^+(x, T, \boldsymbol{D})] \leq K^+(x_j, T_j, \boldsymbol{D}_j) \quad \text{for} \quad j > j_0' + j_0.$$

By the definition of the inferior limit of a sequence, we conclude that

$$M[\mathfrak{S}^+(x, T, \boldsymbol{D})] \leq \liminf K^+(x_j, T_j, \boldsymbol{D}_j).$$

Since this holds for every system $\mathfrak{S}^+(x, T, \boldsymbol{D})$, (48) follows.

Remark 10. Relevant special cases of the preceding theorems include the following.

(i) Given T and $\boldsymbol{D}$ as in **II.3.1**, let $\{\boldsymbol{D}_j\}$ be a sequence of domains such that (a) $\boldsymbol{D}_j \subset \boldsymbol{D}$, (b) $\bar{\boldsymbol{D}}_j \subset \boldsymbol{D}_{j+1}$ for every j, and (c) for every compact set $F \subset \boldsymbol{D}$ there exists an integer $j_0 = j_0(F)$ such that $F \subset \boldsymbol{D}_j$ for $j > j_0$.

Under these conditions, we shall say that the domains $\boldsymbol{D}_j$ fill up $\boldsymbol{D}$ from the interior. On choosing $T_j = T$, $x_j = x$, the assumptions of theorem 4 are clearly satisfied, and hence

$$K(x, T, \boldsymbol{D}) \leq \liminf K(x, T, \boldsymbol{D}_j). \tag{61}$$

On the other hand, theorem 2 yields

$$K(x, T, \boldsymbol{D}) \geq K(x, T, \boldsymbol{D}_j) \quad \text{for every } j.$$

In view of (61) it follows that if the domains $\boldsymbol{D}_j$ fill up $\boldsymbol{D}$ from the interior, then

$$K(x, T, \boldsymbol{D}_j) \to K(x, T, \boldsymbol{D}).$$

An entirely similar argument shows that if the domains $\boldsymbol{D}_j$ fill up $\boldsymbol{D}$ from the interior, then

$$K^+(x, T, \boldsymbol{D}_j) \to K^+(x, T, \boldsymbol{D}), \quad K^-(x, T, \boldsymbol{D}_j) \to K^-(x, T, \boldsymbol{D}).$$

(ii) Choosing $T_j = T$, $\boldsymbol{D}_j = \boldsymbol{D}$, the assumptions of theorem 4 are clearly satisfied, and hence

$$K(x, T, \boldsymbol{D}) \leq \liminf K(x_j, T, \boldsymbol{D}) \quad \text{if} \quad x_j \to x. \tag{62}$$

Thus $K(x, T, \boldsymbol{D})$ is a lower semi-continuous function of x, for fixed T and $\boldsymbol{D}$. Applying the same argument to K^+, K^-, one sees that theorem 3 is implied by theorem 4.

Remark 11. The following observation will be useful later on. Let there be given a system $\mathfrak{S}(x, T, \boldsymbol{D})$ such that

$$M[\mathfrak{S}(x, T, \boldsymbol{D})] = K(x, T, \boldsymbol{D}). \tag{63}$$

The positive and the negative indicator domains comprised in $\mathfrak{S}(x, T, \boldsymbol{D})$ constitute a system $\mathfrak{S}^+(x, T, \boldsymbol{D})$ and a system $\mathfrak{S}^-(x, T, \boldsymbol{D})$ respectively (one or both of which may be empty). We assert that (63) implies the relations

$$M[\mathfrak{S}^+(x, T, \boldsymbol{D})] = K^+(x, T, \boldsymbol{D}), \tag{64}$$

$$M[\mathfrak{S}^-(x, T, \boldsymbol{D})] = K^-(x, T, \boldsymbol{D}). \tag{65}$$

Indeed, if (64) and (65) do not both hold, then [see (11), (12), (63)] it would follow that

$$K(x, T, \boldsymbol{D}) = M[\mathfrak{S}(x, T, \boldsymbol{D})] = M[\mathfrak{S}^+(x, T, \boldsymbol{D})] + \\ + M[\mathfrak{S}^-(x, T, \boldsymbol{D})] < K^+(x, T, \boldsymbol{D}) + K^-(x, T, \boldsymbol{D}),$$

in contradiction with theorem 1.

II.3.3. Essential maximal model continua. *Definition 1.* Given T and $\boldsymbol{D}$ as in **II.3.1**, take a point $x \in R^n$ and consider an m.m.c. C for $(x, T, \boldsymbol{D})$. Then C is termed an *essential maximal model continuum* (abbreviated to e.m.m.c.) for $(x, T, \boldsymbol{D})$ provided that the following condition holds: for every open set O such that $C \subset O \subset \boldsymbol{D}$ there exists an indicator domain D for $(x, T, \boldsymbol{D})$ such that $C \subset D$ and $\overline{D} \subset O$. For given x, the class of all the essential maximal model continua for $(x, T, \boldsymbol{D})$ will be denoted by $\mathfrak{E}(x, T, \boldsymbol{D})$. The number of elements of $\mathfrak{E}(x, T, \boldsymbol{D})$ will be denoted by $k(x, T, \boldsymbol{D})$ (thus k may be equal to zero or ∞).

Definition 2. A continuum $C \in \mathfrak{E}(x, T, \boldsymbol{D})$ is said to be *essentially isolated* if the following condition holds: there exists an open set O such that (i) $C \subset O, \overline{O} \subset \boldsymbol{D}$, and (ii) C is the only element of $\mathfrak{E}(x, T, \boldsymbol{D})$ contained in O. An open set O with these properties (i), (ii) is then termed a *characteristic neighborhood of* C. For given x, the class of essentially isolated essential maximal model continua for $(x, T, \boldsymbol{D})$ will be denoted by $\mathfrak{E}_i(x, T, \boldsymbol{D})$.

Lemma 1. If $C \in \mathfrak{E}_i(x, T, \boldsymbol{D})$, then there exists a domain D such that the following holds: (a) $C \subset D, \overline{D} \subset \boldsymbol{D}$, and (b) if $C' \in \mathfrak{E}(x, T, \boldsymbol{D})$ and $C' \cap D \neq \emptyset$, then $C' = C$.

Proof. Let O be a characteristic neighborhood of C. Since C is an e.m.m.c. for $(x, T, \boldsymbol{D})$ and $C \subset O$, there exists (see definition 1) an indicator domain D for $(x, T, \boldsymbol{D})$ such that $C \subset D, \overline{D} \subset O \subset \boldsymbol{D}$. Consider now any $C' \in \mathfrak{E}(x, T, \boldsymbol{D})$ and suppose that $C' \cap D \neq \emptyset$. By **II.3.1**, lemma 5 it follows that $C' \subset D$ and hence $C' \subset O$. Since C is the only element of $\mathfrak{E}(x, T, \boldsymbol{D})$ contained in O, it follows that $C' = C$.

The following definition will be useful in locating elements of $\mathfrak{E}(x, T, \boldsymbol{D})$.

Definition 3. A determining sequence $\{D_j\}$ for $(x, T, \boldsymbol{D})$ (see **II.3.1**, definition 2) is termed *essential* if $\mu(x, T, D_j) \neq \emptyset$ for every j [or equivalently, if every D_j is an indicator domain for $(x, T, \boldsymbol{D})$].

Lemma 2. If $\{D_j\}$ is an essential determining sequence for $(x, T, \boldsymbol{D})$ then

$$\cap \overline{D}_j = C \in \mathfrak{E}(x, T, \boldsymbol{D}), \tag{1}$$

$$C \subset D_j \quad \text{for every } j. \tag{2}$$

Proof. C is an m.m.c. and (2) holds by **II.3.1**, lemma 6. Take now any open set O such that $C \subset O \subset \boldsymbol{D}$. In view of the definition of C it follows from **I.1.3**, exercise 24, that $\overline{D}_j \subset O$ for j sufficiently large. If j is so chosen, then $C \subset D_j$, $\overline{D}_j \subset O$, and D_j is an indicator domain for $(x, T, \boldsymbol{D})$. Thus (1) follows in view of definition 1.

Lemma 3. If $C \in \mathfrak{E}(x, T, \boldsymbol{D})$, then there exists an essential determining sequence $\{D_j^*\}$ for $(x, T, \boldsymbol{D})$ such that $C = \bigcap \overline{D_j^*}$.

Proof. To begin with, by **II.3.1**, lemma 8 there exists a determining sequence $\{D_j\}$ for $(x, T, \boldsymbol{D})$ such that

$$C = \bigcap \overline{D}_j. \tag{3}$$

Then $C \subset D_1 \subset \boldsymbol{D}$, and D_1 is open. Since C is an e.m.m.c. for $(x, T, \boldsymbol{D})$, there exists therefore an indicator domain D_1^* for $(x, T, \boldsymbol{D})$ such that $C \subset D_1^*$, $\overline{D_1^*} \subset D_1$. Then $C \subset D_2 \cap D_1^* \subset \boldsymbol{D}$. Since $D_2 \cap D_1^*$ is open and C is an e.m.m.c. for $(x, T, \boldsymbol{D})$, there exists an indicator domain D_2^* for $(x, T, \boldsymbol{D})$ such that $C \subset D_2^*, \overline{D_2^*} \subset D_2 \cap D_1^*$. Continuing this process, clearly one obtains a sequence $\{D_j^*\}$ with the desired properties.

Lemma 4. If D is an indicator domain for $(x, T, \boldsymbol{D})$, then D contains at least one e.m.m.c. for $(x, T, \boldsymbol{D})$.

Proof. The assumption implies that

$$x \notin T \, fr\, D, \quad \mu(x, T, D) \neq 0. \tag{4}$$

Let r be any real number such that (see **I.1.4**, definition 7 and exercise 3)

$$0 < r < e(x, T\, fr\, D). \tag{5}$$

Then the spherical neighborhood $\Delta_r(x)$ (see **II.2.1**) is contained in $\complement T\, fr D$. In view of (4) and **II.2.3**, theorem 2, the set $D \cap T^{-1} x$ is nonempty, and hence $\boldsymbol{D} \cap T^{-1} \Delta_r(x) \neq \emptyset$. Set

$$O_r = \boldsymbol{D} \cap T^{-1} \Delta_r(x) \neq \emptyset,$$

and let $\mathfrak{K}_r$ denote the class of those components of O_r that lie in D. By **II.3.1**, lemma 3 [applied with $G = \Delta_r(x)$] it follows that the elements of $\mathfrak{K}_r$ constitute an (x, T, D)-complete sequence. By **II.2.3**, theorem 3 it follows that

$$\mu(x, T, D) = \textstyle\sum' \mu(x, T, D_r), \quad D_r \in \mathfrak{K}_r.$$

In view of (4) we conclude that

$$\mu(x, T, D_r) \neq 0 \quad \text{for some} \quad D_r \in \mathfrak{K}_r. \tag{6}$$

Now take a sequence

$$r_1 > \cdots > r_j > \cdots > 0, \quad 0 < r_j < e(x, T\, fr\, D), \quad r_j \to 0.$$

Applying (6) with $r = r_1$, we obtain a domain D_{r_1} such that

$$D_{r_1} \subset D, \quad \mu(x, T, D_{r_1}) \neq 0, \quad D_{r_1} \in \mathfrak{K}_{r_1}.$$

Replacing D by D_{r_1} and r_1 by r_2, the same argument yields a domain D_{r_2} such that

$$D_{r_2} \subset D_{r_1}, \quad \mu(x, T, D_{r_2}) \neq 0, \quad D_{r_2} \in \mathfrak{K}_{r_2}.$$

Continuation of this process yields a sequence of domains $\{D_{r_j}\}$ such that

$$D_{r_j} \subset D_{r_{j-1}} \subset D, \quad \mu(x, T, D_{r_j}) \neq 0, \quad D_{r_j} \in \mathfrak{K}_{r_j}. \tag{7}$$

We verify now that $\{D_{r_j}\}$ is a determining sequence for $(x, T, \boldsymbol{D})$. Since $\mu(x, T, D_{r_j}) \neq 0$, by **II.2.3**, theorem 2 it follows that

$$x \in T D_{r_j} \subset T \overline{D}_{r_j}.$$

Since $\overline{\Delta_{r_{j+1}}(x)} \subset \Delta_{r_j}(x)$, by **II.3.1**, lemma 4 we have $\overline{D}_{r_{j+1}} \subset D_{r_j}$. As $D_{r_j} \subset T^{-1} \Delta_{r_j}(x)$, it follows that

$$\delta T \overline{D}_{r_j} \leqq \delta T D_{r_{j-1}} \leqq 2 r_{r_{j-1}} \to 0.$$

Thus $\{D_{r_j}\}$ is a determining sequence for $(x, T, \boldsymbol{D})$, and by (7) this determining sequence is essential. By lemma 2 it follows that

$$C = \bigcap \overline{D}_{r_j} \in \mathfrak{C}(x, T, \boldsymbol{D}), \quad C \subset D_{r_j} \subset D,$$

and the lemma is proved.

Remark. In the statement of the preceding lemma 4, let us add the assumption that D is a *positive* indicator domain [that is, $\mu(x, T, \boldsymbol{D}) > 0$]. Inspection of the proof reveals that (6) can be then stated in the form:

$$\mu(x, T, D_r) > 0 \quad \text{for some} \quad D_r \in \mathfrak{K}_r.$$

Similarly, it follows that the domains D_{r_j} can be chosen to satisfy the condition $\mu(x, T, D_{r_j}) > 0$. In summary: if D is a positive indicator domain, then the essential determining sequence $\{D_{r_j}\}$ can be so chosen that each D_{r_j} is a positive indicator domain. Similarly, if D is a negative indicator domain, then each D_{r_j} may be chosen as a negative indicator domain.

Lemma 5. If $C \in \mathfrak{C}_i(x, T, \boldsymbol{D})$ and O is a characteristic neighborhood of C, then the following holds.

(i) If D is any indicator domain for $(x, T, \boldsymbol{D})$ such that $D \subset O$, then $C \subset D$.

(ii) If D_1, D_2 are any two (x, T)-admissible domains such that $C \subset D_1 \subset D_2 \subset O$, then $\mu(x, T, D_1) = \mu(x, T, D_2)$.

(iii) If D is any (x, T)-admissible domain such that $C \subset D \subset O$, then D is an indicator domain for $(x, T, \boldsymbol{D})$.

(iv) If D_1, D_2 are any two indicator domains for $(x, T, \boldsymbol{D})$ such that $C \subset D_1 \subset O$, $C \subset D_2 \subset O$, then $\mu(x, T, D_1) = \mu(x, T, D_2)$.

Proof. Given D as specified in (i), by lemma 4 there exists a continuum $C' \subset D$, $C' \in \mathfrak{C}(x, T, \boldsymbol{D})$. Since then $C' \subset O$, and O is a characteristic neighborhood of C, we must have $C = C'$. Hence $C \subset D$, and (i) is proved. Consider next D_1, D_2 as specified in (ii), and assume that $\mu(x, T, D_1) \neq \mu(x, T, D_2)$. By remark 5 in **II.2.3** there follows the existence of an (x, T)-admissible domain D such that $D \subset D_2$, $D \cap D_1 = \emptyset$, $\mu(x, T, D) \neq 0$. Then D would be an indicator domain for $(x, T, \boldsymbol{D})$ contained in O but not containing C, in contradiction with (i). Thus (ii) is proved. Consider now D as specified in (iii). Since C is an e.m.m.c. for $(x, T, \boldsymbol{D})$ contained in the open set D, there exists an indicator domain D_* for $(x, T, \boldsymbol{D})$ such that $C \subset D_* \subset D$. By (ii) it follows that $\mu(x, T, D) = \mu(x, T, D_*) \neq 0$, and thus D is also an indicator domain for $(x, T, \boldsymbol{D})$. Finally, given D_1, D_2 as specified in (iv), the inclusion $C \subset D_1 \cap D_2$ implies, since C is an e.m.m.c. for $(x, T, \boldsymbol{D})$, the existence of an indicator domain D for $(x, T, \boldsymbol{D})$ such that $C \subset D \subset D_1 \cap D_2$. By (ii) it follows that

$$\mu(x, T, D_1) = \mu(x, T, D) = \mu(x, T, D_2),$$

and (iv) follows.

Lemma 6. $k(x, T, \boldsymbol{D}) \leq K(x, T, \boldsymbol{D})$ [see definition 1 and **II.3.2** (10)].

Proof. The assertion is trivial if $k(x, T, \boldsymbol{D}) = 0$. So assume that $k(x, T, \boldsymbol{D}) > 0$. Let then $C_1, \ldots, C_m$ be any finite set of elements of $\mathfrak{C}(x, T, \boldsymbol{D})$. Since $C_1, \ldots, C_m$ are pair-wise disjoint continua in $\boldsymbol{D}$, we can select open sets $O_1, \ldots, O_m$ such that $C_j \subset O_j \subset \boldsymbol{D}$, $O_j \cap O_i = \emptyset$ for $j \neq i$. Since $C_j \in \mathfrak{C}(x, T, \boldsymbol{D})$, we can select indicator domains D_j for $(x, T, \boldsymbol{D})$ such that $C_j \subset D_j$, $\overline{D}_j \subset O_j$, $j = 1, \ldots, m$. Then the domains D_j constitute an indicator system $\mathfrak{S}(x, T, \boldsymbol{D})$, and by **II.3.2** (7), (10) we conclude that

$$m = N[\mathfrak{S}(x, T, \boldsymbol{D})] \leq M[\mathfrak{S}(x, T, \boldsymbol{D})] \leq K(x, T, \boldsymbol{D}).$$

Thus $\mathfrak{C}(x, T, \boldsymbol{D})$ cannot contain more than $K(x, T, \boldsymbol{D})$ elements, and the lemma is proved.

Lemma 7. $k(x, T, \boldsymbol{D})$ is a lower semi-continuous function of x (for fixed $T, \boldsymbol{D}$).

Proof. Take any point x_0. If $k(x_0, T, \boldsymbol{D}) = 0$, then clearly k is lower semi-continuous at x_0, since $k \geq 0$. So we can assume that $k(x_0, T, \boldsymbol{D}) > 0$. Let any real number $\lambda < k(x_0, T, \boldsymbol{D})$ be assigned. Then the class $\mathfrak{C}(x_0, T, \boldsymbol{D})$, which is now non-empty by assumption, contains a set of distinct continua $C_1, \ldots, C_m$, where $m > \lambda$. As in the proof of lemma 6, there follows the existence of pair-wise disjoint indicator domains $D_1, \ldots, D_m$ for $(x_0, T, \boldsymbol{D})$ such that $C_j \subset D_j$, $j = 1, \ldots, m$. Then $x_0 \in \complement T\, fr\, D_j$, $j = 1, \ldots, m$. The set

$$O = \bigcap_{j=1}^{m} \complement T\, fr\, D_j$$

is then an open set containing x_0. Hence O has a (unique) component Δ containing x_0, and by **I.2.1**, exercise 4 this component Δ is a domain. The lower semi-continuity of k at x_0 will be established if we can show that

$$k(x, T, \boldsymbol{D}) > \lambda \quad \text{for} \quad x \in \Delta. \tag{8}$$

Now Δ is a connected subset of $\complement T\, fr D_j$, $j = 1, \ldots, m$, and hence (by **II.2.3**, remark 1)

$$\mu(x, T, D_j) = \mu(x_0, T, D_j) \quad \text{for} \quad x \in \Delta. \tag{9}$$

Now $\mu(x_0, T, D_j) \neq 0$, since D_j is an indicator domain for $(x_0, T, \boldsymbol{D})$. Thus, in view of (9), $\mu(x, T, D_j) \neq 0$ for $x \in \Delta$. Accordingly, D_j is an indicator domain for $(x, T, \boldsymbol{D})$ if $x \in \Delta$. By lemma 4, there follows the existence of a continuum $C'_j \in \mathfrak{C}(x, T, \boldsymbol{D})$ such that $C'_j \subset D_j$. Since the domains $D_1, \ldots, D_m$ are pair-wise disjoint, the continua $C'_1, \ldots, C'_m$ are all distinct. In summary: for $x \in \Delta$, the class $\mathfrak{C}(x, T, \boldsymbol{D})$ contains (at least) $m > \lambda$ distinct elements, and (8) is verified.

II.3.4. The essential local index $i_e(C, T)$ and the algebraic multiplicity function $\mu_e(x, T, D)$. Given T and $\boldsymbol{D}$ as in **II.3.1**, take a point $x \in R^n$ and consider a continuum $C \in \mathfrak{C}_i(x, T, \boldsymbol{D})$. Let O be a characteristic neighborhood of C. Since C is an e.m.m.c. for $(x, T, \boldsymbol{D})$, there exists an indicator domain D for $(x, T, \boldsymbol{D})$ such that $C \subset D$, $\bar{D} \subset O$.

Definition 1. In terms of O and D as specified above, the *essential local index* $i_e(C, T)$ is defined by the formula

$$i_e(C, T) = \mu(x, T, D). \tag{1}$$

Remark 1. To justify this definition, we have to verify, of course, that we obtain the same value for $i_e(C, T)$ if we select any other characteristic neighborhood O' of C and any indicator domain D' for $(x, T, \boldsymbol{D})$ such that $C \subset D'$, $\bar{D}' \subset O'$. Now since $C \subset D \cap D'$, there exists an indicator domain D_* for $(x, T, \boldsymbol{D})$ such that $C \subset D_*$, $\bar{D}_* \subset D \cap D'$. By **II.3.3**, lemma 5 it follows that

$$\mu(x, T, D) = \mu(x, T, D_*) = \mu(x, T, D').$$

Thus we get the same value for $i_e(C, T)$ whether we use O and D or O' and D'.

Definition 2. For points $x \in R^n$ such that $K(x, T, \boldsymbol{D}) < \infty$, the *algebraic multiplicity function* $\mu_e(x, T, \boldsymbol{D})$ is defined by the formula

$$\mu_e(x, T, \boldsymbol{D}) = \sum i_e(C, T), \quad C \in \mathfrak{C}_i(x, T, \boldsymbol{D}), \tag{2}$$

with the understanding that $\mu_e(x, T, \boldsymbol{D}) = 0$ if $\mathfrak{C}_i(x, T, \boldsymbol{D})$ is empty.

Remark 2. To clarify this definition, let us note that in view of **II.3.3**, lemma 6 the assumption $K(x, T, \boldsymbol{D}) < \infty$ implies that the class $\mathfrak{C}(x, T, \boldsymbol{D})$ is finite. It follows that $\mathfrak{C}(x, T, \boldsymbol{D}) = \mathfrak{C}_i(x, T, \boldsymbol{D})$. Indeed, this is obvious if $\mathfrak{C}(x, T, \boldsymbol{D})$ is empty. On the other hand, if $\mathfrak{C}(x, T, \boldsymbol{D})$ is non-empty and consists of the continua $C_1, \ldots, C_m$, then these continua are pair-wise disjoint, and hence there exist pair-wise disjoint open sets $O_1, \ldots, O_m$ such that $C_j \subset O_j \subset \boldsymbol{D}$. Clearly, each C_j is therefore essentially isolated. One sees now that the summation in (2) is finite (perhaps empty), and hence the definition (2) is certainly meaningful. One also sees that the condition $C \in \mathfrak{C}_i(x, T, \boldsymbol{D})$ could be replaced by the condition $C \in \mathfrak{C}(x, T, \boldsymbol{D})$.

Remark 3. If D is an indicator domain for $(x, T, \boldsymbol{D})$, then $\mu(x, T, D)$ is an integer different from zero. In view of (1) it follows that $i_e(C, T)$ is an integer different from zero. On the other hand, (2) may yield the value zero for $\mu_e(x, T, \boldsymbol{D})$ as a result of cancellations or because $\mathfrak{C}_i(x, T, \boldsymbol{D})$ is empty.

Theorem 1. Given T and $\boldsymbol{D}$ as in **II.3.1**, let D be a domain and x a point such that the following holds. (i) $\bar{D} \subset \boldsymbol{D}$. (ii) D is (x, T)-admissible. (iii) D contains only a finite number $m \geq 0$ of continua $C \in \mathfrak{C}(x, T, \boldsymbol{D})$. Then

$$\mu(x, T, D) = \sum i_e(C, T), \quad C \in \mathfrak{C}_i(x, T, \boldsymbol{D}), \quad C \subset D, \tag{3}$$

with the understanding that an empty summation has the value zero.

Proof. Case 1. $m = 0$. Then the summation in (3) is empty, and we have to show that

$$\mu(x, T, D) = 0. \tag{4}$$

If (4) is denied, then D should be an indicator domain for $(x, T, \boldsymbol{D})$, and by **II.3.3**, lemma 4 there would follow the existence of an e.m.m.c. for $(x, T, \boldsymbol{D})$ contained in D, in contradiction with the assumption $m = 0$.

Case 2. $0 < m < \infty$. Let then $C_1, \ldots, C_m$ be the set of those elements of $\mathfrak{C}(x, T, \boldsymbol{D})$ that lie in D. Since $C_1, \ldots, C_m$ are pair-wise disjoint continua, there exist pair-wise disjoint open sets $O_1, \ldots, O_m$ such that $C_j \subset O_j \subset D$, $j = 1, \ldots, m$. Clearly, each C_j is essentially isolated, and O_j is a characteristic neighborhood of C_j (see **II.3.3**, definition 2). Thus

$$C_j \in \mathfrak{C}_i(x, T, \boldsymbol{D}), \quad j = 1, \ldots, m, \tag{5}$$

and there exists an indicator domain D_j for $(x, T, \boldsymbol{D})$ such that

$$C_j \subset D_j, \quad \overline{D}_j \subset O_j, \quad j = 1, \ldots, m. \tag{6}$$

By definition 1, we have then

$$i_e(C_j, T) = \mu(x, T, D_j), \quad j = 1, \ldots, m,$$

and thus (3) will be established if we can show that

$$\mu(x, T, D) = \sum_{j=1}^{m} \mu(x, T, D_j). \tag{7}$$

Let (7) be denied. Then by **II.2.3**, theorem 4 there follows the existence of a domain D_* such that $D_* \subset D$, $D_* \cap D_j = \emptyset$, $j = 1, \ldots, m$, and $\mu(x, T, D_*) \neq 0$. By **II.3.3**, lemma 4 there follows the existence of a continuum $C_* \in \mathfrak{C}(x, T, \boldsymbol{D})$ contained in D_*. Then C_* would be an element of $\mathfrak{C}(x, T, \boldsymbol{D})$ contained in D and different from $C_1, \ldots, C_m$, in contradiction with the fact that $C_1, \ldots, C_m$ are all the elements of $\mathfrak{C}(x, T, \boldsymbol{D})$ lying in D. Thus (7) must hold, and the proof is complete.

Remark 4. In the preceding theorem, we required the inclusion $\overline{D} \subset \boldsymbol{D}$. Actually, in the proof we needed only the fact that T was defined and continuous on $\overline{D}$. Accordingly, the theorem applies even if $D = \boldsymbol{D}$, provided that T is defined and continuous on $\overline{\boldsymbol{D}}$. This observation yields the following statement.

Theorem 2. Given T and $\boldsymbol{D}$ as in **II.3.1**, assume that the inclusion $\overline{\boldsymbol{D}} \subset U$ holds (thus T is now defined and continuous on $\overline{\boldsymbol{D}}$). Then

$$\mu_e(x, T, \boldsymbol{D}) = \mu(x, T, \boldsymbol{D}) \tag{8}$$

for all points x satisfying the conditions

$$x \notin T \, fr \, \boldsymbol{D}, \quad K(x, T, \boldsymbol{D}) < \infty. \tag{9}$$

Proof. Observe that the conditions (9) characterize precisely those points x where $\mu_e(x, T, \boldsymbol{D})$ and $\mu(x, T, \boldsymbol{D})$ are both defined. As noted in definition 2, the assumption $K(x, T, \boldsymbol{D}) < \infty$ implies that the class $\mathfrak{C}(x, T, \boldsymbol{D})$ is finite. Accordingly (see remark 3) we can apply theorem 1 with $D = \boldsymbol{D}$, obtaining

$$\mu(x, T, \boldsymbol{D}) = \sum i_e(C, T), \quad C \in \mathfrak{C}_i(x, T, \boldsymbol{D}), \tag{10}$$

since now every element of $\mathfrak{C}_i(x, T, \boldsymbol{D})$ is contained in $D = \boldsymbol{D}$. Comparison of (10) and (2) yields (8).

Theorem 3. Given T and $\boldsymbol{D}$ as in **II.3.1**, the following statements hold for the functions $k(x, T, \boldsymbol{D})$, $K(x, T, \boldsymbol{D})$.

(i) $0 \leq k(x, T, \boldsymbol{D}) \leq K(x, T, \boldsymbol{D})$.

(ii) $k(x, T, \boldsymbol{D}) = 0$ if and only if $K(x, T, \boldsymbol{D}) = 0$.

(iii) $k(x, T, \boldsymbol{D}) = \infty$ if and only if $K(x, T, \boldsymbol{D}) = \infty$.

Proof. (i) is merely a re-statement of lemma 6 in **II.3.3**. In view of (i), $K(x, T, \boldsymbol{D}) = 0$ implies that $k(x, T, \boldsymbol{D}) = 0$. Conversely, if $k(x, T, \boldsymbol{D}) = 0$, then by **II.3.3**, lemma 4 it follows that there exists no indicator domain for $(x, T, \boldsymbol{D})$, and hence $K(x, T, \boldsymbol{D}) = 0$ by **II.3.2**, remark 2. Thus (ii) is also proved. In view of (i), $k(x, T, \boldsymbol{D}) = \infty$ implies that $K(x, T, \boldsymbol{D}) = \infty$. There remains to show that if $k(x, T, \boldsymbol{D})$ is finite, then $K(x, T, \boldsymbol{D})$ is also finite. If $k(x, T, \boldsymbol{D}) = 0$, then this follows from (ii). So we assume that

$$0 < k(x, T, \boldsymbol{D}) < \infty. \tag{11}$$

In this case, the class $\mathfrak{E}(x, T, \boldsymbol{D})$ is finite and non-empty. Let $C_1, \ldots, C_m$ be the continua that constitute $\mathfrak{E}(x, T, \boldsymbol{D})$, where $m = k(x, T, \boldsymbol{D})$. As noted in remark 2, each C_j, $j = 1, \ldots, m$, is now essentially isolated, and

$$\mathfrak{E}(x, T, \boldsymbol{D}) = \mathfrak{E}_i(x, T, \boldsymbol{D}). \tag{12}$$

Accordingly, $i_e(C_j, T)$ is defined, $j = 1, \ldots, m$. Put

$$\sum_{j=1}^{m} |i_e(C_j, T)| = N. \tag{13}$$

Then N is a (finite) positive integer. Consider now any indicator system $\mathfrak{S}(x, T, \boldsymbol{D})$, consisting of indicator domains $D_1, \ldots, D_t$. Then the assumptions of theorem 1 hold for each D_k, $k = 1, \ldots, t$, and we obtain the relation

$$\mu(x, T, D_k) = \sum i_e(C_j, T), \quad C_j \subset D_k. \tag{14}$$

Since the domains $D_1, \ldots, D_t$ are pair-wise disjoint, no continuum C_j is utilized in (14) for two different values of k. Accordingly, (14) yields [in view of (13)]

$$M[\mathfrak{S}(x, T, D)] = \sum_{k=1}^{t} |\mu(x, T, D_k)| \leq \sum_{j=1}^{m} |i_e(C_j, T)| = N.$$

Since the system $\mathfrak{S}(x, T, \boldsymbol{D})$ was arbitrary, in view of **II.3.2** (10) it follows that

$$K(x, T, \boldsymbol{D}) \leq \sum_{j=1}^{m} |i_e(C_j, T)| = N. \tag{15}$$

Since N is finite, it follows that $K(x, T, \boldsymbol{D})$ is finite, and the proof is complete.

Theorem 4. Given T and $\boldsymbol{D}$ as in **II.3.1**, let x be a point such that

$$K(x, T, \boldsymbol{D}) < \infty. \tag{16}$$

Then the following formulas hold (with the understanding that an empty summation has the value zero).

$$K(x, T, \boldsymbol{D}) = \sum |i_e(C, T)|, \quad C \in \mathfrak{E}_i(x, T, \boldsymbol{D}), \tag{17}$$

$$\mu_e(x, T, \boldsymbol{D}) = K^+(x, T, \boldsymbol{D}) - K^-(x, T, \boldsymbol{D}). \tag{18}$$

Proof. Case 1. $K(x, T, \boldsymbol{D}) = 0$. By theorem 3 the class $\mathfrak{C}(x, T, \boldsymbol{D})$ and hence also the class $\mathfrak{C}_i(x, T, \boldsymbol{D})$ is empty, and hence $\mu_e(x, T, \boldsymbol{D}) = 0$ by definition, and the summation in (17) has the value zero. Also, K^+ and K^- are both equal to zero by **II.3.2** (13), (14). Thus the theorem is trivial in this case.

Case 2. $K(x, T, \boldsymbol{D}) > 0$. In view of (16) we have then

$$0 < K(x, T, \boldsymbol{D}) < \infty.$$

By theorem 3 it follows that

$$0 < k(x, T, \boldsymbol{D}) < \infty. \tag{19}$$

Since (19) coincides with (11), the conclusions drawn from (11) in the course of the proof of theorem 3 are available. Hence, on denoting again by $C_1, \ldots, C_m$ the continua that constitute the class $\mathfrak{C}(x, T, \boldsymbol{D}) = \mathfrak{C}_i(x, T, \boldsymbol{D})$, the inequality (15) holds again. Since the continua $C_1, \ldots, C_m$ are pair-wise disjoint, we have pair-wise disjoint open sets $O_1, \ldots, O_m$ such that $C_j \subset O_j \subset \boldsymbol{D}$, $j = 1, \ldots, m$. Clearly each O_j is a characteristic neighborhood for the corresponding C_j, and we have indicator domains D_j such that $C_j \subset D_j$, $\bar{D}_j \subset O_j$, $j = 1, \ldots, m$. By definition 1, we have then

$$i_e(C_j, T) = \mu(x, T, D_j), \quad j = 1, \ldots, m. \tag{20}$$

The domains $D_1, \ldots, D_m$ constitute a system $\mathfrak{S}(x, T, \boldsymbol{D})$, and (20) yields

$$M[\mathfrak{S}(x, T, \boldsymbol{D})] = \sum_{j=1}^{m} |i_e(C_j, T)|. \tag{21}$$

In view of **II.3.2** (10) it follows that

$$K(x, T, \boldsymbol{D}) \geq \sum_{j=1}^{m} |i_e(C_j, T)|. \tag{22}$$

From (15) and (22) we conclude that

$$K(x, T, \boldsymbol{D}) = \sum_{j=1}^{m} |i_e(C_j, T)|, \tag{23}$$

and (17) is proved. (21) and (23) yield

$$K(x, T, \boldsymbol{D}) = M[\mathfrak{S}(x, T, \boldsymbol{D})]. \tag{24}$$

In view **II.3.2**, remark 11, we infer from (24) the relations

$$K^+(x, T, \boldsymbol{D}) = \sum i_e(C_j, T), \qquad i_e(C_j, T) > 0, \tag{25}$$

$$K^-(x, T, \boldsymbol{D}) = -\sum i_e(C_j, T), \quad i_e(C_j, T) < 0, \tag{26}$$

which correspond to (64), (65) in **II.3.2**. Subtracting (26) from (25), in view of definition 2 we obtain (18), and the proof is complete.

Remark 5. In view of part (iii) of theorem 3, the condition $K(x,T,\boldsymbol{D})<\infty$ can be replaced by the equivalent condition $k(x,T,\boldsymbol{D})<\infty$ in definition 2, theorem 2 and theorem 4. Furthermore, in view of remark 2, in (17) the condition $C\in\mathfrak{C}_i(x, T, \boldsymbol{D})$ can be replaced by the condition $C\in\mathfrak{C}(x, T, \boldsymbol{D})$.

II.3.5. Local properties. Given T and $\boldsymbol{D}$ as in **II.3.1**, let x be a point in R^n and D a domain such that $\overline{D}\subset\boldsymbol{D}$. Then the property of D being (x, T)-admissible depends only upon the behavior of T on $fr\,D$, since D is (x, T)-admissible if and only if $x\notin T\,fr\,D$. Furthermore, if D is (x, T)-admissible, then $\mu(x, T, D)$ is defined and its value depends only upon the behavior of T on $fr\,D$ (see **II.2.3**, remark 7). Accordingly, the property of D being a positive or negative indicator domain for $(x, T, \boldsymbol{D})$ in the sense of **II.3.2** depends also only upon the behavior of T on $fr\,D$. The preceding observations yield examples of *local properties*, that is properties that depend only upon the behavior of T on certain subsets of $\boldsymbol{D}$. To obtain precise statements of further relevant local properties, let us consider two transformations

$$T_1: U_1\to R^n, \quad T_2: U_2\to R^n, \tag{1}$$

where T_1, T_2 are continuous on the subsets U_1, U_2 of R^n respectively. Let $\boldsymbol{D}_1$, $\boldsymbol{D}_2$ be bounded domains such that $\boldsymbol{D}_1\subset U_1$, $\boldsymbol{D}_2\subset U_2$. As in **II.3.1**, we assume that T_1, T_2 are bounded on $\boldsymbol{D}_1$, $\boldsymbol{D}_2$ respectively.

Theorem. Let x be a point in R^n and C a continuum in $\boldsymbol{D}_1\cap\boldsymbol{D}_2$. Suppose that there exists an open set G such that

$$C\subset G, \quad \overline{G}\subset\boldsymbol{D}_1\cap\boldsymbol{D}_2, \quad T_1|G=T_2|G. \tag{2}$$

Then the following holds.

(i) If C is an m.m.c. for $(x, T_1, \boldsymbol{D}_1)$, then it is also an m.m.c. for $(x, T_2, \boldsymbol{D}_2)$ (see **II.3.1**, definition 1).

(ii) If $C\in\mathfrak{C}(x, T_1, \boldsymbol{D}_1)$, then $C\in\mathfrak{C}(x, T_2, \boldsymbol{D}_2)$ (see **II.3.3**, definition 1).

(iii) If $C\in\mathfrak{C}_i(x, T_1, \boldsymbol{D}_1)$, then $C\in\mathfrak{C}_i(x, T_2, \boldsymbol{D}_2)$ (see **II.3.3**, definition 2).

(iv) If $i_e(C, T_1)$ is defined, then $i_e(C, T_2)$ is also defined, and $i_e(C, T_1)=i_e(C, T_2)$ (see **II.3.4**, definition 1).

Proof of (i). Suppose that C is an m.m.c. for $(x, T_1, \boldsymbol{D}_1)$. Then $T_1C=x$, and hence by (2) also $T_2C=x$. Thus

$$C\subset G\cap T_2^{-1}x\subset\boldsymbol{D}_2\cap T_2^{-1}x.$$

As C is connected, it follows that C is contained in a component C_2 of $\boldsymbol{D}_2\cap T_2^{-1}x$. Clearly (i) will be proved if we can show that $C_2=C$. Now since C is an m.m.c. for $(x, T_1, \boldsymbol{D}_1)$, by **II.3.1**, lemma 8 there exists a determining sequence $\{D_j\}$ for $(x, T_1, \boldsymbol{D}_1)$ such that

$$C = \bigcap \overline{D}_j, \tag{3}$$

$$\overline{D}_j < G \quad \text{for every } j. \tag{4}$$

Since $T_1|G = T_2|G$ and D_j is (x, T_1)-admissible, we infer from (4) that D_j is also (x, T_2)-admissible. As C_2 is a component of $\boldsymbol{D}_2 \cap T_2^{-1} x$, from **II.3.1**, lemma 5 it follows that either $C_2 < D_j$ or $C_2 < \complement \overline{D}_j$. The second alternative is ruled out by the observation that $C_2 \cap \overline{D}_j \neq \emptyset$ since $C_2 > C$ and $C < \overline{D}_j$. Thus

$$C < C_2 < D_j < \overline{D}_j \quad \text{for every } j. \tag{5}$$

From (3) and (5) we conclude that

$$C < C_2 < \bigcap \overline{D}_j = C.$$

Hence $C_2 = C$, and (i) is proved.

Proof of (ii). Suppose that $C \in \mathfrak{C}(x, T_1, \boldsymbol{D}_1)$. Then C is an m.m.c. for $(x, T_1, \boldsymbol{D}_1)$ and hence [by (i)] also for $(x, T_2, \boldsymbol{D}_2)$. Let there be assigned any open set O such that $C < O < \boldsymbol{D}_2$. Clearly (ii) will be proved if we can exhibit an indicator domain D for $(x, T_2, \boldsymbol{D}_2)$ such that $C < D, \overline{D} < O$. Put $O^* = O \cap G$. Then O^* is an open set, and $C < O^* < \boldsymbol{D}_1$ by (2). Since $C \in \mathfrak{C}(x, T_1, \boldsymbol{D}_1)$, there follows the existence of an indicator domain D for $(x, T_1, \boldsymbol{D}_1)$ such that $C < D, \overline{D} < O^*$. As $O^* < G$ and $T_1|G = T_2|G$, we have also $T_1|\overline{D} = T_2|\overline{D}$. Hence (by the initial remarks in the present section) D is an indicator domain for $(x, T_2, \boldsymbol{D}_2)$ also, and $C < D, \overline{D} < O^* < O$. Thus (ii) is proved.

Proof of (iii). Suppose that $C \in \mathfrak{C}_i(x, T_1, \boldsymbol{D}_1)$. Then $C \in \mathfrak{C}(x, T_1, \boldsymbol{D}_1)$ and hence also $C \in \mathfrak{C}(x, T_2, \boldsymbol{D}_2)$ by (ii). To prove (iii), there remains to exhibit a characteristic neighborhood of C (relative to $T_2, \boldsymbol{D}_2$) in the sense of **II.3.3**, definition 2. Now since $C \in \mathfrak{C}_i(x, T_1, \boldsymbol{D}_1)$, there exists a characteristic neighborhood O_1 of C relative to $T_1, \boldsymbol{D}_1$. On setting $O_2 = O_1 \cap G$, it follows readily that O_2 is a characteristic neighborhood of C relative to $T_2, \boldsymbol{D}_2$. Indeed, obviously $C < O_2$, $\overline{O}_2 < \overline{G} < \boldsymbol{D}_2$ by (2). Thus we have to show yet that C is the only element of $\mathfrak{C}(x, T_2, \boldsymbol{D}_2)$ contained in O_2. Now if $C_2 \in \mathfrak{C}(x, T_2, \boldsymbol{D}_2)$ and $C_2 < O_2$, then also $C_2 < G$, and on reversing in (ii) the roles of $T_1|\boldsymbol{D}_1$ and $T_2|\boldsymbol{D}_2$, it follows that $C_2 \in \mathfrak{C}(x, T_1, \boldsymbol{D}_1)$. As $C_2 < O_2 < O_1$ and O_1 is a characteristic neighborhood of C (relative to $T_1, \boldsymbol{D}_1$), C_2 must coincide with C, and thus it is established that C is the only element of $\mathfrak{C}(x, T_2, \boldsymbol{D}_2)$ contained in O_2.

Proof of (iv). Suppose that $i_e(C, T_1)$ is defined. Then $C \in \mathfrak{C}_i(x, T_1, \boldsymbol{D}_1)$ and hence $C \in \mathfrak{C}_i(x, T_2, \boldsymbol{D}_2)$ by (iii). Thus $i_e(C, T_2)$ is defined. Let O_1, O_2 be characteristic neighborhoods of C relative to $T_1, \boldsymbol{D}_1$ and

$T_2, \boldsymbol{D}_2$ respectively. Put $O = O_1 \cap O_2 \cap G$. Then clearly O is a characteristic neighborhood of C, in view of (2), with respect to both $T_1, \boldsymbol{D}_1$ and $T_2, \boldsymbol{D}_2$. Let D be an indicator domain for $(x, T_1, \boldsymbol{D}_1)$ such that $C \subset D, \bar{D} \subset O$. Since $T_1|O = T_2|O$ in view of (2) and $\bar{D} \subset O$, it follows by the initial remarks in the present section that

$$\mu(x, T_1, D) = \mu(x, T_2, D). \tag{6}$$

On the other hand, by definition 1 in **II.3.4**,

$$i_e(C, T_1) = \mu(x, T_1, D), \quad i_e(C, T_2) = \mu(x, T_2, D).$$

By (6) it follows that $i_e(C, T_1) = i_e(C, T_2)$, and the proof is complete.

Remark 1. Given T and $\boldsymbol{D}$ as in **II.3.1**, consider a domain $\boldsymbol{D}^* \subset \boldsymbol{D}$. Put $\boldsymbol{D}_1 = \boldsymbol{D}^*$, $\boldsymbol{D}_2 = \boldsymbol{D}$, $T_1 = T_2 = T$, and let x be a point in R^n and C a continuum in $\boldsymbol{D}^*$. The following statements are then direct consequences of the preceding theorem.

(i) C is an m.m.c. for $(x, T, \boldsymbol{D}^*)$ if and only if it is an m.m.c. for $(x, T, \boldsymbol{D})$.

(ii) $C \in \mathfrak{C}(x, T, \boldsymbol{D}^*)$ if and only if $C \in \mathfrak{C}(x, T, \boldsymbol{D})$.

(iii) $C \in \mathfrak{C}_i(x, T, \boldsymbol{D}^*)$ if and only if $C \in \mathfrak{C}_i(x, T, \boldsymbol{D})$.

(iv) $i_e(C, T)$ is defined relative to $T, \boldsymbol{D}^*$ if and only if it is defined relative to $T, \boldsymbol{D}$ and has the same value in either case.

Remark 2. Given $T, \boldsymbol{D}, \boldsymbol{D}^*$ as in remark 1, take a point x in R^n. Then **II.3.3**, definition 1 applies to T and $\boldsymbol{D}^*$ and yields a multiplicity function $k(x, T, \boldsymbol{D}^*)$. By part (ii) of remark 1 it follows that $k(x, T, \boldsymbol{D}^*)$ is equal to the number of those continua $C \in \mathfrak{C}(x, T, \boldsymbol{D})$ that lie in $\boldsymbol{D}^*$.

Remark 3. Given $T, \boldsymbol{D}, \boldsymbol{D}^*, x$ as in remark 2, suppose that $k(x, T, \boldsymbol{D}^*) < \infty$. By **II.3.4**, theorem 3 (applied to $T, \boldsymbol{D}^*$) we have then also $K(x, T, \boldsymbol{D}^*) < \infty$, and **II.3.4**, definition 2 yields the algebraic multiplicity function

$$\mu_e(x, T, \boldsymbol{D}^*) = \sum i_e(C, T), \quad C \in \mathfrak{C}_i(x, T, \boldsymbol{D}^*).$$

In view of remark 1 we have the equivalent formula

$$\mu_e(x, T, \boldsymbol{D}^*) = \sum i_e(C, T), \quad C \in \mathfrak{C}_i(x, T, \boldsymbol{D}), \quad C \subset \boldsymbol{D}^*.$$

Similarly, remark 1 and **II.3.4**, theorem 4 yield the formula

$$K(x, T, \boldsymbol{D}^*) = \sum |i_e(C, T)|, \quad C \in \mathfrak{C}_i(x, T, \boldsymbol{D}), \quad C \subset \boldsymbol{D}^*.$$

Remark 4. Given T and $\boldsymbol{D}$ as in **II.3.1**, consider a domain $D \subset \boldsymbol{D}$ and a point $x \in R^n$ such that

$$\bar{D} \subset \boldsymbol{D}, \quad x \notin T\, fr\, D. \tag{7}$$

Then $\mu(x, T, D)$ is defined (see **II.2.2**). Assume that

$$\mu(x, T, D) \neq 0. \tag{8}$$

Then

$$K(x, T, D) \geq 1. \tag{9}$$

Indeed, (7) and (8) imply that D is an indicator domain for $(x, T, \boldsymbol{D})$. By **II.3.3**, lemma 4 it follows that D contains at least one e.m.m.c. C for $(x, T, \boldsymbol{D})$. By remark 1, C is then also an e.m.m.c. for (x, T, D), and thus $k(x, T, D) \geq 1$. From lemma 6 in **II.3.3** (applied to D), we conclude that

$$K(x, T, D) \geq k(x, T, D) \geq 1,$$

and (9) is verified.

II.3.6. Essential sets. Given T and $\boldsymbol{D}$ as in **II.3.1**, denote by $E(x, T, \boldsymbol{D})$ the union of all the continua $C \in \mathfrak{C}(x, T, \boldsymbol{D})$ (see **II.3.3**, definition 1). In symbols:

$$E(x, T, \boldsymbol{D}) = \bigcup C, \quad C \in \mathfrak{C}(x, T, \boldsymbol{D}). \tag{1}$$

Next, denote by $E(T, \boldsymbol{D})$ the union of all the sets $E(x, T, \boldsymbol{D})$ corresponding to all the points $x \in R^n$. In symbols:

$$E(T, \boldsymbol{D}) = \bigcup E(x, T, \boldsymbol{D}), \quad x \in R^n. \tag{2}$$

Let $\mathfrak{C}^p(x, T, \boldsymbol{D})$ be the class of all those continua $C \in \mathfrak{C}(x, T, \boldsymbol{D})$ which reduce to single points, and put

$$E^p(x, T, \boldsymbol{D}) = \bigcup C, \quad C \in \mathfrak{C}^p(x, T, \boldsymbol{D}), \tag{3}$$

$$E^p(T, \boldsymbol{D}) = \bigcup E^p(x, T, \boldsymbol{D}), \quad x \in R^n. \tag{4}$$

Similarly, the class $\mathfrak{C}_i(x, T, \boldsymbol{D})$ (see **II.3.3**, definition 2) gives rise to the sets

$$E_i(x, T, \boldsymbol{D}) = \bigcup C, \quad C \in \mathfrak{C}_i(x, T, \boldsymbol{D}), \tag{5}$$

$$E_i(T, \boldsymbol{D}) = \bigcup E_i(x, T, \boldsymbol{D}), \quad x \in R^n. \tag{6}$$

On denoting by $\mathfrak{C}_i^p(x, T, \boldsymbol{D})$ the class of those continua $C \in \mathfrak{C}_i(x, T, \boldsymbol{D})$ which reduce to single points, we obtain [in analogy with (3), (4)] the sets

$$E_i^p(x, T, \boldsymbol{D}) = \bigcup C, \quad C \in \mathfrak{C}_i^p(x, T, \boldsymbol{D}), \tag{7}$$

$$E_i^p(T, \boldsymbol{D}) = \bigcup E_i^p(x, T, \boldsymbol{D}), \quad x \in R^n. \tag{8}$$

Some or all of the sets defined in (1) to (8) may be empty. The following relations are direct consequences of the definitions.

$$E_i(T, \boldsymbol{D}) \subset E(T, \boldsymbol{D}) \subset \boldsymbol{D}, \tag{9}$$

$$E_i^p(T, \boldsymbol{D}) \subset E^p(T, \boldsymbol{D}) \subset \boldsymbol{D}, \tag{10}$$

$$E^p(T, \boldsymbol{D}) \subset E(T, \boldsymbol{D}), \tag{11}$$

$$E_i^p(T, \boldsymbol{D}) \subset E_i(T, \boldsymbol{D}), \tag{12}$$

$$E_i^p(T, \boldsymbol{D}) = E^p(T, \boldsymbol{D}) \cap E_i(T, \boldsymbol{D}). \tag{13}$$

Definition 1. The sets $E(T, \boldsymbol{D})$, $E^p(T, \boldsymbol{D})$, $E_i(T, \boldsymbol{D})$, $E_i^p(T, \boldsymbol{D})$ are termed the *essential sets* for $(T, \boldsymbol{D})$.

Our first objective in this section is to show that these essential sets are BOREL sets (see **I.1.4**, definition 11). In preparation, we shall presently introduce and discuss some auxiliary sets.

Definition 2. Let j and m be integers such that $j \geqq 0$, $m > 0$. Then $S_{jm}(T, \boldsymbol{D})$ will denote the set of those points $u \in \boldsymbol{D}$ for which there exists a domain D (depending upon u, j, m) with the following properties.

(i) $u \in D$, $\bar{D} \subset \boldsymbol{D}$.

(ii) D is an indicator domain for $(Tu, T, \boldsymbol{D})$.

(iii) $\delta T \bar{D} < 1/m$.

(iv) $k(Tu, T, D) > j$ (see **II.3.3**, definition 1 and **II.3.5**, remark 2).

Lemma 1. The set $S_{jm}(T, \boldsymbol{D})$ is open.

Proof. The assertion is obvious if $S_{jm}(T, \boldsymbol{D}) = \emptyset$. So assume that $S_{jm}(T, \boldsymbol{D}) \neq \emptyset$, and consider an arbitrary point $u_0 \in S_{jm}(T, \boldsymbol{D})$. Then there exists a corresponding domain D with the properties (i) to (iv) stated in definition 2. Property (ii) implies that $Tu_0 \notin T\,fr\,D$ and $\mu(Tu_0, T, D) \neq 0$. By **II.2.3**, theorem 1 there follows the existence of an open set G_1 such that

$$T u_0 \in G_1, \tag{14}$$

$$\mu(x, T, D) = \mu(T u_0, T, D) \neq 0 \quad \text{for} \quad x \in G_1. \tag{15}$$

Since $k(x, T, D)$ is a lower semi-continuous function of x (see **II.3.3**, lemma 7), property (iv) implies the existence of an open set G_2 such that

$$T u_0 \in G_2, \quad k(x, T, D) > j \quad \text{for} \quad x \in G_2. \tag{16}$$

Then $G_1 \cap G_2$ is an open set containing Tu_0. Since T is continuous, there exists an open set O such that

$$u_0 \in O \subset D, \quad TO \subset G_1 \cap G_2. \tag{17}$$

Clearly, the lemma will be proved if we can show that $O \subset S_{jm}(T, \boldsymbol{D})$. In turn, this inclusion will be established by showing that the domain D possesses the properties (i) to (iv) stated in definition 2 relative to every point $\tilde{u} \in O$. For the properties (i) and (iii) this is obvious. As regards property (ii), we have $T\tilde{u} \in TO \subset G_1$ by (17), and hence $\mu(T\tilde{u}, T, D) \neq 0$ by (15). Thus D is an indicator domain for $(T\tilde{u}, T, \boldsymbol{D})$, and property (ii) is verified. By (17) we have $T\tilde{u} \in TO \subset G_2$, and hence $k(T\tilde{u}, T, D) > j$ by (16). Thus property (iv) is also verified, and the lemma is proved.

Definition 3. If j is a non-negative integer, then the set $S_j(T, \boldsymbol{D})$ is defined by the formula

$$S_j(T, \boldsymbol{D}) = \bigcap_{m=1}^{\infty} S_{jm}(T, \boldsymbol{D}). \tag{18}$$

Lemma 2. $S_j(T, \boldsymbol{D}) > S_{j+1}(T, \boldsymbol{D})$.

Proof. Obviously $S_{jm}(T, \boldsymbol{D}) > S_{j+1,m}(T, \boldsymbol{D})$, and thus the assertion is a direct consequence of (18).

Lemma 3. If $u_0 \in S_j(T, \boldsymbol{D})$, then there exists a continuum C_0 and a sequence of domains $\{D_m\}$ such that the following holds.

(i) $u_0 \in D_m$, $\overline{D}_m < \boldsymbol{D}$.

(ii) Each D_m is an indicator domain for $(T u_0, T, \boldsymbol{D})$.

(iii) $\delta T \overline{D}_m < 1/m$.

(iv) $k(T u_0, T, D_m) > j$.

(v) $u_0 \in C_0 < D_m$, $C_0 \in \mathfrak{C}(T u_0, T, \boldsymbol{D})$.

Proof. In view of (18), the assumption implies that $u_0 \in S_{jm}(T, \boldsymbol{D})$ fcr every integer $m \geq 1$. Accordingly, for every integer $m \geq 1$ there follows, by definition 2, the existence of a domain D_m with the properties (i) to (iv). There remains to verify the existence of the continuum C_0 satisfying (v). Observe that the properties (i) to (iii) imply, by **II.3.1**, lemma 9, the existence of a subsequence $\{D_{m_r}\}$ which is a determining sequence for $(T u_0, T, \boldsymbol{D})$. In view of property (ii), this determining sequence is essential (see **II.3.3**, definition 3). Hence by **II.3.3**, lemma 2 the set

$$C_0 = \bigcap \overline{D}_{m_r} \tag{19}$$

is an e.m.m.c. for $(T u_0, T, \boldsymbol{D})$. Since $u_0 \in C_0$ by property (i), we have verified property (v) except for the inclusion

$$C_0 < D_m, \quad m = 1, 2, \ldots . \tag{20}$$

Now since D_m is $(T u_0, T)$-admissible by property (ii), we conclude from **II.3.1**, lemma 5 that either (20) holds or else $C_0 < \complement \overline{D}_m$. The second alternative is ruled out by the observation that $C_0 \cap D_m \neq \emptyset$ since $u_0 \in C_0 \cap D_m$. Thus (20) holds, and the lemma is proved.

Lemma 4. $S_0(T, \boldsymbol{D}) = E(T, \boldsymbol{D})$.

Proof. We first verify the inclusion

$$S_0(T, \boldsymbol{D}) < E(T, \boldsymbol{D}). \tag{21}$$

If $S_0(T, \boldsymbol{D}) = \emptyset$, then (21) is obvious. So assume that $S_0(T, \boldsymbol{D}) \neq \emptyset$, and consider an arbitrary point $u_0 \in S_0(T, \boldsymbol{D})$. By part (v) of lemma 3 and in view of (1) and (2) we have

$$u_0 \in E(T u_0, T, \boldsymbol{D}) < E(T, \boldsymbol{D}),$$

and (21) follows. There remains to verify the inclusion

$$E(T, \boldsymbol{D}) \subset S_0(T, \boldsymbol{D}). \tag{22}$$

If $E(T, \boldsymbol{D}) = \emptyset$, then (22) is obvious. So assume that $E(T, \boldsymbol{D}) \neq \emptyset$ and consider an arbitrary point $u_0 \in E(T, \boldsymbol{D})$. In view of (1) and (2) there exists a continuum C_0 such that

$$u_0 \in C_0, \quad C_0 \in \mathfrak{C}(T u_0, T, \boldsymbol{D}). \tag{23}$$

Take now any integer $m \geq 1$ and consider the spherical neighborhood $\Delta(T u_0, 1/2m)$ (see **I.1.4**, definition 2). The set

$$O = \boldsymbol{D} \cap T^{-1} \Delta(T u_0, 1/2m) \tag{24}$$

is then an open set containing C_0. Since C_0 is an e.m.m.c. for $(T u_0, T, \boldsymbol{D})$, there follows the existence of an indicator domain D for $(T u_0, T, \boldsymbol{D})$ such that

$$C_0 \subset D, \quad \bar{D} \subset O. \tag{25}$$

From (24) and (25) we see that

$$T\bar{D} \subset TO \subset \Delta(T u_0, 1/2m),$$

and hence $\delta T\bar{D} < 1/m$. Next we verify that

$$k(T u_0, T, D) > 0. \tag{26}$$

Indeed, D contains the continuum $C_0 \in \mathfrak{C}(T u_0, T, \boldsymbol{D})$, and thus (26) holds by **II.3.5**, remark 2. In summary: the domain D satisfies, relative to u_0, the conditions (i) to (iv) in definition 2 for $j = 0$ and every integer $m \geq 1$. Hence $u_0 \in S_0(T, \boldsymbol{D})$, and (22) follows.

Lemma 5. $E_i(T, \boldsymbol{D}) \cap S_1(T, \boldsymbol{D}) = \emptyset$.

Proof. We show that the assumption of the existence of a point

$$u_0 \in E_i(T, \boldsymbol{D}) \cap S_1(T, \boldsymbol{D}) \tag{27}$$

leads to a contradiction as follows. Since $u_0 \in S_1(T, \boldsymbol{D})$, from lemma 3 we infer the existence of a sequence of domains $\{D_m\}$ and of a continuum C_0 with the properties (i) to (v) stated there, with $j = 1$. In view of property (v), C_0 is then the unique m.m.c. containing u_0, and since $u_0 \in E_i(T, \boldsymbol{D})$ it follows [see (5), (6)] that $C_0 \in \mathfrak{C}_i(T u_0, T, \boldsymbol{D})$. Accordingly (see **II.3.3**, definition 2) there exists a characteristic neighborhood O of C_0. From the definition of such a neighborhood it follows that if a domain D satisfies the conditions

$$C_0 \subset D, \bar{D} \subset O, \tag{28}$$

then C_0 is the only element of the class $\mathfrak{C}(Tu_0, T, \boldsymbol{D})$ that lies in D. Hence, by remark 2 in **II.3.5**, (28) implies that

$$k(Tu_0, T, \boldsymbol{D}) = 1. \tag{29}$$

As noted in the proof of lemma 3, the sequence $\{D_m\}$ contains a subsequence $\{D_{m_r}\}$ which is a determining sequence for $(Tu_0, T, \boldsymbol{D})$ such that

$$C_0 = \bigcap \overline{D}_{m_r}. \tag{30}$$

Since $C_0 \subset O$, it follows by (v) in lemma 3 and **I.1.3**, exercise 24 that

$$C_0 \subset D_{m_r}, \quad \overline{D}_{m_r} \subset O \quad \text{for } r \text{ large}. \tag{31}$$

Let r be so chosen that (31) holds. Then by (28), (29), we have

$$k(Tu_0, T, D_{m_r}) = 1, \tag{32}$$

while by property (iv) in lemma 3 we should have

$$k(Tu_0, T, D_{m_r}) > 1, \tag{33}$$

since now $j = 1$. Thus (27) leads to the contradictory relations (32) and (33), and the lemma is proved.

Lemma 6. $E_i(T, \boldsymbol{D}) = S_0(T, \boldsymbol{D}) - S_1(T, \boldsymbol{D})$.

Proof. By (9) and lemma 4 we have $E_i(T, \boldsymbol{D}) \subset S_0(T, \boldsymbol{D})$, while lemma 5 yields $E_i(T, \boldsymbol{D}) \subset \complement S_1(T, \boldsymbol{D})$. Hence

$$E_i(T, \boldsymbol{D}) \subset S_0(T, \boldsymbol{D}) - S_1(T, \boldsymbol{D}).$$

There remains to verify the inclusion

$$S_0(T, \boldsymbol{D}) - S_1(T, \boldsymbol{D}) \subset E_i(T, \boldsymbol{D}). \tag{34}$$

This inclusion will be proved by showing that the assumption

$$u_0 \in S_0(T, \boldsymbol{D}) - S_1(T, \boldsymbol{D}) \tag{35}$$

implies that

$$u_0 \in E_i(T, \boldsymbol{D}). \tag{36}$$

So assume that (35) holds. Then $u_0 \in S_0(T, \boldsymbol{D})$. By lemma 3, applied with $j = 0$, there follows the existence of a sequence of domains $\{D_m\}$ and of a continuum C_0 with the properties (i) to (v) stated there. In particular

$$k(Tu_0, T, D_m) > 0, \quad m = 1, 2, \ldots. \tag{37}$$

We assert that

$$k(Tu_0, T, D_m) = 1 \quad \text{for some } m. \tag{38}$$

Indeed, if (38) is denied, then in view of (37) we should have $k(Tu_0, T, D_m) > 1$ for every integer $m \geq 1$, and hence it would follow that $u_0 \in S_1(T, \boldsymbol{D})$, in contradiction with (35). Thus (38) is verified. If m is an integer such that (38) holds, then by **II.3.5**, remark 2 it follows that C_0 is the unique element of the class $\mathfrak{C}(Tu_0, T, \boldsymbol{D})$ contained in D_m. As D_m is open, we conclude that $C_0 \in \mathfrak{C}_i(Tu_0, T, \boldsymbol{D})$, and hence [see (5), (6)]

$$u_0 \in C_0 \subset E_i(Tu_0, T, \boldsymbol{D}) \subset E_i(T, \boldsymbol{D}).$$

Thus (36) is established and the lemma is proved.

Definition 4. If m is a positive integer, then $S_m^*(T, \boldsymbol{D})$ denotes the set of all those points $u \in \boldsymbol{D}$ for which there exists a domain D (depending upon m and u) with the following properties.

(i) $u \in D, \overline{D} \subset \boldsymbol{D}$.

(ii) D is (Tu, T)-admissible (see **II.2.1**).

(iii) $\delta\overline{D} < 1/m$ (see **I.1.4**, definition 4).

Lemma 7. The set $S_m^*(T, \boldsymbol{D})$ is open.

The proof is similar to that of lemma 1.

Definition 5. The set $S^*(T, \boldsymbol{D})$ is defined by the formula

$$S^*(T, \boldsymbol{D}) = \bigcap_{m=1}^{\infty} S_m^*(T, \boldsymbol{D}). \tag{39}$$

Lemma 8. If $u_0 \in \boldsymbol{D}$, then $u_0 \in S^*(T, \boldsymbol{D})$ if and only if the point u_0 by itself is an m.m.c. for $(Tu_0, T, \boldsymbol{D})$.

This is a ready consequence of the definitions involved.

Lemma 9. $E^p(T, \boldsymbol{D}) = E(T, \boldsymbol{D}) \cap S^*(T, \boldsymbol{D})$.

Lemma 10. $E_i^p(T, \boldsymbol{D}) = E_i(T, \boldsymbol{D}) \cap S^*(T, \boldsymbol{D})$.

Since the proofs of these two lemmas are entirely similar, we indicate merely the proof of lemma 10. If $u \in E_i^p(T, \boldsymbol{D})$, then u by itself is an m.m.c. (and indeed an e.m.m.c.) for $(Tu, T, \boldsymbol{D})$, and hence $u \in S^*(T, \boldsymbol{D})$ by lemma 8. Thus $E_i^p(T, \boldsymbol{D}) \subset S^*(T, \boldsymbol{D})$. In view of (12) it follows that

$$E_i^p(T, \boldsymbol{D}) \subset E_i(T, \boldsymbol{D}) \cap S^*(T, \boldsymbol{D}). \tag{40}$$

Next, assume that $u \in E_i(T, \boldsymbol{D}) \cap S^*(T, \boldsymbol{D})$. Then $u \in E_i(T, \boldsymbol{D})$, and hence there exists a continuum $C \in \mathfrak{C}_i(Tu, T, \boldsymbol{D})$ such that $u \in C$. But since also $u \in S^*(T, \boldsymbol{D})$, this continuum C must reduce to the single point u by lemma 8, and hence $u \in E_i^p(T, \boldsymbol{D})$. Thus

$$E_i(T, \boldsymbol{D}) \cap S^*(T, \boldsymbol{D}) \subset E_i^p(T, \boldsymbol{D}),$$

and in view of (40) lemma 10 is proved.

Theorem 1. The essential sets $E(T, \boldsymbol{D})$, $E_i(T, \boldsymbol{D})$, $E^p(T, \boldsymbol{D})$, $E_i^p(T, \boldsymbol{D})$ are Borel sets.

Proof. Since the set $S_{jm}(T, \boldsymbol{D})$ is open by lemma 1, the set $S_j(T, \boldsymbol{D})$, as a countable intersection of Borel sets, is a Borel set. Hence $E(T, \boldsymbol{D})$ is a Borel set by lemma 4, and $E_i(T, \boldsymbol{D})$ is a Borel set by lemma 6. As the set $S_m^*(T, \boldsymbol{D})$ is open by lemma 7, the set $S^*(T, \boldsymbol{D})$, as a countable intersection of Borel sets, is a Borel set. In view of lemma 9 and lemma 10, $E^p(T, \boldsymbol{D})$ and $E_i^p(T, \boldsymbol{D})$ are (finite) intersections of Borel sets and hence these two sets are themselves Borel sets.

Our next objective is to study certain auxiliary sets related to the local index $i_e(C, T)$.

Definition 6. If s is an integer different from zero and m is a positive integer, then $F_{sm}(T, \boldsymbol{D})$ will denote the set of those points $u \in \boldsymbol{D}$ for which there exists a domain D (depending upon u, s, m) with the following properties.

(i) $u \in D, \overline{D} \subset \boldsymbol{D}$.

(ii) D is an indicator domain for $(Tu, T, \boldsymbol{D})$.

(iii) $\delta T\overline{D} < 1/m$.

(iv) $\mu(Tu, T, D) = s \neq 0$.

Lemma 11. The set $F_{sm}(T, \boldsymbol{D})$ is open.

The proof is entirely similar to that of lemma 1.

Definition 7. If s is an integer different from zero, then the set $F_s(T, \boldsymbol{D})$ is defined by the formula

$$F_s(T, \boldsymbol{D}) = \bigcap_{m=1}^{\infty} F_{sm}(T, \boldsymbol{D}). \tag{41}$$

Lemma 12. If $u \in F_s(T, \boldsymbol{D})$, then there exists a continuum C and a sequence of domains $\{D^r\}$ such that the following holds.

(i) $u \in C \subset D^r$ for every r.

(ii) $C \in \mathfrak{C}(Tu, T, \boldsymbol{D})$.

(iii) $\{D^r\}$ is a determining sequence for $(Tu, T, \boldsymbol{D})$.

(iv) $C = \bigcap \overline{D}^r$.

(v) $\mu(Tu, T, D^r) = s$ for every r.

Proof. The assumption $u \in F_s(T, \boldsymbol{D})$ implies [see (41) and definition 6] that for every positive integer m there exists a domain D_m such that

$$u \in D_m, \overline{D}_m \subset \boldsymbol{D}, \tag{42}$$

$$\delta T\overline{D}_m < 1/m, \tag{43}$$

$$\mu(Tu, T, D_m) = s, \tag{44}$$

and each D_m is an indicator domain for $(Tu, T, \boldsymbol{D})$. By **II.3.1**, lemma 9 there follows the existence of a subsequence $\{D_{m_r}\}$ which is a determining sequence for $(Tu, T, \boldsymbol{D})$. Since $s \neq 0$, this determining sequence is essential (see **II.3.3**, definition 3). Hence, by lemma 2 in **II.3.3**, the set $C = \bigcap \overline{D}_{m_r}$ is an e.m.m.c. for Tu, T, $\boldsymbol{D}$), and $C \subset D_{m_r}$ for every r. In view (42) and (44) we have

$$u \in C, \quad \mu(Tu, T, D_{m_r}) = s \quad \text{for every } r.$$

Hence on setting $D^r = D_{m_r}$, the conditions (i) to (v) hold, and the lemma is proved.

Definition 8. If s is an integer different from zero, then $F_s^*(T, \boldsymbol{D})$ denotes the union of all those continua C which satisfy the following conditions.

(i) $C \in \mathfrak{C}_i(x, T, \boldsymbol{D})$ for some point $x \in R^n$.

(ii) $i_e(C, T) = s$.

Lemma 13. $F_s^*(T, \boldsymbol{D}) = F_s(T, \boldsymbol{D}) \cap E_i(T, \boldsymbol{D})$.

Proof. We assert the inclusion

$$F_s(T, \boldsymbol{D}) \cap E_i(T, \boldsymbol{D}) \subset F_s^*(T, \boldsymbol{D}), \tag{45}$$

where we can assume that the set on the left in (45) is non-empty, since otherwise the inclusion is trivial. Consider any point

$$u \in F_s(T, \boldsymbol{D}) \cap E_i(T, \boldsymbol{D}).$$

Then $u \in F_s(T, \boldsymbol{D})$, and hence there exists a continuum C and a sequence $\{D^r\}$ with the properties stated in lemma 12. Since also $u \in E_i(T, \boldsymbol{D})$, it follows that $C \in \mathfrak{C}_i(Tu, T, \boldsymbol{D})$, and there remains to show (see definition 8) that

$$i_e(C, T) = s. \tag{46}$$

Let O be a characteristic neighborhood of C (see **II.3.3**, definition 2). In view of (iii), (iv) in lemma 12, it follows by **I.1.3**, exercise 24, that $\overline{D}^r \subset O$ for r sufficiently large. If $\overline{D}^r \subset O$, then (see **II.3.4**, definition 1) $i_e(C, T) = \mu(Tu, T, D^r)$, and (46) follows by (v) in lemma 12. Thus (45) is verified. The complementary inclusion

$$F_s^*(T, \boldsymbol{D}) \subset F_s(T, \boldsymbol{D}) \cap E_i(T, \boldsymbol{D})$$

is established in a similar manner, and the lemma follows.

Theorem 2. The set $F_s^*(T, \boldsymbol{D})$ is a Borel set.

Proof. Since $F_{sm}(T, \boldsymbol{D})$ is open by lemma 11, the set $F_s(T, \boldsymbol{D})$ is a countable intersection of open sets, and hence it is a Borel set. As $E_i(T, \boldsymbol{D})$ is a Borel set by theorem 1, $F_s^*(T, \boldsymbol{D})$ appears (in view of lemma 13) as the intersection of two Borel sets and hence this set is also a Borel set.

The following two lemmas are included in view of applications in the sequel.

Lemma 14. Let u be a point in $\boldsymbol{D}$. Then $u \in E_i^p(T, \boldsymbol{D})$ if and only if for every $\varepsilon > 0$ there exists a domain $D_{u\varepsilon}$ (which depends upon u and ε) such that the following holds.

(i) $u \in D_{u\varepsilon} \subset \boldsymbol{D}$.

(ii) $\delta D_{u\varepsilon} < \varepsilon$.

(iii) $1 \leq K(Tu, T, D_{u\varepsilon}) < \infty$.

(iv) If D is any domain such that $u \in D \subset D_{u\varepsilon}$, then

$$K(Tu, T, D) = K(Tu, T, D_{u\varepsilon}).$$

Proof. Assume first that for every $\varepsilon > 0$ there exists a domain $D_{u\varepsilon}$ satisfying the conditions (i) to (iv). Consider the domain D_{u1} (corresponding to $\varepsilon = 1$). By (iii) and **II.3.4**, theorem 3 (applied to D_{u1}) it follows that $k(Tu, T, D_{u1})$ is finite and different from zero. In view of remark 2 in **II.3.5**, we conclude that D_{u1} contains a finite, non-zero number of essential maximal model continua $C_1, \ldots, C_m$ for $(Tu, T, \boldsymbol{D})$. Clearly (see **II.3.3**, definition 2)

$$C_1, \ldots, C_m \in \mathfrak{C}_i(Tu, T, \boldsymbol{D}). \tag{47}$$

Consider now a domain $D_{u\varepsilon}$. In view of (ii), there exists a number $\eta > 0$ such that

$$u \in D_{u\varepsilon} \subset D_{u1} \quad \text{if} \quad 0 < \varepsilon < \eta. \tag{48}$$

Assume that $0 < \varepsilon < \eta$. In view of (iii) and **II.3.4**, theorem 3 (applied to $D_{u\varepsilon}$) we have

$$k(Tu, T, D_{u\varepsilon}) \geq 1.$$

By remark 2 in **II.3.5**, we conclude that $D_{u\varepsilon}$ contains at least one e.m.m.c. C for $(Tu, T, \boldsymbol{D})$. In view of (48), it follows that $C \subset D_{u1}$, and hence C must coincide with one of the continua $C_1, \ldots, C_m$. In other words, for arbitrarily small values of $\varepsilon > 0$ at least one of the continua $C_1, \ldots, C_m$ must lie in $D_{u\varepsilon}$. In view of (ii) this clearly implies that one of $C_1, \ldots, C_m$ contains u. Let the notations be so chosen that $u \in C_1$. By the preceding remarks, we have then

$$u \in C_1 \subset D_{u\varepsilon} \quad \text{for} \quad 0 < \varepsilon < \eta. \tag{49}$$

In view of (ii), it is clear from (49) that C_1 must reduce to the point u, and by (47) it follows then that $u \in E_i^p(T, \boldsymbol{D})$ [see **II.3.6** (8)]. Suppose, conversely, that $u \in E_i^p(T, \boldsymbol{D})$. Then u itself is an e.m.m.c. for $(Tu, T, \boldsymbol{D})$, and there exists an open set O such that $u \in O$, $\overline{O} \subset \boldsymbol{D}$, and u itself is the only e.m.m.c. for $(Tu, T, \boldsymbol{D})$ contained in O. Assign now $\varepsilon > 0$ arbitrarily. We can select a domain $D_{u\varepsilon}$ such that

$$u \in D_{u\varepsilon} \subset O, \quad \delta D_{u\varepsilon} < \varepsilon. \tag{50}$$

We assert that $D_{u\varepsilon}$ satisfies the conditions (i) to (iv). For (i) and (ii) this is obvious by (50). Take now any domain D such that $u \in D \subset D_{u\varepsilon}$. Observe that since u itself is the only e.m.m.c. for $(Tu, T, \boldsymbol{D})$ contained in $D_{u\varepsilon}$ and hence in D, we have

$$k(Tu, T, D) = k(Tu, T, D_{u\varepsilon}) = 1.$$

By **II.3.5**, remark 3 it follows that

$$K(Tu, T, D) = |i_e(u, T)| = K(Tu, T, D_{u\varepsilon}). \tag{51}$$

Since $i_e(u, T)$ is finite and different from zero (see **II.3.4**, remark 3), (51) implies (iii) and (iv), and the lemma is proved.

Lemma 15. Let x be a point in R^n such that

$$K(x, T, \boldsymbol{D}) < \infty, \tag{52}$$

$$x \in T[E(T, \boldsymbol{D}) - E_i^p(T, \boldsymbol{D})]. \tag{53}$$

Then there exists a number $\eta = \eta(x) > 0$ such that the following holds: if $D_1, \ldots, D_N$ are pair-wise disjoint domains in $\boldsymbol{D}$ such that $\delta D_j < \eta$, $j = 1, \ldots, N$, then

$$\sum_{j=1}^{N} K(x, T, D_j) < K(x, T, \boldsymbol{D}). \tag{54}$$

Proof. (53) implies that there exists at least one e.m.m.c. for $(x, T, \boldsymbol{D})$. In view of (52) and **II.3.4**, theorem 3 it follows that the number of essential maximal model continua for $(x, T, \boldsymbol{D})$ is finite and different from zero. Let $C_1, \ldots, C_m$ be these continua. At least one of these continua must have a positive diameter, since otherwise (53) could not hold. Let the notations be so chosen that

$$\delta C_1 > 0. \tag{55}$$

We assert that the number

$$\eta = \delta C_1 \tag{56}$$

satisfies the requirements of the lemma. Indeed, let $D_1, \ldots, D_N$ be pair-wise disjoint domains in $\boldsymbol{D}$ such that

$$\delta D_j < \eta, \quad j = 1, \ldots, N. \tag{57}$$

By **II.3.4**, theorem 4 and **II.3.5**, remark 3 we have

$$K(x, T, \boldsymbol{D}) = \sum_{k=1}^{m} |i_e(C_k, T)|, \tag{58}$$

$$K(x, T, D_j) = \sum |i_e(C_k, T)|, \quad C_k \subset D_j. \tag{59}$$

In view of (55), (56), (57) it is clear that the term $|i_e(C_1, T)|$ does not occur for any value $j=1, \ldots, N$ in the formula (59). Thus (58) and (59) yield

$$\sum_{j=1}^{N} K(x, T, D_j) \leq K(x, T, \boldsymbol{D}) - |i_e(C_1, T)|.$$

Since $i_e(C_1, T) \neq 0$, the inequality (54) follows.

II.3.7. The local index $i_{\mathrm{B}}(u, T)$. This index function will be needed later on in connection with a theory of continuous transformations initiated by BANACH, and the purpose of this section is to present its definition and some of its properties. The bounded continuous transformation $T: \boldsymbol{D} \to R^n$ is assumed to be given as in **II.3.1**, where we now assume that $U = \boldsymbol{D}$.

Definition 1. If u and u' are two distinct points in $\boldsymbol{D}$ such that $Tu = Tu'$, then u and u' are termed *relatives* of each other.

Definition 2. The symbol $\mathbf{I}(T, \boldsymbol{D})$ denotes the set of those points $u \in \boldsymbol{D}$ which possess some neighborhood clear of relatives. Thus $u \in \mathbf{I}(T, \boldsymbol{D})$ if and only if there exists an open set O_u such that the conditions $u' \in O_u$, $u' \neq u$ imply that $Tu' \neq Tu$.

Lemma 1. $\mathbf{I}(T, \boldsymbol{D})$ is a BOREL set.

Proof. We can assume that $\mathbf{I}(T, \boldsymbol{D}) \neq \emptyset$, since otherwise the assertion is obvious. Given two positive integers m and n, and a point $u \in \boldsymbol{D}$, let $R_{mn}(u)$ denote the set of those points u' which satisfy the inequalities

$$\frac{1}{m+n} \leq \|u' - u\| \leq \frac{1}{m}.$$

Designate by S_{mn} the set of those points $u \in \boldsymbol{D}$ for which the following holds. (i) $R_{mn}(u) \subset \boldsymbol{D}$. (ii) $R_{mn}(u)$ contains no relative *of* u. In other words, $u' \in R_{mn}(u)$ implies that $u' \in \boldsymbol{D}$ and $Tu' \neq Tu$. One sees readily, by means of elementary arguments, that S_{mn} is open and

$$\mathbf{I}(T, \boldsymbol{D}) = \bigcup_{m=1}^{\infty} \bigcap_{n=1}^{\infty} S_{mn}.$$

Thus $\mathbf{I}(T, \boldsymbol{D})$ is a BOREL set.

We now define the local index $i_{\mathrm{B}}(u, T)$, for $u \in \boldsymbol{D}$, as follows. We set

$$i_{\mathrm{B}}(u, T) = 0 \quad \text{if} \quad u \notin \mathbf{I}(T, \boldsymbol{D}). \tag{1}$$

If $u \in \mathbf{I}(T, \boldsymbol{D})$, then clearly u taken by itself is an m.m.c. for $(Tu, T, \boldsymbol{D})$ (see **II.3.1**, definition 1), and it is then either unessential or essential, in the sense of **II.3.3**, definition 1. Using the *essential sets* introduced in **II.3.6**, clearly a point $u \in \mathbf{I}(T, \boldsymbol{D})$ is an e.m.m.c. if and only if $u \in E_i^p(T, \boldsymbol{D})$. We now set

$$i_{\mathrm{B}}(u, T) = 0 \qquad \text{if} \quad u \in \mathbf{I}(T, \boldsymbol{D}) - E_i^p(T, \boldsymbol{D}), \tag{2}$$

$$i_{\mathrm{B}}(u, T) = i_e(u, T) \quad \text{if} \quad u \in E_i^p(T, \boldsymbol{D}) \cap \mathbf{I}(T, \boldsymbol{D}). \tag{3}$$

As regards (3), note that if $u \in E_i^p(T, \boldsymbol{D})$ then u by itself is an essentially isolated maximal model continuum, and hence the essential local index $i_e(u, T)$ is defined (see **II.3.4**). Thus $i_{\mathrm{B}}(u, T)$ is well defined for $u \in \boldsymbol{D}$.

Lemma 2. $i_{\mathrm{B}}(u, T)$ is Borel measurable in $\boldsymbol{D}$.

Proof. Recall that $\mathbf{I}(T, \boldsymbol{D})$ is a Borel set by lemma 1 and $E_i^p(T, \boldsymbol{D})$ is a Borel set by **II.3.6**, theorem 1. Since $i_{\mathrm{B}}(u, T)$ vanishes on the Borel set $\boldsymbol{D} - \mathbf{I}(T, \boldsymbol{D})$, it suffices to show that $i_{\mathrm{B}}(u, T)$ is Borel measurable on $\mathbf{I}(T, \boldsymbol{D})$. Let s be an integer, and denote by $\mathbf{I}_s(T, \boldsymbol{D})$ the set of those points $u \in \mathbf{I}(T, \boldsymbol{D})$ where $i_{\mathrm{B}}(u, T) = s$. As $i_{\mathrm{B}}(u, T)$ is integral-valued, it is sufficient to show that the sets $\mathbf{I}_s(T, \boldsymbol{D})$ are Borel sets. Now since the essential local index i_e is always different from zero (see **II.3.4**, remark 3), clearly

$$\mathbf{I}_0(T, \boldsymbol{D}) = \mathbf{I}(T, \boldsymbol{D}) - E_i^p(T, \boldsymbol{D}),$$

and thus $\mathbf{I}_0(T, \boldsymbol{D})$ is a Borel set. If $s \neq 0$, then consider the set $F_s^*(T, \boldsymbol{D})$ in **II.3.6**, definition 8. Since $F_s^*(T, \boldsymbol{D})$ is a Borel set by **II.3.6**, theorem 2, and clearly [in view of (2) and (3)]

$$\mathbf{I}_s(T, \boldsymbol{D}) = \mathbf{I}(T, \boldsymbol{D}) \cap F_s^*(T, \boldsymbol{D}) \quad \text{for} \quad s \neq 0,$$

it follows that $\mathbf{I}_s(T, \boldsymbol{D})$ is a Borel set, and the lemma is proved.

Remark. If D is any domain in $\boldsymbol{D}$, then the preceding discussion applies to the transformation $T|D$. Then $\mathbf{I}(T, D)$ denotes the set of those points $u \in D$ which possess a neighborhood clear of relatives. Clearly

$$\mathbf{I}(T, D) = D \cap \mathbf{I}(T, \boldsymbol{D}).$$

In view of the comments in **II.3.5** on *local properties* it is also clear that one has

$$i_{\mathrm{B}}(u, T|D) = i_{\mathrm{B}}(u, T) \quad \text{for} \quad u \in D.$$

II.3.8. Topologically similar transformations. In view of the topological character of the definition of the various multiplicity functions and index functions that we have studied so far, it is clear that these functions should possess certain properties of topological invariance. The purpose of this section is to study this point in some detail.

Definition. Let there be given two bounded continuous transformations

$$T_1 : \boldsymbol{D}_1 \to R^n, \tag{1}$$

$$T_2 : \boldsymbol{D}_2 \to R^n, \tag{2}$$

where $\boldsymbol{D}_1, \boldsymbol{D}_2$ are bounded domains in R^n. Then T_1 and T_2 are termed *topologically similar* if there exists a homeomorphism T from $\boldsymbol{D}_1$ onto $\boldsymbol{D}_2$ such that

$$T_1 u = T_2 T u \quad \text{for} \quad u \in \boldsymbol{D}_1. \tag{3}$$

We shall write

$$T_1 \sim T_2 (t\, s) \tag{4}$$

to state that T_1 and T_2 are topologically similar. In case (4) holds, a homeomorphism T from $\boldsymbol{D}_1$ onto $\boldsymbol{D}_2$ which satisfies (3) will be termed a *connecting homeomorphism.*

To justify the terminology just introduced, observe that if T is a homeomorphism from $\boldsymbol{D}_1$ onto $\boldsymbol{D}_2$ for which (3) holds, then T^{-1} is a homeomorphism from $\boldsymbol{D}_2$ onto $\boldsymbol{D}_1$ such that $T_2 u = T_1 T^{-1} u$ for $u \in \boldsymbol{D}_2$. Thus topological similarity is a symmetric relation. It is also evident that topological similarity is reflexive and transitive.

Let there be given two topologically similar transformations T_1, T_2 and a connecting homeomorphism T as in (1), (2), (3). These transformations are thought of as fixed throughout this section.

Lemma 1. Let D be a domain in $\boldsymbol{D}_1$ and x a point in R^n such that

$$\overline{D} \subset \boldsymbol{D}_1, \quad x \notin T_1 \mathit{fr}\, D. \tag{5}$$

Then TD is a domain in $\boldsymbol{D}_2$ such that

$$\overline{TD} \subset \boldsymbol{D}_2, \quad x \notin T_2 \mathit{fr}\, T D. \tag{6}$$

Proof. Since T is defined on $\overline{D}$ in view of (5), TD is a domain (see **II.2.4**). By **II.2.4** (6) we have $T\overline{D} = \overline{TD}$, and hence

$$\overline{TD} = T\overline{D} \subset T\boldsymbol{D}_1 = \boldsymbol{D}_2. \tag{7}$$

Also, in view of (3) and (7),

$$T_2 \overline{TD} = T_2 T\overline{D} = T_1 \overline{D}. \tag{8}$$

By **II.2.4** (7) we have $T\mathit{fr} D = \mathit{fr}\, TD$, and hence [in view of (3)]

$$T_2 \mathit{fr}\, T D = T_2 T \mathit{fr}\, D = T_1 \mathit{fr}\, D, \tag{9}$$

and the relations (6) follow now from (7), (5) and (9).

Lemma 2. Let D be a domain in $\boldsymbol{D}_1$ and x a point in R^n such that (5) holds. Then (see **II.2.2** and **II.2.4** for notations)

$$\mu(x, T_1, D) = \iota(T, D)\,\mu(x, T_2, T D). \tag{10}$$

Proof. Since (5) holds, (6) also holds by lemma 1. Accordingly, $\mu(x, T_1, D)$, $\mu(x, T_2, TD)$ and $\iota(T, D)$ are defined (see **II.2.2** and **II.2.4**). Take now a frame $[A, B, \Delta]$ which is (x, T_1, D)-admissible (see **II.2.1**, definition 3). Then we have the relations

$$x \in \Delta, \tag{11}$$

$$T_1|\overline{D}:(\overline{D}, fr\, D) \to (A, B). \tag{12}$$

From (8), (9) and (12) it follows that

$$T_2\overline{TD} = T_1\overline{D} \subset A, \quad T_2\, fr\, T D = T_1\, fr\, D \subset B.$$

Hence we have the relation

$$T_2|\overline{TD}:(\overline{TD}, fr\, T D) \to (A, B). \tag{13}$$

Thus the frame $[A, B, \Delta]$ is (x, T_2, TD)-admissible in view of (11) and (13). By the definition of the topological index there follow the relations [see **II.2.2** (3)]

$$(T_1|\overline{D})_*\, \mathfrak{g}[A, B, \Delta] = \mu(x, T_1, D)\, \mathfrak{g}[\overline{D}, fr\, D, D], \tag{14}$$

$$(T_2|\overline{TD})_*\, \mathfrak{g}[A, B, \Delta] = \mu(x, T_2, T D)\, \mathfrak{g}[\overline{TD}, fr\, T D, T D]. \tag{15}$$

Since $T|\overline{D}$ is a homeomorphism from $\overline{D}$ onto $\overline{TD}$, by **II.2.4** (8), (10) there follow the relations

$$T|\overline{D}:(\overline{D}, fr\, D) \to (\overline{TD}, fr\, T D), \tag{16}$$

$$(T|\overline{D})_*\, \mathfrak{g}[\overline{TD}, fr\, T D, T D] = \iota(T, D)\, \mathfrak{g}[\overline{D}, fr\, D, D]. \tag{17}$$

In view of (3), (12), (13), (16), the vector law for induced homomorphisms yields (see **I.6.2**)

$$(T_1|\overline{D})_* = (T|\overline{D})_*\, (T_2|\overline{TD})_*, \tag{18}$$

and (10) follows now from (18), (17), (15) and (14).

Lemma 3. Let x be a point in R^n and C a continuum in $\boldsymbol{D}_1$ which is an m.m.c. for $(x, T_1, \boldsymbol{D}_1)$ (see **II.3.1**, definition 1). Then TC is an m.m.c. for $(x, T_2, \boldsymbol{D}_2)$.

Proof. Since C is a compact and connected subset of $\boldsymbol{D}_1$, TC is a compact and connected subset of $\boldsymbol{D}_2$ (see **I.1.5**, exercise 8). Since $T_1 C = x$, by (3) it follows that $T_2 T C = T_1 C = x$. Thus TC is a continuum in $\boldsymbol{D}_2$ whose image under T_2 is the point x. There remains to prove that TC is a component of the set $T_2^{-1}x$. As TC is connected, there exists a component C_2 of the set $T_2^{-1}x$ such that $TC \subset C_2$. Then $T^{-1}C_2$ is a connected set in $\boldsymbol{D}_1$ such that $C \subset T^{-1}C_2$ and $T_1T^{-1}C_2 = T_2C_2 = x$. Since C is a component of $T_1^{-1}x$, it follows that $C = T^{-1}C_2$ and hence $TC = C_2$. Thus TC is a component of $T_2^{-1}x$, and the lemma is proved.

The preceding lemmas 2 and 3 state invariance properties of the concepts of *topological index* and *maximal model continuum* respectively. Since all of our multiplicity functions and index functions, as well as the various *essential sets*, were defined in terms of these two concepts, the lemmas 3 and 2 yield invariance properties for all the basic concepts introduced in Part II. The following lemma 4 is an immediate consequence of this remark.

Lemma 4. Let there be given two topologically similar transformations T_1, T_2 and a connecting homeomorphism T as in (1), (2), (3). Then (see **II.3.2**, **II.3.6** for terminology)

$$K(x, T_1, \boldsymbol{D}_1) = K(x, T_2, \boldsymbol{D}_2),$$
$$TE(T_1, \boldsymbol{D}_1) = E(T_2, \boldsymbol{D}_2), \quad TE_i(T_1, \boldsymbol{D}_1) = E_i(T_2, \boldsymbol{D}_2),$$
$$TE^p(T_1, \boldsymbol{D}_1) = E^p(T_2, \boldsymbol{D}_2), \quad TE_i^p(T_1, \boldsymbol{D}_1) = E_i^p(T_2, \boldsymbol{D}_2).$$

Part III. Background in Analysis.

§ III.1. Survey of functions of real variables.

III.1.1. Lebesgue measure and Lebesgue integral in Euclidean n-space R^n. If $I: a_i \leq x^i \leq b_i$, $i = 1, \ldots, n$, is an interval in R^n (see **I.2.2**), then the *elementary volume* $v(I)$ of I is defined by the formula

$$v(I) = (b_1 - a_1) \ldots (b_n - a_n).$$

If E is a subset of R^n, then the *exterior* Lebesgue *measure* L^*E of E is defined by the formula

$$\mathsf{L}^* E = \text{gr.l.b.} \sum_j v(I_j),$$

where $\{I_j\}$ is any (finite or infinite) sequence of intervals such that $E \subset \bigcup I_j$, and the greatest lower bound is taken with respect to all such sequences $\{I_j\}$. Thus L^*E may be infinite. If it is desirable to display the dimension n of R^n, then one writes $\mathsf{L}_n^* E$ instead of L^*E. However, n is thought of as fixed throughout this section, and hence the notation L^*E is adequate.

A set $E \subset R^n$ is termed L-measurable (LEBESGUE *measurable*) if the relation $\mathsf{L}^*(A \cup B) = \mathsf{L}^* A + \mathsf{L}^* B$ holds for every pair of sets in R^n such that $A \subset E$, $B \subset \mathsf{C} E$. For L-measurable subsets E of R^n the L-measure (LEBESGUE *measure*) $\mathsf{L} E$ of E is defined by the formula $\mathsf{L} E = \mathsf{L}^* E$. Thus $\mathsf{L} E$ is a non-negative set-function defined for those (and only those) sets $E \subset R^n$ which are L-measurable.

The class of L-measurable subsets of R^n is a completely additive class (see **I.1.4**, definition 10), and $\mathsf{L} E$ is *countably additive* on this class. That is, if $\{E_k\}$ is a finite or infinite sequence of pair-wise disjoint L-measurable sets in R^n, then $\mathsf{L}(\bigcup E_j) = \sum \mathsf{L} E_j$.

If a certain statement $S(x)$, involving a variable point $x \in R^n$, is true for all points x of R^n (or for all points x of a set $E \subset R^n$) with the exception of certain points x in a set F such that $\mathsf{L} F = 0$, then we shall say that $S(x)$ is true a.e. (*almost everywhere*) in R^n (or in E).

Further concepts will now be listed, for convenient reference, in numbered definitions.

Definition 1. Let u_0 be a point of R^n. Consider an L-measurable set $E \subset R^n$. Then u_0 is termed a *point of density* of E if for every sequence $\{Q_j\}$ of oriented n-cubes in R^n (see **I.2.2**), such that $u_0 \in Q_j$, $\delta Q_j \to 0$, one has

$$\lim_{j \to \infty} \frac{\mathsf{L}(E \cap Q_j)}{\mathsf{L} Q_j} = 1 .$$

Definition 2. Let $x_0 = (x_0^1, \ldots, x_0^n)$ be a point of the L-measurable set $E \subset R^n$. Those real numbers h for which

$$(x_0^1 + h, x_0^2, \ldots, x_0^n) \in E ,$$

constitute a set of real numbers. Denote by e the set of these real numbers h. If e (as a subset of the real number-line) is L-measurable and if $h = 0$ is a point of density of e (in the sense of definition 1 applied with $n = 1$), then x_0 is termed a *point of linear density* of E in the direction of the x^1-axis. Points of linear density of E in the direction of the axes $x^2, \ldots, x^n$ are defined in a similar manner.

Definition 3. Let E be an L-measurable subset of R^n. Then the *parameter of regularity* $\pi(E)$ of E is defined by the formula

$$\pi(E) = \text{l.u.b.} \frac{\mathsf{L} E}{\mathsf{L} Q} ,$$

where the least upper bound is taken with respect to all oriented n-cubes $Q \supset E$.

Definition 4. A sequence $\{G_m\}$ of closed subsets of R^n is termed *regular* if there exists a number $\lambda > 0$ such that $\pi(G_m) > \lambda$, $m = 1, 2, \ldots$ (see definition 3).

Definition 5. A real-valued function $f(x)$, defined on an L-measurable set $E \subset R^n$, is termed L-measurable on E if for every real number r the set of those points $x \in E$ where $f(x) > r$ is L-measurable.

Consider an L-measurable set $E \subset R^n$, and let $f(x)$ be a real-valued function which is defined and L-measurable on $E - e$, where e is a subset of E such that $\mathsf{L}e = 0$. If one extends the definition of $f(x)$ by assigning its values on e in an entirely arbitrary manner, then obviously the extended function is L-measurable on E as a consequence of the assumption that $\mathsf{L}e = 0$. On the basis of this observation, we adopt (as a matter of convenience) the following extension of definition 5. If E is an L-measurable set in R^n, and $f(x)$ is a real-valued function defined and L-measurable on $E - e$, where e is a subset of E such that $\mathsf{L}e = 0$, then we shall say that $f(x)$ is L-measurable on E.

We proceed to consider the LEBESGUE *integral* in R^n. Let $f(x)$ be a real-valued, non-negative L-measurable function defined on an L-measurable set $E \subset R^n$. Let $E_1, \ldots, E_m$ be any finite sequence of pair-wise disjoint L-measurable subsets of E such that $E = \bigcup E_j$, $j = 1, \ldots, m$, and denote by $\mu(f, E_j)$ the greatest lower bound of $f(x)$ on E_j, $j = 1, \ldots, m$. If the least upper bound of all the sums of the type

$$\sum_{j=1}^{m} \mu(f, E_j)\, \mathsf{L} E_j$$

is finite, then $f(x)$ is termed L-summable (LEBESGUE *summable*) on E, and the LEBESGUE integral of $f(x)$ on E is defined by the formula

$$\int_E f(x)\, d\,\mathsf{L} = \text{l.u.b.} \sum_{j=1}^{m} \mu(f, E_j)\, \mathsf{L} E_j .$$

If $g(x)$ is a real-valued, L-measurable function (of arbitrary sign) on E, then consider the auxiliary functions $g^+(x)$, $g^-(x)$ defined on E as follows.

$$g^+(x) = \begin{cases} g(x) & \text{if } g(x) \geqq 0, \\ 0 & \text{if } g(x) < 0. \end{cases}$$

$$g^-(x) = \begin{cases} -g(x) & \text{if } g(x) \leqq 0, \\ 0 & \text{if } g(x) > 0. \end{cases}$$

Then $g^+(x)$, $g^-(x)$ are non-negative, L-measurable functions on E. If $g^+(x)$, $g^-(x)$ are both L-summable on E, then $g(x)$ is termed L-summable on E, and its LEBESGUE integral on E is defined by the formula

$$\int_E g(x)\, d\,\mathsf{L} = \int_E g^+(x)\, d\,\mathsf{L} - \int_E g^-(x)\, d\,\mathsf{L} .$$

Let $f(x)$ be a real-valued, L-measurable function in R^n, such that $f(x)$ vanishes for every point $x \in \mathsf{C}E$, where E is a bounded L-measurable subset of R^n. Clearly $f(x)$ is L-summable on R^n if and only if it is L-summable on E. If $f(x)$ is L-summable on R^n and F is any bounded L-measurable set in R^n such that $F \supset E$, then clearly

$$\int_{R^n} f(x)\, d\mathsf{L} = \int_F f(x)\, d\mathsf{L} = \int_E f(x)\, d\mathsf{L} \quad \text{if} \quad f(x) = 0 \quad \text{for} \quad x \in \mathsf{C}E.$$

In the situation just described, we shall write $\int f(x)\, d\mathsf{L}$, instead of $\int_{R^n} f(x)\, d\mathsf{L}$, for the integral of $f(x)$ over R^n.

Definition 6. Let F be a family of L-summable functions on an L-measurable set $E \subset R^n$. Then the family F is said to satisfy the condition (V) (the VITALI condition) on E if for every $\varepsilon > 0$ there exists an $\eta = \eta(\varepsilon) > 0$ such that the following holds: If S is any L-measurable subset of E such that $\mathsf{L}S < \eta$, then

$$\int_S |f(x)|\, d\mathsf{L} < \varepsilon \quad \text{for every} \quad f \in \mathsf{F}.$$

Definition 7. Let A be a subset of R^n. A family F of closed subsets of R^n is said to cover A in the VITALI sense if $A \subset \bigcup F$, $F \in \mathsf{F}$, and for every point $x \in A$ there exists a regular sequence $\{F_j\}$ of sets in F (see definition 4) such that $x \in F_j$ for every j and $\delta F_j \to 0$ for $j \to \infty$.

We shall now list, in the form of lemmas, a series of facts needed later on. The proofs are either contained in the SAKS treatise or the MAYRHOFER treatise (see the *Bibliography*) or can be readily supplied in view of the definitions involved.

Lemma 1. If E is an L-measurable subset of R^n such that $\mathsf{L}E > 0$, then E contains a set which is not L-measurable.

Lemma 2. Let E_1, E_2 be any two L-measurable subsets of R^n. Then

$$\mathsf{L}(E_1 \cup E_2) + \mathsf{L}(E_1 \cap E_2) = \mathsf{L}E_1 + \mathsf{L}E_2.$$

Lemma 3. Let E be a bounded L-measurable subset of R^n. Then $\mathsf{L}E < \infty$, and for every open set O containing E and for every $\varepsilon > 0$ there exists an open set G such that

$$E \subset G \subset O, \quad \mathsf{L}G < \mathsf{L}E + \varepsilon.$$

Lemma 4. Let $\{E_j\}$ be a sequence of L-measurable subsets of R^n such that $E_1 \subset E_2 \subset \dots \subset S$, where S is a bounded subset of R^n. Then $E = \bigcup E_j$ is L-measurable, and $\mathsf{L}(E - E_j) \to 0$ for $j \to \infty$.

Lemma 5. BOREL sets and analytic sets in R^n are L-measurable (see **I.1.4**, definitions 11 and 12).

Lemma 6. (*The* VITALI *covering theorem*). Let F be a family of closed sets in R^n which covers an L-measurable subset A of R^n in the VITALI sense (see definition 7). Then there exists in F a (finite or infinite) sequence $\{F_j\}$ of pair-wise disjoint sets such that $\mathsf{L}(A - \bigcup F_j) = 0$.

Lemma 7. Let S be a subset of R^n such that $\mathsf{L}S=0$. Then there exists a BOREL set $B \subset R^n$ such that $S \subset B$, $\mathsf{L}B=0$.

Lemma 8. Let E be an L-measurable subset of R^n. Then there exists a decomposition $E=B \cup V$, such that B is a BOREL set, and $\mathsf{L}V=0$, $B \cap V=\emptyset$.

Lemma 9. If E is an L-measurable subset of R^n, then almost every point of E is a point of density of E (that is, those points of E which are not points of density of E constitute a set of L-measure zero).

Lemma 10. If E is an L-measurable subset of R^n, then almost every point of E is a point of linear density of E in the direction of each one of the axes $x^1, \ldots, x^n$ (see definition 2).

Lemma 11. If a point x is a point of density of each one of the L-measurable subsets $E_1, \ldots, E_m$ of R^n, then x is also a point of density of the intersection of these sets.

Lemma 12. Let c be a real number, and let j be one of the integers $1, \ldots, n$. Then the set of those points $x=(x^1, \ldots, x^n)$ of R^n for which $x^j=c$ is of L-measure zero.

Lemma 13. Let $\{Q_j\}$ be a (finite or infinite) sequence of oriented n-cubes in R^n, such that

$$\operatorname{int} Q_j \cap \operatorname{int} Q_k=\emptyset \quad \text{for} \quad j \neq k.$$

Then $\mathsf{L} \bigcup Q_j=\sum \mathsf{L} Q_j$.

Lemma 14. Let F be a bounded, L-measurable subset of R^n, and let O be an open subset of R^n such that $F \subset O$. Then for every $\varepsilon>0$ there exists a (finite or infinite) sequence $\{Q_j\}$ of oriented n-cubes such that $F \subset \bigcup Q_j \subset O$, $\sum \mathsf{L} Q_j<\mathsf{L}F+\varepsilon$, and $\operatorname{int} Q_j \cap \operatorname{int} Q_k=\emptyset$ for $j \neq k$.

Lemma 15. Given two distinct points u_1, u_2 of R^n, consider the open spherical neighborhoods (see **I.1.4**, definition 2)

$$\Delta_1=\Delta(u_1, \|u_2-u_1\|), \quad \Delta_2=\Delta(u_2, \|u_2-u_1\|).$$

Then the number

$$\alpha=\frac{\mathsf{L}(\bar{\Delta}_1 \cap \bar{\Delta}_2)}{\mathsf{L}\bar{\Delta}_1+\mathsf{L}\bar{\Delta}_2}$$

satisfies the inequalities $0<\alpha<1$, and has the same value for every choice of the distinct points u_1, u_2 in R^n.

Lemma 16. Let P be a parallelotope in R^n with edge-vectors $v_j=(v_j^1, \ldots, v_j^n)$, $j=1, \ldots, n$ (see **I.2.2**). Then $\mathsf{L}P=|\det(v_j^i)|$.

Lemma 17. Let E be a bounded, L-measurable subset of R^2. For each real number r, denote by E_r the set of those real numbers x^1 for which $(x^1, r) \in E$. Assume that there exists a set e of real numbers of (linear) L-measure zero such that for $r \notin e$ the set E_r is of (linear) L-measure zero. Then $\mathsf{L}E=0$ (if E is not L-measurable, then this statement is false).

Lemma 18. If f_1, f_2 are L-measurable functions on an L-measurable set $E \subset R^n$, then f_1+f_2, $f_1 f_2$ are also L-measurable on E. If $\{f_j\}$ is a sequence of L-measurable functions on an L-measurable set $E \subset R^n$, then $\limsup f_j$, $\liminf f_j$ are also L-measurable on E, and if the sequence converges a.e. on E then $\lim f_j$ is L-measurable on E (see the comments following definition 5).

Definition 8. A real-valued function $f(x)$, defined on a BOREL set $B \subset R^n$, is termed BOREL measurable on B if for every real number r the set of those points $x \in B$ where $f(x)>r$ is a BOREL set.

Lemma 19. If the real-valued function $f(x)$ is Borel measurable on the Borel set $B \subset R^n$, then $f(x)$ is L-measurable on B.

Lemma 20. If the real-valued function $f(x)$ is lower semi-continuous (see **I.1.3**, definition 28) on the Borel set $B \subset R^n$, then $f(x)$ is Borel measurable and hence also L-measurable on B.

Lemma 21. Let $f_1(x), f_2(x)$ be two real-valued functions on the L-measurable set $E \subset R^n$. If $f_1(x)$ is L-measurable on E and $f_1(x) = f_2(x)$ a.e. on E, then $f_2(x)$ is also L-measurable on E.

Lemma 22. (Lusin's *Theorem.*) If the real-valued function $f(x)$ is L-measurable and a.e. finite on the L-measurable set $E \subset R^n$, then for every $\varepsilon > 0$ there exists a closed set $E_\varepsilon \subset E$ such that $\mathsf{L}(E - E_\varepsilon) < \varepsilon$ and $f(x)$ is continuous on E_ε (relative to E_ε).

Lemma 23. If $H(x)$ is a real-valued, finite-valued, L-measurable function in R^n, then there exists a finite-valued Borel measurable function $H^*(x)$ in R^n such that $H^*(x) = H(x)$ a.e. in R^n.

Lemma 24. If $f(x)$ is a non-negative, real-valued, L-measurable function on a Borel set $B \subset R^n$, then there exists a Borel measurable function $f_*(x)$ on B such that $0 \leq f_*(x) \leq f(x)$ and $f_*(x) = f(x)$ a.e. on B.

Lemma 25. Let $f_1(x), f_2(x)$ be two real-valued, L-measurable functions on the L-measurable set $E \subset R^n$. If $f_1(x)$ is L-summable on E and $f_1(x) = f_2(x)$ a.e. on E, then $f_2(x)$ is also L-summable on E and

$$\int_E f_2(x)\, d\mathsf{L} = \int_E f_1(x)\, d\mathsf{L}.$$

Lemma 26. If the real-valued, L-measurable function $f(x)$ is L-summable on the L-measurable set $E \subset R^n$, then $|f(x)|$ is also L-measurable and L-summable on E.

Lemma 27. Let $\{f_m\}$ be a sequence of real-valued, non-negative, L-measurable functions on the L-measurable set $E \subset R^n$, and let $f(x)$ be a real-valued, non-negative function on E such that

$$f(x) = \sum_{m=1}^{\infty} f_m(x) \quad \text{a.e. on } E.$$

Then the following holds.

(i) $f(x)$ is L-measurable on E.

(ii) If $f(x)$ is L-summable on E, then $f_m(x)$ is also L-summable on E, $m = 1, 2, \ldots$, and

$$\int_E f(x)\, d\mathsf{L} = \sum_{m=1}^{\infty} \int_E f_m(x)\, d\mathsf{L}.$$

(iii) Conversely, if $f_m(x)$ is L-summable on E, $m = 1, 2, \ldots$, and the series $\sum \int_E f_m(x)\, d\mathsf{L}$ is convergent, then $f(x)$ is L-summable on E [and hence (ii) is applicable].

Lemma 28. (Minkowski's inequality). Given a real number $p \geq 1$, let $f_1(x), \ldots, f_m(x)$ be real-valued, L-measurable functions on the L-measurable set $E \subset R^n$ such that $|f_1|^p, \ldots, |f_m|^p$ are L-summable on E. Then $(|f_1| + \cdots + |f_m|)^p$ is also L-summable on E and

$$\left[\int_E \left(\sum_{j=1}^{m} |f_j(x)|\right)^p d\mathsf{L}\right]^{1/p} \leq \sum_{j=1}^{m} \left[\int_E |f_j(x)|^p\, d\mathsf{L}\right]^{1/p}.$$

Lemma 29. (HÖLDER's inequality). Let p, q be two real numbers such that

$$p > 1, \quad q > 1, \quad \frac{1}{p} + \frac{1}{q} = 1.$$

If $f(x), g(x)$ are real-valued, L-measurable functions on an L-measurable set $E \subset R^n$ such that $|f|^p$ and $|g|^q$ are L-summable on E, then fg is L-summable on E and

$$\Big|\int_E f(x)\, g(x)\, d\mathsf{L}\Big| \leq \Big[\int_E |f(x)|^p\, d\mathsf{L}\Big]^{1/p} \Big[\int_E |g(x)|^q\, d\mathsf{L}\Big]^{1/q}.$$

Lemma 30. (FATOU's *theorem or lemma.*) Let $f(x), f_j(x)$, $j = 1, 2, \ldots$, be real-valued, non-negative, L-measurable functions on an L-measurable set $E \subset R^n$, such that $f_j(x)$ is L-summable on E, $j = 1, 2, \ldots$, and

$$f(x) \leq \liminf f_j(x) \quad \text{a.e. on } E, \quad \liminf \int_E f_j(x)\, d\mathsf{L} < \infty.$$

Then $f(x)$ is L-summable on E, and

$$\int_E f(x)\, d\mathsf{L} \leq \liminf \int_E f_j(x)\, d\mathsf{L}.$$

Lemma 31. Let $f(x)$ be a real-valued, non-negative, L-summable function on an L-measurable set $E \subset R^n$. Then the set of those points $x \in E$ where $f(x) = \infty$ is of L-measure zero.

Lemma 32. If $f(x)$ is a real-valued, non-negative, L-summable function in R^n, then there exists a BOREL set $X \subset R^n$ such that $\mathsf{L}X = 0$ and $f(x) < \infty$ for $x \in \mathsf{C}X$.

Lemma 33. Let $f(x)$ be a real-valued, L-summable function on an L-measurable set $E \subset R^n$. Then for every $\varepsilon > 0$ there exists an $\eta = \eta(\varepsilon) > 0$ such that

$$\int_S |f(x)|\, d\mathsf{L} < \varepsilon$$

whenever S is an L-measurable subset of E such that $\mathsf{L}S < \eta$.

Lemma 34. Let $f(x)$ be a real-valued, L-summable function on an open set $G \subset R^n$. Let E be an L-measurable set in R^n such that $\bar{E} \subset G$. Then there exists for every $\varepsilon > 0$ an open set $O = O(\varepsilon)$ such that

$$\bar{E} \subset O \subset G, \quad \int_O |f(x)|\, d\mathsf{L} < \int_{\bar{E}} |f(x)|\, d\mathsf{L} + \varepsilon.$$

Lemma 35. Let $f(x)$ be a real-valued, non-negative, L-summable function on an L-measurable set $E \subset R^n$ such that $\int_E f(x)\, d\mathsf{L} = 0$. Then $f(x) = 0$ a.e. on E.

Lemma 36. Let f, f_j, $j = 1, 2, \ldots$, be real-valued, L-summable functions on an L-measurable set $E \subset R^n$, such that $f_j \to f$ a.e. on E and the sequence $\{f_j\}$ satisfies the condition (V) on E (see definition 6). Then

$$\int_E f_j(x)\, d\mathsf{L} \to \int_E f(x)\, d\mathsf{L}.$$

Lemma 37. If $f(x)$ is a real-valued, L-summable function on an L-measurable set $E \subset R^n$ and $\{E_j\}$ is a sequence of pair-wise disjoint L-measurable subsets of E such that $E = \bigcup E_j$, then

$$\int_E f(x)\, d\mathsf{L} = \sum \int_{E_j} f(x)\, d\mathsf{L}.$$

Lemma 38. Let f, f_j, $j = 1, 2, \ldots$, be real-valued, L-summable functions on an L-measurable set $E \subset R^n$, such that $f_j \to f$ a.e. on E. If there exists a non-negative,

L-summable function F on E such that $|f_j| \leq F$ a.e. on E, $j = 1, 2, \ldots$, then

$$\int_E f_j(x)\, d\mathsf{L} \to \int_E f(x)\, d\mathsf{L}.$$

Lemma 39. Let $f(x)$ be a non-negative, L-measurable function on an L-measurable set $E \subset R^n$. Let $\{E_j\}$ be a sequence of pair-wise disjoint L-measurable subsets of E such that $E = \bigcup E_j$. If $f(x)$ is L-summable on E_j, $j = 1, 2, \ldots$, and the series

$$\sum \int_{E_j} f(x)\, d\mathsf{L}$$

is convergent, then $f(x)$ is L-summable on E and

$$\int_E f(x)\, d\mathsf{L} = \sum \int_{E_j} f(x)\, d\mathsf{L}.$$

Lemma 40. Let $\{f_j\}$ be a sequence of real-valued, non-negative, L-summable functions on an L-measurable set $E \subset R^n$, such that the series

$$\sum \int_E f_j(x)\, d\mathsf{L}$$

is convergent. Then the series $\sum f_j(x)$ is convergent a.e. on E, the sum of this series is L-summable on E, and

$$\int_E \left(\sum f_j(x)\right) d\mathsf{L} = \sum \int_E f_j(x)\, d\mathsf{L}.$$

Lemma 41. Let $f, f_j, j = 1, 2, \ldots$, be real-valued, L-summable functions on an L-measurable set $E \subset R^n$, such that

$$\int_E |f(x) - f_j(x)|\, d\mathsf{L} \to 0.$$

Then $\int_E |f_j(x)|\, d\mathsf{L} \to \int_E |f(x)|\, d\mathsf{L}$.

Lemma 42. Let $\{Q_j\}$ be a sequence of oriented n-cubes in R^n such that $\operatorname{int} Q_j \cap \operatorname{int} Q_k = \emptyset$ for $j \neq k$, and let f be a real-valued function which is L-summable on the set $E = \bigcup Q_j$. Then

$$\int_E f(x)\, d\mathsf{L} = \sum \int_{Q_j} f(x)\, d\mathsf{L}.$$

Lemma 43. Let f be a real-valued, L-summable function on an open set $O \subset R^n$. Then there exists a subset e of O of L-measure zero such that the following holds for every point $u \in O - e$.

(i) $f(u)$ is finite.

(ii) If $\{G_m\}$ is any regular sequence of closed subsets of O (see definition 4) such that $u \in G_m$ and $\delta G_m \to 0$, then

$$f(u) = \lim \frac{1}{\mathsf{L} G_m} \int_{G_m} f(x)\, d\mathsf{L}.$$

Lemma 44. Let f_1, f_2 be real-valued, L-summable functions on an open set $O \subset R^n$, such that

$$\int_{\operatorname{int} Q} f_1(x)\, d\mathsf{L} = \int_{\operatorname{int} Q} f_2(x)\, d\mathsf{L}$$

for every oriented n-cube $Q \subset O$. Then $f_1 = f_2$ a.e. in O.

III.1.2. Finite measures on BOREL sets in R^n. We shall need only a few basic concepts and facts from the very comprehensive theory of general measures. In this section we collect this minimum amount of information for convenient reference (proofs and further details may be found in the SAKS treatise listed in the *Bibliography*).

A real-valued function $\nu(B)$, defined for every BOREL set B in a fixed BOREL set $B^0 \subset R^n$, is termed a *finite measure* on BOREL sets in B^0 if the following holds. (i) $0 \leq \nu(B) < \infty$ for every BOREL set $B \subset B^0$. (ii) ν is *countably additive* [that is, if $\{B_j\}$ is a finite or infinite sequence of pair-wise disjoint BOREL sets in B^0, then $\nu(\bigcup B_j) = \sum \nu(B_j)$].

A finite measure ν on BOREL sets in B^0 is termed *absolutely continuous* (relative to L-measure in R^n) if for every BOREL set $B \subset B^0$ such that $\mathsf{L} B = 0$ one has $\nu(B) = 0$. This condition is equivalent to the following one: for every $\varepsilon > 0$ there exists an $\eta = \eta(\varepsilon) > 0$ such that $\nu(B) < \varepsilon$ whenever B is a BOREL set in B^0 such that $\mathsf{L} B < \eta$.

A finite measure ν on BOREL sets in B^0 is termed *singular* (relative to L-measure in R^n) if there exists a BOREL set $b \subset B^0$ such that (i) $\mathsf{L}\, b = 0$ and (ii) $\nu(B) = 0$ for every BOREL set $B \subset B^0 - b$. This condition is equivalent to the following one: for every BOREL set $B \subset B^0$ one has $\nu(B) = \nu(B \cap b)$, where b is a BOREL set in B^0 such that $\mathsf{L}\, b = 0$.

Let ν be a finite measure on BOREL sets in B^0. Then ν gives rise to a ν-integral on BOREL sets in B^0 as follows (note the analogy with the L-integral considered in **III.1.1**). Let $f(x)$ be a real-valued, non-negative, BOREL measurable function on a BOREL set $B \subset B^0$. Let $B_1, \ldots, B_m$ be any finite system of pair-wise disjoint BOREL sets such that $B = \bigcup B_j$, $j = 1, \ldots, m$. Denote by $\mu(f, B_j)$ the greatest lower bound of $f(x)$ on B_j, $j = 1, \ldots, m$. Then $f(x)$ is termed ν-summable on B if the least upper bound of all the sums of the type $\sum \mu(f, B_j)\, \nu(B_j)$ is finite, and if this is the case, then the ν-integral of $f(x)$ on B is defined by the formula

$$\int_B f(x)\, d\nu = \text{l.u.b.} \sum \mu(f, B_j)\, \nu(B_j).$$

If $g(x)$ is a real-valued, BOREL measurable function (of arbitrary sign) defined on a BOREL set $B \subset B^0$, then the functions

$$g^+(x) = \begin{cases} g(x) & \text{if } g(x) \geq 0, \\ 0 & \text{if } g(x) < 0, \end{cases} \qquad g^-(x) = \begin{cases} -g(x) & \text{if } g(x) \leq 0, \\ 0 & \text{if } g(x) > 0, \end{cases}$$

are again BOREL measurable (and non-negative) on B. If $g^+(x)$, $g^-(x)$ are both ν-summable on B, then $g(x)$ is termed ν-summable on B, and one defines:

$$\int_B g(x)\, d\nu = \int_B g^+(x)\, d\nu - \int_B g^-(x)\, d\nu.$$

Let ν be a finite measure on BOREL sets B in an open set $G \subset R^n$. Given a BOREL set $B \subset G$ and a real number $\varepsilon > 0$, there exist two sets F, O in G such that F is closed, O is open, $F \subset B \subset O$, and $\nu(O) < \nu(B) + \varepsilon$, $\nu(B) < \nu(F) + \varepsilon$.

Let ν be a finite measure on BOREL sets $B \subset B^0$. A real-valued, BOREL measurable function $f(x)$ defined on a BOREL set $B \subset B^0$ is ν-summable on B if and only if $|f(x)|$ is ν-summable on B. If $f_1(x), \ldots, f_m(x)$ are real-valued, BOREL measurable functions, defined and ν-summable on the BOREL set $B \subset B^0$, and $c_1, \ldots, c_m$ are real constants, then

$$\int_B \left[\sum_{j=1}^{m} c_j f_j(x)\right] d\nu = \sum_{j=1}^{m} c_j \int_B f_j(x)\, d\nu.$$

Let $f(x)$ be a real-valued, non-negative, BOREL measurable function on a BOREL set $B \subset B^0$, and let $\{B_j\}$ be a (finite or infinite) sequence of pair-wise disjoint BOREL sets such that $B = \cup B_j$ and $f(x)$ is ν-summable on B_j, $j = 1, 2, \ldots$. Then $f(x)$ is ν-summable on B if and only if the series

$$\sum \int_{B_j} f(x)\, d\nu$$

is convergent; and if this is the case, then

$$\int_B f(x)\, d\nu = \sum \int_{B_j} f(x)\, d\nu.$$

If ν is a finite measure on BOREL sets $B \subset B^0$, then there exists a unique decomposition of the form $\nu = \nu_a + \nu_s$, where ν_a, ν_s are finite measures on BOREL sets in B^0 such that ν_a is absolutely continuous and ν_s is singular (relative to L-measure). This unique decomposition is termed the LEBESGUE *decomposition* of ν. Furthermore, there exists a real-valued, non-negative, BOREL measurable, L-summable function $F(x)$ on B^0 such that $\nu_a(B) = \int_B F(x)\, d\mathsf{L}$ for every BOREL set $B \subset B^0$. Thus the LEBESGUE decomposition of ν can be written in the form

$$\nu(B) = \int_B F(x)\, d\mathsf{L} + \nu_s(B),$$

where B is any BOREL set in B^0. The uniqueness of the LEBESGUE decomposition implies that ν is absolutely continuous on B^0 if and only if $\nu_s \equiv 0$, and similarly ν is singular on B^0 if and only if $\nu_a \equiv 0$. If $f(x)$ is a real-valued, BOREL measurable function on a BOREL set $B \subset B^0$, then $f(x)$ is ν-summable on B if and only if it is both ν_a-summable and ν_s-summable on B. Furthermore, if $f(x)$ is ν-summable on B, then

$$\int_B f(x)\, d\nu = \int_B f(x) F(x)\, d\mathsf{L} + \int_B f(x)\, d\nu_s,$$

where $F(x)$ is the function occurring in the LEBESGUE decomposition of ν. The preceding formula, which expresses the relation between the ν-integral and the L-integral, is referred to as the theorem on *change of measure*. An important special case arises if ν is absolutely continuous. One has then

$$\int_B f(x)\, d\nu = \int_B f(x)\, F(x)\, d\mathsf{L}.$$

In fact, this formula holds as soon as one of the two integrals involved exists (assuming, of course, that ν is absolutely continuous and $f(x)$ is BOREL measurable on B).

III.1.3. Derivatives and differentials in $\boldsymbol{R}^n$. We shall collect in this section, for convenient reference, a few definitions and facts concerning derivatives and differentials of real-valued functions of several real variables, in the form best suited for applications later on. Proofs and further details may be found in the SAKS treatise listed in the *Bibliography*.

In an open set $O \subset R^n$, let $f(x) = f(x^1, \ldots, x^n)$ be a real-valued continuous function of the point $x = (x^1, \ldots, x^n)$. The first partial derivative of f with respect to x^j will be denoted by $\partial f/\partial x^j$ (assuming that it exists at a certain point in O). If $\partial f/\partial x^j$ exists a.e. in O, then it is L-measurable there (see the comments following definition 5 in **III.1.1**). Let $x_0 = (x_0^1, \ldots, x_0^n)$ be a point of O such that $\partial f/\partial x^1, \ldots, \partial f/\partial x^n$ exist at x_0. Then the expression

$$\sum_{j=1}^{n} \frac{\partial f(x_0)}{\partial x^j}\,(x^j - x_0^j),$$

where $x = (x^1, \ldots, x^n)$ is a variable point in O, is termed the *differential* of f at x_0. If

$$\frac{1}{|x - x_0|} \left| f(x) - f(x_0) - \sum_{j=1}^{n} \frac{\partial f(x_0)}{\partial x^j}\,(x^j - x_0^j) \right| \to 0$$

for $x \neq x_0$, $x \to x_0$, then f is said to possess a *total differential* at x_0. If for every $\varepsilon > 0$ the point x_0 is a point of density of the set

$$E_\varepsilon = \left\{ x \,\middle|\, x \in O,\ \left| f(x) - f(x_0) - \sum_{j=1}^{n} \frac{\partial f(x_0)}{\partial x^j}\,(x^j - x_0^j) \right| \leq \varepsilon \|x - x_0\| \right\},$$

then f is said to possess an *approximate total differential* at x_0. Actually, this concept can be (and usually is) formulated in a more general manner. However, the preceding definition is adequate for our purposes.

A real-valued function $f(x)$, defined on a set $E \subset R^n$, is said to be *Lipschitzian* on E if there exists a (finite) constant $M \geq 0$ such that

$$|f(x_2) - f(x_1)| \leq M \|x_2 - x_1\|$$

for every pair of points x_1, x_2 in E. In terms of this concept, we have the following theorem: if $f(x)$ is a real-valued function on an open set $O \subset R^n$ which is Lipschitzian on O, then $f(x)$ possesses a total differential a.e. on O.

More generally, let $f(x)$ be a real-valued, continuous function on an open set $O \subset R^n$, such that

$$\limsup_{x \to x_0} \frac{|f(x) - f(x_0)|}{\|x - x_0\|} < \infty$$

at all those points $x_0 \in O$ which do not belong to a certain subset e of O such that $\mathsf{L}e = 0$. Then $f(x)$ possesses a total differential a.e. in O (*theorem of* RADEMACHER).

Let $f(x)$ be a real-valued, continuous function in an open set $O \subset R^n$, such that the first partial derivatives $\partial f/\partial x^1, \ldots, \partial f/\partial x^n$ exist a.e. in O. Then $f(x)$ possesses an approximate total differential a.e. in O (this is a special case of a theorem of STEPANOFF).

§ III.2. Functions of open intervals in R^n [1].

III.2.1. Bounded variation. An interval I in R^n is a point set determined by inequalities of the form $a_i \leq u^i \leq b_i$, where $a_i < b_i$, $i = 1, \ldots, n$. An interval I is a compact set (see **I.2.2**, exercise 3) and clearly its interior $\operatorname{int} I$ is given by the inequalities $a_i < u^i < b_i$, $i = 1, \ldots, n$. The set $\operatorname{int} I$ is termed an open interval in R^n.

Definition 1. Let O be a non-empty open set in R^n. A (finite) system of intervals $I_1, \ldots, I_m$ will be termed admissible for O, and will be denoted generically by $\sigma(O)$, if $\operatorname{int} I_k \subset O$, $k = 1, \ldots, m$, and $\operatorname{int} I_j \cap \operatorname{int} I_k = \emptyset$ for $j \neq k$.

Definition 2. Let O be an open set in R^n. If with every open interval $\operatorname{int} I \subset O$ there is associated a (finite) real number $\varphi(\operatorname{int} I)$, then φ is termed a (real-valued) function of open intervals in O.

Of course, one can also consider functions $\psi(I)$ of closed intervals I. If $\varphi(\operatorname{int} I)$ is a function of open intervals, then on setting $\psi(I) = \varphi(\operatorname{int} I)$ one obtains a function ψ of closed intervals. Conversely, if $\psi(I)$ is a function of closed intervals, then on setting $\varphi(\operatorname{int} I) = \psi(I)$ one obtains

[1] For historical comments and further details, see the SAKS treatise listed in the *Bibliography*.

a function φ of open intervals. For our purposes it is more convenient to operate with functions of *open* intervals.

To simplify formulas, let us introduce the following notations. Given $\sigma(O)$ and φ as specified in definitions 1 and 2, put

$$\sum[\varphi, \sigma(O)] = \sum \varphi(\text{int } I), \quad I \in \sigma(O), \tag{1}$$

$$\mathsf{L}[\sigma(O)] = \sum \mathsf{L} I, \quad I \in \sigma(O). \tag{2}$$

Throughout this section, φ will denote a function of open intervals in an open set $O \subset R^n$.

Definition 3. φ is termed bounded in O if there exists a (finite) constant $K \geq 0$ such that $|\varphi(\text{int } I)| \leq K$ for $\text{int } I \subset O$.

Definition 4. The least upper bound of the sums $\sum[|\varphi|, \sigma(O)]$ is termed the *total variation* $V(\varphi, O)$ of φ in O. In symbols:

$$V(\varphi, O) = \underset{\sigma(O)}{\text{l.u.b.}} \sum[|\varphi|, \sigma(O)]. \tag{3}$$

Note that $V(\varphi, O)$ may be infinite. Clearly $V(\varphi, O) \geq 0$.

Definition 5. φ is said to be of *bounded variation* in O if there exists a (finite) constant $M \geq 0$ such that

$$\sum[|\varphi|, \sigma(O)] \leq M \tag{4}$$

for every system $\sigma(O)$.

The following two lemmas are direct consequences of the definitions involved.

Lemma 1. φ is of bounded variation in O if and only if $V(\varphi, O) < \infty$.

Lemma 2. If φ is of bounded variation in O, then it is also of bounded variation in every open set $O^* \subset O$, and $V(\varphi, O^*) \leq V(\varphi, O)$.

III.2.2. Absolute continuity. The notations and definitions introduced in **III.2.1** will be used in this section.

Definition. φ is said to be *absolutely continuous* in O if for every $\varepsilon > 0$ there exists an $\eta(\varepsilon) > 0$ such that

$$\sum[|\varphi|, \sigma(O)] < \varepsilon \quad \text{if} \quad \mathsf{L}[\sigma(O)] < \eta(\varepsilon). \tag{1}$$

Lemma 1. If φ is absolutely continuous in O, then it is also absolutely continuous in every open set $O^* \subset O$.

This is a direct consequence of the definition.

Lemma 2. Assume that φ is absolutely continuous in O. Take a positive number ε, and let $\eta(\varepsilon)$ correspond to ε in the sense of (1). Let O^* be any (non-empty) open set in O such that

$$\mathsf{L}\, O^* < \eta(\varepsilon). \tag{2}$$

Then
$$V(\varphi, O^*) \leq \varepsilon. \tag{3}$$

Proof. Consider any system $\sigma(O^*)$. In view of (2), clearly $\mathsf{L}[\sigma(O^*)] < \eta(\varepsilon)$. Since $\sigma(O^*)$ is also a system $\sigma(O)$, by (1) we conclude that
$$\sum [|\varphi|, \sigma(O^*)] < \varepsilon. \tag{4}$$
As $\sigma(O^*)$ was arbitrary, (4) implies (3).

Lemma 3. Assume that φ is absolutely continuous and bounded in the bounded open set O. Then φ is of bounded variation in O.

Proof. As O is bounded, clearly
$$\mathsf{L}\, O < \infty. \tag{5}$$
Since φ is bounded in O, we have a constant K such that
$$|\varphi(\operatorname{int} I)| < K \quad \text{for} \quad \operatorname{int} I \subset O. \tag{6}$$
On choosing $\varepsilon = 1$, (1) yields
$$\sum [|\varphi|, \sigma(O)] < 1 \quad \text{if} \quad \mathsf{L}[\sigma(O)] < \eta(1). \tag{7}$$
Let us put
$$\lambda = \eta(1)/2. \tag{8}$$
Consider now a system $\sigma_0(O)$, consisting of intervals $I_1, \ldots, I_m$.

Case 1. Suppose that
$$\mathsf{L}\, I_j < \lambda, \quad j = 1, \ldots, m. \tag{9}$$
Let us say that a grouping of the intervals $I_1, \ldots, I_m$ into mutually exclusive systems $\sigma_1(O), \ldots, \sigma_k(O)$ is a λ-grouping if
$$\mathsf{L}[\sigma_s(O)] < 2\lambda, \quad s = 1, \ldots, k. \tag{10}$$
In view of (9), we have at least one λ-grouping [namely, the one obtained by considering each individual interval as constituting a system $\sigma(O)$]. Clearly, there exists a λ-grouping $\sigma_1(O), \ldots, \sigma_k(O)$ for which the number k has a minimum value. Let $\sigma_1(O), \ldots, \sigma_k(O)$ be a λ-grouping with this minimal property. Then the inequality $\mathsf{L}[\sigma_s(O)] < \lambda$ holds for at most one of the integers $s = 1, \ldots, k$. Otherwise two of the systems $\sigma_s(O)$ could be combined into one, and we would obtain a λ-grouping consisting of $k-1$ systems $\sigma(O)$, in contradiction with the minimal property of the λ-grouping $\sigma_1(O), \ldots, \sigma_k(O)$. Thus the sum $\mathsf{L}[\sigma_s(O)]$ has a value $\geq \lambda$ for at least $k-1$ of the integers $s = 1, \ldots, k$. Hence
$$(k-1)\,\lambda \leq \mathsf{L}[\sigma_0(O)] \leq \mathsf{L}\, O,$$
and thus
$$k \leq 1 + \frac{\mathsf{L}\, O}{\lambda}. \tag{11}$$

Observe now that

$$\sum [|\varphi|, \sigma_0(O)] = \sum_{s=1}^{k} \sum [|\varphi|, \sigma_s(O)], \tag{12}$$

and [by (10), (8) and (7)]

$$\sum [|\varphi|, \sigma_s(O)] < 1, \quad s = 1, \ldots, k. \tag{13}$$

From (11) to (13) it follows that

$$\sum [|\varphi|, \sigma_0(O)] < 1 + \frac{\mathsf{L}O}{\lambda}. \tag{14}$$

Case 2. Suppose that $\mathsf{L} I_j \geq \lambda$, $j = 1, \ldots, m$. Then clearly

$$m\lambda \leq \mathsf{L}[\sigma_0(O)] \leq \mathsf{L}O,$$

and thus $m \leq \mathsf{L}O/\lambda$. By (6) we conclude that

$$\sum [|\varphi|, \sigma_0(O)] \leq \frac{K\mathsf{L}O}{\lambda}. \tag{15}$$

In the general case, $\sigma_0(O)$ can be split into two systems $\sigma_0'(O)$ and $\sigma_0''(O)$, one presenting *Case 1* and the other presenting *Case 2*. By (14) and (15), applied to $\sigma_0'(O)$ and $\sigma_0''(O)$ respectively, it follows that

$$\sum [|\varphi|, \sigma_0(O)] < 1 + \frac{\mathsf{L}O}{\lambda} + \frac{K\mathsf{L}O}{\lambda}. \tag{16}$$

Since $\sigma_0(O)$ was arbitrary, (16) shows that φ is of bounded variation in O.

III.2.3. Derivatives. The notations and definitions introduced in **III.2.1** and **III.2.2** will be used in this section.

Definition 1. Let u be a point of O. The *upper derivative* $\overline{D}(u, \varphi)$ of φ at u is the least upper bound of all those numbers λ (including $\pm\infty$) for which there exists a sequence of oriented n-cubes $\{Q_k\}$ such that

$$u \in \operatorname{int} Q_k \subset O, \quad \mathsf{L}Q_k \to 0, \quad \frac{\varphi(\operatorname{int} Q_k)}{\mathsf{L}Q_k} \to \lambda.$$

The *lower derivative* $\underline{D}(u, \varphi)$ is the greatest lower bound of these same numbers λ.

Definition 2. If $\overline{D}(u, \varphi)$ and $\underline{D}(u, \varphi)$ are finite and equal to each other at a point $u \in O$, then their common value is termed the derivative $D(u, \varphi)$ of φ at u.

Theorem 1. $\overline{D}(u, \varphi)$ and $\underline{D}(u, \varphi)$ are Borel measurable in O.

Proof. Since the argument is entirely similar in both cases, we present it only for $\overline{D}(u, \varphi)$. Assign any real number α. Let Q be a generic notation for an oriented n-cube. For each pair of positive integers m, j define a set $E_{\alpha m j}$ by the formula

$$E_{\alpha m j} = \bigcup \operatorname{int} Q, \quad \operatorname{int} Q \subset O, \quad \mathsf{L}\, Q < \frac{1}{j}, \quad \frac{\varphi(\operatorname{int} Q)}{\mathsf{L}\, Q} > \alpha + \frac{1}{m}.$$

If E_α denotes the set of those points $u \in O$ where $\overline{D}(u, \varphi) > \alpha$, then it follows readily from the definition of $\overline{D}(u, \varphi)$ that

$$E_\alpha = \bigcup_m \bigcap_j E_{\alpha m j}. \tag{1}$$

Since $E_{\alpha m j}$ is open, (1) shows that E_α is a BOREL set, and hence $\overline{D}(u, \varphi)$ is BOREL measurable.

Theorem 2. Let E be the set of those points $u \in O$ where $D(u, \varphi)$ exists (see definition 2). Then E is a BOREL set, and $D(u, \varphi)$ is BOREL measurable on E.

Proof. Clearly, E consists of those points $u \in O$ where the following conditions hold simultaneously.

$$-\infty < \overline{D}(u, \varphi) < \infty, \quad -\infty < \underline{D}(u, \varphi) < \infty, \quad \overline{D}(u, \varphi) = \underline{D}(u, \varphi).$$

In view of theorem 1 and **III.1.1**, definition 8, it follows that E is the intersection of three BOREL sets, and hence E is a BOREL set. As $D(u, \varphi) = \overline{D}(u, \varphi)$ on E, by theorem 1 one sees that $D(u, \varphi)$ is BOREL measurable on E.

Theorem 3. If $D(u, \varphi)$ exists a.e. in O, then it is L-measurable in O.

Proof. Denoting again by E the set of those points $u \in O$ where $D(u, \varphi)$ exists, the set $O - E$ is now of L-measure zero by assumption. In view of the conventions stated in **III.1.1**, the present theorem is therefore a consequence of theorem 2, since BOREL measurability implies L-measurability (see **III.1.1**, lemma 19).

Theorem 4. Assume that the set O is bounded, φ is of bounded variation in O, and $D(u, \varphi)$ exists a.e. in O. Then $D(u, \varphi)$ is L-summable in O, and (see **III.2.1**, definition 4)

$$\int_O |D(u, \varphi)|\, d\mathsf{L} \leqq V(\varphi, O). \tag{2}$$

Proof. Take an oriented n-cube Q such that $O \subset Q$. We shall make use now of the subdivision D_{p_j} of Q (see **I.2.3**, definition 3). For each positive integer j, let Q_j be a generic notation for a cube of D_{p_j} for which $\operatorname{int} Q_j \subset O$, and let $\mathfrak{Q}_j$ be the class of all such cubes Q_j. Since O is bounded, clearly $\mathfrak{Q}_j$ is a finite class. For each positive integer j, define in O a function $f_j(u)$ as follows: If there exists a cube Q_j such that $u \in \operatorname{int} Q_j$, then

$$f_j(u) = \frac{\varphi(\operatorname{int} Q_j)}{\mathsf{L}\, Q_j}. \tag{3}$$

while $f_j(u) = 0$ if no such cube Q_j exists. Clearly, $f_j(u)$ is BOREL measurable in O, and

$$\int_O |f_j(u)|\, d\mathsf{L} = \sum_{Q_j} |\varphi(\text{int } Q_j)|. \tag{4}$$

If u is any point in O, then by **I.2.3**, exercise 2 this point will lie for j sufficiently large in the interior of some cube Q_j, and thus (3) will hold. Hence, if $D(u, \varphi)$ exists at u, then $f_j(u) \to D(u, \varphi)$. Thus

$$\lim_{j\to\infty} |f_j(u)| = |D(u, \varphi)| \quad \text{a.e. in } O. \tag{5}$$

Since $\mathfrak{Q}_j$ is clearly a system $\sigma(O)$, we infer from (4) that (see **III.2.1**, definition 4)

$$\int_O |f_j(u)|\, d\mathsf{L} \leq V(\varphi, O) < \infty. \tag{6}$$

By the theorem of FATOU (see **III.1.1**, lemma 30), (5) and (6) imply both the L-summability of $|D(u, \varphi)|$ in O and the inequality (2).

Lemma. Assume that O is bounded and $D(u, \varphi)$ exists a.e. in O. Using the notations Q_j, $\mathfrak{Q}_j$ as in the proof of theorem 4, assume further that

$$\lim_{j\to\infty} \sum_{Q_j} |\varphi(\text{int } Q_j)| = 0. \tag{7}$$

Then

$$D(u, \varphi) = 0 \quad \text{a.e. in } O. \tag{8}$$

Proof. Note that we do not assume now that φ is of bounded variation in O. Still, we can introduce the functions $f_j(u)$ as in the proof of theorem 4, and the relations (4) and (5) are available as before. From (4) and (7) we see that

$$\lim_{j\to\infty} \int_O |f_j(u)|\, d\mathsf{L} = 0. \tag{9}$$

By the theorem of FATOU (see **III.1.1**, lemma 30), (5) and (9) imply that

$$\int_O |D(u, \varphi)|\, d\mathsf{L} = 0,$$

and (8) follows by **III.1.1**, lemma 35.

Definition 3. φ is termed *sub-additive* in O if

$$\sum_{j=1}^{m} \varphi(\text{int } I_j) \leq \varphi(\text{int } I), \tag{10}$$

for every choice of the intervals $I_1, \ldots, I_m, I$ subject to the conditions

$$\text{int } I_j \subset \text{int } I \subset O, \quad j = 1, \ldots, m, \tag{11}$$

$$\text{int } I_j \cap \text{int } I_k = \emptyset \quad \text{for} \quad j \neq k. \tag{12}$$

Theorem 5. If φ is non-negative and sub-additive in O, then $D(u,\varphi)$ exists a.e. in O.

By **I.2.2**, exercise 8, O can be covered by a countable set of open intervals. Hence it is sufficient to show that if I is any interval such that $\operatorname{int} I \subset O$, then $D(u,\varphi)$ exists a.e. in $\operatorname{int} I$. For clarity, we subdivide the proof into several steps.

(a) Let I be an interval such that $\operatorname{int} I \subset O$, and let α be a positive number. Denote by E_α the set of those points $u \in \operatorname{int} I$ where $\overline{D}(u,\varphi) > \alpha$. Then

$$\alpha \, \mathsf{L} E_\alpha \leq \varphi(\operatorname{int} I). \tag{13}$$

Proof. Let F be the family of those oriented n-cubes Q which satisfy the conditions

$$\operatorname{int} Q \subset \operatorname{int} I, \quad \frac{\varphi(\operatorname{int} Q)}{\mathsf{L} Q} > \alpha.$$

Clearly, the cubes $Q \in \mathsf{F}$ cover E_α in the sense required in the VITALI covering theorem (see **III.1.1**, lemma 6). Hence F contains a (finite or infinite) sequence of cubes $\{Q_j\}$ such that $Q_i \cap Q_j = \emptyset$ for $i \neq j$ and

$$\mathsf{L}(E_\alpha - \bigcup Q_j) = 0. \tag{14}$$

Since φ is sub-additive, there follows for every positive integer k the inequality

$$\alpha \sum_{i=1}^{k} \mathsf{L} Q_i < \sum_{i=1}^{k} \varphi(\operatorname{int} Q_i) \leq \varphi(\operatorname{int} I).$$

For $k \to \infty$ we obtain, in view of (14), the inequality (13). If $\{Q_j\}$ is a finite sequence, then the step $k \to \infty$ is not needed in the proof.

(b) Let I be an interval such that $\operatorname{int} I \subset O$. Denote by E^* the set of those points $u \in \operatorname{int} I$ where $\overline{D}(u,\varphi) = \infty$. Then

$$\mathsf{L} E^* = 0. \tag{15}$$

Proof. Clearly $E^* \subset E_\alpha$ for every positive number α, and thus (15) follows from (13).

(c) Let I be an interval such that $\operatorname{int} I \subset O$. Then $\overline{D}(u,\varphi) = \underline{D}(u,\varphi)$ a.e. in $\operatorname{int} I$.

Proof. If $\underline{D}(u,\varphi) < \overline{D}(u,\varphi)$ at a point $u \in \operatorname{int} I$, then there exist two rational numbers r and s such that

$$\underline{D}(u,\varphi) < r < s < \overline{D}(u,\varphi). \tag{16}$$

Denote by E_{rs} the set of those points $u \in \operatorname{int} I$ where (16) holds for a given pair of rational numbers r, s such that $0 < r < s$. Then the set of those points $u \in \operatorname{int} I$ where $\underline{D}(u,\varphi) \neq \overline{D}(u,\varphi)$ is covered by the family of all the sets E_{rs}, corresponding to all possible pairs of rational numbers r, s. Since this family is countable, it is sufficient to show that each set E_{rs} is of L-measure zero. Deny this, and suppose that

$$\mathsf{L}\, E_{rs} > 0, \quad 0 < r < s, \tag{17}$$

for some pair r, s. Take a positive number ε. Since E_{rs} is Borel measurable by theorem 1, we have (see **III.1.1**, lemma 3) an open set G such that

$$E_{rs} \subset G \subset \operatorname{int} I, \quad \mathsf{L}\, G < \mathsf{L}\, E_{rs} + \varepsilon. \tag{18}$$

Let now F denote the family of those oriented cubes Q which satisfy the conditions

$$\operatorname{int} Q \subset G, \quad \frac{\varphi(\operatorname{int} Q)}{\mathsf{L}\, Q} < r. \tag{19}$$

Clearly, the cubes $Q \in \mathsf{F}$ cover E_{rs} in the manner required in the Vitali covering theorem (see **III.1.1**, lemma 6). Hence F contains a (finite or infinite) sequence of cubes $\{Q_j\}$ such that $Q_i \cap Q_j = \emptyset$ for $i \neq j$ and

$$\mathsf{L}(E_{rs} - \cup Q_j) = 0. \tag{20}$$

There follow [by (19) and (18)] the inequalities

$$\sum_j \varphi(\operatorname{int} Q_j) < r \sum_j \mathsf{L}\, Q_j \leqq r\, \mathsf{L}\, G < r(\mathsf{L}\, E_{rs} + \varepsilon). \tag{21}$$

Observe now that $E_{rs} \cap \operatorname{int} Q_j$ is a subset of the set of those points $u \in \operatorname{int} Q_j$ where $\overline{D}(u, \varphi) > s$. In view of (a) above there follows the inequality

$$s\, \mathsf{L}(E_{rs} \cap \operatorname{int} Q_j) \leqq \varphi(\operatorname{int} Q_j). \tag{22}$$

By (20) we have

$$\sum_j \mathsf{L}(E_{rs} \cap \operatorname{int} Q_j) = \mathsf{L}\, E_{rs}. \tag{23}$$

From (23), (22), and (21) we obtain the inequality

$$s\, \mathsf{L}\, E_{rs} < r(\mathsf{L}\, E_{rs} + \varepsilon).$$

Since ε was arbitrary, it follows that

$$s\, \mathsf{L}\, E_{rs} \leqq r\, \mathsf{L}\, E_{rs},$$

and in view of (17) this is a contradiction. Thus (c) is established.

Combining (b) and (c), we see that $\overline{D}(u, \varphi)$ and $\underline{D}(u, \varphi)$ are equal to each other and are finite a.e. in $\operatorname{int} I$. Hence (see definition 2) $D(u, \varphi)$ exists a.e. in $\operatorname{int} I$, and the proof is complete.

III.2.4. Applications. In this section, $\boldsymbol{D}$ denotes a bounded domain in R^n, and F denotes a family of subsets of $\boldsymbol{D}$ such that F contains, in particular, every domain $D \subset \boldsymbol{D}$.

Definition 1. If with every set $F \in \mathsf{F}$ there is associated a (finite) real number $\Phi(F)$, then Φ is termed a (real-valued) function of sets in F.

If I is an interval such that int $I \subset \boldsymbol{D}$, then (since int I is a domain by **I.2.2**, exercise 3) $\Phi(\operatorname{int} I)$ is defined. Thus Φ gives rise to a function Φ^* of open intervals in $\boldsymbol{D}$, by means of the formula

$$\Phi^*(\operatorname{int} I) = \Phi(\operatorname{int} I), \quad \operatorname{int} I \subset \boldsymbol{D}. \tag{1}$$

Accordingly, we can define the derivative $D(u, \Phi)$ of Φ by the formula (see **III.2.3**, definition 2)

$$D(u, \Phi) = D(u, \Phi^*). \tag{2}$$

By **III.2.3**, theorem 2, the set where $D(u, \Phi)$ exists is a Borel set, and $D(u, \Phi)$ is Borel measurable on that set.

Definition 2. A function Φ of sets $F \in \mathsf{F}$ is termed sub-additive on F if

$$\sum_{k=1}^{m} \Phi(F_k) \leqq \Phi(F), \tag{3}$$

whenever $F_1, \ldots, F_m, F$ are sets of the family F such that $F_k \subset F$, $k = 1, \ldots, m$, and $F_i \cap F_j = \emptyset$ for $i \neq j$.

Theorem 1. Assume that Φ is a non-negative, sub-additive function of sets $F \in \mathsf{F}$. Then the derivative $D(u, \Phi)$ exists a.e. in $\boldsymbol{D}$ and is L-summable in $\boldsymbol{D}$. Furthermore, if O is any open set in $\boldsymbol{D}$ such that $O \in \mathsf{F}$, then

$$\int_O D(u, \Phi)\, d\mathsf{L} \leqq \Phi(O). \tag{4}$$

Proof. Clearly, the associated function Φ^* of open intervals [see (1)] is non-negative and sub-additive. Accordingly, in view of (2) and **III.2.3**, theorems 3 and 5, $D(u, \Phi)$ exists a.e. in $\boldsymbol{D}$ and is L-measurable in $\boldsymbol{D}$. Let now O be a (non-empty) open set in $\boldsymbol{D}$ such that $O \in \mathsf{F}$. Consider any system $\sigma(O)$ (see **III.2.1**, definition 1). Using the notations of **III.2.1**, we have by (3) and (1) the inequality

$$\sum [\Phi^*, \sigma(O)] \leqq \Phi(O) < \infty.$$

Since $\sigma(O)$ was arbitrary, it follows that

$$V(\Phi^*, O) \leqq \Phi(O) < \infty. \tag{5}$$

Hence, by **III.2.3**, theorem 4, $D(u, \Phi^*)$ is L-summable in O and

$$\int_O D(u, \Phi^*)\, d\mathsf{L} \leqq V(\Phi^*, O). \tag{6}$$

Now (4) follows from (6), (5) and (2). The L-summability of $D(u, \Phi)$ in $\boldsymbol{D}$ follows if one chooses $O = \boldsymbol{D}$ in the preceding argument.

Let us now consider a finite measure $\nu(B)$ on BOREL sets in $\boldsymbol{D}$ (see **III.1.2**). The family of BOREL sets in $\boldsymbol{D}$ contains all open sets in $\boldsymbol{D}$ and hence also all domains in $\boldsymbol{D}$. If $B_1, \ldots, B_m$, B are BOREL sets in $\boldsymbol{D}$ such that $B_k \subset B$, $k=1, \ldots, m$, and $B_i \cap B_j = \emptyset$ for $i \neq j$, then

$$\nu(B) = \sum_{k=1}^{m} \nu(B_k) + \nu\left(B - \bigcup_{k=1}^{m} B\right).$$

Since ν is non-negative, it follows that

$$\sum_{k=1}^{m} \nu(B_k) \leq \nu(B).$$

Thus ν is non-negative and sub-additive on BOREL sets in $\boldsymbol{D}$. The assumptions of theorem 1 are therefore satisfied (the family of BOREL sets in $\boldsymbol{D}$ playing now the role of the family F), and hence we have the following statement.

Lemma 1. If $\nu(B)$ is a finite measure on BOREL sets in $\boldsymbol{D}$, then the derivative $D(u, \nu)$ of ν exists a.e. in $\boldsymbol{D}$ and is L-summable in $\boldsymbol{D}$. Furthermore, if O is any open set in $\boldsymbol{D}$, then

$$\int_O D(u, \nu)\, d\mathsf{L} \leq \nu(O). \tag{7}$$

Lemma 2. Let $\nu(B)$ be a finite measure on BOREL sets in $\boldsymbol{D}$. If B is a BOREL set in $\boldsymbol{D}$ such that

$$\nu(B) = 0, \tag{8}$$

then

$$D(u, \nu) = 0 \quad \text{a.e. on } B. \tag{9}$$

Proof. Assign $\varepsilon > 0$. In view of **III.1.1**, lemma 3, (8) implies the existence of an open set O such that

$$\nu(O) < \varepsilon, \quad B \subset O \subset \boldsymbol{D}. \tag{10}$$

Since $D(u, \nu) \geq 0$, from (10) and (7) we obtain

$$0 \leq \int_B D(u, \nu)\, d\mathsf{L} \leq \int_O D(u, \nu)\, d\mathsf{L} \leq \nu(O) < \varepsilon.$$

As ε was arbitrary, it follows that

$$\int_B D(u, \nu)\, d\mathsf{L} = 0,$$

and (9) follows by **III.1.1**, lemma 35.

Lemma 3. Let $\nu(B)$ be a finite measure on BOREL sets in $\boldsymbol{D}$. If ν is singular relative to L-measure, then

$$D(u, \nu) = 0 \quad \text{a.e. in } \boldsymbol{D}. \tag{11}$$

Proof. Since ν is singular relative to L-measure, there exists (see **III.1.2**) a BOREL set $B_0 \subset \boldsymbol{D}$ such that $\mathsf{L}B_0 = 0$, $\nu(\boldsymbol{D} - B_0) = 0$. By lemma 2 we conclude that $D(u, \nu) = 0$ a.e. on $\boldsymbol{D} - B_0$. Since $\mathsf{L}B_0 = 0$, the relation (11) follows.

Given a finite measure $\nu(B)$ on BOREL sets in $\boldsymbol{D}$, let us recall that we have at our disposal the LEBESGUE decomposition (see **III.1.2**)

$$\nu(B) = \int_B f(u)\, d\mathsf{L} + \nu_s(B), \tag{12}$$

where $f(u)$ is non-negative and L-summable in $\boldsymbol{D}$, ν_s is a finite singular measure on BOREL sets in $\boldsymbol{D}$, and B is any BOREL set in $\boldsymbol{D}$. In view of lemma 3 and **III.1.1**, lemma 43, we obtain from (12) the relation

$$f(u) = D(u, \nu) \quad \text{a.e. in } \boldsymbol{D}. \tag{13}$$

Accordingly, (12) may be re-written in the form

$$\nu(B) = \int_B D(u, \nu)\, d\mathsf{L} + \nu_s(B). \tag{14}$$

Theorem 2. Let $\nu(B)$ be a finite measure on BOREL sets in $\boldsymbol{D}$ which is absolutely continuous with respect to L-measure. Then

$$\nu(B) = \int_B D(u, \nu)\, d\mathsf{L} \tag{15}$$

for every BOREL set $B \subset \boldsymbol{D}$.

Indeed, in this case $\nu_s \equiv 0$ by **III.1.2**, and thus (15) follows from (14).

Theorem 3. Let $\nu(B)$ be a finite measure on BOREL sets in $\boldsymbol{D}$. Then the derivative $D(u, \nu)$ exists a.e. in $\boldsymbol{D}$ and is L-summable in $\boldsymbol{D}$. Furthermore

$$\int_{\boldsymbol{D}} D(u, \nu)\, d\mathsf{L} \leq \nu(\boldsymbol{D}), \tag{16}$$

where the sign of equality holds if and only if ν is absolutely continuous in $\boldsymbol{D}$ with respect to L-measure.

Proof. In view of lemma 1 and theorem 2, there remains to show that ν is absolutely continuous with respect to L-measure if

$$\int_{\boldsymbol{D}} D(u, \nu)\, d\mathsf{L} = \nu(\boldsymbol{D}). \tag{17}$$

Now (17) implies, in view of (14), that $\nu_s(\boldsymbol{D}) = 0$. Since $0 \leq \nu_s(B) \leq \nu_s(\boldsymbol{D})$ for every BOREL set $B \subset \boldsymbol{D}$, it follows that $\nu_s \equiv 0$, and hence ν is absolutely continuous with respect to L-measure (see **III.1.2**).

Theorem 4. Let $\nu(B)$ be a finite measure on BOREL sets in $\boldsymbol{D}$ which is absolutely continuous with respect to L-measure, and let $h(u)$ be a BOREL measurable function in $\boldsymbol{D}$. If B is any BOREL set in $\boldsymbol{D}$, then

$$\int_B h(u)\, d\nu = \int_B h(u)\, D(u, \nu)\, d\mathsf{L}, \tag{18}$$

as soon as one of the two integrals involved exists.

In view of (13), this statement is a direct consequence of the general theorem on *change of measure* in **III.1.2.**

Part IV.

Bounded variation and absolute continuity in R^n [1].

§ IV.1. Measurability questions[2].

IV.1.1. Basic measurability theorems. From this point on, our general objective is the study of continuous transformations in Euclidean n-space R^n from the point of view of Analysis. As before, we shall deal with transformations given in the form

$$T: \boldsymbol{D} \rightarrow R^n, \tag{1}$$

where $\boldsymbol{D}$ is a bounded domain in R^n and T is continuous and bounded in $\boldsymbol{D}$. The notations and the terminology adopted in **II.3.1** will be followed.

Theorem 1. Given a continuous transformation T as in (1) and a BOREL set $B \subset R^n$, the set $T^{-1}B$ is again a BOREL set.

Proof. Let $\mathfrak{S}$ denote the class of those sets $S \subset R^n$ for which the set $T^{-1}S$ is a BOREL set. The class $\mathfrak{S}$ possesses the following properties

(i) If O is an open set in R^n, then $O \in \mathfrak{S}$.

Indeed, since T is continuous, the set $T^{-1}O$ is open and hence it is a BOREL set.

(ii) If $S \in \mathfrak{S}$, then $\complement S \in \mathfrak{S}$.

Indeed, $T^{-1}\complement S = \boldsymbol{D} - T^{-1}S$. Since $\boldsymbol{D}$ and $T^{-1}S$ are BOREL sets, it follows that $T^{-1}\complement S$ is a BOREL set.

(iii) If $S_1, \ldots, S_m, \ldots$ is a (finite or infinite) sequence of sets of the class $\mathfrak{S}$, then $\bigcup S_m \in \mathfrak{S}$.

[1] Part IV and Part V constitute an elaboration (including various improvements) of the note by T. RADO and P. V. REICHELDERFER, On n-dimensional concepts of bounded variation, absolute continuity, and generalized Jacobian [Proc. Nat. Acad. Sci. U.S.A. **35**, 678–681 (1949)].

[2] For historical comments and further details concerning the role of BOREL sets and analytic sets in connection with transformations, see the HAHN treatise listed in the *Bibliography*. A paper by E. J. MICKLE, Metric foundations of continuous transformations [Trans. Amer. Math. Soc. **63**, 368–391 (1948)], contains a great deal of instructive material.

Indeed, $T^{-1} \cup S_m = \cup T^{-1} S_m$, and hence $T^{-1} \cup S_m$ is a BOREL set, since the sets $T^{-1} S_m$ are BOREL sets.

The properties (i), (ii), (iii) mean that $\mathfrak{S}$ is a completely additive class of sets in R^n, and hence $\mathfrak{S}$ contains all BOREL sets in R^n (see **I.1.4**, definitions 10 and 11). Thus if B is any BOREL set in R^n, then $T^{-1}B$ is a BOREL set.

Theorem 2. If $H(x)$ is a (real-valued) BOREL measurable function in R^n, then the function $H(Tu)$ is BOREL measurable in $\boldsymbol{D}$ (see **III.1.1**, definition 8).

Proof. Let c be any real number, and denote by S_c the set of those points $x \in R^n$ where $H(x) > c$, and by S_c^* the set of those points $u \in \boldsymbol{D}$ where $H(Tu) > c$. Clearly $S_c^* = T^{-1} S_c$. Since S_c is a BOREL set [since $H(x)$ is BOREL measurable], it follows by theorem 1 that S_c^* is also a BOREL set. As c was arbitrary, we conclude that $H(Tu)$ is BOREL measurable.

There exist examples showing that the continuous image of a BOREL set need not be a BOREL set. Due to this discrepancy, one has to utilize analytic sets (see **I.1.4**, definition 12) in deriving certain measurability theorems needed in the sequel.

Theorem 3. If A is an analytic set in $\boldsymbol{D}$, then TA is an analytic set in R^n.

Proof. Since the assertion is trivial if A is empty, we assume that A is non-empty. In view of **I.1.4**, exercise 7, we can further assume that A is given in terms of a regular determining system $\mathfrak{A}$ of closed sets $A(m_1, \ldots, m_k)$ such that

$$\lim_{k\to\infty} \delta A(m_1, \ldots, m_k) = 0. \tag{2}$$

Consider the system $\mathfrak{A}^*$ of the sets

$$A^*(m_1, \ldots, m_k) = \overline{T(\boldsymbol{D} \cap A(m_1, \ldots, m_k))}. \tag{3}$$

Clearly, $\mathfrak{A}^*$ is a regular determining system of closed sets. We first verify that (see **I.1.4**, definition 12)

$$TA \subset \mathsf{A}(\mathfrak{A}^*). \tag{4}$$

Indeed, consider any point $x \in TA$. Then there is a point $u \in A$ such that $x = Tu$. Since $u \in A = \mathsf{A}(\mathfrak{A})$, there exists an infinite sequence $m_1, m_2, \ldots$ of integers such that

$$u \in \boldsymbol{D} \cap A(m_1, \ldots, m_k), \quad k = 1, 2, \ldots .$$

In view of (3) it follows that

$$x = Tu \in A^*(m_1, \ldots, m_k), \quad k = 1, 2, \ldots ,$$

and hence $x \in \mathsf{A}(\mathfrak{A}^*)$. Thus (4) is established. Next we verify that

$$\mathsf{A}(\mathfrak{A}^*) \subset TA. \tag{5}$$

Indeed, take any point $x \in \mathsf{A}(\mathfrak{A}^*)$. Then there exists an infinite sequence of integers $m_1, m_2, \ldots$, such that

$$x \in A^*(m_1, \ldots, m_k), \quad k = 1, 2, \ldots.$$

In view of (3), for each integer k there is a point x_k such that

$$x_k \in T(\boldsymbol{D} \cap A(m_1, \ldots, m_k)), \quad \|x_k - x\| < 1/k. \tag{6}$$

The inclusion (6) implies the existence of a point u_k such that

$$u_k \in \boldsymbol{D} \cap A(m_1, \ldots, m_k), \quad T u_k = x_k. \tag{7}$$

Since the determining system $\mathfrak{A}$ is regular, clearly

$$u_j \in A(m_1, \ldots, m_k) \quad \text{for} \quad j \geq k. \tag{8}$$

In view of (2) it follows that the points $u_1, u_2, \ldots$ constitute a Cauchy sequence (see **I.1.4**, definition 8), and hence they constitute a convergent sequence. Put

$$u = \lim u_k. \tag{9}$$

Since the set $A(m_1, \ldots, m_k)$ is closed, it follows from (8) that

$$u \in A(m_1, \ldots, m_k), \quad k = 1, 2, \ldots,$$

and hence [since $A = \mathsf{A}(\mathfrak{A})$]

$$u \in A \subset \boldsymbol{D}. \tag{10}$$

Since T is continuous, (10), (9), (7), (6) imply that

$$T u = \lim T u_k = \lim x_k = x.$$

By (10) it follows that $x \in TA$, and (5) is established. From (4) and (5) it follows that $TA = \mathsf{A}(\mathfrak{A}^*)$, and the theorem is proved.

Theorem 4. If B is a Borel set in $\boldsymbol{D}$, then TB is L-measurable.

Proof. Since Borel sets are analytic (see **I.1.4**, exercise 7), TB is an analytic set in R^n by theorem 3. But analytic sets in R^n are L-measurable (see **III.1.1**, lemma 5), and the theorem follows.

IV.1.2. The crude multiplicity function. Given T as in **IV.1.1**, let S be a subset of $\boldsymbol{D}$ and x a point in R^n. The *crude multiplicity function* $N(x, T, S)$ is then equal to the number (perhaps ∞) of the points u in the set $S \cap T^{-1}x$. Some elementary properties of $N(x, T, S)$, needed in the sequel, will now be verified.

Since $\boldsymbol{D}$ is a bounded domain in R^n, we have an oriented n-cube $\boldsymbol{Q}$ containing $\boldsymbol{D}$. For each positive integer j, let p_j denote the j-th positive prime, and consider the subdivision of $\boldsymbol{Q}$ into p_j^n congruent oriented n-cubes (see **I.2.3**, definition 3). Let Q_j be a generic notation for a cube occurring in this subdivision. For each j, designate by $\mathfrak{Q}_j$ the class of those cubes Q_j which satisfy the condition $Q_j \subset \boldsymbol{D}$.

If G is a set in R^n, then denote by $c(x, G)$ the characteristic function of G, defined as follows:

$$c(x, G) = \begin{cases} 1 & \text{if} \quad x \in G, \\ 0 & \text{if} \quad x \notin G. \end{cases}$$

Then clearly G is L-measurable if and only if $c(x, G)$ is L-measurable, and G is a Borel set if and only if $c(x, G)$ is Borel measurable (see **III.1.1**, definitions 5 and 8).

We proceed to verify the following relations, in which S stands for an arbitrary subset of $\boldsymbol{D}$.

$$N(x, T, S) \geq \sum N(x, T, S \cap \operatorname{int} Q_j), \qquad Q_j \in \mathfrak{Q}_j. \tag{1}$$

$$N(x, T, S) = \lim_{j \to \infty} \sum N(x, T, S \cap \operatorname{int} Q_j), \quad Q_j \in \mathfrak{Q}_j. \tag{2}$$

$$N(x, T, S) \geq \sum c\big(x, T(S \cap \operatorname{int} Q_j)\big), \qquad Q_j \in \mathfrak{Q}_j. \tag{3}$$

$$N(x, T, S) = \lim_{j \to \infty} \sum c\big(x, T(S \cap \operatorname{int} Q_j)\big), \quad Q_j \in \mathfrak{Q}_j. \tag{4}$$

Note that for fixed j the sets $S \cap \operatorname{int} Q_j$ (corresponding to the various cubes $Q_j \in \mathfrak{Q}_j$) are pair-wise disjoint subsets of S, and thus (1) is obvious. If F is any subset of $\boldsymbol{D}$, then clearly

$$N(x, T, F) \geq c(x, TF) \geq 0, \tag{5}$$

and thus (3) follows from (1). In view of (5) it is evident that (2) is a consequence of (1) and (4) and thus there remains to show that (4) is valid. Fix a point $x \in R^n$. If $N(x, T, S) = 0$, then (4) is obvious, since then every term in the summation appearing in (4) vanishes. So assume that $N(x, T, S) > 0$. Let m be any positive integer such that

$$N(x, T, S) \geq m. \tag{6}$$

The set $S \cap T^{-1} x$ contains then at least m distinct points $u_1, \ldots, u_m$. By **I.2.3**, exercise 2, for j sufficiently large each of these points is interior to a cube Q_j. Let $Q_j^1, \ldots, Q_j^m$ be the cubes Q_j for which

$$u_1 \in \operatorname{int} Q_j^1, \ldots, u_m \in \operatorname{int} Q_j^m.$$

Since the diameters of the cubes Q_j converge to zero for $j \to \infty$ and since $u_1, \ldots, u_m$ are distinct points of the domain $\boldsymbol{D}$, it is clear that for j sufficiently large the cubes $Q_j^1, \ldots, Q_j^m$ are distinct and are contained in $\boldsymbol{D}$, and hence belong to the class $\mathfrak{Q}_j$. Thus, for j sufficiently large, we have

$$m \leq \sum_{i=1}^{m} c\left(x, T(S \cap \operatorname{int} Q_j^i)\right) \leq \sum c\left(x, T(S \cap \operatorname{int} Q_j)\right), \quad Q_j \in \mathfrak{Q}_j.$$

Since m is any integer satisfying (6), it follows that

$$N(x, T, S) \leq \liminf_{j \to \infty} \sum c\left(x, T(S \cap \operatorname{int} Q_j)\right), \quad Q_j \in \mathfrak{Q}_j. \tag{7}$$

Clearly, (7) and (3) imply (4).

Theorem. If S is a Borel set in $\boldsymbol{D}$, then $N(x, T, S)$ is L-measurable in R^n.

Proof. Consider the expression (4) for $N(x, T, S)$. Since now $S \cap \operatorname{int} Q_j$ is a Borel set, the image set $T(S \cap \operatorname{int} Q_j)$ is L-measurable by **IV.1.1**, theorem 4. Hence the characteristic functions $c\left(x, T(S \cap \operatorname{int} Q_j)\right)$ are also L-measurable, and thus (4) shows that $N(x, T, S)$ is L-measurable (see **III.1.1**, lemma 18).

Let now O be an open set in $\boldsymbol{D}$. For each positive integer j, denote by $\mathfrak{Q}_j^*$ the class of those cubes Q_j for which $\operatorname{int} Q_j \subset O$. An argument entirely analogous to that used in verifying (1), (2), (3), (4) yields the following relations.

$$N(x, T, O) \geq \sum N(x, T, \operatorname{int} Q_j), \qquad Q_j \in \mathfrak{Q}_j^*. \tag{8}$$

$$N(x, T, O) = \lim_{j \to \infty} \sum N(x, T, \operatorname{int} Q_j), \quad Q_j \in \mathfrak{Q}_j^*. \tag{9}$$

$$N(x, T, O) \geq \sum c(x, T \operatorname{int} Q_j), \qquad Q_j \in \mathfrak{Q}_j^*. \tag{10}$$

$$N(x, T, O) = \lim_{j \to \infty} \sum c(x, T \operatorname{int} Q_j), \quad Q_j \in \mathfrak{Q}_j^*. \tag{11}$$

IV.1.3. Summatory functions. Given the transformation T as in **IV.1.1**, let $f(u)$ be a real-valued, non-negative function in $\boldsymbol{D}$ (see **I.1.2**). If S is a subset of $\boldsymbol{D}$ and x is point in R^n, put

$$W(x, T, S, f) = \sum f(u), \quad u \in S \cap T^{-1} x. \tag{1}$$

Note that according to the conventions adopted in **I.1.2**, the non-negative function $f(u)$ is permitted to take on the value ∞. Furthermore, the summation in (1) is generally non-countable, since the set $S \cap T^{-1} x$ need not be countable. The general comments in **I.1.2** concerning non-countable summations apply to the present situation. In terms of the notations used there, the defining formula (1) appears in the form

$$W(x, T, S, f) = \sum (S \cap T^{-1} x, f), \tag{2}$$

which yields the precise meaning of the summation in (1).

Definition 1. The quantity $W(x, T, S, f)$ is termed the summatory function induced by T, S, f.

Thus $W(x, T, S, f)$ is a non-negative function defined for all points $x \in R^n$. If the transformation T is thought of as fixed, then the simpler notation $W(x, S, f)$ may be used. In cases where the function $f(u)$ is also fixed, we shall write merely $W(x, S)$, omitting both T and f. In view of (2), the various elementary properties of non-countable summations (listed in **I.1.2**) are available in dealing with the summatory function W. When using such properties in the sequel, the reader will be referred to the general comments in **I.1.2**, since the application to W will be immediate.

Lemma. Let S be a subset of $\boldsymbol{D}$, and consider the characteristic function

$$f_S(u) = \begin{cases} 1 & \text{if} \quad u \in S, \\ 0 & \text{if} \quad u \in \boldsymbol{D} - S, \end{cases}$$

of S in $\boldsymbol{D}$. If F is any set in $\boldsymbol{D}$, then

$$W(x, T, F, f_S) = N(x, T, F \cap S).$$

This statement is an obvious consequence of the definitions involved.

Theorem. If $f(u)$ is a non-negative, Borel measurable function in $\boldsymbol{D}$ and B is any Borel set in $\boldsymbol{D}$, then the summatory function $W(x, T, B, f)$ is L-measurable in R^n.

Proof. Since T is fixed in the course of the argument, we shall use the simplified notation $W(x, B, f)$. For convenience, the proof is given in three steps.

Case 1. Assume that f is the characteristic function in $\boldsymbol{D}$ of a Borel set $B_* \subset \boldsymbol{D}$. That is,

$$f(u) = \begin{cases} 1 & \text{if} \quad u \in B_*, \\ 0 & \text{if} \quad u \in \boldsymbol{D} - B_*. \end{cases}$$

If B is any Borel set in $\boldsymbol{D}$, then (see the preceding lemma)

$$W(x, B, f) = N(x, T, B \cap B_*). \tag{3}$$

Since $B \cap B_*$ is a Borel set, the crude multiplicity function appearing in (3) is L-measurable by the theorem in **IV.1.2**, and thus (3) shows that $W(x, B, f)$ is L-measurable.

Case 2. Assume that $f(u)$ is non-negative, Borel measurable, and *bounded* in $\boldsymbol{D}$. Then we have a constant $M > 0$ such that $0 \leq f(u) \leq M$ in $\boldsymbol{D}$. Take a positive integer m. For each integer j such that $0 \leq j \leq 2^m$ denote by B_{mj} the set of those points $u \in \boldsymbol{D}$ where

$$\frac{j}{2^m} M \leq f(u) < \frac{j+1}{2^m} M.$$

Then the sets B_{mj}, $j=0, 1, \ldots, 2^m$, are pair-wise disjoint BOREL sets whose union is $\boldsymbol{D}$. Let $f_{mj}(u)$ be the characteristic function in $\boldsymbol{D}$ of the set B_{mj}. That is,

$$f_{mj}(u) = \begin{cases} 1 & \text{if} \quad u \in B_{mj}, \\ 0 & \text{if} \quad u \in \boldsymbol{D} - B_{mj}. \end{cases} \tag{4}$$

Introduce the auxiliary function

$$f_m(u) = \sum \frac{j}{2^m} M f_{mj}(u), \quad j = 0, 1, \ldots, 2^m. \tag{4*}$$

Clearly

$$0 \leqq f_1 \leqq f_2 \leqq \cdots, \tag{5}$$

$$0 \leqq f - f_m < \frac{M}{2^m}, \tag{6}$$

and hence

$$\lim f_m = f. \tag{7}$$

Let B be any BOREL set in $\boldsymbol{D}$. The relations (5) and (7) imply (see **I.1.2**) that

$$\lim W(x, B, f_m) = W(x, B, f). \tag{8}$$

In view of (4), (4*) and *Case 1*, $W(x, B, f_m)$ is L-measurable, and thus (8) shows that $W(x, B, f)$ is also L-measurable (see **III.1.1**, lemma 18).

Case 3. Let finally $f(u)$ be an arbitrary non-negative BOREL measurable function in $\boldsymbol{D}$. For each positive integer m, put

$$g_m(u) = \begin{cases} f(u) & \text{if} \quad 0 \leqq f(u) \leqq m, \\ m & \text{if} \quad f(u) > m. \end{cases} \tag{9}$$

Then clearly $g_m(u)$ is BOREL measurable, and

$$0 \leqq g_1 \leqq g_2 \leqq \cdots, \tag{10}$$

$$\lim g_m = f. \tag{11}$$

In view of **I.1.2**, the relations (10) and (11) imply that

$$\lim W(x, B, g_m) = W(x, B, f), \tag{12}$$

where B is any BOREL set in $\boldsymbol{D}$. Since g_m is non-negative, BOREL measurable, and bounded, it follows by *Case 2* that $W(x, B, g_m)$ is L-measurable, and thus the L-measurability of $W(x, B, f)$ follows in view of (12).

IV.1.4. The condition (N). Again, $T: \boldsymbol{D} \to R^n$ is a bounded continuous transformation, as in **IV.1.1**.

Definition. If U is a subset of $\boldsymbol{D}$, then the transformation T is said to satisfy condition (N) on U if every subset of U of LEBESGUE measure zero is transformed by T into a set of LEBESGUE measure zero. In symbols: $V \subset U$ and $\mathsf{L}V=0$ imply $\mathsf{L}TV=0$.

Lemma 1. (i) If T satisfies condition (N) on U, then it also satisfies condition (N) on every subset of U. (ii) If U is a subset of $\boldsymbol{D}$ such that $\mathsf{L}TU=0$, then T satisfies condition (N) on U. (iii) If $\{U_m\}$ is a sequence of sets in $\boldsymbol{D}$ and T satisfies condition (N) on every U_m, then it also satisfies condition (N) on $\bigcup U_m$.

These statements are obvious consequences of the definitions involved.

Lemma 2. Let f be a non-negative BOREL measurable function on $\boldsymbol{D}$, and L an L-measurable subset of $\boldsymbol{D}$ such that T satisfies condition (N) on the set L_f consisting of those points $u \in L$ where $f(u)>0$. Then the summatory function $W(x, L, f)$ is L-measurable in R^n.

Proof. The set L_f is L-measurable and hence it admits of a decomposition (see **III.1.1**, lemma 8)

$$L_f = B \cup V, \quad B \cap V = \emptyset, \quad \mathsf{L}V = 0, \tag{1}$$

where B is a BOREL set. Since T satisfies condition (N) on L_f, it follows from (1) that

$$\mathsf{L}TV = 0. \tag{2}$$

From (1) one infers, by **I.1.2**, that

$$W(x, L, f) = W(x, L_f, f) = W(x, B, f) + W(x, V, f). \tag{3}$$

Clearly $W(x, V, f) = 0$ for $x \in \complement TV$, so (2) and (3) imply that

$$W(x, L, f) = W(x, B, f) \quad \text{a.e. in } R^n. \tag{4}$$

Since $W(x, B, f)$ is L-measurable in R^n by the theorem in **IV.1.3**, it follows from (4) that $W(x, L, f)$ is also L-measurable in R^n, and the lemma is proved.

Corollary 1. If L is an L-measurable subset of $\boldsymbol{D}$ such that T satisfies condition (N) on L, then the crude multiplicity function $N(x, T, L)$ is L-measurable in R^n.

Proof. The characteristic function $f_{\boldsymbol{D}}$ of the domain $\boldsymbol{D}$ is clearly BOREL measurable in $\boldsymbol{D}$, and by the lemma in **IV.1.3** one has

$$W(x, L, f_{\boldsymbol{D}}) = N(x, T, L).$$

Thus corollary 1 is an immediate consequence of lemma 2.

Corollary 2. If L is an L-measurable subset of $\boldsymbol{D}$ such that T satisfies condition (N) on L, then TL is L-measurable.

Proof. Since TL coincides with the set of points x in R^n where $N(x, T, L) \geq 1$, corollary 2 is a direct consequence of corollary 1.

Theorem. Let f be a non-negative L-measurable function on $\boldsymbol{D}$ such that T satisfies condition (N) on the set $\boldsymbol{D}_f$ of points $u \in \boldsymbol{D}$ where $f(u) > 0$. Then for every L-measurable set L in $\boldsymbol{D}$ the summatory function $W(x, L, f)$ is L-measurable in R^n.

Proof. By **III.1.1**, lemma 24 there exists a non-negative Borel measurable function f_* on $\boldsymbol{D}$ such that

$$0 \leq f_*(u) \leq f(u), \quad u \in \boldsymbol{D}, \tag{5}$$

$$f_*(u) = f(u) \quad \text{a.e. in } \boldsymbol{D}. \tag{6}$$

Let V denote the set of points $u \in \boldsymbol{D}$ where $f_*(u) \neq f(u)$. From (5) and (6) it follows that $V \subset \boldsymbol{D}_f$ and $\mathsf{L} V = 0$. Since T satisfies condition (N) on $\boldsymbol{D}_f$, one concludes that

$$\mathsf{L} T V = 0. \tag{7}$$

If $x \in \mathsf{C} T V$, it is clear that $f_*(u) = f(u)$ for $u \in L \cap T^{-1} x$, and consequently $W(x, L, f_*) = W(x, L, f)$ if $x \in \mathsf{C} T V$. Using (7) one concludes that

$$W(x, L, f_*) = W(x, L, f) \quad \text{a.e. in } R^n. \tag{8}$$

In view of (5) the set L_{f_*} of points $u \in L$ where $f_*(u) > 0$ is a subset of $\boldsymbol{D}_f$. Since T satisfies condition (N) on $\boldsymbol{D}_f$, it follows that it also satisfies condition (N) on L_{f_*}. Thus the summatory function $W(x, L, f_*)$ is L-measurable in R^n by lemma 2, and the conclusion of the theorem now follows from (8).

Lemma 3. Let $\mathfrak{F}$ be a family of *connected* sets C in $\boldsymbol{D}$, such that the following conditions are satisfied. (i) No set $C \in \mathfrak{F}$ reduces to a single point. (ii) T is constant on each set $C \in \mathfrak{F}$ (that is, TC is a single point). (iii) T satisfies the condition (N) on the set

$$S = \bigcup C, \quad C \in \mathfrak{F}. \tag{9}$$

Then

$$\mathsf{L} T S = 0. \tag{10}$$

Proof. Let k be an integer, m a positive integer, and let j be one of the integers $1, \ldots, n$. Denote by G_{kmj} the set of those points $u = (u^1, \ldots, u^n) \in \boldsymbol{D}$ for which $u_j = k/2^m$. By **III.1.1**, lemma 12 we have then $\mathsf{L} G_{kmj} = 0$. In view of condition (iii) it follows that

$$\mathsf{L} T (S \cap G_{kmj}) = 0.$$

Since the collection of the sets G_{kmj} (corresponding to all admissible choices for k, m, j) is countable, it follows that on setting

$$G = \bigcup G_{kmj} \tag{11}$$

we have

$$\mathsf{L}\, T(S \cap G) = 0. \tag{12}$$

We assert that

$$C \cap G \neq \emptyset \quad \text{if} \quad C \in \mathfrak{F}. \tag{13}$$

Indeed, by condition (i), the set C contains at least two distinct points $u_1 = (u_1^1, \ldots, u_1^n)$, $u_2 = (u_2^1, \ldots, u_2^n)$. For some integer $j = 1, \ldots, n$ we must have then $u_1^j \neq u_2^j$, say $u_1^j < u_2^j$. It follows that for some integer k and some positive integer m we shall have the inequalities

$$u_1^j < k/2^m < u_2^j.$$

Since C is connected, by **I.1.3**, exercise 37 there follows the existence of a point $u = (u^1, \ldots, u^j, \ldots, u^n) \in C$ such that $u^j = k/2^m$. For this point we have then

$$u \in C \cap G_{kmj} \subset C \cap G,$$

and (13) is verified. Next we assert that

$$TC = T(C \cap G) \quad \text{if} \quad C \in \mathfrak{F}. \tag{14}$$

Indeed, by (13) there exists a point $u \in C \cap G$. Since T is constant on C, we conclude that $TC = Tu \in T(C \cap G)$, and (14) follows, since obviously $T(C \cap G) \subset TC$. From (14) and (9) we conclude that $TS = T(S \cap G)$, and (10) follows in view of (12).

§ IV.2. Bounded variation and absolute continuity with respect to a base-function.

IV.2.1. Base-functions. Let there be given a bounded continuous transformation $T: D \to R^n$ as in **IV.1.1.**

Definition 1. A function $b(u)$, defined in $\boldsymbol{D}$, is termed a *base-function* in $\boldsymbol{D}$ if it possesses the following properties. (i) At each point $u \in \boldsymbol{D}$, $b(u)$ is either a non-negative integer or $+\infty$. (ii) $b(u)$ is Borel measurable in $\boldsymbol{D}$.

The base-functions actually utilized in the sequel will be selected on the basis of topological considerations. Our present objective is the development of a metric theory of base-functions which is of sufficient generality to cover the situations actually arising in the sequel. This theory could be generalized in various ways. For example, the requirement that the finite values of a base-function should be integers could be dropped, as far as the topics discussed in **§ IV.2** are concerned.

Definition 2. Given a base-function $b(u)$ in $\boldsymbol{D}$, the set of those points $u \in \boldsymbol{D}$ where $b(u) > 0$ will be denoted by $\boldsymbol{D}_b$ and will be termed the *base-set* for b. The set of those points $u \in \boldsymbol{D}$ where $0 < b(u) < \infty$ will be denoted by $\boldsymbol{D}_b^r$ and will be termed the *reduced base-set* for b.

Since $b(u)$ is BOREL measurable in $\boldsymbol{D}$, it is clear that $\boldsymbol{D}_b$ and $\boldsymbol{D}_b^r$ are BOREL sets. Obviously

$$\boldsymbol{D}_b^r \subset \boldsymbol{D}_b. \tag{1}$$

If S is any subset of $\boldsymbol{D}$, then there arises the summatory function

$$W(x, T, S, b) = \sum b(u), \quad u \in S \cap T^{-1} x, \tag{2}$$

in the sense of **IV.1.3**. As T and b will be kept fixed throughout the rest of **§ IV.2**, we shall write $W(x, S)$ instead of $W(x, T, S, b)$. Since $b(u)$ is BOREL measurable, by the theorem in **IV.1.3** we have the following statement (which is formulated as a lemma merely for the sake of convenient reference).

Lemma 1. If B is a BOREL set in $\boldsymbol{D}$, then the summatory function $W(x, B)$ is L-measurable (as a function of x in R^n).

Let S be a subset of $\boldsymbol{D}$. If $x \notin T S$, then $S \cap T^{-1} x = \emptyset$, and hence clearly

$$W(x, S) = 0 \quad \text{if} \quad x \in \complement T S. \tag{3}$$

The following lemmas are concerned with the case when $x \in T S$.

Lemma 2. If $S \subset \boldsymbol{D} - \boldsymbol{D}_b$, then $W(x, S) = 0$ even for $x \in T S$.

Proof. Since now $S \cap T^{-1} x \subset \boldsymbol{D} - \boldsymbol{D}_b$, and $b(u) = 0$ for $u \in \boldsymbol{D} - \boldsymbol{D}_b$, the assertion is obvious in view of (2).

Lemma 3. If $S \subset \boldsymbol{D}_b - \boldsymbol{D}_b^r$, then $W(x, S) = +\infty$, and $W(x, \boldsymbol{D}) = +\infty$ for $x \in T S$.

Proof. The assertion is obvious, since $b(u) = +\infty$ for $u \in \boldsymbol{D}_b - \boldsymbol{D}_b^r$ and $W(x, \boldsymbol{D}) \geq W(x, S)$.

Lemma 4. If $S \subset \boldsymbol{D}_b^r$, then $W(x, S) \geq 1$ for $x \in T S$.

Proof. If $x \in T S$, then there exists at least one point $u_0 \in S \subset \boldsymbol{D}_b^r$ such that $T u_0 = x$. Then $u_0 \in S \cap T^{-1} x$ and $b(u_0) \geq 1$, and hence

$$W(x, S) \geq b(u_0) \geq 1.$$

IV.2.2. Bounded variation with respect to a base-function.

Definition 1. The transformation T, given as in **IV.1.1**, is said to be of bounded variation in $\boldsymbol{D}$ with respect to the base-function b (briefly, BVb in $\boldsymbol{D}$) if the corresponding summatory function $W(x, \boldsymbol{D})$ is L-summable in R^n (see **IV.2.1**).

Observe that $W(x, \boldsymbol{D})$ is L-measurable by **IV.2.1**, lemma 1, and hence the preceding definition is meaningful. If B is any BOREL set in $\boldsymbol{D}$, then clearly

$$0 \leq W(x, B) \leq W(x, \boldsymbol{D}). \tag{1}$$

Thus the L-summability of $W(x, \boldsymbol{D})$ implies the L-summability of $W(x, B)$. This remark yields the following statement.

Lemma 1. If T is BVb in $\boldsymbol{D}$, then it is also BVb in every domain $D \subset \boldsymbol{D}$.

Assume now that T is BVb in $\boldsymbol{D}$. If B is any Borel set in $\boldsymbol{D}$, then $W(x, B)$ is L-summable in R^n as a consequence of (1). On setting

$$\nu(B) = \int W(x, B)\, d\mathsf{L}, \tag{2}$$

we obtain therefore a finite-valued, non-negative function ν of Borel sets in $\boldsymbol{D}$.

As regards the definition of $\nu(B)$, observe that $W(x, B) = 0$ for $x \in \mathsf{C}TB$ by **IV.2.1** (3), and hence (since T is bounded) $W(x, B)$ vanishes outside of a properly selected n-cube in R^n. The integration in (2) is taken over R^n, as explained in **III.1.1**. Note that ν depends also upon T and b, and hence a fully descriptive notation for ν should be $\nu(B, T, b)$. Since T and b are thought of as fixed in the present context, the concise notation $\nu(B)$ is adequate. Observing that $W(x, B) = 0$ for $x \in \mathsf{C}TB$ by **IV.2.1** (3), we see that (2) may be written in the equivalent form

$$\nu(B) = \int_{TB} W(x, B)\, d\mathsf{L}. \tag{3}$$

Lemma 2. If T is BVb in $\boldsymbol{D}$, then $\nu(B)$ is a finite measure on Borel sets in $\boldsymbol{D}$.

Proof. Let $B_1, \ldots, B_m, \ldots$ be a (finite or infinite) sequence of pair-wise disjoint Borel sets in $\boldsymbol{D}$. Put $B = \bigcup B_m$. Then (see **I.1.2**)

$$W(x, B) = \sum W(x, B_m).$$

Integration yields (see **III.1.1**, lemma 27)

$$\nu(B) = \sum \nu(B_m).$$

Thus ν is countably additive on Borel sets in $\boldsymbol{D}$, and we already noted above that ν is finite-valued and non-negative. Thus ν is a finite measure on Borel sets in $\boldsymbol{D}$ (see **III.1.2**).

Definition 2. The measure ν is termed the measure induced in $\boldsymbol{D}$ by the transformation T and the base-function b, where T is assumed to be BVb in $\boldsymbol{D}$. The derivative of ν is denoted by $D(u, \nu)$.

Observe that $D(u, \nu)$ exists a.e. in $\boldsymbol{D}$ and is L-summable in $\boldsymbol{D}$ by **III.2.4**, lemma 1. From the point of view of applications, it is important to obtain information concerning points $u \in \boldsymbol{D}$ where $D(u, \nu) = 0$. If B is a Borel set in $\boldsymbol{D}$, then (by **III.2.4**, lemma 2)

$$\nu(B) = 0 \quad \text{implies} \quad D(u, \nu) = 0 \quad \text{a.e. on } B. \tag{4}$$

Accordingly, we proceed to determine BOREL sets $B \subset \boldsymbol{D}$ on which ν vanishes. *It will be assumed throughout that T is BVb in $\boldsymbol{D}$.*

Lemma 3. $\nu(\boldsymbol{D}-\boldsymbol{D}_b)=0$ (see **IV.2.1**, definition 2).

Proof. Since $W(x, \boldsymbol{D}-\boldsymbol{D}_b)=0$ for $x \in T(\boldsymbol{D}-\boldsymbol{D}_b)$ by **IV.2.1**, lemma 2, the assertion follows directly from (3) with $B=\boldsymbol{D}-\boldsymbol{D}_b$.

Lemma 4. If B is a BOREL set in $\boldsymbol{D}$ such that $\mathsf{L}TB=0$, then $\nu(B)=0$.

This is a direct consequence of (3).

Lemma 5. $\nu(\boldsymbol{D}_b-\boldsymbol{D}_b^r)=0$ (see **IV.2.1**, definition 2).

Proof. In view of **IV.2.1**, lemma 3, applied with $S=\boldsymbol{D}_b-\boldsymbol{D}_b^r$, the set $T(\boldsymbol{D}_b-\boldsymbol{D}_b^r)$ is contained in the set where $W(x, \boldsymbol{D})=+\infty$. This latter set is of L-measure zero since $W(x, \boldsymbol{D})$ is L-summable in R^n (see **III.1.1**, lemma 31). Thus $\mathsf{L}T(\boldsymbol{D}_b-\boldsymbol{D}_b^r)=0$, and hence $\nu(\boldsymbol{D}_b-\boldsymbol{D}_b^r)=0$ by lemma 4.

Lemma 6. If B is a BOREL subset of $\boldsymbol{D}_b^r$, then $\nu(B)=0$ if and only if $\mathsf{L}TB=0$.

Proof. By **IV.2.1**, lemma 4, one has $W(x,B)\geqq 1$ for $x\in TB$. Hence, by (3), $\nu(B)\geqq \mathsf{L}TB$. Thus $\nu(B)=0$ implies that $\mathsf{L}TB=0$. Conversely, if $\mathsf{L}TB=0$, then $\nu(B)=0$ by (3).

Lemma 7. $\nu(\boldsymbol{D}-\boldsymbol{D}_b^r)=0$.

This is a direct consequence of lemmas 3 and 5.

Lemma 8. If B is any BOREL set in $\boldsymbol{D}$, then $\nu(B)=\nu(B\cap \boldsymbol{D}_b^r)$.

Proof. B is the union of the disjoint BOREL sets $B_1=B\cap(\boldsymbol{D}-\boldsymbol{D}_b^r)$ and $B_2=B\cap \boldsymbol{D}_b^r$. Hence

$$\nu(B) = \nu(B_1) + \nu(B_2).$$

In view of lemma 7, we have

$$\nu(B_1) \leqq \nu(\boldsymbol{D}-\boldsymbol{D}_b^r) = 0$$

and thus $\nu(B_1)=0$. Hence $\nu(B)=\nu(B_2)$, and the lemma is proved.

The preceding lemmas will now be applied to obtain statements concerning the vanishing of the derivative $D(u, \nu)$ (see definition 2), it being assumed throughout that T is BVb in $\boldsymbol{D}$.

Lemma 9. $D(u, \nu)=0$ a.e. on $\boldsymbol{D}-\boldsymbol{D}_b^r$.

Proof. Since $\nu(\boldsymbol{D}-\boldsymbol{D}_b^r)=0$ by lemma 7, the assertion follows from (4).

Lemma 10. If S is a set in R^n such that $\mathsf{L}S=0$, then $D(u, \nu)=0$ a.e. on $T^{-1}S$.

Proof. We can select (see **III.1.1**, lemma 7) a BOREL set $B\subset R^n$ such that $S\subset B$ and $\mathsf{L}B=0$. Then $T^{-1}B$ is a BOREL set in $\boldsymbol{D}$ (see **IV.1.1**, theorem 1), and $TT^{-1}B \subset B$. Thus $\mathsf{L}TT^{-1}B=0$, and hence $\nu(T^{-1}B)=0$ by lemma 4. In view of (4) it follows that $D(u, v)=0$ a.e. on $T^{-1}B$, and hence also on $T^{-1}S$, since $T^{-1}S \subset T^{-1}B$.

Lemma 11. If F is any set in R^n such that $\mathsf{L}(F \cap T\boldsymbol{D}_b^r)=0$, then

$$D(u, v) = 0 \quad \text{a.e. on} \quad T^{-1}F.$$

Proof. F is the union of the disjoint sets

$$F_1 = F \cap \complement T\boldsymbol{D}_b^r \quad \text{and} \quad F_2 = F \cap T\boldsymbol{D}_b^r.$$

Since $\mathsf{L}F_2=0$ by assumption, we have by lemma 10

$$D(u, v) = 0 \quad \text{a.e. on} \quad T^{-1}F_2. \tag{5}$$

On the other hand, clearly $T^{-1}F_1 \subset \boldsymbol{D}-\boldsymbol{D}_b^r$. As $D(u, v)=0$ a.e. on $\boldsymbol{D}-\boldsymbol{D}_b^r$ by lemma 9, it follows that

$$D(u, v) = 0 \quad \text{a.e. on} \quad T^{-1}F_1. \tag{6}$$

Since $T^{-1}F$ is the union of $T^{-1}F_1$ and $T^{-1}F_2$, (5) and (6) imply that

$$D(u, v) = 0 \quad \text{a.e. on} \quad T^{-1}F.$$

Lemma 12. If G is any set in $\boldsymbol{D}$ such that $\mathsf{L}TG=0$, then $D(u, v)=0$ a.e. on G.

Proof. On setting $S=TG$, and noting that $G \subset T^{-1}TG=T^{-1}S$, one sees that lemma 12 is an immediate consequence of lemma 10.

IV.2.3. Absolute continuity with respect to a base-function. Let there be given a bounded continuous transformation $T: \boldsymbol{D} \to R^n$ and a base-function $b(u)$ in $\boldsymbol{D}$, as in **IV.2.2**.

Definition. T is said to be absolutely continuous in $\boldsymbol{D}$ with respect to the base-function b (briefly, ACb in $\boldsymbol{D}$) if T is BVb in $\boldsymbol{D}$ and the measure ν generated by T and b is absolutely continuous with respect to L-measure in $\boldsymbol{D}$ (see **IV.2.2** and **III.1.2**).

Explicitly, T is ACb in $\boldsymbol{D}$ if the following holds. (i) The summatory function $W(x, \boldsymbol{D})$ corresponding to b (see **IV.2.1**) is L-summable in R^n. (ii) $\nu(B)=0$ for every Borel set $B \subset \boldsymbol{D}$ such that $\mathsf{L}B=0$. Suppose now that these conditions are satisfied, and consider any domain $D \subset \boldsymbol{D}$. Then $0 \leq W(x, D) \leq W(x, \boldsymbol{D})$, and hence $W(x, D)$ is also L-summable. Thus condition (i) still holds if $\boldsymbol{D}$ is replaced by D, and the same is obvious for condition (ii). Hence the following statement.

Lemma 1. If T is ACb in $\boldsymbol{D}$, then T is also ACb in every domain $D \subset \boldsymbol{D}$.

Lemma 2. Suppose that T is BVb in $\boldsymbol{D}$. Then T is ACb in $\boldsymbol{D}$ if and only if T satisfies the condition (N) on the reduced base-set $\boldsymbol{D}_b^r$ (see **IV.2.1**, definition 2 and **IV.1.4**).

Proof. Suppose first that T is ACb in $\boldsymbol{D}$, and consider any set $S \subset \boldsymbol{D}_b^r$ such that $\mathsf{L}S = 0$. There exists then a BOREL set B such that $S \subset B$ and $\mathsf{L}B = 0$. Since $\boldsymbol{D}_b^r$ itself is a BOREL set, the set $B \cap \boldsymbol{D}_b^r$ is a BOREL set such that

$$S \subset B \cap \boldsymbol{D}_b^r \subset \boldsymbol{D}_b^r, \quad \mathsf{L}(B \cap \boldsymbol{D}_b^r) = 0.$$

As ν is absolutely continuous, it follows that $\nu(B \cap \boldsymbol{D}_b^r) = 0$, and hence $\mathsf{L}T(B \cap \boldsymbol{D}_b^r) = 0$ by **IV.2.2**, lemma 6. In view of the inclusion $TS \subset T(B \cap \boldsymbol{D}_b^r)$ we conclude that $\mathsf{L}TS = 0$. Thus it is established that T satisfies the condition (N) on $\boldsymbol{D}_b^r$. Suppose, conversely, that T is BVb in $\boldsymbol{D}$ and satisfies the condition (N) on $\boldsymbol{D}_b^r$. Consider any BOREL set $B \subset \boldsymbol{D}$ such that $\mathsf{L}B = 0$. Then $B \cap \boldsymbol{D}_b^r$ is also a BOREL set and $\mathsf{L}(B \cap \boldsymbol{D}_b^r) = 0$. Since T satisfies the condition (N) on $\boldsymbol{D}_b^r$, it follows that

$$\mathsf{L}T(B \cap D_b^r) = 0. \tag{1}$$

In view of **IV.2.2**, lemma 6, we infer from (1) that

$$\nu(B \cap \boldsymbol{D}_b^r) = 0. \tag{2}$$

By **IV.2.2**, lemma 8 we have

$$\nu(B) = \nu(B \cap \boldsymbol{D}_b^r). \tag{3}$$

(2) and (3) imply that $\nu(B) = 0$. Thus ν is absolutely continuous, and the lemma is proved.

Lemma 3. If T is ACb in $\boldsymbol{D}$, then T satisfies the condition (N) on the base-set $\boldsymbol{D}_b$.

Proof. Let S be any set in $\boldsymbol{D}_b$ such that $\mathsf{L}S = 0$. As S is the union of the sets

$$S_1 = S \cap (\boldsymbol{D}_b - \boldsymbol{D}_b^r) \quad \text{and} \quad S_2 = S \cap \boldsymbol{D}_b^r,$$

we have $TS = TS_1 \cup TS_2$, and hence the lemma will be proved if we can establish the relations

$$\mathsf{L}TS_1 = 0, \tag{4}$$

$$\mathsf{L}TS_2 = 0. \tag{5}$$

Now since $S_2 \subset \boldsymbol{D}_b^r$ and $\mathsf{L}S_2 = 0$, (5) is a direct consequence of lemma 2. As $S_1 \subset \boldsymbol{D}_b - \boldsymbol{D}_b^r$, by **IV.2.1**, lemma 3 the set TS_1 is a subset of the set where $W(x, \boldsymbol{D}) = +\infty$. This latter set is however of L-measure zero, since $W(x, \boldsymbol{D})$ is L-summable, and thus (4) follows.

Lemma 4. If T is ACb in $\boldsymbol{D}$ and L is any L-measurable set in $\boldsymbol{D}$, then $W(x, L)$ is L-measurable and L-summable in R^n.

Proof. By lemma 3, T satisfies the condition (N) on $\boldsymbol{D}_b$, and thus $W(x, L)$ is L-measurable by **IV.1.4**, lemma 2 [recall that the base function $b(u)$ is Borel measurable]. Since $W(x, \boldsymbol{D})$ is L-summable and clearly $0 \leq W(x, L) \leq W(x, \boldsymbol{D})$, it follows that $W(x, L)$ is L-summable also.

Lemma 5. Suppose that T is ACb in $\boldsymbol{D}$, and let S be a set in $\boldsymbol{D}$ such that $\mathsf{L}S = 0$. Then $W(x, S) = 0$ a.e. in R^n.

Proof. There exists (see **III.1.1**, lemma 7) a Borel set B such that $S \subset B \subset \boldsymbol{D}$ and $\mathsf{L}B = 0$. Since T is ACb in $\boldsymbol{D}$, we have $\nu(B) = 0$. As $W(x, B) \geq 0$, in view of **IV.2.2** (2) it follows that $W(x, B) = 0$ a.e. in R^n. Since $0 \leq W(x, S) \leq W(x, B)$, we conclude that $W(x, S) = 0$ a.e. in R^n.

IV.2.4. Transformation formulas for definite integrals. Let there be given a bounded continuous transformation $T: \boldsymbol{D} \to R^n$ and a base-function $b(u)$, as in **IV.2.1**.

Theorem 1. Assume that T is BVb in $\boldsymbol{D}$, and let ν be the measure induced by T and b (see **IV.2.2**, definition 2). If $h(u)$ is a (finite-valued) non-negative, Borel measurable function in $\boldsymbol{D}$ and B is a Borel set in $\boldsymbol{D}$, then

$$\int_B h(u)\, d\nu = \int W(x, B, h b)\, d\mathsf{L}, \tag{1}$$

as soon as one of the two integrals involved exists. More explicitly: if either of the two integrals exists, then so does the other, and (1) holds.

Proof. The function hb is Borel measurable in $\boldsymbol{D}$, and hence the summatory function $W(x, B, hb)$ is L-measurable in R^n by **IV.1.3.** Since h is Borel measurable, the measurability conditions necessary for the consideration of the integrals appearing in (1) are fulfilled. Take a positive number ε. For each non-negative integer j denote by B_j the set of those points $u \in B$ where

$$\varepsilon j \leq h(u) < \varepsilon (j + 1). \tag{2}$$

Evidently the sets B_j, $j = 0, 1, 2, \ldots$ are pair-wise disjoint Borel sets whose union is B. Integration of the inequalities (2) over B_j with respect to the measure ν yields

$$\varepsilon j\, \nu(B_j) \leq \int_{B_j} h(u)\, d\nu \leq \varepsilon (j + 1)\, \nu(B_j). \tag{3}$$

Since the base function $b(u)$ is non-negative, multiplication of the inequalities (2) by $b(u)$ yields

$$\varepsilon j\, b(u) \leq h(u)\, b(u) \leq \varepsilon (j + 1)\, b(u), \quad u \in B_j. \tag{4}$$

For the summatory functions $W(x, B_j, b)$, $W(x, B_j, h\,b)$ we obtain from (4) the inequalities (see **I.1.2**)

$$\varepsilon j\, W(x, B_j, b) \leq W(x, B_j, h\,b) \leq \varepsilon (j+1)\, W(x, B_j, b). \tag{5}$$

Since T is BVb in $\boldsymbol{D}$, $W(x, B_j, b)$ is L-summable (see **IV.2.2**). Thus (5) shows that $W(x, B_j, h\,b)$ is also L-summable. Integration of (5) yields

$$\varepsilon j\, \nu(B_j) \leq \int W(x, B_j, h\,b)\, d\mathsf{L} \leq \varepsilon (j+1)\, \nu(B_j). \tag{6}$$

From (3) and (6) we infer that

$$\left| \int_{B_j} h(u)\, d\nu - \int W(x, B_j, h\,b)\, d\mathsf{L} \right| \leq \varepsilon\, \nu(B_j).$$

Hence for every integer $m \geq 0$ we have the inequality

$$\left| \sum_{j=0}^{m} \int_{B_j} h(u)\, d\nu - \sum_{j=0}^{m} \int W(x, B_j, h\,b)\, d\mathsf{L} \right| \leq \varepsilon \sum_{j=0}^{m} \nu(B_j) \leq \varepsilon\, \nu(B). \tag{7}$$

From (7) it is evident that the two series of non-negative terms

$$\sum_{j=0}^{\infty} \int_{B_j} h(u)\, d\nu \quad \text{and} \quad \sum_{j=0}^{\infty} \int W(x, B_j, h\,b)\, d\mathsf{L} \tag{8}$$

are either both convergent or both divergent. Since B is the union of the pair-wise disjoint BOREL sets B_j, and $h(u)$ is non-negative, it follows (see **III.1.2**) that h is ν-summable on B if and only if the first series in (8) is convergent, and that

$$\int_B h(u)\, d\nu = \sum_{j=0}^{\infty} \int_{B_j} h(u)\, d\nu \tag{9}$$

if h is ν-summable on B. On the other hand, by **I.1.2** we have

$$W(x, B, h\,b) = \sum_{j=0}^{\infty} W(x, B_j, h\,b).$$

So it follows (see **III.1.2**) that $W(x, B, hb)$ is L-summable if and only if the second series in (8) is convergent, and that

$$\int W(x, B, h\,b)\, d\mathsf{L} = \sum_{j=0}^{\infty} \int W(x, B_j, h\,b)\, d\mathsf{L} \tag{10}$$

if $W(x, B, hb)$ is L-summable. We conclude from the preceding observations that the first integral in (1) exists if and only if the second one exists. Assume now the existence of these two integrals. From (7), (9), (10) we obtain the inequality

$$\left| \int_B h(u)\, d\nu - \int W(x, B, h\,b)\, d\mathsf{L} \right| \leq \varepsilon\, \nu(B).$$

Since $\varepsilon > 0$ was arbitrary, the relation (1) follows.

Lemma 1. Assume that T is BVb in $\boldsymbol{D}$, and let ν be the measure induced by T and b. If $H(x)$ is a (finite-valued) non-negative, BOREL measurable function in R^n and B is a BOREL set in $\boldsymbol{D}$, then

$$\int_B H(Tu)\, d\nu = \int H(x)\, W(x, B)\, d\mathsf{L}, \tag{11}$$

as soon as one of the two integrals involved exists.

Proof. Put

$$h(u) = H(Tu), \quad u \in \boldsymbol{D}. \tag{12}$$

By **IV.1.1**, theorem 2, it follows that $h(u)$ is a finite-valued, non-negative, BOREL measurable function in $\boldsymbol{D}$, and clearly

$$W(x, B, hb) = H(x)\, W(x, B). \tag{13}$$

In view of (12) and (13), the present lemma appears as a direct consequence of theorem 1.

Theorem 2. Assume that T is BVb in $\boldsymbol{D}$, and let ν be the measure induced by T and b. If $H(x)$ is a (finite-valued) real-valued, BOREL measurable function in R^n, and B is a BOREL set in $\boldsymbol{D}$, then

$$\int_B H(Tu)\, d\nu = \int H(x)\, W(x, B)\, d\mathsf{L}, \tag{14}$$

as soon as one of the two integrals involved exists.

Proof. Consider the auxiliary functions

$$H_1(x) = |H(x)|, \quad H_2(x) = |H(x)| - H(x). \tag{15}$$

Observe that $H_1(x), H_2(x)$ are finite-valued, non-negative, BOREL measurable functions. Hence, by lemma 1,

$$\int_B H_1(Tu)\, d\nu = \int H_1(x)\, W(x, B)\, d\mathsf{L}, \tag{16}$$

as soon as one of the two integrals involved exists, and similarly

$$\int_B H_2(Tu)\, d\nu = \int H_2(x)\, W(x, B)\, d\mathsf{L}, \tag{17}$$

as soon as one of the two integrals involved exists. Now suppose that the first integral in (14) exists. By (15) and **III.1.1**, lemma 26 it follows that the first integral in (16) also exists. Since clearly $0 \leq H_2 \leq 2|H| = 2H_1$, it follows further that the first integral in (17) exists. Hence (16) and (17) both hold, and on subtracting (17) from (16) we obtain (14), in view of (15). Suppose next that the second integral in (14) exists. The same reasoning as before [applied to the second integrals in (16) and (17)] yields again (14).

We proceed to consider the case when T is ACb in $\boldsymbol{D}$.

Lemma 2. Assume that T is ACb in $\boldsymbol{D}$, and let $D(u, v)$ denote the derivative of the measure v induced by T and b. If $H(x)$ is an L-measurable (finite-valued) function in R^n and B is any Borel set in $\boldsymbol{D}$, then

$$\int_B H(Tu)\, D(u, v)\, d\mathsf{L} = \int H(x)\, W(x, B)\, d\mathsf{L}, \tag{18}$$

as soon as one of the two integrals involved exists.

Proof. There exists (see **III.1.1**, lemma 23) a (finite-valued) Borel measurable function $H^*(x)$ in R^n such that $H^*(x) = H(x)$ a.e. in R^n. On denoting by e the exceptional set where $H^*(x) \neq H(x)$, we have

$$H(x) = H^*(x) \quad \text{if} \quad x \notin e, \quad \mathsf{L}\, e = 0. \tag{19}$$

We assert that

$$H(Tu)\, D(u, v) = H^*(Tu)\, D(u, v) \quad \text{a.e. in } \boldsymbol{D}. \tag{20}$$

Indeed, if $u \in T^{-1}\mathsf{C}e$, then $H(Tu) = H^*(Tu)$ by (19). Since $D(u, v)$ exists a.e. in $\boldsymbol{D}$ (see **IV.2.2**), it follows that

$$H(Tu)\, D(u, v) = H^*(Tu)\, D(u, v) \quad \text{a.e. on } T^{-1}\mathsf{C}\, e. \tag{21}$$

On the other hand, since $D(u, v) = 0$ a.e. on $T^{-1}e$ by **IV.2.2**, lemma 10, we have

$$H(Tu)\, D(u, v) = H^*(Tu)\, D(u, v) \quad \text{a.e. on } T^{-1}e. \tag{22}$$

As $\boldsymbol{D} = T^{-1}\mathsf{C}e \cup T^{-1}e$, clearly (20) follows from (21) and (22). Now since $H^*(Tu)$ is Borel measurable by **IV.1.1**, theorem 2, and $D(u, v)$ is L-measurable by **III.2.3**, theorem 3, it follows from (20) that $H(Tu)D(u, v)$ is L-measurable in $\boldsymbol{D}$ (see **III.1.1**, lemma 21). Also, $W(x, B)$ is L-measurable by **IV.1.3**. Thus the measurability conditions necessary for the consideration of the two integrals in (18) are fulfilled. We proceed to discuss the existence and equality of those two integrals. In view of (20) it follows that

$$\int_B H(Tu)\, D(u, v)\, d\mathsf{L} = \int_B H^*(Tu)\, D(u, v)\, d\mathsf{L}, \tag{23}$$

as soon as one of the two integrals involved exists. By **III.2.4**, theorem 4, we have

$$\int_B H^*(Tu)\, D(u, v)\, d\mathsf{L} = \int_B H^*(Tu)\, dv, \tag{24}$$

as soon as one of the two integrals involved exists. By theorem 2 we have

$$\int_B H^*(Tu)\, dv = \int H^*(x)\, W(x, B)\, d\mathsf{L}, \tag{25}$$

as soon as one of the two integrals involved exists. Finally, in view of (19) it follows that

$$\int H^*(x)\, W(x, B)\, d\mathsf{L} = \int H(x)\, W(x, B)\, d\mathsf{L}, \tag{26}$$

as soon as one of the two integrals involved exists. Clearly, it follows that if one of the two integrals in (18) exists, then all the integrals appearing in (23) to (26) exist and have the same value. Thus the lemma is proved.

Theorem 3. Assume that T is ACb in $\boldsymbol{D}$, and let $D(u, v)$ denote the derivative of the measure v induced by T and b. If $H(x)$ is an L-measurable (finite-valued) function in R^n and L is any L-measurable set in $\boldsymbol{D}$, then

$$\int_L H(T\,u)\, D(u, v)\, d\mathsf{L} = \int H(x)\, W(x, L)\, d\mathsf{L}, \tag{27}$$

as soon as one of the two integrals involved exists.

Proof. Note first that $W(x, L)$ is L-measurable in R^n by **IV.2.3**, lemma 4, and $H(T u)\, D(u, v)$ is L-measurable in $\boldsymbol{D}$, as it has been shown in the course of the proof of lemma 2. Note next that there exists a BOREL set B such that

$$L \subset B \subset \boldsymbol{D}, \quad \mathsf{L}(B - L) = 0. \tag{28}$$

Clearly, therefore,

$$\int_L H(T\,u)\, D(u, v)\, d\,\mathsf{L} = \int_B H(T\,u)\, D(u, v)\, d\,\mathsf{L}, \tag{29}$$

as soon as one of the two integrals involved exists. By lemma 2 we have

$$\int_B H(T\,u)\, D(u, v)\, d\mathsf{L} = \int H(x)\, W(x, B)\, d\mathsf{L}, \tag{30}$$

as soon as one of the two integrals involved exists. We assert that

$$W(x, B) = W(x, L) \quad \text{a.e. in } R^n. \tag{31}$$

Indeed, since the sets L and $B - L$ are disjoint, we have

$$W(x, B) = W(x, L) + W(x, B - L). \tag{32}$$

By **IV.2.3**, lemma 5, we have in view of (28)

$$W(x, B - L) = 0 \quad \text{a.e. in } R^n. \tag{33}$$

Clearly, (31) follows from (32) and (33). From (31) we infer that

$$\int H(x)\, W(x, B)\, d\mathsf{L} = \int H(x)\, W(x, L)\, d\mathsf{L}, \tag{34}$$

as soon as one of the two integrals involved exists. It is now clear that if one of the two integrals appearing in (27) exists, then all the integrals appearing in (29), (30), (34) exist and have the same value, and thus the theorem is proved.

§ IV.3. Bounded variation and absolute continuity with respect to a multiplicity function.

IV.3.1. Admissible multiplicity functions. Let $T: \boldsymbol{D} \to R^n$ be a bounded continuous transformation, where $\boldsymbol{D}$ is a bounded domain in R^n. The term *multiplicity function* has been already used in various contexts. In **§ II.3**, we considered the multiplicity functions $K(x, T, D)$, $K^+(x, T, D)$, $K^-(x, T, D)$, $k(x, T, D)$, and the crude multiplicity function $N(x, T, S)$ entered the picture on numerous occasions. Our objective in the present **§ IV.3** is the development of a general theory applicable to all those multiplicity functions that are relevant for our purposes. The transformation T is assumed given as above.

Definition 1. Let $M(x, T, D)$ be a function defined for every point $x \in R^n$ and every domain $D \subset \boldsymbol{D}$ (and depending also upon T), such that each value of $M(x, T, D)$ is either a non-negative integer or $+\infty$. Then $M(x, T, D)$ is termed a *non-negative multiplicity function* for the given transformation T.

The examples mentioned above are non-negative multiplicity functions in this sense. A useful theory of such multiplicity functions will naturally depend upon additional properties that are available in relevant special cases. Before stating such properties, let us introduce a function $b(u)$ associated with a given non-negative multiplicity function $M(x, T, D)$ in the following manner. Let u be a point in $\boldsymbol{D}$, and consider all those domains D which satisfy the conditions $u \in D \subset \boldsymbol{D}$. If D is such a domain, then $M(Tu, T, D)$ is defined and is either a non-negative integer or $+\infty$. We define $b(u)$ as the greatest lower bound of all these values $M(Tu, T, D)$. In symbols

$$b(u) = \text{gr.l.b.}\, M(Tu, T, D), \quad u \in D \subset \boldsymbol{D}. \tag{1}$$

Clearly, each value of $b(u)$ is either a non-negative integer or $+\infty$. Since $b(u)$ depends upon M, a more descriptive notation like $b(u, M)$ may be used if M is not clearly identified by the context.

Definition 2. A non-negative multiplicity function $M(x, T, D)$ is termed *admissible* (for the given transformation T) if the following conditions are satisfied.

(i) If $x \notin TD$, then $M(x, T, D) = 0$.

(ii) For fixed x, M is *sub-additive* with respect to D. Explicitly: if $D_1, \ldots, D_m, D$ are domains in $\boldsymbol{D}$ such that $D_k \subset D$, $k = 1, \ldots, m$, and $D_j \cap D_k = \emptyset$ if $j \neq k$, then

$$\sum_{k=1}^{m} M(x, T, D_k) \leqq M(x, T, D). \tag{2}$$

(iii) If D is a domain in $\boldsymbol{D}$ and $\{D_m\}$ is an infinite sequence of domains filling up D from the interior (see **II.3.2**, remark 10), then

$$M(x, T, D) = \lim_{m\to\infty} M(x, T, D_m).$$

(iv) For fixed D, $M(x, T, D)$ is L-measurable as a function of x in R^n, and the function $b(u)$, associated with M by means of (1), is Borel measurable in $\boldsymbol{D}$.

We shall verify later on that all the multiplicity functions relevant for our purposes are admissible in the sense of the preceding definition.

Definition 3. The function $b(u)$ defined by (1) is termed the *base-function generated by the admissible multiplicity function M.*

As noted above, each value of $b(u)$ is clearly a non-negative integer or $+\infty$. Also, $b(u)$ is Borel measurable by (iv). Thus the term *base-function* is used in the preceding definition in conformity with **IV.2.1**, definition 1. The base-function $b(u)$ determines a summatory function [see **IV.2.1** (2)], given by the formula

$$W(x, T, S) = \sum b(u), \quad u \in S \cap T^{-1} x, \tag{3}$$

where S is any set in $\boldsymbol{D}$. This summatory function depends upon M, and hence a more descriptive notation like $W(x, T, S, M)$ may be used if the multiplicity function M is not clearly identified by the context.

Definition 4. The function W, defined by (3), is termed the summatory function generated by the multiplicity function M.

Throughout this section, the transformation T and the admissible non-negative multiplicity function M are thought of as fixed. The symbols b, W are used in the sense of (1) and (3) respectively.

Lemma 1. If D is any domain in $\boldsymbol{D}$ and x is any point in R^n, then

$$0 \leqq W(x, T, D) \leqq M(x, T, D). \tag{4}$$

Proof. The assertion is obvious if $W(x, T, D) = 0$. So assume that $W(x, T, D) > 0$, and take any real number α such that

$$W(x, T, D) > \alpha \geqq 0. \tag{5}$$

In view of the definition of W, there follows that existence of a finite sequence $u_1, \ldots, u_m$ of distinct points in $D \cap T^{-1} x$ such that

$$\sum_{k=1}^{m} b(u_k) > \alpha. \tag{6}$$

Select pair-wise disjoint domains $D_1, \ldots, D_m$ such that $u_k \in D_k \subset D$, $k = 1, \ldots, m$. By (ii), (1), (6) it follows that

$$M(x, T, D) \geq \sum_{k=1}^{m} M(x, T, D_k) \geq \sum_{k=1}^{m} b(u_k) > \alpha. \tag{7}$$

Since α was any number satisfying (5), clearly (7) implies (4).

Lemma 2. Given a point $u \in \boldsymbol{D}$, there exists a domain D_u such that

$$u \in D_u \subset \boldsymbol{D}, \tag{8}$$

$$b(u) = W(T u, T, D_u) = M(T u, T, D_u). \tag{9}$$

Proof. Since each value of M is either a non-negative integer or $+\infty$, it is clear from the defining formula (1) that there exists a domain D_u such that $u \in D_u \subset \boldsymbol{D}$ and

$$b(u) = M(T u, T, D_u). \tag{10}$$

On the other hand, by the definition of W and by lemma 1 we have

$$b(u) \leq W(T u, T, D_u) \leq M(T u, T, D_u),$$

and (9) follows in view of (10).

Lemma 3. Let x be a point in R^n and D a domain in $\boldsymbol{D}$ such that

$$M(x, T, D) < \infty. \tag{11}$$

Assign a compact set F in D. Then there exists a domain D_0 such that

$$F \subset D_0, \quad \bar{D}_0 \subset D, \quad M(x, T, D_0) = M(x, T, D). \tag{12}$$

Proof. According to **I.2.3**, exercise 9, we can select a sequence of domains $\{D_m\}$ filling up D from the interior. By the definition of such a sequence we have then

$$\bar{D}_m \subset D, \quad m = 1, 2, \ldots, \tag{13}$$

$$F \subset D_m \quad \text{for } m \text{ large}, \tag{14}$$

and by (iii)

$$\lim_{m \to \infty} M(x, T, D_m) = M(x, T, D). \tag{15}$$

Since the finite values of M are integers, (15) and (11) imply that

$$M(x, T, D_m) = M(x, T, D) \quad \text{for } m \text{ large}. \tag{16}$$

On choosing m sufficiently large and setting $D_0 = D_m$, (12) will hold in view of (13), (14), (16).

Lemma 4. Let x be a point in R^n and D_* a domain in $\boldsymbol{D}$ such that

$$W(x, T, D_*) = M(x, T, D_*) < \infty. \tag{17}$$

Then for every domain $D \subset D_*$ we have

$$W(x, T, D) = M(x, T, D). \tag{18}$$

Proof. (18) is obvious, in view of lemma 1, if $M(x, T, D) = 0$. So assume that $M(x, T, D) > 0$. By (ii) we have then

$$M(x, T, D_*) \geq M(x, T, D) > 0. \tag{19}$$

From (19) and (17) we see that

$$0 < W(x, T, D_*) < \infty. \tag{20}$$

Let F_* be the set of those points $u \in D_* \cap T^{-1}x$ where $b(u) > 0$. Then [see (3)]

$$W(x, T, D_*) = \sum b(u), \quad u \in F_*. \tag{21}$$

If D' is any domain in D_*, then clearly [see (3)]

$$W(x, T, D') = \sum b(u), \quad u \in D' \cap F_*. \tag{22}$$

Since $b(u)$ is integral-valued, we conclude from (20) and (21) that F_* is a finite, non-empty subset of D_*. Returning to the domain D, we have by (22) (applied with $D' = D$)

$$W(x, T, D) = \sum b(u), \quad u \in D \cap F_*. \tag{23}$$

Since $D \cap F_*$ is a finite set, it is compact (perhaps empty). By lemma 3 we have therefore a domain D_0 such that

$$D \cap F_* \subset D_0, \quad \overline{D}_0 \subset D, \quad M(x, T, D_0) = M(x, T, D). \tag{24}$$

In view of (22) (applied with $D' = D_0$), the first two relations in (24) imply that

$$W(x, T, D_0) = \sum b(u), \quad u \in D_0 \cap F_* = D \cap F_*,$$

and hence by (23)

$$W(x, T, D_0) = W(x, T, D). \tag{25}$$

Consider now the open set $D_* - \overline{D}_0$. In view of (24), this set is non-empty, and it contains the set $F_* - D \cap F_*$. This latter set is finite (perhaps empty). Accordingly, we can select a finite number of components $D_1, \ldots, D_m$ of $D_* - \overline{D}_0$ such that

$$F_* - D \cap F_* \subset \bigcup_{k=1}^{m} D_k. \tag{26}$$

Note that each D_k is a domain, since $D_* - \overline{D}_0$ is open. The domains $D_0, D_1, \ldots, D_m$ are pair-wise disjoint sub-domains of D_*, and hence by (ii)

$$\sum_{k=0}^{m} M(x, T, D_k) \leqq M(x, T, D_*). \tag{27}$$

Furthermore, F_* is contained in the union of the domains $D_0, D_1, \ldots, D_m$ by (24) and (26). Hence [applying (22) with $D' = D_0, D_1, \ldots, D_m, D_*$]

$$\sum_{k=0}^{m} W(x, T, D_k) = W(x, T, D_*). \tag{28}$$

Let us put

$$\gamma_k = M(x, T, D_k) - W(x, T, D_k), \quad k = 0, 1, \ldots, m. \tag{29}$$

Subtraction of (28) from (27) yields, in view of (17),

$$\sum_{k=0}^{m} \gamma_k \leqq 0.$$

Since $\gamma_k \geqq 0$ by lemma 1, it follows that $\gamma_k = 0$, $k = 0, 1, \ldots, m$. In view of (29), (25), (24), the relation $\gamma_0 = 0$ is equivalent to (18), and the lemma is proved.

Lemma 5. Let x be a point in R^n such that

$$M(x, T, \boldsymbol{D}) < \infty. \tag{30}$$

Then there exists a positive number $\zeta = \zeta(x)$ such that the following holds: if D is any domain such that (see **I.1.4**, definition 4)

$$D \subset \boldsymbol{D}, \quad \delta D < \zeta(x), \tag{31}$$

then

$$W(x, T, D) = M(x, T, D). \tag{32}$$

Proof. According to lemma 3, we can select domains D', D'' such that

$$\overline{D}' \subset D'', \quad \overline{D}'' \subset \boldsymbol{D}, \tag{33}$$

$$M(x, T, D') = M(x, T, \boldsymbol{D}). \tag{34}$$

If $T^{-1}x = \emptyset$, then it follows from (i) and lemma 1 that $M(x, T, D) = W(x, T, D) = 0$ for every domain $D \subset \boldsymbol{D}$. Thus $\zeta(x)$ may be chosen as any positive number in this case. So suppose that $T^{-1}x \neq \emptyset$. For each point $u \in T^{-1}x$ there exists, by lemma 2, a domain D_u such that $u \in D_u \subset \boldsymbol{D}$ and

$$W(x, T, D_u) = M(x, T, D_u). \tag{35}$$

Note that since $\boldsymbol{D}$ is bounded, the set $\bar{D}'' \cap T^{-1}x$ is bounded and closed, and hence compact. Clearly the domains D_u, corresponding to the points $u \in T^{-1}x$, constitute an open covering of the compact set $\bar{D}'' \cap T^{-1}x$. Hence there exists (see **I.1.4**, exercise 6) a positive number η such that the following holds: if S is any set satisfying the conditions

$$S \cap \bar{D}'' \cap T^{-1}x \neq \emptyset, \quad \delta S < \eta,$$

then $S \subset D_u$ for some point $u \in \bar{D}'' \cap T^{-1}x$. Observe next that (33) yields $e(D', \complement D'') > 0$ (see **I.1.4**, definition 7). Choose now $\zeta(x)$ as the smaller one of the two positive numbers η and $e(D', \complement D'')$. We proceed to show that this choice of $\zeta(x)$ is adequate for the purpose of the lemma. Take any domain D satisfying (31). It is convenient to consider three cases.

Case 1. Assume that $D \cap D' = \emptyset$. Then D and D' are disjoint domains in $\boldsymbol{D}$, and hence by (ii)

$$0 \leq M(x, T, D) + M(x, T, D') \leq M(x, T, \boldsymbol{D}). \tag{36}$$

From (36) and (34) we conclude that $M(x, T, D) = 0$, and thus (32) follows by lemma 1.

Case 2. Assume that $D \cap T^{-1}x = \emptyset$. By (i) we have then $M(x, T, D) = 0$, and (32) follows by lemma 1.

Case 3. There remains to discuss the situation when

$$D \cap D' \neq \emptyset \quad \text{and} \quad D \cap T^{-1}x \neq \emptyset. \tag{37}$$

Since $\delta D < \zeta(x) \leq e(D', \complement D'')$, the first relation in (37) implies that

$$D \subset D'' \subset \bar{D}''. \tag{38}$$

From (38) and the second relation in (37) one concludes that

$$D \cap \bar{D}'' \cap T^{-1}x \neq \emptyset. \tag{39}$$

Since $\delta D < \zeta(x) \leq \eta$, we infer from (39) that there exists a point u such that

$$u \in \bar{D}'' \cap T^{-1}x, \quad D \subset D_u. \tag{40}$$

In view of (35), (30), and lemma 4, applied with $D_* = D_u$, (32) follows from the second relation in (40), and the proof is complete.

Our next lemma is concerned with the reduced base-set $\boldsymbol{D}_b^r$ (see **IV.2.1**, definition 2) corresponding to the base-function $b(u)$ given by (1). Let us recall that

$$\boldsymbol{D}_b^r = \{u \mid u \in \boldsymbol{D},\ 0 < b(u) < \infty\}. \tag{41}$$

Lemma 6. A point $u \in \boldsymbol{D}$ belongs to the reduced base-set $\boldsymbol{D}_b^r$ if and only if for every $\varepsilon > 0$ there exists a domain $D_{u\varepsilon}$ (which depends upon u and ε) such that the following holds.

(i) $u \in D_{u\varepsilon} \subset \boldsymbol{D}$.

(ii) $\delta D_{u\varepsilon} < \varepsilon$.

(iii) $1 \leq M(Tu, T, D_{u\varepsilon}) < \infty$.

(iv) If D is any domain such that $u \in D \subset D_{u\varepsilon}$, then $M(Tu, T, D) = M(Tu, T, D_{u\varepsilon})$.

Proof. Assume first that a domain $D_{u\varepsilon}$ with the properties (i) to (iv) exists for every $\varepsilon > 0$. From (1), (i) and (iii) we infer that

$$b(u) < \infty. \tag{42}$$

Let us take now any domain D such that $u \in D \subset \boldsymbol{D}$. Since D is open, from (i) and (ii) it follows that we shall have the inclusions $u \in D_{u\varepsilon} \subset D$ if ε is sufficiently small, and hence (if ε is sufficiently small)

$$M(Tu, T, D_{u\varepsilon}) \leq M(Tu, T, D).$$

In view of (iii) it follows that

$$M(Tu, T, D) \geq 1 \quad \text{if} \quad u \in D \subset \boldsymbol{D},$$

and thus [see (1)]

$$b(u) \geq 1. \tag{43}$$

From (42), (43), (41) we see that $u \in \boldsymbol{D}_b^r$. Suppose, conversely, that $u \in \boldsymbol{D}_b^r$. Then

$$1 \leq b(u) < \infty. \tag{44}$$

By lemma 2, there exists a domain D_u such that

$$u \in D_u \subset \boldsymbol{D}, \quad b(u) = M(Tu, T, D_u). \tag{45}$$

Now assign $\varepsilon > 0$ arbitrarily. Since D_u is open and $u \in D_u$, we can select a domain $D_{u\varepsilon}$ such that

$$u \in D_{u\varepsilon} \subset D_u, \quad \delta D_{u\varepsilon} < \varepsilon. \tag{46}$$

We assert that $D_{u\varepsilon}$ satisfies the conditions (i), (ii), (iii), (iv). For (i) and (ii) this is clear in view of (46). Also, since $u \in D_{u\varepsilon} \subset D_u$, we have

$$b(u) \leq M(Tu, T, D_{u\varepsilon}) \leq M(Tu, T, D_u),$$

and hence (iii) is also satisfied in view of (44) and (45). To verify (iv), consider any domain D such that $u \in D \subset D_{u\varepsilon}$. Then [see (45) and (46)]

$$b(u) \leq M(T u, T, D) \leq M(T u, T, D_{u\varepsilon}) \leq M(T u, T, D_u) = b(u),$$

and hence the sign of equality must hold throughout. In particular

$$M(T u, T, D) = M(T u, T, D_{u\varepsilon}).$$

Thus (iv) is verified, and the lemma is proved.

IV.3.2. Bounded variation. Throughout this section, T denotes a transformation given as in **IV.3.1**, and we assume that there is assigned an admissible (non-negative) multiplicity function M in the sense of **IV.3.1**, definition 2. The symbols b and W will be used in the sense of the definitions 3 and 4 in **IV.3.1**. The transformation T and the admissible (non-negative) multiplicity function M are thought of as fixed, and hence the corresponding base-function b and the summatory function W are also considered as fixed.

Definition 1. T is said to be of bounded variation in $\boldsymbol{D}$ with respect to the admissible (non-negative) multiplicity function $M(x, T, D)$ (briefly, BVM in $\boldsymbol{D}$) if $M(x, T, \boldsymbol{D})$ is L-summable in R^n.

Lemma 1. If T is BVM in $\boldsymbol{D}$, then it is also BVM in every domain $D \subset \boldsymbol{D}$.

Proof. If D is a domain in $\boldsymbol{D}$, then

$$0 \leq M(x, T, D) \leq M(x, T, \boldsymbol{D})$$

by the condition (ii) in **IV.3.1**, definition 2. Thus the L-summability of $M(x, T, \boldsymbol{D})$ implies the L-summability of $M(x, T, D)$, and hence T is BVM in D.

Assume that T is BVM in $\boldsymbol{D}$. If D is any domain in $\boldsymbol{D}$, then $M(x, T, D)$ is L-summable, and hence we can define a function $\mathfrak{M}(D)$ of domains $D \subset \boldsymbol{D}$ by the formula

$$\mathfrak{M}(D) = \int M(x, T, D)\, d\mathsf{L}, \quad D \subset \boldsymbol{D}. \tag{1}$$

Clearly

$$0 \leq \mathfrak{M}(D) < \infty, \quad D \subset \boldsymbol{D}. \tag{2}$$

Since $\mathfrak{M}(D)$ depends upon T and M, the concise notation $\mathfrak{M}(D)$ is adequate only if T and M are thought of as fixed, which is the case at present.

Definition 2. The function of domains $\mathfrak{M}(D)$, given by (1), is said to be generated by the admissible multiplicity function M.

Lemma 2. Assume that T is BVM in $\boldsymbol{D}$, and let $\mathfrak{M}(D)$ be the function of domains generated by M. Then the following holds.

(i) $\mathfrak{M}(D)$ is sub-additive on the family of domains in $\boldsymbol{D}$ (see **III.2.4**, definition 2).

(ii) The derivative $D(u, \mathfrak{M})$ of $\mathfrak{M}$ [see **III.2.4** (2)] exists a.e. in $\boldsymbol{D}$ and is L-summable in $\boldsymbol{D}$.

(iii) If D is any domain in $\boldsymbol{D}$, then

$$\int_D D(u, \mathfrak{M})\, d\,\mathsf{L} \leq \mathfrak{M}(D). \tag{3}$$

Proof. Take domains $D_1, \ldots, D_m, D$ in $\boldsymbol{D}$, such that $D_k \subset D$, $k=1, \ldots, m$, and $D_j \cap D_k = \emptyset$ for $j \neq k$. By condition (ii) in **IV.3.1**, definition 2, we have then

$$\sum_{k=1}^{m} M(x, T, D_k) \leq M(x, T, D).$$

Integration yields the inequality

$$\sum_{k=1}^{m} \mathfrak{M}(D_k) \leq \mathfrak{M}(D),$$

and (i) is proved. In view of (2) and (i), $\mathfrak{M}$ is a non-negative, finite-valued, sub-additive function of domains $D \subset \boldsymbol{D}$. Thus (ii) and (iii) follow directly from **III.2.4**, theorem 1.

Lemma 3. Assume that T is BVM in $\boldsymbol{D}$, and let b denote the base-function generated by M (see **IV.3.1**, definition 3). Then T is BVb in $\boldsymbol{D}$ (in the sense of **IV.2.2**, definition 1).

Proof. By **IV.3.1**, lemma 1, we have

$$0 \leq W(x, T, \boldsymbol{D}) \leq M(x, T, \boldsymbol{D}).$$

Since $M(x, T, \boldsymbol{D})$ is now L-summable (see definition 1), it follows that $W(x, T, \boldsymbol{D})$ is also L-summable, and thus T is BVb in $\boldsymbol{D}$.

Once it is known that T is BVb in $\boldsymbol{D}$, we have at our disposal the finite measure [see **IV.2.2** (2)]

$$\nu(B) = \int W(x, T, B)\, d\,\mathsf{L}, \tag{4}$$

defined for all Borel sets $B \subset \boldsymbol{D}$, as well as the derivative $D(u, \nu)$ of ν (see **IV.2.2**, definition 2).

Summing up: if T is BVM in $\boldsymbol{D}$, then M generates a non-negative, finite-valued, sub-additive function $\mathfrak{M}(D)$ of domains $D \subset \boldsymbol{D}$ which possesses a derivative $D(u, \mathfrak{M})$ a.e. in $\boldsymbol{D}$. On the other hand, M generates a base-function $b(u)$ and a corresponding summatory function W, in the sense of **IV.3.1**, definitions 3 and 4. In turn, W yields the finite measure ν, given by (4), which possesses a derivative $D(u, \nu)$ a.e. in $\boldsymbol{D}$. The relationships amongst these various functions will now be discussed.

Lemma 4. Assume that T is BVM in $\boldsymbol{D}$. Then the following statements hold.

(i) If D is any domain in $\boldsymbol{D}$, then

$$\nu(D) \leq \mathfrak{M}(D). \tag{5}$$

(ii) The sign of equality holds in (5) if and only if

$$W(x, T, D) = M(x, T, D) \quad \text{a.e. in } R^n. \tag{6}$$

(iii) If $\nu(\boldsymbol{D}) = \mathfrak{M}(\boldsymbol{D})$, then $\nu(D) = \mathfrak{M}(D)$ for every domain $D \subset \boldsymbol{D}$.

Proof. If D is a domain in $\boldsymbol{D}$, then

$$0 \leq W(x, T, D) \leq M(x, T, D) \tag{7}$$

by **IV.3.1**, lemma 1. In view of (1) and (4), integration of (7) yields (5). By **III.1.1**, lemma 35, (7) implies that the sign of equality holds in (5) if and only if (6) holds. Thus (i) and (ii) are proved. Suppose now that $\nu(\boldsymbol{D}) = \mathfrak{M}(\boldsymbol{D})$. By (ii) it follows that

$$W(x, T, \boldsymbol{D}) = M(x, T, \boldsymbol{D}) \quad \text{a.e. in } R^n. \tag{8}$$

Since $M(x, T, \boldsymbol{D})$ is now L-summable, we have (see **III.1.1**, lemma 31)

$$M(x, T, \boldsymbol{D}) < \infty \quad \text{a.e. in } R^n. \tag{9}$$

Take now any domain $D \subset \boldsymbol{D}$. By **IV.3.1**, lemma 4 we conclude from (8) and (9) that

$$W(x, T, D) = M(x, T, D) \quad \text{a.e. in } R^n,$$

and the relation $\nu(D) = \mathfrak{M}(D)$ follows by (ii).

Lemma 5. Assume that T is BVM in $\boldsymbol{D}$. Then

$$D(u, \mathfrak{M}) = D(u, \nu) \quad \text{a.e. in } \boldsymbol{D}. \tag{10}$$

Proof. As noted above, $D(u, \mathfrak{M})$ and $D(u, \nu)$ exist a.e. in $\boldsymbol{D}$. To prove (10) we consider an auxiliary function $\varphi(\operatorname{int} I)$ of open intervals in $\boldsymbol{D}$, defined by the formula

$$\varphi(\operatorname{int} I) = \mathfrak{M}(\operatorname{int} I) - \nu(\operatorname{int} I), \quad \operatorname{int} I \subset \boldsymbol{D}. \tag{11}$$

Note that φ is non-negative by (5). Since $D(u, \mathfrak{M})$ and $D(u, \nu)$ exist a.e. in $\boldsymbol{D}$, it follows from (11) that $D(u, \varphi)$ also exists a.e. in $\boldsymbol{D}$ and

$$D(u, \varphi) = D(u, \mathfrak{M}) - D(u, \nu) \quad \text{a.e. in } \boldsymbol{D}.$$

Thus (10) will be established if we prove that

$$D(u, \varphi) = 0 \quad \text{a.e. in } \boldsymbol{D}. \tag{12}$$

The proof of (12) will be based on the lemma in **III.2.3**, which will be applied with $O=\boldsymbol{D}$. Accordingly, Q_j will now be a generic notation for a cube whose interior lies in $\boldsymbol{D}$ and which belongs to the subdivision D_{p_j}, and $\boldsymbol{Q}_j$ denotes the class of all such cubes Q_j (for a fixed positive integer j). According to the lemma in **III.2.3**, (12) will be established if we show that

$$\lim_{j\to\infty} \sum_{Q_j} \varphi(\operatorname{int} Q_j) = 0. \tag{13}$$

To prove (13), let us introduce for each positive integer j the function

$$g_j(x) = \sum_{Q_j}{}' [M(x, T, \operatorname{int} Q_j) - W(x, T, \operatorname{int} Q_j)], \quad x \in R^n, \tag{14}$$

where the symbol $\sum'$ is used to indicate that the summation is extended only over those terms which are not of the indeterminate form $\infty - \infty$. By lemma 1 in **IV.3.1** and by condition (ii) in **IV.3.1**, definition 2 we have then

$$0 \leq g_j(x) \leq M(x, T, \boldsymbol{D}). \tag{15}$$

Since T is BVM in $\boldsymbol{D}$, $M(x, T, \boldsymbol{D})$ is L-summable. Hence, on denoting by X_∞ the set of those points $x \in R^n$ where $M(x, T, \boldsymbol{D}) = \infty$, we have

$$\mathsf{L} X_\infty = 0. \tag{16}$$

Consider now a point x such that

$$x \in \mathsf{C} X_\infty. \tag{17}$$

Then $M(x, T, \boldsymbol{D}) < \infty$, and hence by **IV.3.1**, lemma 5 there exists a positive number $\zeta(x)$ such that if D is any domain in $\boldsymbol{D}$ satisfying the condition $\delta D < \zeta(x)$, then

$$M(x, T, D) = W(x, T, D).$$

On the other hand, clearly $\delta Q_j < \zeta(x)$ for all the cubes Q_j if j is sufficiently large, say $j > j_0(x)$. Hence

$$M(x, T, \operatorname{int} Q_j) - W(x, T, \operatorname{int} Q_j) = 0 \quad \text{if} \quad j > j_0(x). \tag{18}$$

From (14), (17) and (18) we conclude that

$$g_j(x) \to 0 \quad \text{a.e. in } R^n. \tag{19}$$

Since $M(x, T, \boldsymbol{D})$ is L-summable, (15) and (19) imply (see **III.1.1**, lemma 38) that

$$\lim_{j\to\infty} \int g_j(x)\, d\,\mathsf{L} = 0. \tag{20}$$

In view of (14), (1), (4) and (11), the relation (20) is equivalent to (13), and the proof is complete.

Lemma 6. Assume that T is BVM in $\boldsymbol{D}$. Then

$$D(u, \mathfrak{M}) = 0 \quad \text{a.e. on} \quad \boldsymbol{D} - \boldsymbol{D}_b^r, \tag{21}$$

where $\boldsymbol{D}_b^r$ is the reduced base-set (see **IV.2.1**, definition 2) corresponding to the base-function $b(u)$ generated by the multiplicity function M.

Proof. We have $D(u, \nu) = 0$ a.e. on $\boldsymbol{D} - \boldsymbol{D}_b^r$ by **IV.2.2**, lemma 9, and $D(u, \nu) = D(u, \mathfrak{M})$ a.e. in $\boldsymbol{D}$ by the preceding lemma 5, and thus (21) follows.

Lemma 7. Assume that T is BVM in $\boldsymbol{D}$, and let D be any domain in $\boldsymbol{D}$. Then

$$\int_D D(u, \mathfrak{M})\, d\mathsf{L} = \int_D D(u, \nu)\, d\mathsf{L} \leq \nu(D) \leq \mathfrak{M}(D). \tag{22}$$

Proof. Note that T is BVb in $\boldsymbol{D}$ by lemma 3, and hence the measure ν is available. The relation (22) follows then by direct application of lemma 5, part (i) of lemma 4, and **III.2.4**, lemma 1.

Lemma 8. Assume that T is BVM in $\boldsymbol{D}$, and let G be any set in $\boldsymbol{D}$ such that $\mathsf{L}TG = 0$. Then

$$D(u, \mathfrak{M}) = 0 \quad \text{a.e. on } G. \tag{23}$$

Proof. Note that T is BVb in $\boldsymbol{D}$ by lemma 3 and hence the measure ν [given by (4)] is available. We have $D(u, \nu) = 0$ a.e. on G by **IV.2.2**, lemma 12, and $D(u, \mathfrak{M}) = D(u, \nu)$ a.e. in $\boldsymbol{D}$ by lemma 5, and thus (23) follows.

IV.3.3. Absolute continuity. In this section, we assume that a transformation T and an admissible (non-negative) multiplicity function M are given as in **IV.3.2**. The symbols $b, W, \mathfrak{M}, \nu, D(u, \mathfrak{M}), D(u, \nu)$ are used in the same sense as in **IV.3.2**.

Definition. T is said to be *absolutely continuous* in $\boldsymbol{D}$ with respect to the admissible (non-negative) multiplicity function M (briefly, ACM in $\boldsymbol{D}$) if T is BVM in $\boldsymbol{D}$ and

$$\int_{\boldsymbol{D}} D(u, \mathfrak{M})\, d\mathsf{L} = \mathfrak{M}(\boldsymbol{D}). \tag{1}$$

To justify and motivate this definition, let us recall that if T is BVM in $\boldsymbol{D}$, then

$$\int_{\boldsymbol{D}} D(u, \mathfrak{M})\, d\mathsf{L} \leq \mathfrak{M}(\boldsymbol{D}) \tag{2}$$

by **IV.3.2**, lemma 2. Thus the property ACM in $\boldsymbol{D}$ requires that the sign of equality should hold in (2), or equivalently, that $\mathfrak{M}(\boldsymbol{D})$ should be expressible as the L-integral over $\boldsymbol{D}$ of its derivative $D(u, \mathfrak{M})$.

Lemma 1. Suppose that T is ACM in $\boldsymbol{D}$. Then the following holds.

(i) T is ACb in $\boldsymbol{D}$.

(ii) $\nu(\boldsymbol{D}) = \mathfrak{M}(\boldsymbol{D})$.

Proof. Since T is ACM in $\boldsymbol{D}$, it is also BVM in $\boldsymbol{D}$ by definition. By **IV.3.2**, lemma 3, T is therefore BVb in $\boldsymbol{D}$. To prove (i), we have to show yet that the measure $\nu(B)$ is absolutely continuous on Borel sets in $\boldsymbol{D}$ (see **IV.2.3**). Now since T is BVM in $\boldsymbol{D}$, by lemma 7 in **IV.3.2** we have

$$\int_{\boldsymbol{D}} D(u, \mathfrak{M})\, d\mathsf{L} = \int_{\boldsymbol{D}} D(u, \nu)\, d\mathsf{L} \leq \nu(\boldsymbol{D}) \leq \mathfrak{M}(\boldsymbol{D}). \tag{3}$$

In view of (1), the sign of equality must hold throughout in (3). This yields (ii), as well as the relation

$$\int_{\boldsymbol{D}} D(u, \nu)\, d\mathsf{L} = \nu(\boldsymbol{D}).$$

By **III.2.4**, theorem 3 it follows that ν is absolutely continuous on Borel sets in $\boldsymbol{D}$, and the proof is complete.

Lemma 2. Assume that T is BVM in $\boldsymbol{D}$. Then T is ACM in $\boldsymbol{D}$ if and only if the following two conditions hold simultaneously.

(i) T is ACb in $\boldsymbol{D}$.

(ii) $\nu(\boldsymbol{D}) = \mathfrak{M}(\boldsymbol{D})$.

Proof. If T is ACM in $\boldsymbol{D}$, then (i) and (ii) hold by lemma 1. Suppose conversely that T is BVM in $\boldsymbol{D}$ and (i) and (ii) hold. By **IV.2.3**, (i) implies that ν is absolutely continuous on Borel sets in $\boldsymbol{D}$, and hence

$$\int_{\boldsymbol{D}} D(u, \nu)\, d\mathsf{L} = \nu(\boldsymbol{D}), \tag{4}$$

by **III.2.4**, theorem 3. Also, since T is BVM in $\boldsymbol{D}$, we have $D(u, \nu) = D(u, \mathfrak{M})$ a.e. in $\boldsymbol{D}$ by **IV.3.2**, lemma 5. Hence

$$\int_{\boldsymbol{D}} D(u, \nu)\, d\mathsf{L} = \int_{\boldsymbol{D}} D(u, \mathfrak{M})\, d\mathsf{L}. \tag{5}$$

Now (ii), (4) and (5) imply (1), and thus T is ACM in $\boldsymbol{D}$.

Lemma 3. Suppose that T is ACM in $\boldsymbol{D}$. Let D be any domain in $\boldsymbol{D}$. Then T is ACM in D.

Proof. Note that T is BVM in $\boldsymbol{D}$, and hence also BVM in D, by **IV.3.2**, lemma 1. Also, T is ACb in $\boldsymbol{D}$ by lemma 1, and hence T is ACb in D by **IV.2.3**, lemma 1. Since T is ACM in $\boldsymbol{D}$, we have $\nu(\boldsymbol{D}) = \mathfrak{M}(\boldsymbol{D})$ by lemma 1, and by part (iii) of lemma 4 in **IV.3.2** we conclude that $\nu(D) = \mathfrak{M}(D)$. On applying lemma 2 (with $\boldsymbol{D}$ replaced by D) it follows that T is ACM in D.

Lemma 4. Assume that T is BVM in $\boldsymbol{D}$. Then T is ACM in $\boldsymbol{D}$ if and only if

$$\int_D D(u, \mathfrak{M})\, d\mathsf{L} = \mathfrak{M}(D) \tag{6}$$

for every domain $D \subset \boldsymbol{D}$.

Proof. Suppose first that (6) holds for every domain $D \subset \boldsymbol{D}$. Then (6) holds, in particular, for $D = \boldsymbol{D}$, and thus T is ACM in $\boldsymbol{D}$, by definition. Suppose, conversely, that T is ACM in $\boldsymbol{D}$. Then T is ACM in every domain $D \subset \boldsymbol{D}$ by lemma 3, and hence (6) holds by the definition of the ACM property.

Theorem. Assume that T is ACM in $\boldsymbol{D}$. Let $H(x)$ be a finite-valued, real-valued, L-measurable function of $x \in R^n$ and let D be any domain in $\boldsymbol{D}$. Then

$$\int_D H(T u)\, D(u, \mathfrak{M})\, d\mathsf{L} = \int H(x)\, M(x, T, D)\, d\mathsf{L}, \tag{7}$$

as soon as one of the two integrals involved exists.

Proof. By lemma 1, T is ACb in $\boldsymbol{D}$. Hence, by **IV.2.4**, theorem 3,

$$\int_D H(T u)\, D(u, v)\, d\mathsf{L} = \int H(x)\, W(x, T, D)\, d\mathsf{L}, \tag{8}$$

as soon as one of the two integrals involved exists. By lemma 2, $v(\boldsymbol{D}) = \mathfrak{M}(\boldsymbol{D})$, and hence $v(D) = \mathfrak{M}(D)$ by part (iii) of lemma 4 in **IV.3.2**. From part (ii) of that same lemma it follows that

$$W(x, T, D) = M(x, T, D) \quad \text{a.e. in } R^n,$$

and hence also

$$H(x)\, W(x, T, D) = H(x)\, M(x, T, D) \quad \text{a.e. in } R^n.$$

In view of **III.1.1**, lemma 25 we have therefore

$$\int H(x)\, W(x, T, D)\, d\mathsf{L} = \int H(x)\, M(x, T, D)\, d\mathsf{L}, \tag{9}$$

as soon as one the two integrals involved exists. By **IV.3.2**, lemma 5, $D(u, v) = D(u, \mathfrak{M})$ a.e. in $\boldsymbol{D}$ and hence a.e. in D, and therefore also

$$H(T u)\, D(u, v) = H(T u)\, D(u, \mathfrak{M}) \quad \text{a.e. in } D.$$

Consequently

$$\int_D H(T u)\, D(u, v)\, d\mathsf{L} = \int_D H(T u)\, D(u, \mathfrak{M})\, d\mathsf{L}, \tag{10}$$

as soon as one of the two integrals involved exists. Combining these facts one sees that if one of the two integrals in (7) exists, then all the integrals appearing in (8), (9), (10) exist and have the same value, and (7) follows.

IV.3.4. Some relevant examples of admissible multiplicity functions. Let there be given a bounded continuous transformation $T: \boldsymbol{D} \to R^n$ as in **IV.3.1**, and a non-negative multiplicity function $M(x, T, D)$ (see **IV.3.1**, definition 1).

Theorem 1. Assume that $M(x, T, D)$ satisfies the conditions (i), (ii), (iii) stated in **IV.3.1**, definition 2, and the additional condition:

(iv)* For each domain $D \subset \boldsymbol{D}$, $M(x, T, D)$ is a lower semi-continuous function of $x \in R^n$ (see **I.1.3**, definition 28).

Then $M(x, T, D)$ is admissible, in the sense of **IV.3.1**, definition 2.

Proof. For fixed $D \subset \boldsymbol{D}$, $M(x, T, D)$ is lower semi-continuous with respect to $x \in R^n$ by assumption, and hence it is L-measurable in R^n (see **III.1.1**, lemma 20). Inspection of definition 2 in **IV.3.1** reveals that there remains only to show that the base-function $b(u)$ generated by M is Borel measurable in $\boldsymbol{D}$. To verify this point, take a real number α and a positive integer j, and denote by $S_{\alpha j}$ the set of those points $u \in \boldsymbol{D}$ which satisfy the following condition: There exists a domain D such that

$$u \in D \subset \boldsymbol{D}, \quad \delta D < 1/j, \quad M(Tu, T, D) > \alpha. \tag{1}$$

We show presently that $S_{\alpha j}$ is an open set. This is obvious if $S_{\alpha j}$ is empty. So assume that $S_{\alpha j}$ is non-empty. Take a point $u \in S_{\alpha j}$ and a domain D for which (1) holds. From the lower semi-continuity of $M(x, T, D)$ it follows [in view of (1)] that there is an open set $O^* \subset R^n$ such that

$$Tu \in O^*, \tag{2}$$

$$x \in O^* \quad \text{implies} \quad M(x, T, D) > \alpha. \tag{3}$$

Since T is continuous, it follows from (1) and (2) that there is an open set O such that

$$u \in O \subset D, \tag{4}$$

$$u' \in O \quad \text{implies} \quad Tu' \in O^*. \tag{5}$$

Consider any point $u' \in O$. From (5) and (3) we infer that

$$M(Tu', T, D) > \alpha. \tag{6}$$

Combining (1), (4) and (6), we see that

$$u' \in D \subset \boldsymbol{D}, \quad \delta D < 1/j, \quad M(Tu', T, D) > \alpha.$$

Thus $u' \in S_{\alpha j}$. Since u' was an arbitrary point of O, it follows that $O \subset S_{\alpha j}$. Summing up: if $u \in S_{\alpha j}$, then there exists an open set O such that $u \in O \subset S_{\alpha j}$. Thus $S_{\alpha j}$ is open. Now let S_α be the set of those points $u \in \boldsymbol{D}$ where the base-function $b(u)$ generated by M satisfies the inequality $b(u) > \alpha$. We assert that

$$S_\alpha = \bigcap_{j=1}^{\infty} S_{\alpha j}. \tag{7}$$

Indeed, if $u \in S_\alpha$, then [see **IV.3.1** (1)]

$$\alpha < b(u) \leq M(T u, T, D) \tag{8}$$

for every domain D such that $u \in D \subset \boldsymbol{D}$. Given any positive integer j, we can select a domain D such that

$$u \in D \subset \boldsymbol{D}, \quad \delta D < 1/j.$$

By (8) it follows then that $u \in S_{\alpha j}$. Since u was an arbitrary point of S_α, we conclude that

$$S_\alpha \subset S_{\alpha j}$$

for every positive integer j, and hence

$$S_\alpha \subset \bigcap_{j=1}^{\infty} S_{\alpha j}. \tag{9}$$

Consider next a point

$$u \in \bigcap_{j=1}^{\infty} S_{\alpha j}. \tag{10}$$

Then for every positive integer j there is a domain D_j such that

$$u \in D_j \subset \boldsymbol{D}, \quad \delta D_j < 1/j, \quad M(T u, T, D_j) > \alpha. \tag{11}$$

Consider any domain D such that $u \in D \subset \boldsymbol{D}$. From (11) it follows that

$$u \in D_j \subset D \tag{12}$$

for j sufficiently large. If j is so chosen that (12) holds, then in view of (12) and condition (ii) in **IV.3.1**, definition 2 we have

$$M(T u, T, D) \geq M(T u, T, D_j) > \alpha.$$

Thus

$$u \in D \subset \boldsymbol{D} \quad \text{implies} \quad M(T u, T, D) > \alpha. \tag{13}$$

Since the finite values of M are integers, (13) implies [in view of **IV.3.1**(1)] that $b(u) > \alpha$. Hence $u \in S_\alpha$. Since u was an arbitrary point satisfying (10), it follows that

$$\bigcap_{j=1}^{\infty} S_{\alpha j} \subset S_\alpha. \tag{14}$$

Now (7) follows from (9) and (14). Since $S_{\alpha j}$ is an open set, (7) shows that S_α is a Borel set, and hence $b(u)$ is Borel measurable.

Remark 1. The preceding theorem will be applied below to the multiplicity functions $K(x, T, D)$, $K^+(x, T, D)$, $K^-(x, T, D)$, $k(x, T, D)$ discussed in § **II.3**. In general, the statements in § **II.3** are formulated for the case when $D = \boldsymbol{D}$. However, since $\boldsymbol{D}$ was any bounded domain in which T is continuous and bounded, the concepts and results discussed in § **II.3** apply for any domain $D \subset \boldsymbol{D}$. As regards situations involving *local properties*, section **II.3.5** contains the facts that should be kept in mind when one operates with subdomains of $\boldsymbol{D}$.

Theorem 2. The multiplicity functions $K(x, T, D)$, $K^+(x, T, D)$, $K^-(x, T, D)$ are admissible (see **IV.3.1**, definition 2).

Proof. From the definitions given in § **II.3** it is evident that each value of K, K^+, K^- is either a non-negative integer or $+\infty$. The properties (i), (ii), (iii), (iv)* required in theorem 1 have been established for K, K^+, K^- in remark 2, theorem 2, remark 10, and theorem 3 of section **II.3.2**, and thus the admissibility of K, K^+, K^- follows.

Remark 2. While theorem 2 covers the cases actually needed in the sequel, it is of interest to observe that there are various other multiplicity functions whose admissibility follows from theorem 1. For example, the multiplicity function $k(x, T, D)$ (see **II.3.3**) is readily seen to satisfy the conditions (i), (ii), (iii) required in theorem 1 as an immediate consequence of its definition, and it also satisfies condition (iv)* by **II.3.3**, lemma 7. Thus $k(x, T, D)$ is admissible. A further example is obtained as follows. Take a point $x \in R^n$ and a domain $D \subset \boldsymbol{D}$. Denote by $k^+(x, T, D)$ the number (possibly infinite) of those essential maximal model continua C for (x, T, D) which satisfy the following condition: if O is any open set such that $C \subset O \subset D$, then there exists a positive indicator domain D_* for (x, T, D) (see **II.3.2**) such that $C \subset D_* \subset O$. One sees readily that $k^+(x, T, D)$ satisfies the conditions required in theorem 1, and hence it is admissible. The same holds for the multiplicity function $k^-(x, T, D)$ obtained by replacing the term *positive indicator domain* by the term *negative indicator domain* in the definition of $k^+(x, T, D)$. Inspection of the literature reveals further examples of multiplicity functions that are admissible by virtue of theorem 1.

Theorem 3. The crude multiplicity function $N(x, T, D)$ is admissible.

Proof. Note that $N(x, T, D)$ is not lower semi-continuous with respect to x, and hence theorem 1 does not apply. However, $N(x, T, D)$ obviously satisfies conditions (i), (ii), (iii) in **IV.3.1**, definition 2, and it is L-measurable as a function of x by **IV.1.2**. There remains to show that the base-function $b(u)$ generated by $N(x, T, D)$ is Borel measurable. Consider the set $\mathbf{I}(T, \boldsymbol{D})$ in **II.3.7**, definition 2. If $u \in \boldsymbol{D} - \mathbf{I}(T, \boldsymbol{D})$, then every neighborhood of u contains some point

$u' \neq u$ such that $Tu' = Tu$. It obviously follows that if D is any domain such that $u \in D \subset \boldsymbol{D}$, then $N(Tu, T, D) = \infty$. Hence

$$b(u) = \infty \quad \text{if} \quad u \in \boldsymbol{D} - \mathbf{I}(T, \boldsymbol{D}). \tag{15}$$

On the other hand, if $u \in \mathbf{I}(T, \boldsymbol{D})$, then there exists an open set O such that $u \in O \subset \boldsymbol{D}$ and $Tu' \neq Tu$ for $u' \in O$, $u' \neq u$. Thus $N(Tu, T, O) = 1$. Take any domain D such that $u \in D \subset O$. It follows that

$$N(Tu, T, D) \leqq N(Tu, T, O) = 1,$$

and thus $b(u) \leqq 1$. On the other hand, if D is any domain such that $u \in D \subset \boldsymbol{D}$, then $N(Tu, T, D) \geqq 1$ and hence $b(u) \geqq 1$. Thus we see that

$$b(u) = 1 \quad \text{if} \quad u \in \mathbf{I}(T, \boldsymbol{D}). \tag{16}$$

Since $\mathbf{I}(T, D)$ is a Borel set by **II.3.7**, lemma 1, it follows from (15) and (16) that $b(u)$ is Borel measurable.

§ IV.4. Essential bounded variation and absolute continuity.

IV.4.1. Essential bounded variation. Let there be given a bounded continuous transformation $T: \boldsymbol{D} \to R^n$, where $\boldsymbol{D}$ is a bounded domain in R^n. Throughout the present § **IV.4**, constant use will be made of the results obtained in § **II.3** about the essential multiplicity functions $K(x, T, D)$, $K^+(x, T, D)$, $K^-(x, T, D)$, $k(x, T, D)$ (see **II.3.2**, **II.3.3**) and the essential sets $E(T, \boldsymbol{D})$, $E_i^p(T, \boldsymbol{D})$ (see **II.3.6**). The reader should turn to the sections just referred to when in doubt about the meaning of terms or symbols.

Definition 1. T is said to be *essentially of bounded variation* in $\boldsymbol{D}$ (briefly, eBV in $\boldsymbol{D}$) if the essential multiplicity function $K(x, T, \boldsymbol{D})$ is L-summable in R^n.

Observe that $K(x, T, \boldsymbol{D})$ is a lower semi-continuous function of x (see **II.3.2**, theorem 3). Thus $K(x, T, \boldsymbol{D})$ is Borel measurable and hence also L-measurable in R^n (see **III.1.1**, lemma 20). As noted in **IV.3.4**, $K(x, T, D)$ is defined for every domain $D \subset \boldsymbol{D}$, and is in fact an admissible multiplicity function in the sense of **IV.3.1**, according to **IV.3.4**, theorem 2. Comparison of definition 1 above with definition 1 in **IV.3.2** reveals that the property eBV is equivalent to bounded variation with respect to the essential multiplicity function $K(x, T, D)$. Hence the theory developed in § **IV.3** applies to the present situation. To simplify formulas, we introduce the following notations. The function of domains generated by K, in the sense of **IV.3.2**, definition 2, will be denoted by $\mathfrak{K}(D)$. Thus explicitly

$$\mathfrak{K}(D) = \int K(x, T, D)\, d\mathsf{L}, \quad D \subset \boldsymbol{D}. \tag{1}$$

The base-function generated by K, in the sense of **IV.3.1**, definition 3, will be denoted by $b_e(u)$. Thus

$$b_e(u) = \text{gr.l.b.}\, K(Tu, T, D), \quad u \in D \subset \boldsymbol{D}. \tag{2}$$

The corresponding summatory function [see **IV.3.1** (3)] will be denoted by W_e. Thus, if S is a set in $\boldsymbol{D}$, then

$$W_e(x, T, S) = \sum b_e(u), \quad u \in S \cap T^{-1} x. \tag{3}$$

The corresponding measure [see **IV.3.2** (4)] will be denoted by ν_e (assuming that T is eBV in $\boldsymbol{D}$). Thus, if B is a Borel set in $\boldsymbol{D}$, then

$$\nu_e(B) = \int W_e(x, T, B)\, d\mathsf{L}. \tag{4}$$

The derivative of the function of domains $\mathfrak{K}(D)$ will be denoted by $D_e(u, T)$. Thus

$$D_e(u, T) = D(u, \mathfrak{K}), \tag{5}$$

where the symbol $D(u, \mathfrak{K})$ is used in the sense of **III.2.4** (2). The reduced base-set (see **IV.2.1**, definition 2) corresponding to the base-function b_e will be denoted by $\mathfrak{B}_e^r$. Thus

$$\mathfrak{B}_e^r = \{u \mid u \in \boldsymbol{D}, 0 < b_e(u) < \infty\}. \tag{6}$$

Since T and $\boldsymbol{D}$ are thought of as fixed in the present section, the preceding notations are adequate.

Lemma 1. $\mathfrak{B}_e^r = E_i^p(T, \boldsymbol{D})$.

Proof. This is a direct consequence of **II.3.6**, lemma 14 and **IV.3.1**, lemma 6.

Lemma 2. If T is eBV in $\boldsymbol{D}$, then T is also eBV in every domain $D \subset \boldsymbol{D}$.

Proof. This is a direct consequence of **IV.3.2**, lemma 1.

Theorem 1. Assume that T is eBV in $\boldsymbol{D}$. Then the following holds.

(i) The derivative $D_e(u, T)$ exists a.e. in $\boldsymbol{D}$ and is L-summable there.

(ii) $D_e(u, T) = 0$ a.e. on $\boldsymbol{D} - E_i^p(T, \boldsymbol{D})$.

(iii) If D is any domain in $\boldsymbol{D}$, then [see (1)]

$$\int_D D_e(u, T)\, d\mathsf{L} \leqq \mathfrak{K}(D).$$

(iv) If G is any set in $\boldsymbol{D}$ such that $\mathsf{L}TG = 0$, then $D_e(u, T) = 0$ a.e. on G.

Proof. (i) and (iii) follow directly from **IV.3.2**, lemma 2, and (ii) follows from lemma 1 and **IV.3.2**, lemma 6. Finally (iv) follows directly from **IV.3.2**, lemma 8.

Definition 2. Let D be any domain in $\boldsymbol{D}$. Then the *essential total variation* $V_e(T, D)$ is given by the formula $V_e(T, D) = \mathfrak{K}(D)$ [see (1)] if T is eBV in D, while $V_e(T, D) = \infty$ if T is not eBV in D.

Lemma 3. $V_e(T, D)$ is finite if and only if T is eBV in D.

This is a formal consequence of the definitions involved.

Theorem 2. Given T as above, let $T_j: \boldsymbol{D}_j \to R^n$, $j = 1, 2, \ldots$, be a sequence of bounded continuous transformations which converge uniformly to T on compact subsets of $\boldsymbol{D}$ (see **II.3.2**, remark 9). Then the following holds.

(i) $V_e(T, \boldsymbol{D}) \leq \liminf V_e(T_j, \boldsymbol{D}_j)$.

(ii) If $\liminf\limits_{j\to\infty} V_e(T_j, \boldsymbol{D}_j) < \infty$, then T is eBV in $\boldsymbol{D}$.

Proof. If the lim inf in (i) is infinite, the assertion is obvious. So we can assume that the lim inf is finite. Passing on to a subsequence if necessary, we can further assume that $V_e(T_j, \boldsymbol{D}_j) < \infty$ for every j. By definition 2 we have then

$$V_e(T_j, \boldsymbol{D}_j) = \int K(x, T_j, \boldsymbol{D}_j)\, d\mathsf{L}.$$

By **II.3.2**, theorem 4 we have

$$K(x, T, \boldsymbol{D}) \leq \liminf_{j\to\infty} K(x, T_j, \boldsymbol{D}_j).$$

By the lemma of FATOU (see **III.1.1**, lemma 30) we conclude that $K(x, T, \boldsymbol{D})$ is L-summable in R^n and

$$\int K(x, T, \boldsymbol{D})\, d\mathsf{L} \leq \liminf_{j\to\infty} \int K(x, T_j, \boldsymbol{D}_j)\, d\mathsf{L}. \tag{7}$$

In view of definition 2, the relation (7) is equivalent to (i). Finally, (ii) is a direct consequence of (i) and lemma 3.

IV.4.2. Essential absolute continuity. The terminology of **IV.4.1** will be used in the present section. The concept of *essential absolute continuity*, to be introduced presently, means *absolute continuity with respect to the admissible multiplicity function* $K(x, T, D)$, in the sense of **IV.3.3**. The explicit wording is as follows [see **IV.4.1** (1), (5)].

Definition. T is termed essentially absolutely continuous in $\boldsymbol{D}$ (briefly, eAC in $\boldsymbol{D}$) if it is eBV in $\boldsymbol{D}$ and

$$\int_{\boldsymbol{D}} D_e(u, T)\, d\mathsf{L} = \int K(x, T, \boldsymbol{D})\, d\mathsf{L}. \tag{1}$$

Note that if T is eBV in $\boldsymbol{D}$, then both integrals appearing in (1) exist by **IV.4.1**, definition 1 and theorem 1, and thus the preceding definition is meaningful. As regards motivation, note that if T is eBV in $\boldsymbol{D}$, then

$$\int_{\boldsymbol{D}} D_e(u, T)\, d\mathsf{L} \leq \int K(x, T, \boldsymbol{D})\, d\mathsf{L}$$

by **IV.4.1**, theorem 1, and thus the eAC property requires that the sign of equality should hold in the preceding relation.

Theorem 1. If T is eAC in $\boldsymbol{D}$, then it is also eAC in every domain $D \subset \boldsymbol{D}$, and hence

$$\int_D D_e(u, T)\, d\mathsf{L} = \int K(x, T, D)\, d\mathsf{L}, \quad D \subset \boldsymbol{D}. \tag{2}$$

This is a direct consequence of **IV.3.3**, lemma 3.

Lemma 1. Let x be a point in R^n such that

$$K(x, T, \boldsymbol{D}) < \infty, \tag{3}$$

$$x \in T[E(T, \boldsymbol{D}) - E_i^p(T, \boldsymbol{D})]. \tag{4}$$

Then

$$W_e(x, T, \boldsymbol{D}) < K(x, T, \boldsymbol{D}). \tag{5}$$

Proof. Since $x \in TE(T, \boldsymbol{D})$ by (4), there exists at least one e.m.m.c. for $(x, T, \boldsymbol{D})$, and thus $k(x, T, \boldsymbol{D}) \geqq 1$. By **II.3.4**, theorem 3 it follows that $K(x, T, \boldsymbol{D}) \geqq 1$. Hence, in view of (3),

$$1 \leqq K(x, T, \boldsymbol{D}) < \infty. \tag{6}$$

Note that

$$0 \leqq W_e(x, T, \boldsymbol{D}) \leqq K(x, T, \boldsymbol{D}) \tag{7}$$

by **IV.3.1**, lemma 1. Now if $W_e(x, T, \boldsymbol{D}) = 0$, then (5) is obvious in view of (6). Hence we can assume [see (6) and (7)] that

$$0 < W_e(x, T, \boldsymbol{D}) < \infty.$$

By the definition of W_e it follows that there exists a finite (non-zero) number of points $u \in T^{-1}x$ where $b_e(u) \neq 0$. Let $u_1, \ldots, u_m$ be these points. Note that if D is any domain in $\boldsymbol{D}$, then

$$W_e(x, T, D) = \sum b_e(u_j), \quad u_j \in D. \tag{8}$$

Let $D_1, \ldots, D_m$ be pair-wise disjoint domains in $\boldsymbol{D}$ such that

$$u_j \in D_j, \quad j = 1, \ldots, m.$$

By (8) we have then

$$W_e(x, T, \boldsymbol{D}) = \sum_{j=1}^{m} W_e(x, T, D_j). \tag{9}$$

By **II.3.6**, lemma 15 the assumptions (3) and (4) imply the existence of a number $\eta > 0$ such that if the domains D_j above satisfy the condition

$$\delta D_j < \eta, \quad j = 1, \ldots, m, \tag{10}$$

then

$$\sum_{j=1}^{m} K(x, T, D_j) < K(x, T, \boldsymbol{D}). \tag{11}$$

Now since the number of the points $u_1, \ldots, u_m$ is finite, we can clearly comply with (10) in selecting the domains $D_1, \ldots, D_m$. Since

$$W_e(x, T, D_j) \leq K(x, T, D_j)$$

by **IV.3.1**, lemma 1, the relations (9) and (11) yield

$$W_e(x, T, \boldsymbol{D}) = \sum_{j=1}^{m} W_e(x, T, D_j) \leq \sum_{j=1}^{m} K(x, T, D_j) < K(x, T, \boldsymbol{D}),$$

and (5) is proved.

Lemma 2. Let x be a point in R^n such that

$$K(x, T, \boldsymbol{D}) < \infty, \tag{12}$$

$$x \notin T[E(T, \boldsymbol{D}) - E_i^p(T, \boldsymbol{D})]. \tag{13}$$

Then

$$W_e(x, T, \boldsymbol{D}) = K(x, T, \boldsymbol{D}). \tag{14}$$

Proof. (14) is obvious if $K(x, T, \boldsymbol{D}) = 0$, since

$$0 \leq W_e(x, T, \boldsymbol{D}) \leq K(x, T, \boldsymbol{D})$$

by **IV.3.1**, lemma 1. So we can assume, in view of (12), that

$$0 < K(x, T, \boldsymbol{D}) < \infty.$$

By **II.3.4**, theorem 3 it follows that

$$0 < k(x, T, \boldsymbol{D}) < \infty.$$

In other words, the number of essential maximal model continua for $(x, T, \boldsymbol{D})$ is finite and non-zero. Furthermore, each of these continua reduces to a single point by (13). Let the points $u_1, \ldots, u_m$ be the essential maximal model continua for $(x, T, \boldsymbol{D})$. Let us select pair-wise disjoint domains $D_1, \ldots, D_m$ in $\boldsymbol{D}$ such that $u_j \in D_j$, $j = 1, \ldots, m$. By **II.3.4**, theorem 4 and **II.3.5**, remark 3 we have then

$$K(x, T, \boldsymbol{D}) = \sum_{j=1}^{m} |i_e(u_j, T)|,$$

$$K(x, T, D_j) = |i_e(u_j, T)|, \quad j = 1, \ldots, m,$$

and hence

$$K(x, T, \boldsymbol{D}) = \sum_{j=1}^{m} K(x, T, D_j). \tag{15}$$

By **IV.3.1**, lemma 5, the domains D_j can be so chosen that

$$W_e(x, T, D_j) = K(x, T, D_j), \quad j = 1, \ldots, m. \tag{16}$$

Since the domains $D_1, \ldots, D_m$ are pair-wise disjoint, clearly

$$\sum_{j=1}^{m} W_e(x, T, D_j) \leq W_e(x, T, \boldsymbol{D}). \tag{17}$$

Also, by **IV.3.1**, lemma 1,

$$W_e(x, T, \boldsymbol{D}) \leq K(x, T, \boldsymbol{D}). \tag{18}$$

Combining (15), (16), (17), (18), we obtain

$$K(x, T, \boldsymbol{D}) \leq W_e(x, T, \boldsymbol{D}) \leq K(x, T, \boldsymbol{D}),$$

and (14) follows.

Lemma 3. Let x be a point in R^n such that $K(x, T, \boldsymbol{D}) < \infty$. Then the relation

$$W_e(x, T, \boldsymbol{D}) = K(x, T, \boldsymbol{D})$$

holds if and only if

$$x \notin T[E(T, \boldsymbol{D}) - E_i^p(T, \boldsymbol{D})].$$

This is a formal consequence of lemmas 1 and 2.

Theorem 2. Assume that T is eAC in $\boldsymbol{D}$. Then

$$\mathsf{L}\, T[E(T, \boldsymbol{D}) - E_i^p(T, \boldsymbol{D})] = 0. \tag{19}$$

Proof. Let X_∞ be the set of those points in R^n where $K(x, T, \boldsymbol{D}) = \infty$. Since T is also eBV in $\boldsymbol{D}$, $K(x, T, \boldsymbol{D})$ is L-summable in R^n, and hence

$$\mathsf{L}\, X_\infty = 0. \tag{20}$$

Since T is eAC in $\boldsymbol{D}$, by **IV.3.3**, lemma 1 we have $\nu_e(\boldsymbol{D}) = \mathfrak{K}(\boldsymbol{D})$. By **IV.3.2**, lemma 4 it follows that

$$W_e(x, T, \boldsymbol{D}) = K(x, T, \boldsymbol{D}) \quad \text{a.e. in } R^n.$$

Hence, on setting

$$X = \{x \mid W_e(x, T, \boldsymbol{D}) \neq K(x, T, \boldsymbol{D})\}, \tag{21}$$

we have $\mathsf{L}X = 0$, and hence also [see (20)]

$$\mathsf{L}(X \cup X_\infty) = 0. \tag{22}$$

Consider now a point $x \in R^n$ such that

$$x \in \mathsf{C}(X \cup X_\infty). \tag{23}$$

Then

$$K(x, T, \boldsymbol{D}) < \infty, \quad K(x, T, \boldsymbol{D}) = W_e(x, T, \boldsymbol{D}),$$

and hence by lemma 3

$$x \notin T[E(T, \boldsymbol{D}) - E_i^p(T, \boldsymbol{D})]. \tag{24}$$

Thus (23) implies (24), and consequently

$$T[E(T, \boldsymbol{D}) - E_i^p(T, \boldsymbol{D})] \subset X \cup X_\infty. \tag{25}$$

The relation (19) follows now from (25) and (22).

Lemma 4. Assume that T is eBV in $\boldsymbol{D}$ and satisfies the condition (N) on the essential set $E(T, \boldsymbol{D})$. Then

$$\mathsf{L}\, T[E(T, \boldsymbol{D}) - E_i^p(T, \boldsymbol{D})] = 0. \tag{26}$$

Proof. Since T satisfies the condition (N) on $E(T, \boldsymbol{D})$, it also satisfies the condition (N) on the set $E(T, \boldsymbol{D}) - E^p(T, \boldsymbol{D})$. But this latter set is either empty or else it is a union of non-degenerate maximal model continua for T. By **IV.1.4**, lemma 3 it follows that

$$\mathsf{L}\, T[E(T, \boldsymbol{D}) - E^p(T, \boldsymbol{D})] = 0. \tag{27}$$

Now consider any point

$$u \in E^p(T, \boldsymbol{D}) - E_i^p(T, \boldsymbol{D}). \tag{28}$$

Then u itself is an e.m.m.c. for $(Tu, T, \boldsymbol{D})$ which is *not* essentially isolated. Hence $K(Tu, T, \boldsymbol{D}) = \infty$. Thus $Tu \in X_\infty$, where X_∞ denotes again the set of those points $x \in R^n$ where $K(x, T, \boldsymbol{D}) = \infty$. Since u was any point satisfying (28), it follows that

$$T[E^p(T, \boldsymbol{D}) - E_i^p(T, \boldsymbol{D})] \subset X_\infty. \tag{29}$$

But $\mathsf{L} X_\infty = 0$ since $K(x, T, \boldsymbol{D})$ is L-summable. Hence (29) implies that

$$\mathsf{L}\, T[E^p(T, \boldsymbol{D}) - E_i^p(T, \boldsymbol{D})] = 0,$$

and (26) follows in view of (27).

Theorem 3. Assume that T is eBV in $\boldsymbol{D}$. Then T is eAC in $\boldsymbol{D}$ if and only if it satisfies the condition (N) on the essential set $E(T, \boldsymbol{D})$.

Proof. Assume first that T is eAC in $\boldsymbol{D}$. In view of theorem 2, T obviously satisfies the condition (N) on the set $E(T, \boldsymbol{D}) - E_i^p(T, \boldsymbol{D})$. Also, T is ACb_e in $\boldsymbol{D}$ by **IV.3.3**, lemma 1, and hence by **IV.2.3**, lemma 2 it satisfies the condition (N) on the reduced base-set $\mathfrak{B}_e'$ which corresponds to the base-function b_e (see **IV.4.1**). Since $\mathfrak{B}_e' = E_i^p(T, \boldsymbol{D})$ by **IV.4.1**, lemma 1, it follows that T satisfies the condition (N) on the set $E_i^p(T, \boldsymbol{D})$ also. Combining these facts, we see that T satisfies the condition (N) on $E(T, \boldsymbol{D})$. Suppose, conversely, that T satisfies the condition (N) on $E(T, \boldsymbol{D})$. Make the following observations.

(i) T is eBV in $\boldsymbol{D}$ by assumption, and hence it is also BVb_e in $\boldsymbol{D}$ by **IV.3.2**, lemma 3.

(ii) Since $E_i^p(T, \boldsymbol{D}) \subset E(T, \boldsymbol{D})$, T satisfies the condition (N) on $E_i^p(T, \boldsymbol{D})$.

(iii) Since $E_i^p(T, \boldsymbol{D})$ coincides, by lemma 1 in **IV.4.1**, with the reduced base-set corresponding to b_e, (i) and (ii) imply (by **IV.2.3**, lemma 2) that T is ACb_e in $\boldsymbol{D}$.

(iv) Since T is eBV in $\boldsymbol{D}$ and satisfies the condition (N) on $E(T, \boldsymbol{D})$, we have

$$\mathsf{L}\, T[E(T, \boldsymbol{D}) - E_i^p(T, \boldsymbol{D})] = 0$$

by lemma 4.

(v) On setting again

$$X_\infty = \{x \mid K(x, T, \boldsymbol{D}) = \infty\},$$

we have $\mathsf{L} X_\infty = 0$ since T is eBV in $\boldsymbol{D}$ and hence $K(x, T, \boldsymbol{D})$ is L-summable in R^n.

Note now that (iv) and (v) yield

$$K(x, T, \boldsymbol{D}) < \infty, \quad x \notin T[E(T, \boldsymbol{D}) - E_i^p(T, \boldsymbol{D})] \quad \text{a.e. in } R^n.$$

By lemma 3 it follows that

$$W_e(x, T, \boldsymbol{D}) = K(x, T, \boldsymbol{D}) \quad \text{a.e. in } R^n.$$

Integration yields [see **IV.4.1**, (1), (4)]

$$\nu_e(\boldsymbol{D}) = \mathfrak{K}(\boldsymbol{D}). \tag{30}$$

From (iii) and (30) we conclude by **IV.3.3**, lemma 2 that T is absolutely continuous in $\boldsymbol{D}$ with respect to the multiplicity function K. By definition, this means that T is eAC in $\boldsymbol{D}$, and the theorem is proved.

IV.4.3. The essential generalized Jacobian $J_e(u, T)$. The terminology of **IV.4.1**, **IV.4.2** will be used in the present section. Let us recall the relations (see **II.3.2**, theorem 1 and **II.3.4**, as well as remark 1 in **IV.3.4**).

$$0 \leqq K^+(x, T, D) \leqq K(x, T, D), \tag{1}$$

$$0 \leqq K^-(x, T, D) \leqq K(x, T, D), \tag{2}$$

$$K(x, T, D) = K^+(x, T, D) + K^-(x, T, D), \tag{3}$$

$$\mu_e(x, T, D) = K^+(x, T, D) - K^-(x, T, D) \quad \text{if} \quad K(x, T, D) < \infty, \tag{4}$$

where D is any domain in $\boldsymbol{D}$. Let us also recall that $K^+(x, T, D)$ and $K^-(x, T, D)$ are admissible multiplicity functions (see **IV.3.4**, theorem 2). Accordingly, it is legitimate to consider the properties BVK^+, BVK^-, ACK^+, ACK^- in the sense of **IV.3.2** and **IV.3.3**.

Lemma 1. T is eBV in $\boldsymbol{D}$ if and only if it is both BVK^+ and BVK^- in $\boldsymbol{D}$.

Proof. If T is eBV in $\boldsymbol{D}$, then $K(x, T, \boldsymbol{D})$ is L-summable in R^n. In view of (1) and (2) it follows that $K^+(x, T, \boldsymbol{D})$ and $K^-(x, T, \boldsymbol{D})$ are also L-summable in R^n, and hence T is both BVK^+ and BVK^- in $\boldsymbol{D}$ by definition. Conversely, if T is both BVK^+ and BVK^- in $\boldsymbol{D}$, then $K^+(x, T, \boldsymbol{D})$ and $K^-(x, T, \boldsymbol{D})$ are L-summable in R^n. By (3) it follows that $K(x, T, \boldsymbol{D})$ is also L-summable in R^n, and hence T is eBV in $\boldsymbol{D}$ by definition.

Assuming that T is eBV in $\boldsymbol{D}$, we introduce the following notations. Since T is BVK^+ in $\boldsymbol{D}$ by lemma 1, there arises a corresponding function of domains (see **IV.3.2**, definition 2) which will be denoted by $\mathfrak{K}^+(D)$. Thus explicitly

$$\mathfrak{K}^+(D) = \int K^+(x, T, D)\, d\mathsf{L}, \quad D \subset \boldsymbol{D}. \tag{5}$$

The base-function generated by K^+ (see **IV.3.1**, definition 3) will be denoted by $b_e^+(u)$. Thus explicitly

$$b_e^+(u) = \text{gr.l.b.}\, K^+(Tu, T, D), \quad u \in D \subset \boldsymbol{D}. \tag{6}$$

By **IV.3.2**, lemma 2, the derivative $D(u, \mathfrak{K}^+)$ exists a.e. in $\boldsymbol{D}$. We put

$$D_e^+(u, T) = D(u, \mathfrak{K}^+). \tag{7}$$

The reduced base-set (see **IV.2.1**, definition 2) corresponding to $b_e^+(u)$ will be denoted by B_e^{r+}. Thus

$$B_e^{r+} = \{u \mid u \in \boldsymbol{D}, 0 < b_e^+(u) < \infty\}. \tag{8}$$

Similarly, since T is also BVK^- in $\boldsymbol{D}$ by lemma 1, we obtain analogous definitions relative to K^-, as follows.

$$\mathfrak{K}^-(D) = \int K^-(x, T, D)\, d\mathsf{L}, \quad D \subset \boldsymbol{D}, \tag{9}$$

$$b_e^-(u) = \text{gr.l.b.}\, K^-(Tu, T, D), \quad u \in D \subset \boldsymbol{D}, \tag{10}$$

$$D_e^-(u, T) = D(u, \mathfrak{K}^-), \tag{11}$$

$$\mathfrak{B}_e^{r-} = \{u \mid u \in \boldsymbol{D}, 0 < b_e^-(u) < \infty\}. \tag{12}$$

For convenient reference, let us recall the analogous definitions relative to K itself (see **IV.4.1**).

$$\mathfrak{K}(D) = \int K(x, T, \boldsymbol{D})\, d\mathsf{L}, \quad D \subset \boldsymbol{D}, \tag{13}$$

$$b_e(u) = \text{gr.l.b.}\, K(T\,u, T, D), \quad u \in D \subset \boldsymbol{D}, \tag{14}$$

$$D_e(u, T) = D(u, \mathfrak{K}), \tag{15}$$

$$\mathfrak{B}_e^r = \{u \mid u \in \boldsymbol{D}, 0 < b_e(u) < \infty\}. \tag{16}$$

The set of those points x where $K(x, T, \boldsymbol{D}) = \infty$ will be denoted by X_∞. Thus

$$X_\infty = \{x \mid K(x, T, \boldsymbol{D}) = \infty\}. \tag{17}$$

Let us introduce a new function of domains $\mathfrak{A}_e(D)$ by the formula

$$\mathfrak{A}_e(D) = \mathfrak{K}^+(D) - \mathfrak{K}^-(D), \quad D \subset \boldsymbol{D}. \tag{18}$$

In view of the geometrical interpretation obtained for $\mathfrak{K}^+$ and $\mathfrak{K}^-$ from (5) and (9), $\mathfrak{A}_e(D)$ may be thought of as the *essential algebraic area* of the image of D under T. Observe that the derivatives $D_e^+(u, T)$, $D_e^-(u, T)$ exist a.e. in $\boldsymbol{D}$ by **IV.3.2**, lemma 2, and hence in view of (18) the derivative $D(u, \mathfrak{A}_e)$ of $\mathfrak{A}_e$ also exists a.e. in $\boldsymbol{D}$. We put

$$J_e(u, T) = D(u, \mathfrak{A}_e). \tag{19}$$

The following definition is motivated by the geometrical interpretation of $\mathfrak{A}_e(D)$.

Definition. $J_e(u, T)$ is termed the *essential generalized Jacobian* for T (where T is assumed to be eBV in $\boldsymbol{D}$).

Theorem 1. If T is eBV in $\boldsymbol{D}$, then the essential generalized Jacobian $J_e(u, T)$ exists a.e. in $\boldsymbol{D}$ and is L-summable in $\boldsymbol{D}$. Furthermore,

$$J_e(u, T) = D_e^+(u, T) - D_e^-(u, T) \quad \text{a.e. in } \boldsymbol{D}. \tag{20}$$

Proof. Since T is both BVK^+ and BVK^- in $\boldsymbol{D}$, by **IV.3.2**, lemma 2 the derivatives $D_e^+(u, T)$ and $D_e^-(u, T)$ exist a.e. in $\boldsymbol{D}$ and are L-summable there. The existence of $J_e(u, T)$ a.e. in $\boldsymbol{D}$ as well as the formula (20) are thus direct consequences of (18) and (19). Finally, (20) yields the L-summability of $J_e(u, T)$ in $\boldsymbol{D}$.

In preparation for the following discussion, some additional comments about the essential set $E_i^p(T, \boldsymbol{D})$ will now be made (see **II.3.6**). If u is a point of $E_i^p(T, \boldsymbol{D})$, then u itself is an e.m.m.c., and it is essentially isolated, by definition. Accordingly, the essential local index $i_e(u, T)$

is defined and its value is an integer different from zero. Let us put

$$E_i^{p+}(T, \boldsymbol{D}) = \{u | u \in E_i^p(T, \boldsymbol{D}),\ i_e(u, T) > 0\}, \tag{21}$$

$$E_i^{p-}(T, \boldsymbol{D}) = \{u | u \in E_i^p(T, \boldsymbol{D}),\ i_e(u, T) < 0\}. \tag{22}$$

Clearly

$$E_i^{p+}(T, \boldsymbol{D}) \cup E_i^{p-}(T, \boldsymbol{D}) = E_i^p(T, \boldsymbol{D}), \tag{23}$$

$$E_i^{p+}(T, \boldsymbol{D}) \cap E_i^{p-}(T, \boldsymbol{D}) = \emptyset. \tag{24}$$

Lemma 2. The following inclusions hold [see (8), (12), (21), (22)].

$$E_i^{p-}(T, \boldsymbol{D}) < C\,\mathfrak{B}_e^{r+}, \tag{25}$$

$$E_i^{p+}(T, \boldsymbol{D}) < C\,\mathfrak{B}_e^{r-}. \tag{26}$$

Proof. Consider a point

$$u \in E_i^{p-}(T, \boldsymbol{D}). \tag{27}$$

Then u by itself is an essentially isolated e.m.m.c. for $(Tu, T, \boldsymbol{D})$. Let O be a characteristic neighborhood of u (see **II.3.3**, definition 2), and denote by D the component of O containing u. Then u is the only e.m.m.c. for $(Tu, T, \boldsymbol{D})$ in the domain D, and $i_e(u, T) < 0$ by (27). By **II.3.4** (25), applied to the domain D, it follows that $K^+(Tu, T, D) = 0$. Since [see (6)]

$$0 \leq b_e^+(u) \leq K^+(Tu, T, D),$$

we conclude that $b_e^+(u) = 0$, and hence $u \notin \mathfrak{B}_e^{r+}$. As u was an arbitrary point satisfying (27), the relation (25) follows. The proof of (26) is entirely similar.

Lemma 3. Assume that T is eBV in $\boldsymbol{D}$. Then

$$D_e^+(u, T) = 0 \quad \text{a.e. on} \quad E_i^{p-}(T, \boldsymbol{D}), \tag{28}$$

$$D_e^-(u, T) = 0 \quad \text{a.e. on} \quad E_i^{p+}(T, \boldsymbol{D}). \tag{29}$$

Proof. Since the proof is entirely similar in either case, we discuss only (28) in detail. As T is BVK^+ in $\boldsymbol{D}$ by lemma 1, in view of **IV.3.2**, lemma 6 we have the relation

$$D_e^+(u, T) = 0 \quad \text{a.e. on} \quad \boldsymbol{D} - \mathfrak{B}_e^{r+}, \tag{30}$$

and (28) follows by (25).

Theorem 2. If T is eBV in $\boldsymbol{D}$, then

$$|J_e(u, T)| = D_e(u, T) \quad \text{a.e. in } \boldsymbol{D}. \tag{31}$$

Proof. Let D be any domain in $\boldsymbol{D}$. From (1), (2), (3), (13), (5), (9) it follows that

$$0 \leqq \mathfrak{K}^+(D) \leqq \mathfrak{K}(D), \quad 0 \leqq \mathfrak{K}^-(D) \leqq \mathfrak{K}(D), \quad \mathfrak{K}(D) = \mathfrak{K}^+(D) + \mathfrak{K}^-(D),$$

and hence clearly

$$0 \leqq D_e^+(u, T) \leqq D_e(u, T) \quad \text{a.e. in } \boldsymbol{D}, \tag{32}$$

$$0 \leqq D_e^-(u, T) \leqq D_e(u, T) \quad \text{a.e. in } \boldsymbol{D}, \tag{33}$$

$$D_e(u, T) = D_e^+(u, T) + D_e^-(u, T) \quad \text{a.e. in } \boldsymbol{D}. \tag{34}$$

Note that

$$D_e(u, T) = 0 \quad \text{a.e. on} \quad \boldsymbol{D} - E_i^p(T, \boldsymbol{D}) \tag{35}$$

by **IV.4.1**, theorem 1, and

$$J_e(u, T) = D_e^+(u, T) - D_e^-(u, T) \quad \text{a.e. in } \boldsymbol{D} \tag{36}$$

by (20). From (35), (32), (33), (36), (34) we see that

$$|J_e(u, T)| = 0 = D_e(u, T) \quad \text{a.e. on} \quad \boldsymbol{D} - E_i^p(T, \boldsymbol{D}). \tag{37}$$

From (34), (36), (28) we obtain

$$|J_e(u, T)| = D_e^-(u, T) = D_e(u, T) \quad \text{a.e. on} \quad E_i^{p-}(T, \boldsymbol{D}). \tag{38}$$

Similarly, from (34), (36), (29) we conclude that

$$|J_e(u, T)| = D_e^+(u, T) = D_e(u, T) \quad \text{a.e. on} \quad E_i^{p+}(T, \boldsymbol{D}). \tag{39}$$

Since by (23)

$$\boldsymbol{D} = [\boldsymbol{D} - E_i^p(T, \boldsymbol{D})] \cup E_i^{p+}(T, \boldsymbol{D}) \cup E_i^{p-}(T, \boldsymbol{D}),$$

the relations (37), (38), (39) imply (31).

Theorem 3. Assume that T is eBV in $\boldsymbol{D}$. Then

$$\int_{\boldsymbol{D}} |J_e(u, T)|\, d\mathsf{L} \leqq \int K(x, T, \boldsymbol{D})\, d\mathsf{L}, \tag{40}$$

and the sign of equality holds if and only if T is eAC in $\boldsymbol{D}$.

Proof. Since $|J_e(u,T)| = D_e(u,T)$ a.e. in $\boldsymbol{D}$ by theorem 2, the present theorem is a direct consequence of **IV.4.1**, theorem 1 and of the definition of essential absolute continuity (see **IV.4.2**).

Lemma 4. Assume that T is defined and continuous in $\overline{\boldsymbol{D}}$ and eBV in $\boldsymbol{D}$. Let D be a domain such that

$$\overline{D} \subset \boldsymbol{D}, \quad \mathsf{L}\, T\, fr\, D = 0. \tag{41}$$

Then

$$\mu_e(x, T, D) = \mu(x, T, D) \quad \text{a.e. in } R^n, \tag{42}$$

$$\mathfrak{A}_e(D) = \int \mu(x, T, D)\, d\mathsf{L}. \tag{43}$$

Proof. By **II.3.4**, theorem 2 we have

$$\mu_e(x, T, D) = \mu(x, T, D)$$

at those points $x \in R^n$ which satisfy the conditions

$$x \notin T\, fr\, D, \quad K(x, T, D) < \infty. \tag{44}$$

Now since T is eBV in $\boldsymbol{D}$, $K(x, T, D)$ is L-summable in R^n for every domain $D \subset \boldsymbol{D}$, and hence $K(x, T, D) < \infty$ a.e. in R^n. Thus in view of (41) the conditions (44) hold a.e. in R^n, and (42) follows. Furthermore, since $K(x, T, D) < \infty$ a.e. in R^n, we have by (4)

$$\mu_e(x, T, D) = K^+(x, T, D) - K^-(x, T, D) \quad \text{a.e. in } R^n,$$

and (43) follows by integration in view of (42), (5), (9), (18).

Lemma 5. Assume that T is eBV in $\boldsymbol{D}$. Let G be any set in $\boldsymbol{D}$ such that $\mathsf{L}TG = 0$. Then

$$J_e(u, T) = 0 \quad \text{a.e. on } G. \tag{45}$$

Proof. We have $|J_e(u, T)| = D_e(u, T)$ a.e. in $\boldsymbol{D}$ by theorem 2, and $D_e(u, T) = 0$ a.e. on G by **IV.4.1**, theorem 1. Thus (45) follows.

IV.4.4. Transformation formulas for definite integrals. The terminology of **IV.4.1, IV.4.2, IV.4.3** will be used in the present section.

Theorem 1. Assume that T is eAC in $\boldsymbol{D}$. Let $H(x)$ be a finite-valued L-measurable function in R^n. Then

$$\int_{\boldsymbol{D}} H(Tu)\, |J_e(u, T)|\, d\mathsf{L} = \int H(x)\, K(x, T, \boldsymbol{D})\, d\mathsf{L},$$

as soon as one of the two integrals involved exists.

Proof. Since the property eAC is equivalent to the property ACK (see **IV.4.2**), by the theorem in **IV.3.3** we know that

$$\int_{\boldsymbol{D}} H(Tu)\, D(u, \mathfrak{R})\, d\mathsf{L} = \int H(x)\, K(x, T, \boldsymbol{D})\, d\mathsf{L},$$

as soon as one of the two integrals involved exists. Now $D(u, \mathfrak{R}) = D_e(u, T)$ as a matter of notation [see **IV.4.1** (5)]. Since T is also eBV in $\boldsymbol{D}$ by definition, we have $D_e(u, T) = |J_e(u, T)|$ a.e. in $\boldsymbol{D}$ by **IV.4.3**, theorem 2, and the present theorem follows.

Our next objective is the derivation of a transformation formula which involves $J_e(u, T)$ instead of $|J_e(u, T)|$. In preparation, we establish the following

Lemma. T is eAC in $\boldsymbol{D}$ if and only if it is both ACK^+ and ACK^- in $\boldsymbol{D}$.

Proof. Assume that T is both ACK^+ and ACK^- in $\boldsymbol{D}$. Then T is also both BVK^+ and BVK^- in $\boldsymbol{D}$, and hence it is also eBV in $\boldsymbol{D}$ by **IV.4.3**, lemma 1. On setting (see **IV.4.3**)

$$\lambda = \mathfrak{R}(\boldsymbol{D}) - \int_{\boldsymbol{D}} D(u, \mathfrak{R})\, d\mathsf{L}, \tag{1}$$

$$\lambda^+ = \mathfrak{R}^+(\boldsymbol{D}) - \int_{\boldsymbol{D}} D(u, \mathfrak{R}^+)\, d\mathsf{L}, \tag{2}$$

$$\lambda^- = \mathfrak{R}^-(\boldsymbol{D}) - \int_{\boldsymbol{D}} D(u, \mathfrak{R}^-)\, d\mathsf{L}, \tag{3}$$

we have

$$\lambda = \lambda^+ + \lambda^- \tag{4}$$

by **IV.4.3** (3), (7), (11), (15), (34), and

$$\lambda \geqq 0, \quad \lambda^+ \geqq 0, \quad \lambda^- \geqq 0 \tag{5}$$

by **IV.3.2**, lemma 2. Now since T is both ACK^+ and ACK^- by assumption, we have $\lambda^+ = 0$, $\lambda^- = 0$ (see **IV.3.3**), and hence also $\lambda = 0$ by (4). Accordingly, T is eAC in $\boldsymbol{D}$ (see **IV.4.2**). Suppose, conversely, that T is eAC in $\boldsymbol{D}$. Then T is eBV in $\boldsymbol{D}$, and hence also BVK^+ and BVK^- in $\boldsymbol{D}$ by **IV.4.3**, lemma 1. Thus we can introduce again λ, λ^+, λ^- by means of the formulas (1), (2), (3), and we have again at our disposal the relations (4) and (5). Since T is now eAC in $\boldsymbol{D}$ by assumption, we have $\lambda = 0$, and hence also $\lambda^+ = 0$, $\lambda^- = 0$ in view of (4) and (5). Thus T is both ACK^+ and ACK^- in $\boldsymbol{D}$ (see **IV.3.3**).

Theorem 2. Assume that T is eAC in $\boldsymbol{D}$. Let $H(x)$ be a finite-valued L-measurable function in R^n. Then (see **II.3.4**)

$$\int_{\boldsymbol{D}} H(T\,u)\, J_e(u, T)\, d\mathsf{L} = \int H(x)\, \mu_e(x, T, \boldsymbol{D})\, d\mathsf{L}, \tag{6}$$

as soon as the integral on the left exists. More explicitly: if the product

$$H(T\,u)\, J_e(u, T) \tag{7}$$

is L-summable in $\boldsymbol{D}$, then the product

$$H(x)\, \mu_e(x, T, \boldsymbol{D}) \tag{8}$$

is L-summable in R^n, and (6) holds.

Proof. Assume that the product (7) is L-summable in $\boldsymbol{D}$. Then the product

$$H(T\,u)\, |J_e(u, T)|$$

is also L-summable in $\boldsymbol{D}$ (see **III.1.1**, lemma 26). By theorem 1 it follows that the product

$$H(x)\, K(x, T, \boldsymbol{D})$$

is L-summable in R^n. Since

$$0 \leq K^+(x, T, \boldsymbol{D}) \leq K(x, T, \boldsymbol{D}), \quad 0 \leq K^-(x, T, \boldsymbol{D}) \leq K(x, T, \boldsymbol{D})$$

by **IV.4.3** (1), (2), it follows further that the products

$$H(x)\, K^+(x, T, \boldsymbol{D}), \quad H(x)\, K^-(x, T, \boldsymbol{D})$$

are also L-summable in R^n (note that H, K^+, K^- are known to be L-measurable in R^n). As T is both ACK^+ and ACK^- in $\boldsymbol{D}$ by the preceding lemma, we conclude from the theorem in **IV.3.3** that the products

$$H(T\,u)\, \boldsymbol{D}(u, \mathfrak{R}^+), \quad H(T\,u)\, \boldsymbol{D}(u, \mathfrak{R}^-)$$

are L-summable in $\boldsymbol{D}$, and

$$\int_{\boldsymbol{D}} H(T\,u)\, D(u, \mathfrak{R}^+)\, d\mathsf{L} = \int H(x)\, K^+(x, T, \boldsymbol{D})\, d\mathsf{L}, \tag{9}$$

$$\int_{\boldsymbol{D}} H(T\,u)\, D(u, \mathfrak{R}^-)\, d\mathsf{L} = \int H(x)\, K^-(x, T, \boldsymbol{D})\, d\mathsf{L}. \tag{10}$$

Since T is eBV in $\boldsymbol{D}$, $K(x, T, \boldsymbol{D})$ is L-summable in R^n, and hence $K(x, T, \boldsymbol{D}) < \infty$ a.e. in R^n. Accordingly, (see **II.3.4**, definition 2) the algebraic multiplicity function $\mu_e(x, T, \boldsymbol{D})$ is defined a.e. in R^n, and

$$\mu_e(x, T, \boldsymbol{D}) = K^+(x, T, \boldsymbol{D}) - K^-(x, T, \boldsymbol{D}) \quad \text{a.e. in } R^n, \tag{11}$$

by **II.3.4**, theorem 4. Also

$$J_e(u, T) = D(u, \mathfrak{R}^+) - D(u, \mathfrak{R}^-) \quad \text{a.e. in } \boldsymbol{D}, \tag{12}$$

by **IV.4.3** (20), (7), (11). Thus on subtracting (10) from (9) one obtains (6) in view of (11) and (12), and the theorem is proved.

Theorem 3. Assume that T is defined and continuous on $\overline{\boldsymbol{D}}$ and eAC in $\boldsymbol{D}$. Assume further that

$$\mathsf{L}\, T\, \mathit{fr}\, \boldsymbol{D} = 0. \tag{13}$$

Let $H(x)$ be a finite-valued L-measurable function in R^n. Then (see **II.3.4**, theorem 2)

$$\int_{\boldsymbol{D}} H(T\,u)\, J_e(u, T)\, d\mathsf{L} = \int H(x)\, \mu(x, T, \boldsymbol{D})\, d\mathsf{L}, \tag{14}$$

as soon as the integral on the left exists.

Proof. By **IV.4.3**, lemma 4 (applied with $D=\boldsymbol{D}$) we have

$$\mu_e(x, T, \boldsymbol{D}) = \mu(x, T, \boldsymbol{D}) \quad \text{a.e. in } R^n.$$

Thus the present theorem appears as a direct consequence of theorem 2.

The special case when $H(x) \equiv 1$ yields the following statement as a direct application of the preceding theorems.

Theorem 4. Assume that T is eAC in $\boldsymbol{D}$. Then

$$\int_{\boldsymbol{D}} |J_e(u, T)|\, d\mathsf{L} = \int K(x, T, \boldsymbol{D})\, d\mathsf{L}, \tag{15}$$

$$\int_{\boldsymbol{D}} J_e(u, T)\, d\mathsf{L} = \int \mu_e(x, T, \boldsymbol{D})\, d\mathsf{L}. \tag{16}$$

Furthermore, if T is defined and continuous on $\bar{\boldsymbol{D}}$ and $\mathsf{L}T\,fr\,\boldsymbol{D}=0$, then

$$\int_{\boldsymbol{D}} J_e(u, T)\, d\mathsf{L} = \int \mu(x, T, \boldsymbol{D})\, d\mathsf{L}. \tag{17}$$

Note that $J_e(u,T)$ is L-summable in $\boldsymbol{D}$ by **IV.4.3**, theorem 1, and hence theorems 1, 2, 3 are applicable. The formulas (15), (16), (17) may be interpreted as yielding expressions, in terms of the essential generalized Jacobian, for the *essential absolute area* and the *essential algebraic area* of the image of $\boldsymbol{D}$ under T.

Remark. If T is eAC in $\boldsymbol{D}$, then it is also eAC in every domain $D \subset \boldsymbol{D}$, by **IV.4.2**, theorem 1. Accordingly, the preceding theorems 1, 2, 3, 4 remain valid if $\boldsymbol{D}$ is replaced by any domain $D \subset \boldsymbol{D}$.

IV.4.5. Closure theorems. Let there be given bounded continuous transformations

$$T: \boldsymbol{D} \to R^n, \tag{1}$$

$$T_j: \boldsymbol{D}_j \to R^n, \quad j = 1, 2, \ldots, \tag{2}$$

where $\boldsymbol{D}, \boldsymbol{D}_j$ are bounded domains in R^n. In the present context, by a *closure theorem* we mean a statement to the effect that if each one of the transformations T_j possesses the eAC property and the transformations T_j converge to T in an appropriate manner, then T also possesses the eAC property. In other words, a closure theorem states that the class of eAC transformations is closed under certain passages to the limit.

Lemma 1. Given T and T_j as in (1) and (2), assume that the following conditions are satisfied.

(i) The sequence $\{T_j\}$ converges to T uniformly on compact subsets of $\boldsymbol{D}$ (see **II.3.2**, remark 9).

(ii) T_j is eAC in $\boldsymbol{D}_j$, $j=1, 2, \ldots$.

(iii) $\liminf_{j\to\infty} \int_{\boldsymbol{D}_j} |J_e(u, T_j)|\, d\mathsf{L} < \infty$.

Then T is eBV in $\boldsymbol{D}$, and

$$\int_{\boldsymbol{D}} |J_e(u, T)|\, d\mathsf{L} \leq \int K(x, T, \boldsymbol{D})\, d\mathsf{L} \leq \liminf_{j\to\infty} \int_{\boldsymbol{D}_j} |J_e(u, T_j)|\, d\mathsf{L}. \tag{3}$$

Proof. Condition (ii) and **IV.4.3**, theorem 3 yield

$$\int K(x, T_j, \boldsymbol{D}_j)\, d\mathsf{L} = \int_{\boldsymbol{D}_j} |J_e(u, T_j)|\, d\mathsf{L}.$$

In view of (iii) and **IV.4.1**, theorem 2 it follows that T is eBV in $\boldsymbol{D}$ and

$$\int K(x, T, \boldsymbol{D})\, d\mathsf{L} \leq \liminf_{j\to\infty} \int K(x, T_j, \boldsymbol{D}_j)\, d\mathsf{L} = \liminf_{j\to\infty} \int_{\boldsymbol{D}_j} |J_e(u, T_j)|\, d\mathsf{L}.$$

Since (see **IV.4.3**, theorem 3)

$$\int_{\boldsymbol{D}} |J_e(u, T)|\, d\mathsf{L} \leq \int K(x, T, \boldsymbol{D})\, d\mathsf{L},$$

the relations (3) follow.

Theorem 1. Given T and T_j as in (1) and (2), assume that the following conditions are satisfied.

(i) The sequence $\{T_j\}$ converges to T uniformly on compact subsets of $\boldsymbol{D}$ (see **II.3.2**, remark 9).

(ii) T_j is eAC in $\boldsymbol{D}_j$, $j=1, 2, \ldots$.

(iii) T is eBV in $\boldsymbol{D}$.

(iv) $\liminf_{j\to\infty} \int_{\boldsymbol{D}_j} |J_e(u, T_j)|\, d\mathsf{L} \leq \int_{\boldsymbol{D}} |J_e(u, T)|\, d\mathsf{L}$.

Then T is eAC in $\boldsymbol{D}$.

Proof. Since T is eBV in $\boldsymbol{D}$, $J_e(u, T)$ is L-summable in $\boldsymbol{D}$ by **IV.4.3**, theorem 1. Thus, in view of (i), (ii), (iv), the assumptions of lemma 1 are satisfied and hence (3) holds. Combining (3) and (iv) we obtain the inequalities

$$\int_{\boldsymbol{D}} |J_e(u, T)|\, d\mathsf{L} \leq \int K(x, T, \boldsymbol{D})\, d\mathsf{L} \leq \int_{\boldsymbol{D}} |J_e(u, T)|\, d\mathsf{L}.$$

It follows that

$$\int_{\boldsymbol{D}} |J_e(u, T)|\, d\mathsf{L} = \int K(x, T, \boldsymbol{D})\, d\mathsf{L},$$

and hence T is eAC in $\boldsymbol{D}$ by **IV.4.3**, theorem 3.

Theorem 2. Given T as in (1), assume that there exists a sequence of domains $\boldsymbol{D}_j$ which fill up $\boldsymbol{D}$ from the interior (see **II.3.2**, remark 10), such that T is eAC in $\boldsymbol{D}_j$, $j=1, 2, \ldots$. Assume further that the sequence of integrals

$$\int_{D_j} |J_e(u, T)| \, d\mathsf{L}, \quad j = 1, 2, \ldots, \tag{4}$$

is bounded. Then T is eAC in $\boldsymbol{D}$.

Proof. Since T is eAC in $\boldsymbol{D}_j$, $J_e(u, T)$ exists a.e. in $\boldsymbol{D}_j$ and is L-summable there (see **IV.4.3**, theorem 1). On setting, for clarity, $T_j = T|\boldsymbol{D}_j$, clearly the sequence $\{T_j\}$ converges to T uniformly on compact subsets of $\boldsymbol{D}$, in the sense of **II.3.2**, remark 9. Thus, in view of the boundedness of the sequence (4), the assumptions of lemma 1 are satisfied, and hence T is eBV in $\boldsymbol{D}$. Accordingly, $J_e(u, T)$ is L-summable in $\boldsymbol{D}$ (see **IV.4.3**, theorem 1), and clearly $J_e(u, T_j) = J_e(u, T)$ a.e. in $\boldsymbol{D}_j$. Hence

$$\int_{D_j} |J_e(u, T_j)| \, d\mathsf{L} = \int_{D_j} |J_e(u, T)| \, d\mathsf{L} \leq \int_{D} |J_e(u, T)| \, d\mathsf{L}.$$

Thus the assumptions of theorem 1 are obviously satisfied, and hence T is eAC in $\boldsymbol{D}$.

Lemma 2. Given T as in (1), assume that the following conditions are satisfied.

(i) T is eBV in $\boldsymbol{D}$.

(ii) T is eAC in every domain D such that

$$\overline{D} \subset \boldsymbol{D}, \quad \mathsf{L}\, fr\, D = 0. \tag{5}$$

Then T is eAC in $\boldsymbol{D}$.

Proof. There exists (see **I.2.3**, exercise 9) a sequence of domains $\boldsymbol{D}_j$ which fill up $\boldsymbol{D}$ from the interior, such that each $\boldsymbol{D}_j$ satisfies the conditions (5). In view of (i), $J_e(u, T)$ is L-summable in $\boldsymbol{D}$ by **IV.4.3**, theorem 1. Hence

$$\int_{D_j} |J_e(u, T)| \, d\mathsf{L} \leq \int_{D} |J_e(u, T)| \, d\mathsf{L} < \infty.$$

Thus, in view of (ii), the assumptions of theorem 2 are satisfied, and hence T is eAC in $\boldsymbol{D}$.

Theorem 3. Given T and T_j as in (1) and (2), assume that the following conditions are satisfied.

(i) The sequence $\{T_j\}$ converges to T uniformly on compact subsets of $\boldsymbol{D}$ (see **II.3.2**, remark 9).

(ii) T_j is eAC in $\boldsymbol{D}_j$, $j = 1, 2, \ldots$.

(iii) T is eBV in $\boldsymbol{D}$.

(iv) For every oriented n-cube $Q \subset \boldsymbol{D}$ we have

$$\lim_{j \to \infty} \int_{Q} |J_e(u, T_j)| \, d\mathsf{L} = \int_{Q} |J_e(u, T)| \, d\mathsf{L}.$$

Then T is eAC in $\boldsymbol{D}$.

Proof. In view of lemma 2 it is sufficient to prove that T is eAC in every domain D such that

$$\overline{D} \subset \boldsymbol{D}, \quad \mathsf{L}\, fr\, D = 0. \tag{6}$$

In view of (iii), $J_e(u, T)$ exists a.e. in $\boldsymbol{D}$ and is L-summable there (see **IV.4.3**, theorem 1). In view of (6) we have

$$\int_{\overline{D}} |J_e(u, T)|\, d\mathsf{L} = \int_D |J_e(u, T)|\, d\mathsf{L}. \tag{7}$$

Given a positive number ε, it follows from (6) and (7) that there exists (see **III.1.1**, lemma 34) an open set $O = O(\varepsilon)$ such that

$$\overline{D} \subset O \subset \boldsymbol{D}, \tag{8}$$

$$\int_O |J_e(u, T)|\, d\mathsf{L} < \int_D |J_e(u, T)|\, d\mathsf{L} + \varepsilon. \tag{9}$$

In view of (8), we can select a finite system of oriented n-cubes $Q_1, \ldots, Q_k$ such that

$$\operatorname{int} Q_i \cap \operatorname{int} Q_j = \emptyset \quad \text{for} \quad i \neq j, \tag{10}$$

$$\overline{D} \subset \bigcup_{i=1}^{k} Q_i \subset O. \tag{11}$$

From (10), (11) and (9) we find that

$$\sum_{i=1}^{k} \int_{Q_i} |J_e(u, T)|\, d\mathsf{L} < \int_D |J_e(u, T)|\, d\mathsf{L} + \varepsilon. \tag{12}$$

Using (11), (8), and condition (iv), we obtain

$$\lim_{j\to\infty} \sum_{i=1}^{k} \int_{Q_i} |J_e(u, T_j)|\, d\mathsf{L} = \sum_{i=1}^{k} \int_{Q_i} |J_e(u, T)|\, d\mathsf{L}. \tag{13}$$

In view of condition (i) we have

$$\bigcup_{i=1}^{k} Q_i \subset \boldsymbol{D}_j \quad \text{for } j \text{ large}.$$

Thus (11) implies that

$$\int_D |J_e(u, T_j)|\, d\mathsf{L} \leq \sum_{i=1}^{k} \int_{Q_i} |J_e(u, T_j)|\, d\mathsf{L} \quad \text{for } j \text{ large}. \tag{14}$$

From (14), (13), (12) we conclude that

$$\liminf_{j\to\infty} \int_D |J_e(u, T_j)|\, d\mathsf{L} < \int_D |J_e(u, T)|\, d\mathsf{L} + \varepsilon.$$

Since $\varepsilon > 0$ was arbitrary, it follows that

$$\liminf_{j\to\infty} \int_D |J_e(u, T_j)|\, d\mathsf{L} \leq \int_D |J_e(u, T)|\, d\mathsf{L}. \tag{15}$$

In view of (8) and condition (i) we have $D \subset \boldsymbol{D}_j$ for j large, say $j > j_0$. Set

$$D_j^* = D, \quad T_j^* = T_j | D \quad \text{for} \quad j > j_0. \tag{16}$$

Clearly condition (i) implies that the sequence $\{T_j^*\}$ converges to $T|D$ uniformly on compact subsets of D. From condition (ii) and **IV.4.2**, theorem 1 it follows that T_j^* is eAC in D. From condition (iii) and **IV.4.1**, lemma 2 it follows that T is eBV in D. Hence, in view of (15) and (16), the transformations $T|D$ and T_j^* satisfy the assumptions of theorem 1, with $\boldsymbol{D}, \boldsymbol{D}_j$ replaced by D. Accordingly, T is eAC in D, and the proof is complete.

Theorem 4. Given T and T_j as in (1) and (2), assume that the following conditions are satisfied.

(i) The sequence $\{T_j\}$ converges to T uniformly on compact subsets of $\boldsymbol{D}$.

(ii) T_j is eAC in $\boldsymbol{D}_j$, $j = 1, 2, \ldots$.

(iii) T is eBV in $\boldsymbol{D}$.

(iv) For every oriented n-cube $Q \subset \boldsymbol{D}$ we have

$$\lim_{j \to \infty} \int_Q |J_e(u, T) - J_e(u, T_j)| \, d\mathsf{L} = 0.$$

Then T is eAC in $\boldsymbol{D}$.

Proof. By **III.1.1**, lemma 41, the present condition (iv) implies condition (iv) in theorem 3. In view of this fact, theorem 4 is an immediate consequence of theorem 3.

IV.4.6. Topological transformations. Let there be given a bounded continuous transformation

$$T : \boldsymbol{D} \to R^n, \tag{1}$$

which is one-to-one in the bounded domain $\boldsymbol{D} \subset R^n$. Thus we assume now that if u_1, u_2 are any two distinct points in $\boldsymbol{D}$, then $T u_1 \neq T u_2$. By **II.2.4**, theorem 5, the image $T\boldsymbol{D}$ of $\boldsymbol{D}$ is then again a domain. Let us put

$$\boldsymbol{D}^* = T\boldsymbol{D}. \tag{2}$$

Since T is bounded, $\boldsymbol{D}^*$ is a bounded domain. Furthermore (see **II.2.4**), T is a homeomorphism from $\boldsymbol{D}$ onto $\boldsymbol{D}^*$, and its inverse

$$T^{-1} : \boldsymbol{D}^* \to R^n \tag{3}$$

is a homeomorphism from $\boldsymbol{D}^*$ onto $\boldsymbol{D}$. Under the circumstances just described, we shall term T a *bounded homeomorphism.*

Consider now a domain D such that $\bar{D} \subset \boldsymbol{D}$. Then T possesses an index $\iota(T, D)$ relative to D (see **II.2.4**). This index $\iota(T, D)$ has the value $+1$ or -1, and is independent of the choice of D in the following sense: if D_1, D_2 are any two domains such that $\bar{D}_1 \subset \boldsymbol{D}$, $\bar{D}_2 \subset \boldsymbol{D}$, then $\iota(T, D_1) = \iota(T, D_2)$ (see **II.2.4**, theorem 5). This fact justifies the following definition.

Definition 1. The bounded homeomorphism T, given as in (1), is termed *even* or *odd* in $\boldsymbol{D}$ according as $\iota(T, D) = +1$ or $\iota(T, D) = -1$ for every domain D such that $\bar{D} \subset \boldsymbol{D}$.

Lemma 1. Given a bounded homeomorphism T and its inverse T^{-1} as in (1) and (3), T and T^{-1} are both even or both odd (in $\boldsymbol{D}$ and in $\boldsymbol{D}^*$ respectively).

Proof. Take a domain D such that $\bar{D} \subset \boldsymbol{D}$. Put $\Delta = TD$. By **II.2.4** (6) we have then $T\bar{D} = \overline{TD} = \bar{\Delta}$, and thus (see **II.2.4**) Δ is a domain such that $\bar{\Delta} \subset \boldsymbol{D}^*$. Hence $\iota(T^{-1}, \Delta)$ is defined. Also [see **II.2.4** (7)], $T\,\mathit{fr}\,D = \mathit{fr}\,\Delta$. Thus $T|\bar{D}$ is a homeomorphism from the pair $(\bar{D}, \mathit{fr}\,D)$ onto the pair $(\bar{\Delta}, \mathit{fr}\,\Delta)$, and $T^{-1}|\bar{\Delta}$ is a homeomorphism from the pair $(\bar{\Delta}, \mathit{fr}\,\Delta)$ onto the pair $(\bar{D}, \mathit{fr}\,D)$. By **II.2.4**, theorem 4 we conclude that

$$\iota(T^{-1}T, D) = \iota(T, D)\,\iota(T^{-1}, \Delta).$$

But $T^{-1}T$ is the identity mapping in $\boldsymbol{D}$. Hence (see **II.2.4**, theorem 2)

$$\iota(T^{-1}T, D) = 1.$$

It follows that

$$\iota(T, D)\,\iota(T^{-1}, \Delta) = 1. \tag{4}$$

Now if T is even in $\boldsymbol{D}$, then $\iota(T, D) = 1$, and (4) yields $\iota(T^{-1}, \Delta) = 1$. Thus T^{-1} is even in $\boldsymbol{D}^*$ (see definition 1). Similarly, if T is odd in $\boldsymbol{D}$, then $\iota(T, D) = -1$, and (4) yields $\iota(T^{-1}, \Delta) = -1$. Thus T^{-1} is odd in $\boldsymbol{D}^*$.

Lemma 2. Given a bounded homeomorphism T as in (1), let x be a point in R^n and D a domain such $\bar{D} \subset \boldsymbol{D}$ and $x \in TD$. Then $\mu(x, T, D) = +1$ or $\mu(x, T, D) = -1$ according as T is even or odd in $\boldsymbol{D}$.

Proof. By **II.2.4**, theorem 1 we have

$$\mu(x, T, D) = \iota(T, D) \quad \text{if} \quad x \in TD.$$

Thus the present lemma is obvious in view of definition 1.

Lemma 3. Given a bounded homeomorphism T as in (1), the following holds (see **II.3.2, II.3.4** for terminology).

(i) If $x \notin T\boldsymbol{D}$, then

$$K(x, T, \boldsymbol{D}) = 0, \quad K^+(x, T, \boldsymbol{D}) = 0, \quad K^-(x, T, \boldsymbol{D}) = 0, \tag{5}$$

$$\mu_e(x, T, \boldsymbol{D}) = 0. \tag{6}$$

(ii) If $x \in T\boldsymbol{D}$ and T is even in $\boldsymbol{D}$, then

$$K(x, T, \boldsymbol{D}) = K^+(x, T, \boldsymbol{D}) = 1, \tag{7}$$

$$K^-(x, T, \boldsymbol{D}) = 0, \tag{8}$$

$$\mu_e(x, T, \boldsymbol{D}) = 1. \tag{9}$$

(iii) If $x \in T\boldsymbol{D}$ and T is odd in $\boldsymbol{D}$, then

$$K(x, T, \boldsymbol{D}) = K^-(x, T, \boldsymbol{D}) = 1,$$

$$K^+(x, T, \boldsymbol{D}) = 0, \quad \mu_e(x, T, \boldsymbol{D}) = -1.$$

(iv) If u is any point in $\boldsymbol{D}$, then u is an e.m.m.c. for $(Tu, T, \boldsymbol{D})$ (see **II.3.3**, definition 1).

(v) $E(T, \boldsymbol{D}) = E_i(T, \boldsymbol{D}) = E_i^p(T, \boldsymbol{D}) = \boldsymbol{D}$ (see **II.3.6**).

(vi) $i_e u, T) = +1$ or $i_e(u, T) = -1$ for $u \in \boldsymbol{D}$, according as T is even or odd in $\boldsymbol{D}$.

Proof. (5) follows directly from **II.3.2**, remark 2, and (6) is a direct consequence of (5) and **II.3.4**, theorem 4. Thus (i) is verified. The proofs for (ii) and (iii) are entirely similar, and so we discuss only (ii) explicitly. Assume that T is even in $\boldsymbol{D}$. Consider a point $x_0 \in T\boldsymbol{D}$. Since T is a homeomorphism, there exists a unique point $u_0 \in \boldsymbol{D}$ such that $Tu_0 = x_0$. Let D be a domain such that $\bar{D} \subset \boldsymbol{D}$. We assert that D is an indicator domain for $(x_0, T, \boldsymbol{D})$ if and only if $u_0 \in D$. Indeed, if $u_0 \in D$, then $x_0 = Tu_0 \in TD$, and hence $\mu(x_0, T, D) = +1$ by lemma 2. Thus D is an indicator domain for $(x_0, T, \boldsymbol{D})$. Conversely, if D is an indicator domain for $(x_0, T, \boldsymbol{D})$, then $\mu(x_0, T, D) \neq 0$ and hence $x_0 \in TD$ by **II.2.3**, theorem 2. Since T is a homeomorphism, it follows that $u_0 \in D$. From the preceding remarks we draw the following conclusions (it is assumed that T is even in $\boldsymbol{D}$).

(a) There exist no negative indicator domains for $(x_0, T, \boldsymbol{D})$ and hence (see **II.3.2**, remark 2)

$$K^-(x_0, T, \boldsymbol{D}) = 0. \tag{10}$$

(b) If D is a domain such that $\bar{D} \subset \boldsymbol{D}$ and $u_0 \in D$, then D is an indicator domain for $(x_0, T, \boldsymbol{D})$, and $\mu(x_0, T, D) = 1$.

(c) If $\mathfrak{S}(x_0, T, \boldsymbol{D})$ is any non-empty indicator system for $(x_0, T, \boldsymbol{D})$, then this system consists of a single indicator domain. Indeed, if D_1, D_2 were two different indicator domains of the system, then we should have $u_0 \in D_1, u_0 \in D_2$ by the remarks above, in contradiction with the requirement that the domains in an indicator system must be disjoint.

Let D be the unique indicator domain of the system $\mathfrak{S}(x_0, T, \boldsymbol{D})$. Then $\mu(x_0, T, D) = 1$, as noted above, and hence [see **II.3.2** (1)]

$$M[\mathfrak{S}(x_0, T, \boldsymbol{D})] = 1. \tag{11}$$

Since (11) holds for every non-empty indicator system $\mathfrak{S}(x_0, T, \boldsymbol{D})$, it follows [see **II.3.2** (10)] that

$$K(x_0, T, \boldsymbol{D}) = 1. \tag{12}$$

Note that (see **II.3.2**, theorem 1)

$$K(x_0, T, \boldsymbol{D}) = K^+(x_0, T, \boldsymbol{D}) + K^-(x_0, T, \boldsymbol{D}). \tag{13}$$

In view (10), (12), (13), the relations (7) and (8) follow. By (12) and **II.3.4**, theorem 4 we have

$$\mu_e(x_0, T, \boldsymbol{D}) = K^+(x_0, T, \boldsymbol{D}) - K^-(x_0, T, \boldsymbol{D}),$$

and (9) follows in view of (7) and (8). Thus (ii) is proved. Consider now any point $u \in \boldsymbol{D}$. Since T is a homeomorphism, it is clear that u is an m.m.c. for $(Tu, T, \boldsymbol{D})$, and in fact u is the only m.m.c. for $(Tu, T, \boldsymbol{D})$. Thus (iv) and (v) will be proved if we show that u is an e.m.m.c. for $(Tu, T, \boldsymbol{D})$. Assign any open set O such that $u \in O \subset \boldsymbol{D}$, and select any domain D such that $u \in D, \bar{D} \subset O$. By lemma 2 we have then

$$\mu(Tu, T, D) = +1 \quad \text{if } T \text{ is even in } \boldsymbol{D}, \tag{14}$$

$$\mu(Tu, T, D) = -1 \quad \text{if } T \text{ is odd in } \boldsymbol{D}. \tag{15}$$

Thus D is an indicator domain for $(Tu, T, \boldsymbol{D})$ such that $u \in D, \bar{D} \subset O$. Since O was any open set such that $u \in O \subset \boldsymbol{D}$, it follows that u is an e.m.m.c. for $(Tu, T, \boldsymbol{D})$. Finally, (14) and (15) show that $i_e(u, T) = +1$ or $i_e(u, T) = -1$ according as T is even or odd in $\boldsymbol{D}$ (see **II.3.4**, definition 1), and the proof is complete.

Theorem 1. Given a bounded homeomorphism T as in (1), the following holds.

(i) T is eBV in $\boldsymbol{D}$.

(ii) The essential generalized Jacobian $J_e(u, T)$ exists a.e. in $\boldsymbol{D}$ and is L-summable in $\boldsymbol{D}$.

(iii) $J_e(u, T) \geq 0$ a.e. in $\boldsymbol{D}$ if T is even in $\boldsymbol{D}$, and $J_e(u, T) \leq 0$ a.e. in $\boldsymbol{D}$ if T is odd in $\boldsymbol{D}$.

(iv) T is eAC in $\boldsymbol{D}$ if and only if it satisfies the condition (N) in $\boldsymbol{D}$ (see **IV.1.4**).

Proof. By lemma 3, $K(x, T, \boldsymbol{D})$ coincides with the characteristic function of the set $T\boldsymbol{D}$. Since T is bounded, clearly $\mathsf{L}T\boldsymbol{D} < \infty$, and thus $K(x, T, \boldsymbol{D})$ is L-summable in R^n. Hence T is eBV in $\boldsymbol{D}$. (ii) follows now directly from **IV.4.3**, theorem 1. By that same theorem, we have

$$J_e(u, T) = D_e^+(u, T) - D_e^-(u, T) \quad \text{a.e. in } \boldsymbol{D}. \tag{16}$$

Assume that T is even in $\boldsymbol{D}$. If D is any domain in $\boldsymbol{D}$, then

$$0 \leq K^-(x, T, D) \leq K^-(x, T, \boldsymbol{D}) = 0$$

in view of lemma 3. Thus $K^-(x, T, D) = 0$ for every domain $D \subset \boldsymbol{D}$ and for every point $x \in R^n$, and hence obviously [see **IV.4.3** (9), (11)]

$$D_e^-(u, T) = 0 \quad \text{a.e. in } \boldsymbol{D}. \tag{17}$$

Since $D_e^+(u, T)$ is non-negative, it follows from (16) and (17) that $J_e(u, T) \geq 0$ a.e. in $\boldsymbol{D}$ if T is even in $\boldsymbol{D}$. An entirely similar argument shows that $J_e(u, T) \leq 0$ a.e. in $\boldsymbol{D}$ if T is odd in $\boldsymbol{D}$. Finally, (iv) is a direct consequence of (i), part (v) of lemma 3, and **IV.4.2**, theorem 3.

Definition 2. A bounded homeomorphism T, given as in (1), is termed *measurable* in $\boldsymbol{D}$ if for every L-measurable set $S \subset \boldsymbol{D}$ the image set TS is L-measurable in R^n.

Lemma 4. A bounded homeomorphism T, given as in (1), is measurable in $\boldsymbol{D}$ if and only if it satisfies the condition (N) in $\boldsymbol{D}$ (see **IV.1.4**).

Proof. Suppose first that T satisfies the condition (N) in $\boldsymbol{D}$. By **IV.1.4**, corollary 2, for every L-measurable set $L \subset \boldsymbol{D}$ the image set TL is L-measurable, and thus T is measurable in $\boldsymbol{D}$. Suppose, conversely, that T is measurable in $\boldsymbol{D}$. Consider a set $S \subset \boldsymbol{D}$ such that $\mathsf{L}S = 0$. Deny that $\mathsf{L}TS = 0$. Observe that TS is L-measurable (since T is measurable). By **III.1.1**, lemma 1 there follows the existence of a set $S^* \subset TS$ such that S^* is not L-measurable. The set $T^{-1}S^*$ is then a subset of S, and hence $\mathsf{L}T^{-1}S^* = 0$. Thus $T^{-1}S^*$ is L-measurable, and yet its image $TT^{-1}S^* = S^*$ is not L-measurable. This contradicts the assumption that T is measurable in $\boldsymbol{D}$, and the proof is complete.

Remark 1. Definition 2 remains meaningful and lemma 4 remains valid for general continuous transformations in R^n. However, this fact is not needed for our purposes.

Theorem 2. Suppose that the bounded homeomorphism T, given as in (1), is measurable in $\boldsymbol{D}$. Let $H(x)$ be a finite-valued L-measurable function in R^n, and let D be any domain in $\boldsymbol{D}$. Then

$$\int_D H(Tu)\,|J_e(u, T)|\,d\mathsf{L} = \int_{TD} H(x)\,d\mathsf{L},$$

as soon as one of the two integrals involved exists, and

$$\int_D |J_e(u, T)|\,d\mathsf{L} = \mathsf{L}TD. \tag{18}$$

Proof. Observe that T is eAC in $\boldsymbol{D}$ by lemma 4 and part (iv) of theorem 1. Accordingly, by theorem 1 in **IV.4.4**, we have (in view of the remark at the end of **IV.4.4**)

$$\int_D H(Tu)\,|J_e(u, T)|\,d\mathsf{L} = \int H(x)\,K(x, T, D)\,d\mathsf{L},$$

as soon as one of the two integrals involved exists. Since the transformation $T|D$ is one-to-one, we can apply lemma 3 with $\boldsymbol{D}$ replaced by D. Hence, by that lemma, $K(x, T, D) = 0$ if $x \notin TD$ and $K(x, T, D) = 1$ if $x \in TD$. Thus

$$\int H(x)\,K(x, T, D)\,d\mathsf{L} = \int_{TD} H(x)\,d\mathsf{L},$$

as soon as one of the two integrals involved exists, and the first part of the theorem follows. On choosing $H(x) \equiv 1$, one obtains the formula (18).

Definition 3. A bounded homeomorphism T, given as in (1), is termed *bi-measurable* in $\boldsymbol{D}$ if it is measurable in $\boldsymbol{D}$ and its inverse [given as in (3)] is measurable in the image domain $T\boldsymbol{D}$.

Theorem 3. Assume that the bounded homeomorphism T, given as in (1), is bi-measurable in $\boldsymbol{D}$. Let $h(x)$ be a finite-valued L-measurable function defined a.e. in the image domain $T\boldsymbol{D}$, and let D be any domain in $\boldsymbol{D}$. Then the following holds.

(i) $h(Tu)$ is defined a.e. in $\boldsymbol{D}$.

(ii) $\int_D h(Tu)\,|J_e(u, T)|\,d\mathsf{L} = \int_{TD} h(x)\,d\mathsf{L}$, as soon as one of the two integrals involved exists.

Proof. Let S^* be the set of those points $x \in T\boldsymbol{D}$ where $h(x)$ is not defined. Then $\mathsf{L}S^* = 0$ by assumption, and hence $\mathsf{L}T^{-1}S^* = 0$, since T^{-1} is measurable in $T\boldsymbol{D}$ (see lemma 4). As $h(Tu)$ is defined for $u \in \boldsymbol{D} - T^{-1}S^*$, (i) follows. Introduce now an auxiliary function $H(x)$ in R^n as follows: $H(x) = 0$ if $x \notin T\boldsymbol{D} - S^*$, and $H(x) = h(x)$ if $x \in T\boldsymbol{D} - S^*$. Then $H(x)$ is clearly finite-valued and L-measurable in R^n, and hence, by theorem 2,

$$\int_D H(Tu)\,|J_e(u, T)|\,d\mathsf{L} = \int_{TD} H(x)\,d\mathsf{L},$$

as soon as one of the two integrals involved exists. Since obviously $h(x) = H(x)$ a.e. in TD and $h(Tu) = H(Tu)$ a.e. in D, the theorem follows.

Theorem 4. Given a bounded homeomorphism T as in (1), let

$$T^*: T\boldsymbol{D} \to R^n \tag{19}$$

be a bounded continuous transformation from the image domain $T\boldsymbol{D}$ into R^n. Assume that T is bi-measurable in $\boldsymbol{D}$ and T^* is eAC in $T\boldsymbol{D}$. Then the following holds.

(i) T^*T is eAC in $\boldsymbol{D}$.

(ii) $J_e(u, T^*T) = J_e(u, T)\, J_e(Tu, T^*)$ a.e. in $\boldsymbol{D}$.

Proof of (i). Clearly, the transformations T^*T and T^* are topologically similar in the sense of **II.3.8**. Accordingly, by **II.3.8**, lemma 4 we have the relations

$$K(x, T^*T, \boldsymbol{D}) = K(x, T^*, T\,\boldsymbol{D}), \quad x \in R^n, \tag{20}$$

$$T\,E(T^*T, \boldsymbol{D}) = E(T^*, T\,\boldsymbol{D}). \tag{21}$$

Now since T^* is eAC in $T\boldsymbol{D}$ and hence $K(x, T^*, T\boldsymbol{D})$ is L-summable in R^n, (20) shows that $K(x, T^*T, \boldsymbol{D})$ is L-summable in R^n. Thus T^*T is eBV in $\boldsymbol{D}$, and hence (in view of **IV.4.2**, theorem 3) (i) will be proved if we show that T^*T satisfies the condition (N) on the essential set $E(T^*T, \boldsymbol{D})$. So take any set S such that

$$S \subset E(T^*T, \boldsymbol{D}), \quad \mathsf{L}\, S = 0. \tag{22}$$

In view of (21) we have then

$$T\,S \subset E(T^*, T\,\boldsymbol{D}). \tag{23}$$

Since T is measurable in $\boldsymbol{D}$, the second relation in (22) implies that

$$\mathsf{L}\, T\,S = 0. \tag{24}$$

By assumption, T^* is eAC in $T\boldsymbol{D}$, and hence (see **IV.4.2**, theorem 3) T^* satisfies the condition (N) on the essential set $E(T^*, T\,\boldsymbol{D})$. Accordingly, (23) and (24) imply that

$$\mathsf{L}\, T^*\,T\,S = 0. \tag{25}$$

Thus (22) implies (25). In other words, T^*T satisfies the condition (N) on the essential set $E(T^*T, \boldsymbol{D})$, and (i) is proved.

Proof of (ii). There arise two cases, according as T is even or odd in $\boldsymbol{D}$. Since the argument is the same in either case, we discuss only the case when T is odd in $\boldsymbol{D}$. By theorem 1 we have then

$$J_e(u, T) \leqq 0 \quad \text{a.e. in } \boldsymbol{D}. \tag{26}$$

Furthermore, if D is any domain such that $\overline{D} \subset \boldsymbol{D}$, then (see definition 1)

$$\iota(T, D) = -1. \tag{27}$$

Let x be any point in R^n. We proceed to verify the relations

$$K^+(x, T^*T, \boldsymbol{D}) = K^-(x, T^*, T\boldsymbol{D}), \tag{28}$$

$$K^-(x, T^*T, \boldsymbol{D}) = K^+(x, T^*, T\boldsymbol{D}). \tag{29}$$

Let D be an indicator domain for $(x, T^*T, \boldsymbol{D})$. Then

$$\overline{D} \subset \boldsymbol{D}, \quad x \notin T^*T \, fr \, D, \quad \mu(x, T^*T, D) \neq 0. \tag{29a}$$

By lemma 1 and lemma 2 in **II.3.8** it follows that [see (27)]

$$\overline{TD} \subset T\boldsymbol{D}, \quad x \notin T^* \, fr \, TD, \quad \mu(x, T^*, TD) = -\mu(x, T^*T, D). \tag{30}$$

From (29a) and (30) we conclude that if D is a positive indicator domain for $(x, T^*T, \boldsymbol{D})$, then TD is a negative indicator domain for $(x, T^*, T\boldsymbol{D})$. Accordingly, if $D_1, \ldots, D_m$ is a positive indicator system for $(x, T^*T, \boldsymbol{D})$, then $TD_1, \ldots, TD_m$ is a negative indicator system for $(x, T^*, T\boldsymbol{D})$ (see **II.3.2**), and

$$\sum_{j=1}^{m} \mu(x, T^*T, D_j) = -\sum_{j=1}^{m} \mu(x, T^*, TD_j).$$

By the definition of $K^-(x, T^*, T\boldsymbol{D})$ (see **II.3.2**), the summation on the right does not exceed $K^-(x, T^*, T\boldsymbol{D})$. Hence

$$\sum_{j=1}^{m} \mu(x, T^*T, D_j) \leq K^-(x, T^*, T\boldsymbol{D}).$$

Since $D_1, \ldots, D_m$ was an arbitrary positive indicator system for $(x, T^*T, \boldsymbol{D})$, there follows the inequality

$$K^+(x, T^*T, \boldsymbol{D}) \leq K^-(x, T^*, T\boldsymbol{D}). \tag{31}$$

Starting with negative indicator systems for $(x, T^*T, \boldsymbol{D})$, a similar argument yields the inequality

$$K^-(x, T^*T, \boldsymbol{D}) \leq K^+(x, T^*, T\boldsymbol{D}). \tag{32}$$

Note that in deriving (31) and (32) we used only the fact that T^*T and T^* are topologically similar. Since this relationship is symmetric, an entirely similar argument shows that

$$K^+(x, T^*, T\boldsymbol{D}) \leq K^-(x, T^*T, \boldsymbol{D}),$$

$$K^-(x, T^*, T\boldsymbol{D}) \leq K^+(x, T^*T, \boldsymbol{D}),$$

and (28) and (29) follow in view of (31) and (32). Observe now that if D is any domain in $\boldsymbol{D}$, then $T|D$ is a homeomorphism from D onto

TD and $T^*|TD$ is defined and continuous. Accordingly, in the argument leading to (28) and (29) we can replace $\boldsymbol{D}$ by D, and hence

$$K^+(x, T^*T, D) = K^-(x, T^*, T\,D), \tag{33}$$

$$K^-(x, T^*T, D) = K^+(x, T^*, T\,D), \tag{34}$$

for every domain $D \subset \boldsymbol{D}$. Using the remark at the end of **IV.4.4**, let us apply now **IV.4.4**, theorem 4 to T^*T in D. In view of **IV.4.4** (11), applied to T^*T in D, there follows the formula

$$\int_D J_e(u, T^*T)\, d\mathsf{L} = \int K^+(x, T^*T, D)\, d\mathsf{L} - \int K^-(x, T^*T, D)\, d\mathsf{L}. \tag{35}$$

Applying the same references to T^* in TD, we obtain

$$\int_{TD} J_e(x, T^*)\, d\mathsf{L} = \int K^+(x, T^*, T\,D)\, d\mathsf{L} - \int K^-(x, T^*, T\,D)\, d\mathsf{L}. \tag{36}$$

The relations (33) to (36) yield

$$\int_D J_e(u, T^*T)\, d\mathsf{L} = -\int_{TD} J_e(x, T^*)\, d\mathsf{L}. \tag{37}$$

Application of theorem 3, with $h(x) = J_e(x, T^*)$, yields

$$\int_D J_e(T\,u, T^*)\, |J_e(u, T)|\, d\mathsf{L} = \int_{TD} J_e(x, T^*)\, d\mathsf{L}.$$

In view of (26), this last formula is equivalent to

$$\int_D J_e(T\,u, T^*)\, J_e(u, T)\, d\mathsf{L} = -\int_{TD} J_e(x, T^*)\, d\mathsf{L}. \tag{38}$$

Combining (37) and (38) we obtain

$$\int_D J_e(u, T^*T)\, d\mathsf{L} = \int_D J_e(T\,u, T^*)\, J_e(u, T)\, d\mathsf{L}. \tag{39}$$

Since D was any domain in $\boldsymbol{D}$, (39) implies (ii) by **III.1.1**, lemma 44, and the theorem is proved.

Theorem 5. Given a bounded homeomorphism T as in (1), let $T^*: T\boldsymbol{D} \to R^n$ be a bounded homeomorphism defined in the image domain $T\boldsymbol{D}$. Assume that T is bi-measurable in $\boldsymbol{D}$ and T^* is bi-measurable in $T\boldsymbol{D}$. Then the following holds.

(i) T^*T is a bounded bi-measurable homeomorphism.

(ii) $J_e(u, T^*T) = J_e(u, T)\, J_e(T\,u, T^*)$ a.e. in $\boldsymbol{D}$.

Proof. (i) is a direct consequence of the definition of the terms involved. Noting that T^* is eAC in $T\boldsymbol{D}$ by theorem 1 and lemma 4, (ii) appears as a direct consequence of theorem 4.

Remark 2. Assume that T, given as in (1), is the *identity mapping* in $\boldsymbol{D}$ (that is, $Tu=u$ for every point $u\in\boldsymbol{D}$). Clearly, T is then a bounded bi-measurable homeomorphism in $\boldsymbol{D}$. Take any domain D such that $\bar{D}\subset\boldsymbol{D}$. Then $T|\bar{D}$ is the identity mapping in $\bar{D}$, and hence (see **II.2.4**, theorem 2) $\iota(T, D)=1$. Thus T is even in $\boldsymbol{D}$ (see definition 1). Consider now any domain $D\subset\boldsymbol{D}$. Then $T|D$ is the identity mapping in D. We can now apply lemma 3 with $\boldsymbol{D}$ replaced by D. Noting that $T|D$ is even in D and $TD=D$, we obtain from lemma 3 the following information. $K^+(x, T, D)=1$ or 0 according as $x\in D$ or $x\notin D$, and $K^-(x, T, D)\equiv 0$. Accordingly [see **IV.4.3** (5), (9)] $\mathfrak{K}^+(D)=\mathsf{L}D$ and $\mathfrak{K}^-(D)=0$, and hence $\mathfrak{A}_e(D)=\mathfrak{K}^+(D)-\mathfrak{K}^-(D)=\mathsf{L}D$ [see **IV.4.3** (18)]. Since this holds for every domain $D\subset\boldsymbol{D}$, it follows that $J_e(u, T)=D(u, \mathfrak{A}_e)$ exists and $J_e(u, T)=1$ at every point $u\in\boldsymbol{D}$ [see **IV.4.3** (19)].

Theorem 6. Given a bounded homeomorphism T and its inverse as in (1), (2), (3), assume that T is bi-measurable in $\boldsymbol{D}$. Then

$$J_e(u, T)\, J_e(Tu, T^{-1}) = 1 \quad \text{a.e. in } \boldsymbol{D}. \tag{40}$$

Proof. In view of definition 3, T^{-1} is clearly bi-measurable in the image domain $T\boldsymbol{D}$. Thus we can apply theorem 5 with $T^*=T^{-1}$, obtaining

$$J_e(u, T)\, J_e(Tu, T^{-1}) = J_e(u, T^{-1}T) \quad \text{a.e. in } \boldsymbol{D}.$$

Since $T^{-1}T$ is the identity mapping, $J_e(u, T^{-1}T)\equiv 1$ in $\boldsymbol{D}$ by remark 2, and (40) follows.

The theorems 5 and 6 correspond to familiar theorems in classical Analysis, where of course one assumes that the transformations involved are continuously differentiable and one uses the ordinary Jacobian. An important special case of theorem 4 arises if the homeomorphism T stands for a *change of the coordinate system.* Theorem 4 suggests the question whether a similar *product theorem* holds if one drops the assumption that T is a homeomorphism. While several remarkable special results are available in the literature, ultimate clarification of this point seems to be remote as yet.

§ IV.5. Bounded variation and absolute continuity in the BANACH sense.

IV.5.1. Bounded variation. Let there be given a bounded continuous transformation $T: \boldsymbol{D}\to R^n$, where $\boldsymbol{D}$ is a bounded domain in R^n. If I is an n-interval such that $\operatorname{int} I\subset\boldsymbol{D}$, then the set $T(\operatorname{int} I)$ is bounded and L-measurable (see **IV.1.1**, theorem 4). Accordingly, we can define a function $G(\operatorname{int} I)$ of open intervals in $\boldsymbol{D}$ by the formula

$$G(\operatorname{int} I) = \mathsf{L}\, T(\operatorname{int} I), \quad \operatorname{int} I\subset\boldsymbol{D}. \tag{1}$$

Since G depends upon T, a more descriptive notation like $G(\text{int } I, T)$ may be used if T is not clearly identified by the context. An equivalent formula for G is obtained as follows. If F is a set in R^n, then let us denote again by $c(x, F)$ the characteristic function of F. That is,

$$c(x, F) = \begin{Bmatrix} 1 & \text{if} & x \in F, \\ 0 & \text{if} & x \in \complement F. \end{Bmatrix} \tag{2}$$

Then (1) may be re-written in the form

$$G(\text{int } I) = \int c(x, T \text{ int } I)\, d\mathsf{L}. \tag{3}$$

Since T is bounded by assumption, we have $\mathsf{L} T \boldsymbol{D} < \infty$. Thus clearly

$$G(\text{int } I) \leqq \mathsf{L} T \boldsymbol{D} < \infty, \quad \text{int } I \subset \boldsymbol{D}. \tag{4}$$

From (4) we see that G is bounded in $\boldsymbol{D}$.

Definition 1. T is said to be of bounded variation in $\boldsymbol{D}$ in the BANACH sense (briefly, $BV\mathsf{B}$ in $\boldsymbol{D}$) if $G(\text{int } I)$ is of bounded variation in $\boldsymbol{D}$ (see **III.2.1**, definition 5).

We shall show presently that this concept of bounded variation is equivalent to that of bounded variation with respect to the special base-function $b(u) \equiv 1$. In anticipation of this result, we denote the base-function $b(u) \equiv 1$ by $b_{\mathsf{B}}(u)$. Thus

$$b_{\mathsf{B}}(u) \equiv 1, \quad u \in \boldsymbol{D}. \tag{5}$$

The corresponding summatory function W [see **IV.2.1** (2)] will be denoted by W_{B}. Clearly, if S is any set in $\boldsymbol{D}$, then

$$W_{\mathsf{B}}(x, T, S) = N(x, T, S), \tag{6}$$

where N is the *crude multiplicity function* (see **IV.1.2**). By **IV.2.2**, definition 1, we have therefore the following statement.

Lemma 1. T is $BV b_{\mathsf{B}}$ in $\boldsymbol{D}$ if and only if the crude multiplicity function $N(x, T, \boldsymbol{D})$ is L-summable in R^n.

Assuming that T is $BV b_{\mathsf{B}}$ in $\boldsymbol{D}$, we denote by ν_{B} the measure ν induced by T and b_{B} (see **IV.2.2**, definition 2). If B is any BOREL set in $\boldsymbol{D}$, in view of (6) we have then

$$\nu_{\mathsf{B}}(B) = \int N(x, T, B)\, d\mathsf{L}. \tag{7}$$

To simplify notations, we put (see **IV.2.2**, definition 2)

$$D_{\mathsf{B}}(u, T) = D(u, \nu_{\mathsf{B}}). \tag{8}$$

If O is any open set in $\boldsymbol{D}$, we put [see (1) and **III.3.1**, definition 4]

$$V_{\mathsf{B}}(T, O) = V(G, O). \tag{9}$$

Definition 2. $V_{\mathsf{B}}(T, O)$ is termed the *total variation* of T, in the BANACH sense, on the open set $O \subset \boldsymbol{D}$ [observe that $V_{\mathsf{B}}(T, O)$ may be equal to $+\infty$].

Lemma 2. Let O be an open set in $\boldsymbol{D}$. Then $V_{\mathsf{B}}(T, O)$ is finite if and only if the crude multiplicity function $N(x, T, O)$ is L-summable in R^n. Furthermore

$$V_{\mathsf{B}}(T, O) = \int N(x, T, O)\, d\mathsf{L} \quad \text{if} \quad V_{\mathsf{B}}(T, O) < \infty. \tag{10}$$

Proof. Take an oriented n-cube Q such that $\boldsymbol{D} \subset Q$. Using the subdivision D_{p_j} of Q introduced in **I.2.3**, definition 3, for each positive integer j let $\mathfrak{Q}_j^*$ denote the class of those cubes Q of D_{p_j} for which $\operatorname{int} Q \subset O$. Then $\mathfrak{Q}_j^*$ is clearly a system $\sigma(O)$ in the sense of **III.2.1**, definition 1. In terms of the terminology in **III.2.1** we have then [in view of (9)]

$$\sum [G, \mathfrak{Q}_j^*] \leq V_{\mathsf{B}}(T, O). \tag{11}$$

Assume first that

$$V_{\mathsf{B}}(T, O) < \infty. \tag{12}$$

Introduce, for each positive integer j, the auxiliary function [see (2)]

$$f_j(x) = \sum c(x, T \operatorname{int} Q), \quad Q \in \mathfrak{Q}_j^*, \quad x \in R^n. \tag{13}$$

By **IV.1.2** (10), (11) we have then the relations

$$0 \leq f_j(x) \leq N(x, T, O), \tag{14}$$

$$\lim_{j \to \infty} f_j(x) = N(x, T, O). \tag{15}$$

Integration of (13) yields, in view of (3), (11), and (12),

$$\int f_j(x)\, d\mathsf{L} \leq V_{\mathsf{B}}(T, O) < \infty. \tag{16}$$

By the lemma of FATOU (see **III.1.1**, lemma 30) we conclude from (14), (15), (16) that $N(x, T, O)$ is L-summable and

$$\int N(x, T, O)\, d\mathsf{L} \leq V_{\mathsf{B}}(T, O). \tag{17}$$

Suppose, conversely, that $N(x, T, O)$ is L-summable in R^n. Take any system $\sigma(O)$ (see **III.2.1**, definition 1). Clearly

$$N(x, T, O) \geq \sum c(x, T \operatorname{int} I), \quad I \in \sigma(O).$$

Integration yields, in view of (3), the inequality

$$\sum [G, \sigma(O)] \leq \int N(x, T, O)\, d\mathsf{L} < \infty.$$

Since $\sigma(O)$ was arbitrary, it follows that

$$V_{\mathsf{B}}(T, O) \leq \int N(x, T, O)\, d\mathsf{L} < \infty. \tag{18}$$

Thus $V_{\mathsf{B}}(T, O)$ is finite. Finally, (10) follows by (17) and (18).

Theorem 1. T is $BV\mathsf{B}$ in $\boldsymbol{D}$ if and only if the crude multiplicity function $N(x, T, \boldsymbol{D})$ is L-summable in R^n.

Proof. In view of **III.2.1**, lemma 1, T is $BV\mathsf{B}$ in $\boldsymbol{D}$ if and only if $V_{\mathsf{B}}(T, \boldsymbol{D}) < \infty$. Thus the theorem follows from lemma 2 on choosing there $O = \boldsymbol{D}$.

Theorem 2. T is $BV\mathsf{B}$ in $\boldsymbol{D}$ if and only if it is BVb_{B} in $\boldsymbol{D}$.

This is a formal consequence of lemma 1 and theorem 1.

Theorem 3. Assume that T is $BV\mathsf{B}$ in $\boldsymbol{D}$. Then the derivative $D_{\mathsf{B}}(u, T)$ [see (8)] exists a.e. in $\boldsymbol{D}$ and is L-summable there. Furthermore, if O is any open set in $\boldsymbol{D}$, then

$$\int_O D_{\mathsf{B}}(u, T)\, d\mathsf{L} \leq \int N(x, T, O)\, d\mathsf{L} < \infty. \tag{19}$$

Proof. Since T is BVb_{B} in $\boldsymbol{D}$ by theorem 2, the present theorem is a direct consequence of **III.2.4**, lemma 1, in view of (8) and (7).

Assume that T is $BV\mathsf{B}$ in $\boldsymbol{D}$. A remarkable feature of the BANACH theory is the fact that the derivative $D_{\mathsf{B}}(u, T)$ can be replaced by the derivative $D(u, G)$ of G [see (1)]. Let us note that G is generally not sub-additive, and hence theorem 5 in **III.2.3** cannot be applied directly to G. The following lemma will enable us to apply that existence theorem in an indirect manner.

Lemma 3. If $S_1, \ldots, S_m$ are pair-wise disjoint subsets of a set $S \subset \boldsymbol{D}$, then [see (2)]

$$\sum_{k=1}^{m} [N(x, T, S_k) - c(x, T S_k)] \leq N(x, T, S) - c(x, T S). \tag{20}$$

Proof. (20) is obvious if $N(x, T, S) = \infty$. So assume that $N(x, T, S) < \infty$. It is convenient to consider two cases.

Case 1. $c(x, T S_k) = 0$, $k = 1, \ldots, m$. Then $x \notin T S_k$, and hence $N(x, T S_k) = 0$, $k = 1, \ldots, m$. Thus the summation on the left in (20) has the value zero, while clearly

$$N(x, T, S) - c(x, T S) \geq 0.$$

Thus (20) holds in this case.

Case 2. $c(x, T S_k) = 1$ for some k. Then

$$\sum_{k=1}^{m} [N(x, T, S_k) - c(x, T S_k)] \leq \left[\sum_{k=1}^{m} N(x, T, S_k)\right] - 1. \tag{21}$$

As $S_1, \ldots, S_m$ are pair-wise disjoint subsets of S, clearly

$$\sum_{k=1}^{m} N(x, T, S_k) \leq N(x, T, S). \tag{22}$$

As $c(x, T S) \leq 1$, we have

$$N(x, T, S) - 1 \leq N(x, T, S) - c(x, T S), \tag{23}$$

and (20) follows now by (21), (22), (23).

Theorem 4. Assume that T is $BV\mathsf{B}$ in $\boldsymbol{D}$. Then the function G of open intervals [see (1)] possesses a derivative $D(u, G)$ a.e. in $\boldsymbol{D}$, and [see (8)]

$$D(u, G) = D_{\mathsf{B}}(u, T) \quad \text{a.e. in } \boldsymbol{D}. \tag{24}$$

Proof. Note first that $D_{\mathsf{B}}(u, T)$ exists a.e. in $\boldsymbol{D}$ by theorem 3. Introduce now an auxiliary function φ of open intervals in $\boldsymbol{D}$ by the formula

$$\varphi(\text{int } I) = \nu_{\mathsf{B}}(\text{int } I) - G(\text{int } I), \quad \text{int } I \subset \boldsymbol{D}. \tag{25}$$

In view of (3) and (7) we have explicitly

$$\varphi(\text{int } I) = \int [N(x, T, \text{int } I) - c(x, T \text{ int } I)]\, d\mathsf{L}. \tag{26}$$

Thus clearly φ is non-negative. Furthermore, from the representation (26) we infer immediately, by means of lemma 3, that φ is sub-additive in the sense of **III.2.3**, definition 3. By **III.2.3**, theorem 5 we conclude that the derivative $D(u, \varphi)$ of φ exists a.e. in $\boldsymbol{D}$. Since the derivative of ν_{B} also exists a.e. in $\boldsymbol{D}$, (25) yields the existence of the derivative $D(u, G)$ a.e. in $\boldsymbol{D}$, as well as the relation [see (8)]

$$D(u, \varphi) = D_{\mathsf{B}}(u, T) - D(u, G) \quad \text{a.e. in } \boldsymbol{D}.$$

Thus the proof will be complete if we show that

$$D(u, \varphi) = 0 \quad \text{a.e. in } \boldsymbol{D}. \tag{27}$$

For this purpose, we shall use the lemma in **III.2.3**. Taking $O = \boldsymbol{D}$ in that lemma, it follows that (27) is established if we prove that

$$\lim_{j \to \infty} \sum_{Q_j} \varphi(\text{int } Q_j) = 0, \tag{28}$$

where Q_j is used in the same sense as in the lemma in **III.2.3**. For each positive integer j we introduce an auxiliary function $g_j(x)$, $x \in R^n$, by the formula

$$g_j(x) = \sum_{Q_j} [N(x, T, \operatorname{int} Q_j) - c(x, T \operatorname{int} Q_j)]. \tag{29}$$

By **IV.1.2** (8), (9), (10), (11) we have then

$$0 \leq g_j(x) \leq N(x, T, \boldsymbol{D}), \tag{30}$$

$$\lim_{j \to \infty} g_j(x) = 0 \quad \text{if} \quad N(x, T, \boldsymbol{D}) < \infty. \tag{31}$$

Now since T is $BV\mathsf{B}$ in $\boldsymbol{D}$, by theorem 1 we know that $N(x, T, \boldsymbol{D})$ is L-summable in R^n. Hence $N(x, T, \boldsymbol{D}) < \infty$ a.e. in R^n (see **III.1.1**, lemma 31). Thus (31) holds a.e. in R^n, and in view of (30) we conclude from **III.1.1**, lemma 38 that

$$\lim_{j \to \infty} \int g_j(x)\, d\mathsf{L} = 0. \tag{32}$$

By (26) and (29) the relation (32) is equivalent to (28), and the proof is complete.

In the sequel, it will be a matter of importance to have information about points $u \in \boldsymbol{D}$ where $D_{\mathsf{B}}(u, T)$ vanishes. Such information is contained in the following statements.

Theorem 5. Assume that T is $BV\mathsf{B}$ in $\boldsymbol{D}$, and let S be a set in $\boldsymbol{D}$ such that

$$\mathsf{L}\, T\, S = 0. \tag{33}$$

Then

$$D_{\mathsf{B}}(u, T) = 0 \quad \text{a.e. on } S. \tag{34}$$

Proof. Note that T is BVb_{B} in $\boldsymbol{D}$ by theorem 2. Accordingly, by **IV.2.2**, lemma 12 the relation (33) implies that $D(u, \nu_{\mathsf{B}}) = 0$ a.e. on S, and thus (34) follows in view of (8).

Theorem 6. Assume that T is $BV\mathsf{B}$ in $\boldsymbol{D}$, and consider the set $\mathbf{I}(T, \boldsymbol{D})$ of those points $u \in \boldsymbol{D}$ which possess a neighborhood clear of relatives (see **II.3.7**, definitions 1 and 2). Then

$$D_{\mathsf{B}}(u, T) = 0 \quad \text{a.e. on} \quad \boldsymbol{D} - \mathbf{I}(T, \boldsymbol{D}). \tag{35}$$

Proof. Denote by X_∞ the set of those points $x \in R^n$ where $N(x, T, \boldsymbol{D}) = \infty$. Since $N(x, T, \boldsymbol{D})$ is L-summable in R^n by theorem 1, from **III.1.1**, lemma 31 we obtain the relation

$$\mathsf{L} X_\infty = 0. \tag{36}$$

If $u \in \boldsymbol{D} - \mathbf{I}(T, \boldsymbol{D})$, then by definition every open set O containing u must contain a relative of u (distinct from u). There follows the existence of an infinite sequence of distinct points $u_k \in \boldsymbol{D}$, $k=1, 2, \ldots$, such that $u_k \neq u$ and $T u_k = T u$. Thus $N(Tu, T, \boldsymbol{D}) = \infty$ and hence $Tu \in X_\infty$. Since u was an arbitrary point of $\boldsymbol{D} - \mathbf{I}(T, \boldsymbol{D})$, we conclude that

$$T[\boldsymbol{D} - \mathbf{I}(T, \boldsymbol{D})] \subset X_\infty, \tag{37}$$

and (35) follows from theorem 5 in view of (37) and (36).

IV.5.2. Absolute continuity. The terminology of **IV.5.1** will be used in this section.

Definition. T is said to be absolutely continuous in $\boldsymbol{D}$ in the BANACH sense (briefly, $AC\mathsf{B}$ in $\boldsymbol{D}$) if the function $G(\operatorname{int} I)$ is absolutely continuous in $\boldsymbol{D}$ [see **IV.5.1** (1) and **III.2.2**).

Theorem 1. If T is $AC\mathsf{B}$ in $\boldsymbol{D}$, then it is also $BV\mathsf{B}$ in $\boldsymbol{D}$.

Proof. Noting that G is bounded [see **IV.5.1** (4)], we see that the assertion follows directly from **III.2.2**, lemma 3.

Theorem 2. T is $AC\mathsf{B}$ in $\boldsymbol{D}$ if and only if it is ACb_{B} in $\boldsymbol{D}$.

Proof. Assume first that T is $AC\mathsf{B}$ in $\boldsymbol{D}$. By theorem 1 and **IV.5.1**, theorem 2, T is then also BVb_{B} in $\boldsymbol{D}$. There remains to show (see **IV.2.3**) that the measure ν_{B} induced by T and b_{B} is absolutely continuous in $\boldsymbol{D}$ with respect to L-measure. So let us consider any BOREL set $B \subset \boldsymbol{D}$ such that

$$\mathsf{L}\, B = 0. \tag{1}$$

We have to verify that

$$\nu_{\mathsf{B}}(B) = 0. \tag{2}$$

Assign $\varepsilon > 0$. By **III.2.2**, lemma 2 there exists an $\eta(\varepsilon) > 0$ such that if O is any open set satisfying the conditions

$$O \subset \boldsymbol{D}, \quad \mathsf{L}\, O < \eta(\varepsilon) \tag{3}$$

then [see **IV.5.1** (9)]

$$V_{\mathsf{B}}(T, O) < \varepsilon. \tag{4}$$

On the other hand, (1) implies (see **III.1.1**, lemma 3) the existence of an open set O_0 such that

$$B \subset O_0 \subset \boldsymbol{D}, \quad \mathsf{L}\, O_0 < \eta(\varepsilon). \tag{5}$$

From (3), (4), (5) it follows that

$$V_{\mathsf{B}}(T, O_0) < \varepsilon. \tag{6}$$

In view of **IV.5.1** (10) the relation (6) implies that

$$\int N(x, T, O_0)\, d\mathsf{L} < \varepsilon,$$

and by (5) and **IV.5.1** (7) we conclude that

$$\nu_{\mathsf{B}}(B) \leqq \nu_{\mathsf{B}}(O_0) = \int N(x, T, O_0)\, d\mathsf{L} < \varepsilon.$$

Since ε was arbitrary, (2) follows. Assume, conversely, that T is $AC b_{\mathsf{B}}$ in $\boldsymbol{D}$. Then, by definition (see **IV.2.3**), the measure ν_{B} is available and absolutely continuous in $\boldsymbol{D}$ with respect to L-measure. Hence, if $\varepsilon > 0$ is assigned, then there exists an $\eta(\varepsilon) > 0$ such that

$$\nu_{\mathsf{B}}(B) < \varepsilon \quad \text{if} \quad \mathsf{L}\, B < \eta(\varepsilon), \tag{7}$$

where B is a BOREL set in $\boldsymbol{D}$. Take now (see **III.2.1** for notations) any system $\sigma(\boldsymbol{D})$ such that

$$\mathsf{L}[\sigma(\boldsymbol{D})] < \eta(\varepsilon). \tag{8}$$

On setting

$$B_0 = \bigcup \operatorname{int} I, \quad I \in \sigma(\boldsymbol{D}), \tag{9}$$

clearly (8) implies that $\mathsf{L}B_0 < \eta(\varepsilon)$. From (7) it follows then that $\nu_{\mathsf{B}}(B_0) < \varepsilon$. Hence [see (9) and **IV.5.1** (3), (7)]

$$\sum [G, \sigma(\boldsymbol{D})] = \sum_{I \in \sigma(\boldsymbol{D})} \int c(x, T \operatorname{int} I)\, d\mathsf{L} \leqq \int N(x, T, B_0)\, d\mathsf{L} = \nu_{\mathsf{B}}(B_0) < \varepsilon.$$

In summary, (8) implies that

$$\sum [G, \sigma(\boldsymbol{D})] < \varepsilon.$$

Thus $G(\operatorname{int} I)$ is absolutely continuous in $\boldsymbol{D}$ (see **III.2.2**), and hence T is $AC\mathsf{B}$ in $\boldsymbol{D}$ by definition.

From the preceding theorem and from theorem 2 in **IV.5.1** it appears that the BANACH theory of bounded variation and absolute continuity corresponds to the special case of the general theory in § **IV.2** when one selects the base-function $b_{\mathsf{B}}(u) \equiv 1$. This relationship will be exploited in proving the following statements.

Theorem 3. Assume that T is $BV\mathsf{B}$ in $\boldsymbol{D}$. Then T is $AC\mathsf{B}$ in $\boldsymbol{D}$ if and only if it satisfies the condition (N) in $\boldsymbol{D}$ (see **IV.1.4**).

Proof. In view of theorem 2 and **IV.5.1**, theorem 2, the terms $BV\mathsf{B}$ and $AC\mathsf{B}$ can be replaced by the terms $BV b_{\mathsf{B}}$ and $AC b_{\mathsf{B}}$ respectively. Assuming that T is $BV b_{\mathsf{B}}$ in $\boldsymbol{D}$, T is $AC b_{\mathsf{B}}$ in $\boldsymbol{D}$ if and only if it satisfies the condition (N) on the *reduced base-set* where $0 < b_{\mathsf{B}}(u) < \infty$ (see **IV.2.3**, lemma 2). Since $b_{\mathsf{B}}(u) \equiv 1$ in $\boldsymbol{D}$, clearly the reduced base-set coincides with $\boldsymbol{D}$, and the theorem follows.

Theorem 4. Assume that T is $AC\mathsf{B}$ in $\boldsymbol{D}$. If $H(x)$ is an L-measurable, finite-valued function in R^n and L is any L-measurable set in $\boldsymbol{D}$, then

$$\int_L H(T u)\, D_{\mathsf{B}}(u, T)\, d\mathsf{L} = \int H(x)\, N(x, T, L)\, d\mathsf{L},$$

as soon as one of the two integrals involved exists.

Proof. Note that T is ACb_{B} in $\boldsymbol{D}$ by theorem 2, and that $D_{\mathsf{B}}(u, T)$ and $N(x, T, L)$ agree with the derivative $D(u, v_{\mathsf{B}})$ and the summatory function W corresponding to b_{B} respectively [see **IV.5.1** (8), (6)]. Thus the present theorem is a direct consequence of theorem 3 in **IV.2.4**.

Theorem 5. Assume that T is $AC\mathsf{B}$ in $\boldsymbol{D}$, and let L be any L-measurable subset of $\boldsymbol{D}$. Then $N(x, T, L)$ is L-summable in R^n, and

$$\int N(x, T, L)\, d\mathsf{L} = \int_L D_{\mathsf{B}}(u, T)\, d\mathsf{L}.$$

Proof. Note that T is $BV\mathsf{B}$ in $\boldsymbol{D}$ by theorem 1, and hence $N(x, T, \boldsymbol{D})$ is L-summable in R^n by **IV.5.1**, theorem 1. Since

$$N(x, T, L) \leqq N(x, T, \boldsymbol{D}),$$

it follows that $N(x, T, L)$ is also L-summable in R^n. The present theorem appears thus a consequence of theorem 4, corresponding to the special case $H(x) \equiv 1$.

Theorem 6. Assume that T is $BV\mathsf{B}$ in $\boldsymbol{D}$. Then

$$\int_{\boldsymbol{D}} D_{\mathsf{B}}(u, T)\, d\mathsf{L} \leqq \int N(x, T, \boldsymbol{D})\, d\mathsf{L} < \infty, \tag{10}$$

and the sign of equality holds if and only if T is $AC\mathsf{B}$ in $\boldsymbol{D}$.

Proof. Note that (10) holds by **IV.5.1**, theorem 3. Note also that by **IV.5.1** (7), (8) the inequality (10) is equivalent to the inequality

$$\int_{\boldsymbol{D}} D(u, v_{\mathsf{B}})\, d\mathsf{L} \leqq v_{\mathsf{B}}(\boldsymbol{D}). \tag{11}$$

By **III.2.4**, theorem 3 the sign of equality holds in (11) if and only if v_{B} is absolutely continuous in $\boldsymbol{D}$ with respect to L-measure, and hence (see the definition in **IV.2.3**) if and only if T is ACb_{B} in $\boldsymbol{D}$. Since (by theorem 2) this latter condition is equivalent to T being $AC\mathsf{B}$ in $\boldsymbol{D}$, the proof is complete.

IV.5.3. The generalized Jacobian $J_{\mathsf{B}}(u, T)$. Comparison of the multiplicity functions $K(x, T, \boldsymbol{D})$ and $N(x, T, \boldsymbol{D})$. The terminology of **IV.5.1** and **IV.5.2** will be used in this section. Assume that T is $BV\mathsf{B}$ in $\boldsymbol{D}$. Then the derivative $D_{\mathsf{B}}(u, T)$ exists a.e. in $\boldsymbol{D}$ (see **IV.5.1**, theorem 3). Making use of the local index $i_{\mathsf{B}}(u, T)$ discussed in **II.3.7**, a *generalized*

Jacobian $J_{\mathsf{B}}(u,T)$ can be introduced in the BANACH theory by means of the formula

$$J_{\mathsf{B}}(u, T) = i_{\mathsf{B}}(u, T)\, D_{\mathsf{B}}(u, T). \tag{1}$$

In view of **II.3.7**, lemma 2 and **III.2.3**, theorem 3 it is clear that this generalized Jacobian is L-measurable in $\boldsymbol{D}$. The question of the L-summability of $J_{\mathsf{B}}(u,T)$ gives rise to the following comments. While the property eBV implies the L-summability of the generalized Jacobian $J_e(u,T)$ (see **IV.4.3**, theorem 1), the property $BV\mathsf{B}$ generally does not imply the L-summability of the generalized Jacobian $J_{\mathsf{B}}(u,T)$. If T is $BV\mathsf{B}$ in $\boldsymbol{D}$, then $D_{\mathsf{B}}(u,T)$ is L-summable in $\boldsymbol{D}$ (see **IV.5.1**, theorem 3). But since the local index $i_{\mathsf{B}}(u,T)$ may assume arbitrarily large integral values, the L-summability of $D_{\mathsf{B}}(u,T)$ does not necessarily imply the L-summability of $J_{\mathsf{B}}(u,T)$. It is interesting to observe that the situation in this respect depends essentially upon the dimension n of the Euclidean space R^n in which one operates. We shall see later on that in the cases $n=1, 2$ the discrepancy referred to above does not arise. On the other hand, in the general case the fruitfulness of the BANACH theory seems to be limited to those particular situations where the generalized Jacobian $J_{\mathsf{B}}(u,T)$ may be shown to be L-summable by virtue of further assumptions. These observations may serve to motivate the following definitions.

Definition 1. T is termed $sBV\mathsf{B}$ (strongly of bounded variation in the BANACH sense) in $\boldsymbol{D}$ if it is $BV\mathsf{B}$ in $\boldsymbol{D}$ and $J_{\mathsf{B}}(u,T)$ is L-summable in $\boldsymbol{D}$.

Definition 2. T is termed $sAC\mathsf{B}$ (strongly absolutely continuous in the BANACH sense) in $\boldsymbol{D}$ if it is $AC\mathsf{B}$ in $\boldsymbol{D}$ and $J_{\mathsf{B}}(u,T)$ is L-summable in $\boldsymbol{D}$.

Our main objective in this section is to show that the $sAC\mathsf{B}$ transformations are included in the class of eAC transformations. In preparation, various preliminary facts will be first established. Essential use will be made of the set $\mathbf{I}(T, \boldsymbol{D})$ of those points $u \in \boldsymbol{D}$ which have some neighborhood clear of relatives (see **II.3.7**). For each integer s, we denote by $\mathbf{I}_s(T, \boldsymbol{D})$ the set of those points $u \in \mathbf{I}(T, \boldsymbol{D})$ where $i_{\mathsf{B}}(u,T) = s$. To simplify formulas, we introduce the notations

$$\mathbf{I} = \mathbf{I}(T, \boldsymbol{D}), \quad \mathbf{I}_s = \mathbf{I}_s(T, \boldsymbol{D}), \tag{2}$$

which are adequate as long as T and $\boldsymbol{D}$ are kept fixed. Clearly

$$\mathbf{I} = \cup\, \mathbf{I}_s, \quad s = 0, \pm 1, \pm 2, \ldots, \tag{3}$$

$$\mathbf{I}_r \cap \mathbf{I}_s = \emptyset \quad \text{if} \quad r \neq s. \tag{4}$$

Let us introduce further auxiliary sets by the formulas

$$\mathbf{I}_+ = \bigcup \mathbf{I}_s, \quad s = 1, 2, \ldots, \tag{5}$$

$$\mathbf{I}_- = \bigcup \mathbf{I}_{-s}, \quad s = 1, 2, \ldots, \tag{6}$$

$$\mathbf{I}^* = \bigcup \mathbf{I}_s, \quad |s| \neq 1. \tag{7}$$

The set $\mathbf{I}$ is a BOREL set by **II.3.7**, lemma 1. As we noted in the proof of lemma 2 in **II.3.7**, the sets $\mathbf{I}_s$ are also BOREL sets. It follows that the sets $\mathbf{I}_+, \mathbf{I}_-, \mathbf{I}^*$ are also BOREL sets.

Lemma 1. Let x be a point in R^n such that

$$N(x, T, \boldsymbol{D}) < \infty. \tag{8}$$

Then the following relations hold (see **II.3.2**).

$$\sum_{s=1}^{\infty} s\, N(x, T, \mathbf{I}_s) = K^+(x, T, \boldsymbol{D}), \tag{9}$$

$$\sum_{s=1}^{\infty} s\, N(x, T, \mathbf{I}_{-s}) = K^-(x, T, \boldsymbol{D}). \tag{10}$$

Proof. Assume first that $N(x, T, \boldsymbol{D}) = 0$. Then $x \notin T\boldsymbol{D}$, and thus the summations in (9) and (10) have the value zero, and $K^+(x, T, \boldsymbol{D}) = 0$, $K^-(x, T, \boldsymbol{D}) = 0$ by **II.3.2**, remark 2. Thus (9) and (10) trivially hold in this case. So we can assume that

$$0 < N(x, T, \boldsymbol{D}) < \infty.$$

Then the set $T^{-1}x$ is non-empty and finite. Let $u_1, \ldots, u_m$ be the points of $T^{-1}x$. Each of these points is then clearly in the set $\mathbf{I}$ [see (2)]. It is convenient to consider now two cases.

Case 1. $u_j \in \mathbf{I}_0$, $j = 1, \ldots, m$ [see (2)]. Then clearly the summations in (9) and (10) have the value zero. Inspection of the definition of the local index $i_{\mathrm{B}}(u, T)$ reveals (see **II.3.7**) that there exist now no essential maximal model continua for $(x, T, \boldsymbol{D})$, and hence (see **II.3.4**, theorem 3) we have $K(x, T, \boldsymbol{D}) = 0$ and thus *a fortiori* $K^+(x, T, \boldsymbol{D}) = 0$, $K^-(x, T, \boldsymbol{D}) = 0$. Thus (9) and (10) trivially hold in this case also.

Case 2. $u_j \in \mathbf{I}_+ \cup \mathbf{I}_-$ for some j. By the definition of $i_{\mathrm{B}}(u, T)$ and $\mathbf{I}_s$ we have clearly

$$\sum_{s=1}^{\infty} s\, N(x, T, \mathbf{I}_s) = \sum i_{\mathrm{B}}(u_j, T), \quad u_j \in \mathbf{I}_+, \tag{11}$$

$$\sum_{s=1}^{\infty} s\, N(x, T, \mathbf{I}_{-s}) = -\sum i_{\mathrm{B}}(u_j, T), \quad u_j \in \mathbf{I}_-. \tag{12}$$

Now since $u_j \in \mathbf{I}_+ \cup \mathbf{I}_-$ if and only if u_j by itself is an e.m.m.c. for $(x, T, \boldsymbol{D})$, we have [see the definition of $i_{\mathsf{B}}(u, T)$ in **II.3.7**]

$$i_{\mathsf{B}}(u_j, T) = i_e(u_j, T) \quad \text{if} \quad u_j \in \mathbf{I}_+ \cup \mathbf{I}_- . \tag{13}$$

As those points u_j which lie in $\mathbf{I}_+ \cup \mathbf{I}_-$ are now precisely the essential maximal model continua for $(x, T, \boldsymbol{D})$, from **II.3.4** (25), (26) we obtain the relations

$$K^+(x, T, \boldsymbol{D}) = \sum i_e(u_j, T), \quad u_j \in \mathbf{I}_+,$$
$$K^-(x, T, \boldsymbol{D}) = -\sum i_e(u_j, T), \quad u_j \in \mathbf{I}_-,$$

and (9) and (10) follow in view of (11), (12) and (13).

Lemma 2. Assume that T is $BV\mathsf{B}$ in $\boldsymbol{D}$. Then

$$J_{\mathsf{B}}(u, T) = 0 \quad \text{a.e. on} \quad (\boldsymbol{D} - \mathbf{I}) \cup \mathbf{I}_0 . \tag{14}$$

Proof. On the set $(\boldsymbol{D} - \mathbf{I}) \cup \mathbf{I}_0$, $i_{\mathsf{B}}(u, T) = 0$ by definition [see (2) and **II.3.7**]. Since $D_{\mathsf{B}}(u, T)$ exists (and is finite) a.e. in $\boldsymbol{D}$, the relation (14) follows in view of (1).

Lemma 3. Assume that T is $AC\mathsf{B}$ in $\boldsymbol{D}$. If s is any integer, then $J_{\mathsf{B}}(u, T)$ is L-summable on the set $\mathbf{I}_s$, and

$$\int_{\mathbf{I}_s} J_{\mathsf{B}}(u, T)\, d\mathsf{L} = s \int N(x, T, \mathbf{I}_s)\, d\mathsf{L}. \tag{15}$$

Proof. By **IV.5.2**, theorem 5 (applied with $L = \mathbf{I}_s$) we have

$$\int_{\mathbf{I}_s} D_{\mathsf{B}}(u, T)\, d\mathsf{L} = \int N(x, T, \mathbf{I}_s)\, d\mathsf{L}. \tag{16}$$

On multiplying (16) by the constant s, we obtain (15), since $i_{\mathsf{B}}(u, T) = s$ on $\mathbf{I}_s$.

Theorem 1. Assume that T is $sAC\mathsf{B}$ in $\boldsymbol{D}$ (see definition 2). Then the following holds.

(a) T is eAC in $\boldsymbol{D}$.

(b) $J_{\mathsf{B}}(u, T) = J_e(u, T)$ a.e. in $\boldsymbol{D}$.

Proof. By assumption, $J_{\mathsf{B}}(u, T)$ is L-summable in $\boldsymbol{D}$ and hence also on every L-measurable set in $\boldsymbol{D}$. For the set $\mathbf{I}_+$ [see (5)] we obtain in view of lemma 3 the relation (see **III.1.1**, lemma 39)

$$\int_{\mathbf{I}_+} J_{\mathsf{B}}(u, T)\, d\mathsf{L} = \sum_{s=1}^{\infty} \int s\, N(x, T, \mathbf{I}_s)\, d\mathsf{L}. \tag{17}$$

Since $s N(x, T, \mathbf{I}_s) \geq 0$ in (17), by **III.1.1**, lemma 40 we conclude from (17) that

$$\sum_{s=1}^{\infty} s\, N(x, T, \mathbf{I}_s) \tag{18}$$

is an L-summable function of x in R^n. Observe now that T is $BV\mathsf{B}$ in $\boldsymbol{D}$ by **IV.5.2**, theorem 1, and hence $N(x, T, \boldsymbol{D})$ is L-summable in R^n by **IV.5.1**, theorem 1. Accordingly (see **III.1.1**, lemma 31)

$$N(x, T, \boldsymbol{D}) < \infty \quad \text{a.e. in } R^n.$$

In view of lemma 1 it follows that

$$\sum_{s=1}^{\infty} s\, N(x, T, \mathbf{I}_s) = K^+(x, T, \boldsymbol{D}) \quad \text{a.e. in } R^n. \tag{19}$$

Since the function (18) is L-summable in R^n, it follows from (19) that $K^+(x, T, \boldsymbol{D})$ is L-summable in R^n. Since all the terms in (19) are non-negative, integration of (19) yields (see **III.1.1**, lemma 27)

$$\sum_{s=1}^{\infty} \int s\, N(x, T, \mathbf{I}_s)\, d\mathsf{L} = \int K^+(x, T, \boldsymbol{D})\, d\mathsf{L}. \tag{20}$$

From (17) and (20) we obtain

$$\int_{\mathbf{I}_+} J_{\mathsf{B}}(u, T)\, d\mathsf{L} = \int K^+(x, T, \boldsymbol{D})\, d\mathsf{L}. \tag{21}$$

An entirely similar reasoning, applied to the set $\mathbf{I}_-$ [see (6)] yields the conclusion that $K^-(x, T, \boldsymbol{D})$ is L-summable in R^n and

$$\int_{\mathbf{I}_-} J_{\mathsf{B}}(u, T)\, d\mathsf{L} = -\int K^-(x, T, \boldsymbol{D})\, d\mathsf{L}. \tag{22}$$

Now since

$$K(x, T, \boldsymbol{D}) = K^+(x, T, \boldsymbol{D}) + K^-(x, T, \boldsymbol{D})$$

by **II.3.2**, theorem 1, the L-summability in R^n of $K(x, T, \boldsymbol{D})$ follows. Thus T is eBV in $\boldsymbol{D}$ (see **IV.4.1**). Observe now that T satisfies the condition (N) in $\boldsymbol{D}$ by **IV.5.2**, theorem 3. Hence T also satisfies the condition (N) on every subset of $\boldsymbol{D}$, and in particular on the essential set $E(T, \boldsymbol{D})$. Thus T is eAC in $\boldsymbol{D}$ by **IV.4.2**, theorem 3, and part (a) of the theorem is proved. Turning now to part (b) of the theorem, note that since $J_{\mathsf{B}}(u, T) = 0$ a.e. on the set $(\boldsymbol{D} - \mathbf{I}) \cup \mathbf{I}_0$ by lemma 2, we have

$$\int_{\boldsymbol{D}} J_{\mathsf{B}}(u, T)\, d\mathsf{L} = \int_{\mathbf{I}_+} J_{\mathsf{B}}(u, T)\, d\mathsf{L} + \int_{\mathbf{I}_-} J_{\mathsf{B}}(u, T)\, d\mathsf{L}.$$

In view of (21) and (22) there follows the relation

$$\int_{\boldsymbol{D}} J_{\mathsf{B}}(u, T)\, d\mathsf{L} = \int K^+(x, T, \boldsymbol{D})\, d\mathsf{L} - \int K^-(x, T, \boldsymbol{D})\, d\mathsf{L}. \tag{23}$$

Observe now that since clearly T is also $sAC\mathsf{B}$ in every domain $D \subset \boldsymbol{D}$, the preceding conclusions apply if $\boldsymbol{D}$ is replaced by any domain $D \subset \boldsymbol{D}$. In particular, application of this remark to (23) yields the relation

$$\int_{D} J_{\mathsf{B}}(u, T)\, d\mathsf{L} = \int K^+(x, T, D)\, d\mathsf{L} - \int K^-(x, T, D)\, d\mathsf{L}, \quad D \subset \boldsymbol{D}. \tag{24}$$

Since we already know that T is eAC in $\boldsymbol{D}$, we have by **IV.4.4**, theorem 4,

$$\int_D J_e(u, T)\, d\mathsf{L} = \int K^+(x, T, D)\, d\mathsf{L} - \int K^-(x, T, D)\, d\mathsf{L}, \quad D \subset \boldsymbol{D}. \tag{25}$$

(24) and (25) yield

$$\int_D J_{\mathsf{B}}(u, T)\, d\mathsf{L} = \int_D J_e(u, T)\, d\mathsf{L} \tag{26}$$

for every domain $D \subset \boldsymbol{D}$, and part (b) of the theorem follows by **III.1.1**, lemma 44.

The preceding theorem shows that the theory of $sAC\mathsf{B}$ transformations is a special case of the theory of eAC transformations studied in § **IV.4**. There naturally arises the question of the precise relationship between the crude multiplicity function N and the essential multiplicity function K. The following discussion throws some light upon this issue. More detailed information will be obtained, for the case of differentiable transformations, in Part V.

Lemma 4. Let x be a point in R^n such that

$$N(x, T, \boldsymbol{D}) < \infty. \tag{27}$$

Then [see (2)]

$$K(x, T, \boldsymbol{D}) = \sum_{s=1}^{\infty} s\, N(x, T, \mathbf{I}_s) + \sum_{s=1}^{\infty} s\, N(x, T, \mathbf{I}_{-s}), \tag{28}$$

$$N(x, T, \boldsymbol{D}) = N(x, T, \mathbf{I}_0) + \sum_{s=1}^{\infty} N(x, T, \mathbf{I}_s) + \sum_{s=1}^{\infty} N(x, T, \mathbf{I}_{-s}). \tag{29}$$

Proof. Since

$$K(x, T, \boldsymbol{D}) = K^+(x, T, \boldsymbol{D}) + K^-(x, T, \boldsymbol{D})$$

by **II.3.2**, theorem 1, the relation (28) is a direct consequence of lemma 1. As regards (29), note that $\boldsymbol{D}$ is the union of the pair-wise disjoint sets

$$\boldsymbol{D} - \mathbf{I}, \mathbf{I}_0, \mathbf{I}_1, \mathbf{I}_{-1}, \mathbf{I}_2, \mathbf{I}_{-2}, \ldots,$$

and thus (29) will be proved if we show that

$$N(x, T, \boldsymbol{D} - \mathbf{I}) = 0. \tag{30}$$

In view of the definition of the set $\mathbf{I} = \mathbf{I}(T, \boldsymbol{D})$ (see **II.3.7**), the relation (30) is an immediate consequence of (27), and the proof is complete.

Lemma 5. Denote by X_∞ the set of those points $x \in R^n$ where $N(x, T, \boldsymbol{D}) = \infty$, and let $\mathbf{I}^*$ be the set defined by (7). Let x be a point such that

$$x \notin X_\infty \cup T\,\mathbf{I}^*. \tag{31}$$

Then
$$N(x, T, \boldsymbol{D}) = K(x, T, \boldsymbol{D}). \tag{32}$$

Proof. Since $x \notin X_\infty$, the formulas (28) and (29) hold by lemma 4. Furthermore, since $x \notin T\mathbf{I}^*$, it follows that $x \notin T\mathbf{I}_s$ if $|s| \neq 1$, and hence $N(x, T, \mathbf{I}_s) = 0$ if $|s| \neq 1$. Thus (28) and (29) yield now (32).

Theorem 2. Assume that T satisfies the following conditions.

(i) T is $BV\mathsf{B}$ in $\boldsymbol{D}$.

(ii) $\mathsf{L}T\mathbf{I}^* = 0$ [see (7)].

Then the following holds.

(a) $N(x, T, \boldsymbol{D}) = K(x, T, \boldsymbol{D})$ a.e. in R^n.

(b) T is eBV in $\boldsymbol{D}$.

(c) T is $sBV\mathsf{B}$ in $\boldsymbol{D}$.

Proof. Let X_∞ denote again the set of those points $x \in R^n$ where $N(x, T, \boldsymbol{D}) = \infty$. Since $N(x, T, \boldsymbol{D})$ is L-summable in R^n by **IV.5.1**, theorem 1, the set X_∞ is of L-measure zero. In view of the assumption (ii) we have therefore
$$\mathsf{L}(X_\infty \cup T\mathbf{I}^*) = 0,$$

and (a) follows by lemma 5. As $N(x, T, \boldsymbol{D})$ is L-summable in R^n, (a) implies that $K(x, T, \boldsymbol{D})$ is also L-summable in R^n, and (b) is established. To prove (c), there remains to show that $J_{\mathsf{B}}(u, T)$ is L-summable in $\boldsymbol{D}$. Now
$$J_{\mathsf{B}}(u, T) = 0 \quad \text{a.e. on} \quad \boldsymbol{D} - \mathbf{I} \tag{33}$$

by **IV.5.3**, lemma 2. Furthermore, the assumption (ii) implies (by **IV.5.1**, theorem 5) that $D_{\mathsf{B}}(u, T) = 0$ a.e. on $\mathbf{I}^*$, and in view of (1) it follows that
$$J_{\mathsf{B}}(u, T) = 0 \quad \text{a.e. on } \mathbf{I}^*. \tag{34}$$

By (1) and (2) we have
$$|J_{\mathsf{B}}(u, T)| = D_{\mathsf{B}}(u, T) \quad \text{a.e. on} \quad \mathbf{I}_1 \cup \mathbf{I}_{-1}. \tag{35}$$

Now since [see (3) and (7)]
$$\boldsymbol{D} = (\boldsymbol{D} - \mathbf{I}) \cup \mathbf{I}^* \cup \mathbf{I}_1 \cup \mathbf{I}_{-1},$$

we conclude from (33), (34), (35) that
$$|J_{\mathsf{B}}(u, T)| \leq D_{\mathsf{B}}(u, T) \quad \text{a.e. in } \boldsymbol{D},$$

and as $D_{\mathsf{B}}(u, T)$ is L-summable in $\boldsymbol{D}$ by **IV.5.1**, theorem 3, the L-summability of $J_{\mathsf{B}}(u, T)$ in $\boldsymbol{D}$ follows, and the theorem is proved.

Lemma 6. Denote by $\tilde{\mathbf{I}}$ the set of those points $u \in \boldsymbol{D}$ where $|i_{\mathrm{B}}(u,T)| > 1$. Then

$$K(x, T, \boldsymbol{D}) \leq N(x, T, \boldsymbol{D}) \quad \text{for} \quad x \notin T\tilde{\mathbf{I}}. \tag{36}$$

Proof. Note that

$$\tilde{\mathbf{I}} = \bigcup \mathbf{I}_s, \quad |s| \geq 2.$$

Hence, if x is a point in R^n such that $x \notin T\tilde{\mathbf{I}}$, then $x \notin T\mathbf{I}_s$ if $|s| \geq 2$, and thus

$$N(x, T, \mathbf{I}_s) = 0 \quad \text{for} \quad |s| \geq 2. \tag{37}$$

Observe now that (36) is obvious if $N(x, T, \boldsymbol{D}) = \infty$. Hence we can assume that

$$N(x, T, \boldsymbol{D}) < \infty. \tag{38}$$

Lemma 4 yields, in view of (38) and (37),

$$K(x, T, \boldsymbol{D}) = N(x, T, \mathbf{I}_1) + N(x, T, \mathbf{I}_{-1}) \leq N(x, T, \boldsymbol{D}),$$

and (36) is thus proved.

Part V. Differentiable transformations in $\boldsymbol{R}^n$.

§ V.1. Continuous transformations in $\boldsymbol{R}^1$.

V.1.1. Preliminaries. A point of Euclidean 1-space R^1 has just one coordinate x, and thus it should be denoted by (x), in accordance with the general terminology introduced in **I.2.1**. However, as a matter of convenience, we shall speak of the point x instead of the point (x). If this terminology is adopted, then the points of R^1 are merely the real numbers, and if x', x'' are two real numbers then their (Euclidean) distance is equal to $|x'' - x'|$, the absolute value of their difference (see **I.2.1**). While, in general, there exists a significant conceptual difference between an object x and the set consisting of the single object x, it should be noted that in the present case we could operate directly with the space whose points are the real numbers, the distance of two real numbers x', x'' being defined as $|x'' - x'|$. The space so obtained is the *real number-line* of Analysis. In other words, the notational agreement of writing x instead of (x) has the effect that R^1 becomes formally identical with the real number-line. Let now U be a non-empty subset of R^1, and consider a transformation $T: U \to R^1$. Then T associates with every real number $u \in U$ a definite real number $x = T(u)$. Thus T may be thought of as a real-valued function $x = T(u)$ of the real variable $u \in U$. Accordingly, the general theory developed in Part IV may be interpreted in the case $n = 1$ as a geometrical approach to the theory of real-valued functions of a real variable. Our main objective in the present § **V.1** is to show that the various concepts of bounded variation, absolute continuity, generalized Jacobian (introduced in Part IV for general Euclidean n-space R^n) reduce in the case $n=1$ to the classical concepts of bounded variation, absolute continuity, derivative in the theory of real-valued functions of a real variable. In preparation, we shall review in this section a few elementary facts relating to real-valued functions of a real variable.

We have noted earlier (see **I.2.1**, exercise 8) that in R^1 a bounded domain is merely an open (bounded) interval. It will be useful, therefore, to introduce convenient notations for intervals in R^1. If a, b are real numbers, then we put

$$(a, b) = \{x \mid a < x < b\} \quad \text{if} \quad a < b, \tag{1}$$

$$(a, b) = \{x \mid b < x < a\} \quad \text{if} \quad b < a, \tag{2}$$

$$(a, b) = \emptyset \quad \text{if} \quad a = b. \tag{3}$$

Thus if $a \neq b$, then (a, b) is the open interval with end points a, b. Next we put

$$[a, b] = \{x \mid a \leqq x \leqq b\} \quad \text{if} \quad a \leqq b, \tag{4}$$

$$[a, b] = \{x \mid b \leqq x \leqq a\} \quad \text{if} \quad b \leqq a. \tag{5}$$

Thus if $a \neq b$, then $[a, b]$ is the closed interval with end points a, b. For emphasis, we shall speak of non-degenerate and degenerate open and closed intervals (a, b), $[a, b]$ according as $a \neq b$ or $a = b$. The following relations are obvious.

$$(a, b) = (b, a), \quad [a, b] = [b, a], \quad (a, a) = \emptyset, \quad [a, a] = a,$$

$$\operatorname{int}[a, b] = (a, b), \quad \overline{(a, b)} = [a, b] \quad \text{if} \quad a \neq b.$$

Definition 1. Let $S \neq \emptyset$ be a subset of R^1 which does not reduce to a single point, and $f(x)$ a real-valued finite function defined for $x \in S$. The *total variation* $V(f, S)$ of f on S is defined as the least upper bound of all the sums

$$\sum_i |f(b_i) - f(a_i)|,$$

where $[a_1, b_1]$, $[a_2, b_2]$, ... is any finite sequence of non-degenerate closed intervals whose interiors are pair-wise disjoint and whose end points belong to S. If $V(f, S) < \infty$, then f is said to be of bounded variation on S (briefly, BV on S).

Definition 2. Given S and f as in definition 1, f is said to be absolutely continuous on S (briefly, AC on S) if for every $\varepsilon > 0$ there exists an

$\eta = \eta(\varepsilon) > 0$ such that the following holds: if $[a_1, b_1], [a_2, b_2], \ldots$ is any finite sequence of non-degenerate closed intervals whose interiors are pair-wise disjoint and whose end points belong to S, then the inequality $\sum |b_i - a_i| < \eta$ implies the inequality $\sum |f(b_i) - f(a_i)| < \varepsilon$.

If $[a, a]$ is a degenerate closed interval such that $a \in S$, then $|f(a) - f(a)| = 0$, and thus clearly we would obtain equivalent definitions if we would allow some of the intervals $[a_i, b_i]$ to be degenerate.

Definition 3. Given S and f as in definition 1, the *oscillation* $\omega(f, S)$ of f on S is the least upper bound of all the numbers $|f(b) - f(a)|$, where a, b are any two numbers belonging to S.

The following lemmas 1 through 12 are easy consequences of the definitions involved (the proofs of most of these lemmas may be found in the SAKS treatise listed in the *Bibliography*).

Lemma 1. Given S and f as in definition 1, let S^* be a subset of S. Then the following holds.

(i) $V(f, S^*) \leq V(f, S)$.

(ii) If f is BV on S, then f is also BV on S^*.

Lemma 2. Given S and f as in definition 1, assume that f is AC on S. Then f is also AC on every subset of S.

Lemma 3. Given S and f as in definition 1, assume that f is BV on S. Then f is bounded on S.

Lemma 4. Given S and f as in definition 1, the total variation $V(f, S)$ is lower semi-continuous with respect to f, in the following sense: if $f_n(x)$, $n = 1, 2, \ldots$, is any sequence of real-valued, finite functions on S such that $f_n(x) \to f(x)$ on S, then $V(f, S) \leq \liminf V(f_n, S)$.

Lemma 5. Assume that the real-valued function $f(x)$ is bounded, continuous, and AC in the open (bounded) interval $(\boldsymbol{a}, \boldsymbol{b})$. Then f is BV in $(\boldsymbol{a}, \boldsymbol{b})$.

Lemma 6. Let $f(x)$ be a real-valued continuous function in the closed interval $[\boldsymbol{a}, \boldsymbol{b}]$, where $\boldsymbol{a} \neq \boldsymbol{b}$. Then

$$V(f, [\boldsymbol{a}, \boldsymbol{b}]) = V(f, (\boldsymbol{a}, \boldsymbol{b})).$$

Lemma 7. Let $f(x)$ be a real-valued continuous function in the closed interval $[\boldsymbol{a}, \boldsymbol{b}]$, where $\boldsymbol{a} \neq \boldsymbol{b}$. Then f is BV in $[\boldsymbol{a}, \boldsymbol{b}]$ if and only if it is BV in $(\boldsymbol{a}, \boldsymbol{b})$.

Lemma 8. Let $f(x)$ be a real-valued continuous function in the closed interval $[\boldsymbol{a}, \boldsymbol{b}]$, where $\boldsymbol{a} \neq \boldsymbol{b}$. Then f is AC in $[\boldsymbol{a}, \boldsymbol{b}]$ if and only if it is AC in $(\boldsymbol{a}, \boldsymbol{b})$.

Lemma 9. Assume that the real-valued function $f(x)$ is AC in $[\boldsymbol{a}, \boldsymbol{b}]$, where $\boldsymbol{a} \neq \boldsymbol{b}$. Then f is BV in $[\boldsymbol{a}, \boldsymbol{b}]$.

Lemma 10. Assume that the real-valued function $f(x)$ is BV in the open interval $(\boldsymbol{a}, \boldsymbol{b})$, where $\boldsymbol{a} \neq \boldsymbol{b}$. Then $f(x)$ approaches definite (finite) limits as x approaches the end points $\boldsymbol{a}, \boldsymbol{b}$.

Lemma 11. Let $f(x)$ be a real-valued, bounded, continuous function defined in a non-degenerate (bounded) open interval $(\boldsymbol{a}, \boldsymbol{b})$. Its oscillation $\omega(f, (a, b))$ on open intervals $(a, b) \subset (\boldsymbol{a}, \boldsymbol{b})$ is then a function of open intervals in $(\boldsymbol{a}, \boldsymbol{b})$, and (see **III.2.1**, definition 4)

$$V(f, (\boldsymbol{a}, \boldsymbol{b})) = V(\omega, (\boldsymbol{a}, \boldsymbol{b})).$$

Lemma 12. Let $f(x)$ be a real-valued, bounded, continuous function defined in a non-degenerate (bounded) open interval $(\boldsymbol{a}, \boldsymbol{b})$. Then f is BV in $(\boldsymbol{a}, \boldsymbol{b})$ if and only if its oscillation $\omega(f, (a, b))$, considered as a function of open intervals $(a, b) \subset (\boldsymbol{a}, \boldsymbol{b})$, is of bounded variation in $(\boldsymbol{a}, \boldsymbol{b})$.

We shall now make some elementary comments about the derivative $f'(x)$ of a (real-valued) function $f(x)$, defined in an open interval $(\boldsymbol{a}, \boldsymbol{b})$. The statement that the derivative $f'(x)$ exists and has the value $f'(x_0)$ at a point $x_0 \in (\boldsymbol{a}, \boldsymbol{b})$ is equivalent to the following statement: for every $\varepsilon > 0$ there exists a $\delta = \delta(x_0, \varepsilon) > 0$ such that if

$$0 < |h| < \delta, \tag{6}$$

then

$$x_0 + h \in (\boldsymbol{a}, \boldsymbol{b}) \quad \text{and} \quad \left| \frac{f(x_0 + h) - f(x_0)}{h} - f'(x_0) \right| < \varepsilon. \tag{7}$$

Consider now the function $\Delta(a, b)$ of open intervals $(a, b) \subset (\boldsymbol{a}, \boldsymbol{b})$ defined as follows. The notations being so chosen that $a < b$, we set $\Delta(a, b) = f(b) - f(a)$ if a, b both lie in $(\boldsymbol{a}, \boldsymbol{b})$, and $\Delta(a, b) = 0$ otherwise. Then the statement that the derivative $D(x, \Delta)$ of Δ exists and has the value $D(x_0, \Delta)$ at a point $x_0 \in (\boldsymbol{a}, \boldsymbol{b})$ is equivalent (see **III.2.3**, definition 2) to the following statement: for every $\varepsilon > 0$ there exists a $\delta^* = \delta^*(x_0, \varepsilon) > 0$ such that if

$$a < x_0 < b \quad \text{and} \quad b - a < \delta^*, \tag{8}$$

then

$$a, b \in (\boldsymbol{a}, \boldsymbol{b}) \quad \text{and} \quad \left| \frac{f(b) - f(a)}{b - a} - D(x_0, \Delta) \right| < \varepsilon. \tag{9}$$

The following lemma describes the relationship between the derivatives $f'(x)$ and $D(x, \Delta)$.

Lemma 13. Assume that $f(x)$ is real-valued and continuous in the open interval $(\boldsymbol{a}, \boldsymbol{b})$, and let x_0 be a point of $(\boldsymbol{a}, \boldsymbol{b})$. Then $f'(x)$ exists at x_0 if and only if $D(x, \Delta)$ exists there, and

$$f'(x_0) = D(x_0, \Delta) \tag{10}$$

if both exist.

Proof. Assume first that $f'(x)$ exists at x_0. Assign $\varepsilon > 0$, and let $\delta > 0$ correspond to this ε in the sense of (6) and (7). Consider any two numbers a, b such that

$$\mathbf{a} < a < x_0 < b < \mathbf{b}, \quad b - a < \delta. \tag{11}$$

Then $0 < b - x_0 < \delta$, $0 < x_0 - a < \delta$, and hence (7) yields (for $h = b - x_0$ and $h = a - x_0$ respectively)

$$f'(x_0) - \varepsilon < \frac{f(b) - f(x_0)}{b - x_0} < f'(x_0) + \varepsilon, \tag{12}$$

$$f'(x_0) - \varepsilon < \frac{f(a) - f(x_0)}{a - x_0} < f'(x_0) + \varepsilon. \tag{13}$$

Consider now the identity

$$\frac{f(b) - f(a)}{b - a} = \frac{f(a) - f(x_0)}{a - x_0} \, \frac{x_0 - a}{b - a} + \frac{f(b) - f(x_0)}{b - x_0} \, \frac{b - x_0}{b - a}. \tag{14}$$

Noting that

$$\frac{x_0 - a}{b - a} > 0, \qquad \frac{b - x_0}{b - a} > 0, \qquad \frac{x_0 - a}{b - a} + \frac{b - x_0}{b - a} = 1,$$

we obtain from (12), (13), (14) the inequalities

$$f'(x_0) - \varepsilon < \frac{f(b) - f(a)}{b - a} < f'(x_0) + \varepsilon.$$

It follows that

$$\left| \frac{f(b) - f(a)}{b - a} - f'(x_0) \right| < \varepsilon$$

whenever (11) holds. Thus $D(x, \varDelta)$ exists at x_0 and is equal there to $f'(x_0)$. Assume, conversely, that $D(x, \varDelta)$ exists at x_0. Assign $\varepsilon > 0$, and let $\delta^* > 0$ correspond to this ε in the sense of (8) and (9). Denote by δ the smallest one of the numbers δ^*, $\mathbf{b} - x_0$, $x_0 - \mathbf{a}$, and consider any number h such that

$$0 < |h| < \delta. \tag{15}$$

Assume first that $h > 0$. Then (15) implies that

$$x_0 < x_0 + h < \mathbf{b}, \quad 0 < h < \delta^*. \tag{16}$$

Let a be any number such that $\mathbf{a} < a < x_0$. If a is sufficiently close to x_0, then in view of (16) we shall have $x_0 + h - a < \delta^*$. By (9), applied with $b = x_0 + h$, we have then the inequality

$$\left| \frac{f(x_0 + h) - f(a)}{x_0 + h - a} - D(x_0, \varDelta) \right| < \varepsilon.$$

Since f is continuous, for $a \to x_0$ there follows the inequality

$$\left| \frac{f(x_0 + h) - f(x_0)}{h} - D(x_0, \varDelta) \right| \leq \varepsilon. \tag{17}$$

An entirely similar reasoning yields (17) in the case when $h<0$. Thus (17) holds whenever (15) holds. It follows that $f'(x)$ exists at x_0 and is equal there to $D(x_0, \Delta)$, and the lemma is proved.

Remark 1. Let $f(x)$ be a real-valued, bounded, continuous function which is AC in the open interval $(\boldsymbol{a}, \boldsymbol{b})$. By lemma 5, $f(x)$ is then BV in $(\boldsymbol{a}, \boldsymbol{b})$, and hence by lemma 10 the limits

$$\alpha = \lim_{x \to \boldsymbol{a}} f(x), \quad \beta = \lim_{x \to \boldsymbol{b}} f(x),$$

exist. Define now a function $f^*(x)$, $x \in [\boldsymbol{a}, \boldsymbol{b}]$, as follows: $f^*(x) = f(x)$ if $\boldsymbol{a} < x < \boldsymbol{b}$, and $f^*(\boldsymbol{a}) = \alpha$, $f^*(\boldsymbol{b}) = \beta$. Clearly, $f^*(x)$ is then continuous in the closed interval $[\boldsymbol{a}, \boldsymbol{b}]$, and by lemma 8 is follows that $f^*(x)$ is AC in $[\boldsymbol{a}, \boldsymbol{b}]$. Briefly, $f(x)$ admits of a continuous extension to the closed interval $[\boldsymbol{a}, \boldsymbol{b}]$, and this extension is AC in $[\boldsymbol{a}, \boldsymbol{b}]$.

Remark 2. Let $f(x)$ be real-valued, continuous, and BV in the open interval $(\boldsymbol{a}, \boldsymbol{b})$. If (a, b) is any non-degenerate open interval in $(\boldsymbol{a}, \boldsymbol{b})$, then f is BV in (a, b) by lemma 1. Thus $V(f, (a, b))$ is a function of open intervals $(a, b) \subset (\boldsymbol{a}, \boldsymbol{b})$. It can be shown that there exists a non-negative, countably additive function $W(f, B)$ of Borel sets $B \subset (\boldsymbol{a}, \boldsymbol{b})$ such that $W(f, (a, b)) = V(f, (a, b))$ for every open interval in $(\boldsymbol{a}, \boldsymbol{b})$. Briefly, the function $V(f, (a, b))$ of open intervals admits of a non-negative, countably additive extension $W(f, B)$ to Borel sets $B \subset (\boldsymbol{a}, \boldsymbol{b})$. On the other hand, if B is a Borel set in $(\boldsymbol{a}, \boldsymbol{b})$, then there arises the total variation $V(f, B)$ in the sense of definition 1. Simple examples show that generally $V(f, B) \neq W(f, B)$. In the literature, frequently $W(f, B)$ is termed the total variation of f on the Borel set B. The quantity $V(f, B)$, described in definition 1, is frequently referred to as the weak total variation of f on B.

V.1.2. Index functions and multiplicity functions in $\boldsymbol{R}^1$. Let there be given a bounded continuous transformation $T: \boldsymbol{D} \to R^1$, where $\boldsymbol{D}$ is a bounded domain in R^1. As noted above, $\boldsymbol{D}$ is then merely a non-degenerate (bounded) open interval $(\boldsymbol{a}, \boldsymbol{b})$. Accordingly, we write

$$T: (\boldsymbol{a}, \boldsymbol{b}) \to R^1, \tag{1}$$

where $\boldsymbol{a} \neq \boldsymbol{b}$. Then $x = T(u)$ is a real-valued, bounded, continuous function of the variable $u \in (\boldsymbol{a}, \boldsymbol{b})$. Throughout the present section $\boldsymbol{a}, \boldsymbol{b}, T$ are considered as fixed. Let (a, b) be an open interval in $(\boldsymbol{a}, \boldsymbol{b})$ such that $[a, b] \subset (\boldsymbol{a}, \boldsymbol{b})$, $a \neq b$. Consider a point $x \in R^1$. For the topological index $\mu(x, T, (a, b))$ to be defined (see **II.2.2**), it is then necessary and sufficient that $x \notin T\,fr\,(a, b)$. Now $T\,fr\,(a, b)$ consists of the points $T(a)$, $T(b)$ (which may or may not be distinct). Accordingly, we have the following statement.

Lemma 1. $\mu(x, T, (a, b))$ is defined if and only if

$$a \neq b, \quad [a, b] \subset (\boldsymbol{a}, \boldsymbol{b}), \quad x \neq T(a), T(b). \tag{2}$$

The next lemma yields the explicit value of $\mu(x, T, (a, b))$, whenever this quantity is defined.

Lemma 2. Assume that $a<b$ and $\mu(x, T, (a, b))$ is defined. Then the following holds.

$$a < b, \quad [a, b] \subset (\boldsymbol{a}, \boldsymbol{b}), \quad x \neq T(a), T(b). \tag{3}$$

$$\mu(x, T, (a, b)) = 0 \quad \text{if} \quad T(a) = T(b). \tag{4}$$

$$\mu(x, T, (a, b)) = 0 \quad \text{if} \quad x \in \complement[T(a), T(b)]. \tag{5}$$

$$\mu(x, T, (a, b)) = \operatorname{sgn}(T(b) - T(a)) \quad \text{if} \quad x \in (T(a), T(b)). \tag{6}$$

Proof. In view of lemma 1, the relations (3) are direct consequences of the assumptions made. To verify (4), (5), and (6), consider the (unique) linear transformation

$$x = \mathfrak{L}(u) = \alpha u + \beta \tag{7}$$

which satisfies the conditions

$$\mathfrak{L}(a) = T(a), \quad \mathfrak{L}(b) = T(b). \tag{8}$$

Clearly (see **II.2.5**)

$$\det \mathfrak{L} = \alpha = \frac{T(b) - T(a)}{b - a}. \tag{9}$$

Since the points a, b constitute the frontier of (a, b), the conditions (8) mean that $\mathfrak{L} = T$ on $fr(a, b)$. By **II.2.3**, remark 7 it follows that $\mu(x, \mathfrak{L}, (a, b))$ is also defined and

$$\mu(x, T, (a, b)) = \mu(x, \mathfrak{L}, (a, b)). \tag{10}$$

Assume now first that

$$T(a) \neq T(b). \tag{11}$$

In view of (9), $\mathfrak{L}$ is then non-singular. From (8) we see that

$$\mathfrak{L}(a, b) = (T(a), T(b)), \quad \mathfrak{L}[a, b] = [T(a), T(b)].$$

By **II.2.5**, theorem 4 it follows that

$$\mu(x, \mathfrak{L}, (a, b)) = \operatorname{sgn} \det \mathfrak{L} \quad \text{if} \quad x \in (T(a), T(b)), \tag{12}$$

$$\mu(x, \mathfrak{L}, (a, b)) = 0 \quad \text{if} \quad x \in \complement[T(a), T(b)]. \tag{13}$$

Thus (5) and (6) follow from (9), (10), (12), (13), while (4) is of course now vacuous in view of (11). Assume next that

$$T(a) = T(b). \tag{14}$$

Note that (6) is now vacuously true, since $(T(a), T(b)) = \emptyset$ in view of (14). To verify (4) and (5), observe that (14) implies the relations

$$\mathfrak{L}(a, b) = \mathfrak{L}(a) = T(a) = T(b).$$

Since $x \neq T(a), T(b)$ by (2), it follows that we have now the relation $x \in \mathsf{C}\,\mathfrak{L}(a, b)$, and hence $\mu(x, \mathfrak{L}, (a, b)) = 0$ by **II.2.3**, theorem 2. Thus (4) and (5) follow in view of (10), and the proof is complete.

Lemma 3. Let (a, b) be an open interval such that $a < b$ and $[a, b] \subset (\boldsymbol{a}, \boldsymbol{b})$. Then

$$\int \mu(x, T, (a, b))\, d\mathsf{L} = T(b) - T(a). \tag{15}$$

Proof. Assume first that

$$T(a) \neq T(b). \tag{16}$$

By lemma 1, $\mu(x, T, (a, b))$ is then defined for $x \neq T(a), T(b)$ and by lemma 2 we have $\mu(x, T, (a, b)) = 0$ if $x \notin [T(a), T(b)]$ and $\mu(x, T, (a, b)) = \operatorname{sgn}(T(b) - T(a))$ if $x \in (T(a), T(b))$. Thus clearly

$$\int \mu(x, T, (a, b))\, d\mathsf{L} = |T(b) - T(a)| \operatorname{sgn}(T(b) - T(a)) = T(b) - T(a).$$

Thus (15) is proved under the assumption (16). Suppose next that

$$T(a) = T(b). \tag{17}$$

By lemma 1 and lemma 2, $\mu(x, T, (a, b))$ is defined and has the value zero for $x \neq T(a) = T(b)$. Thus now clearly

$$\int \mu(x, T, (a, b))\, d\mathsf{L} = 0,$$

and (15) follows in view of (17).

Lemma 4. If the essential local index $i_e(C, T)$ is defined (see **II.3.4**), then $i_e(C, T) = \pm 1$.

Proof. If $i_e(C, T)$ is defined, then C is an essentially isolated e.m.m.c. for some point $x \in R^1$, and there exists an open interval $(a, b) \subset (\boldsymbol{a}, \boldsymbol{b})$ such that

$$i_e(C, T) = \mu(x, T, (a, b)). \tag{18}$$

If the notation is so chosen that $a < b$, then lemma 2 yields the fact that

$$\mu(x, T, (a, b)) = \pm 1 \quad \text{or} \quad 0. \tag{19}$$

Since $i_e(C, T)$ is always a non-zero integer (see **II.3.4**, remark 3), the relations (18) and (19) imply that $i_e(C, T) = \pm 1$, and the lemma is proved.

Lemma 5. If x is any point in R^1, then (see **II.3.2**, **II.3.3**)

$$K\big(x, T, (\boldsymbol{a}, \boldsymbol{b})\big) = k\big(x, T, (\boldsymbol{a}, \boldsymbol{b})\big). \tag{20}$$

Proof. If $k\big(x, T, (\boldsymbol{a}, \boldsymbol{b})\big)$ is equal to either zero or ∞, then (20) holds by **II.3.4**, theorem 3, and so we can assume that $0 < k\big(x, T, (\boldsymbol{a}, \boldsymbol{b})\big) < \infty$. By **II.3.4**, theorem 4, remark 2, and remark 5 we have then the formula

$$K\big(x, T, (\boldsymbol{a}, \boldsymbol{b})\big) = \sum |i_e(C, T)|, \quad C \in \mathfrak{C}\big(x, T, (\boldsymbol{a}, \boldsymbol{b})\big).$$

Since $|i_e(C, T)| = 1$ by lemma 4, it follows that $K\big(x, T, (\boldsymbol{a}, \boldsymbol{b})\big)$ is equal to the number of the continua $C \in \mathfrak{C}\big(x, T, (\boldsymbol{a}, \boldsymbol{b})\big)$. As this number is equal to $k\big(x, T, (\boldsymbol{a}, \boldsymbol{b})\big)$ by definition (see **II.3.3**, definition 1), the relation (20) follows.

Lemma 6. The local index $i_{\mathrm{B}}(u, T)$ (see **II.3.7**) takes on no value different from $0, 1, -1$.

Proof. By its definition, $i_{\mathrm{B}}(u, T)$ is equal either to zero or else to $i_e(u, T)$, and in the latter case its value is either 1 or -1 by lemma 4.

Lemma 7. If (a, b) is an open interval in $(\boldsymbol{a}, \boldsymbol{b})$ and x is a point in R^1, then (a, b) is an indicator domain (see **II.3.2**) for $\big(x, T, (\boldsymbol{a}, \boldsymbol{b})\big)$ if and only if

$$[a, b] \subset (\boldsymbol{a}, \boldsymbol{b}) \quad \text{and} \quad \big(T(a) - x\big)\big(T(b) - x\big) < 0. \tag{21}$$

Proof. Assume that (a, b) is an indicator domain for $\big(x, T, (\boldsymbol{a}, \boldsymbol{b})\big)$. Then $\mu\big(x, T, (\boldsymbol{a}, \boldsymbol{b})\big)$ must be defined and different from zero. By lemma 1 it follows that $[a, b] \subset (\boldsymbol{a}, \boldsymbol{b})$, and by lemma 2 it follows further that $x \in \big(T(a), T(b)\big)$. This last inclusion is clearly equivalent to the second relation in (21). Assume, conversely, that (21) holds. The second relation in (21) is clearly equivalent to the inclusion $x \in \big(T(a), T(b)\big)$. Thus we have now at our disposal the relations

$$[a, b] \subset (\boldsymbol{a}, \boldsymbol{b}), \quad x \in \big(T(a), T(b)\big). \tag{22}$$

It follows that

$$a \neq b, \tag{23}$$

since otherwise we would have $T(a) = T(b)$, and the second relation in (21) could not hold. Furthermore

$$x \neq T(a), T(b), \tag{24}$$

since otherwise we would have $\big(T(a) - x\big)\big(T(b) - x\big) = 0$, in contradiction with (21). From (22), (23), (24) we conclude, by lemma 1 and lemma 2, that $\mu\big(x, T, (a, b)\big)$ is defined and is equal to $\operatorname{sgn}\big(T(b) - T(a)\big)$

or $\operatorname{sgn}(T(a)-T(b))$, according as $a<b$ or $b<a$ [see (23)]. In either case, $\mu(x, T, (a, b)) = \pm 1$ [note that the second relation (21) implies that $T(a) \neq T(b)$, a fact which we already used above]. Thus $\mu(x, T, (a, b))$ is defined and different from zero, and hence (a, b) is an indicator domain for $(x, T, (\boldsymbol{a}, \boldsymbol{b}))$.

Our next objective is to compare the multiplicity functions $N(x, T, (\boldsymbol{a}, \boldsymbol{b}))$ and $K(x, T, (\boldsymbol{a}, \boldsymbol{b}))$ (see **I.1.2** and **II.3.2**). In preparation, we consider two auxiliary sets $S_{\max}$, $S_{\min}$ defined as follows. A point $u \in (\boldsymbol{a}, \boldsymbol{b})$ belongs to $S_{\max}$ if and only if there exists a number $h = h(u) > 0$ such that

$$[u-h, u+h] \subset (\boldsymbol{a}, \boldsymbol{b})$$

and

$$T(u') \leq T(u) \quad \text{if} \quad u' \in [u-h, u+h].$$

Similarly, a point $u \in (\boldsymbol{a}, \boldsymbol{b})$ belongs to $S_{\min}$ if and only if there exists a number $h_* = h_*(u) > 0$ such that

$$[u-h_*, u+h_*] \subset (\boldsymbol{a}, \boldsymbol{b})$$

and

$$T(u') \geq T(u) \quad \text{if} \quad u' \in [u-h_*, u+h_*].$$

Thus $S_{\max}$ is the set of those points $u \in (\boldsymbol{a}, \boldsymbol{b})$ at which the function $x = T(u)$ has a weak local maximum. Similarly, $S_{\min}$ is the set of those points $u \in (\boldsymbol{a}, \boldsymbol{b})$ at which the function $x = T(u)$ has a weak local minimum.

Lemma 8. The set $T(S_{\max} \cup S_{\min})$ is countable (perhaps empty).

Proof. Clearly it is sufficient to show that the sets $TS_{\max}$ and $TS_{\min}$ are countable. Since the proof is entirely similar in both cases, we consider explicitly only $TS_{\max}$. For each positive integer n, let S^n denote the set of those points $u \in (\boldsymbol{a}, \boldsymbol{b})$ for which it is true that

$$\left[u-\frac{1}{n}, u+\frac{1}{n}\right] \subset (\boldsymbol{a}, \boldsymbol{b}) \tag{25}$$

and

$$T(u') \leq T(u) \quad \text{if} \quad u' \in \left[u-\frac{1}{n}, u+\frac{1}{n}\right]. \tag{26}$$

Clearly $S_{\max} = \bigcup S^n$ and hence $TS_{\max} = \bigcup TS^n$. Thus it is sufficient to prove that TS^n is countable. In fact, we shall find that TS^n is a finite set. If $TS^n = \emptyset$, this is obvious. So assume that $TS^n \neq \emptyset$, and consider a point $x \in TS^n$. Then there exists at least one point $u \in S^n$ such that $T(u) = x$. For each $x \in TS^n$, select such a point u and denote it by $u(x)$. Then

$$u(x) \in S^n, \quad T(u(x)) = x. \tag{27}$$

Consider now any two points x_1, x_2 such that

$$x_1 \neq x_2, \quad x_1, x_2 \in TS^n. \tag{28}$$

We assert that

$$|u(x_2) - u(x_1)| > \frac{1}{n}. \tag{29}$$

Assume, indeed, that $|u(x_2) - u(x_1)| \leq 1/n$. We can then apply (26) with $u = u(x_1)$, $u' = u(x_2)$, obtaining

$$x_2 = T(u(x_2)) \leq T(u(x_1)) = x_1. \tag{30}$$

Also, we can apply (26) with $u = u(x_2)$, $u' = u(x_1)$, obtaining

$$x_1 = T(u(x_1)) \leq T(u(x_2)) = x_2. \tag{31}$$

From (30), (31) it follows that $x_1 = x_2$, in contradiction with (28), and thus (29) is verified. Denote by U the set of the points $u(x)$ corresponding to the points $x \in TS^n$. Then (29) shows that the distance between any two distinct points of U exceeds $1/n$. Since n is now a fixed positive integer and U is a subset of the (bounded) open interval $(\boldsymbol{a}, \boldsymbol{b})$, it follows that U is a finite set. As $TU = TS^n$ by (27), it follows finally that S^n is a finite set, and the lemma is proved.

Lemma 9. Let x be a point in R^1 such that

$$x \notin T(S_{\max} \cup S_{\min}). \tag{32}$$

Then (see **I.1.2** and **II.3.2**)

$$K(x, T, (\boldsymbol{a}, \boldsymbol{b})) = N(x, T, (\boldsymbol{a}, \boldsymbol{b})). \tag{33}$$

Proof. In view of lemma 5, (33) is equivalent to

$$k(x, T, (\boldsymbol{a}, \boldsymbol{b})) = N(x, T, (\boldsymbol{a}, \boldsymbol{b})). \tag{34}$$

To prove (34) we first note that

$$k(x, T, (\boldsymbol{a}, \boldsymbol{b})) \leq N(x, T, (\boldsymbol{a}, \boldsymbol{b})). \tag{35}$$

Indeed, (35) is obvious if $k(x, T, (\boldsymbol{a}, \boldsymbol{b})) = 0$. If $k(x, T, (\boldsymbol{a}, \boldsymbol{b})) > 0$, then select any integer m such that

$$k(x, T, (\boldsymbol{a}, \boldsymbol{b})) \geq m \geq 1. \tag{36}$$

Then there exist, by the definition of k, distinct continua $C_1, \ldots, C_m$ in $(\boldsymbol{a}, \boldsymbol{b})$ such that $TC_j = x$, $j = 1, \ldots, m$. Since each C_j contains at least one point, it follows that $N(x, T, (\boldsymbol{a}, \boldsymbol{b})) \geq m$. As m was any integer satisfying (36), the inequality (35) follows [observe that (32) is not needed to establish (35)]. We show next that

$$N(x, T, (\boldsymbol{a}, \boldsymbol{b})) \leq k(x, T, (\boldsymbol{a}, \boldsymbol{b})) \tag{37}$$

if (32) holds. If $N(x, T, (\boldsymbol{a}, \boldsymbol{b}))=0$, then (37) is obvious. If $N(x, T, (\boldsymbol{a}, \boldsymbol{b}))>0$, then select any integer m such that

$$N(x, T, (\boldsymbol{a}, \boldsymbol{b})) \geq m \geq 1. \tag{38}$$

Then there exist distinct points $u_1, \ldots, u_m$ in $(\boldsymbol{a}, \boldsymbol{b})$ such that

$$T(u_j) = x, \quad j = 1, \ldots, m, \tag{39}$$

$$u_j \notin S_{\max} \cup S_{\min}. \tag{40}$$

Since $u_1, \ldots, u_m$ are distinct points in $(\boldsymbol{a}, \boldsymbol{b})$, we can select a number $h>0$ such that on setting

$$I_j = [u_j - h, u_j + h], \quad j = 1, \ldots, m, \tag{41}$$

we have the relations

$$I_j \subset (\boldsymbol{a}, \boldsymbol{b}), \quad j = 1, \ldots, m, \tag{42}$$

$$I_i \cap I_j = \emptyset \quad \text{if} \quad i \neq j. \tag{43}$$

In view of (40) there follows, for each $j=1, \ldots, m$, the existence of points a_j, b_j such that $a_j, b_j \in I_j$, $T(a_j) > T(u_j) = x$, $T(b_j) < T(u_j) = x$. Clearly, these relations imply that

$$[a_j, b_j] \subset I_j \subset (\boldsymbol{a}, \boldsymbol{b}), \quad (T(a_j) - x)(T(b_j) - x) < 0. \tag{44}$$

By (42), (43), (44) and lemma 7 it follows that $(a_1, b_1), \ldots, (a_m, b_m)$ are pair-wise disjoint indicator domains for $(x, T, (\boldsymbol{a}, \boldsymbol{b}))$. Since each of these indicator domains contains (by **II.3.3**, lemma 4) at least one e.m.m.c. for $(x, T, (\boldsymbol{a}, \boldsymbol{b}))$, it follows that there exist at least m distinct essential maximal model continua for $(x, T, (\boldsymbol{a}, \boldsymbol{b}))$, and hence $k(x, T, (\boldsymbol{a}, \boldsymbol{b})) \geq m$. As m was any integer satisfying (38), the inequality (37) follows. Since (37) and (35) imply (34), the lemma is proved.

Observe that the set $T(S_{\max} \cup S_{\min})$ is countable by lemma 8, and hence this set is of L-measure zero. Thus lemma 9 implies that (33) holds a.e. in R^1. Since we only assumed that T is continuous and bounded in $(\boldsymbol{a}, \boldsymbol{b})$, it is clear that $(\boldsymbol{a}, \boldsymbol{b})$ can be replaced by any non-degenerate open interval $(a, b) \subset (\boldsymbol{a}, \boldsymbol{b})$. Accordingly, we have the following statement.

Lemma 10. If (a, b) is any non-degenerate open interval in $(\boldsymbol{a}, \boldsymbol{b})$, then

$$K(x, T, (a, b)) = N(x, T, (a, b)) \quad \text{a.e. in } R^1.$$

Let us recall that T may be thought of as a real-valued function $x=T(u)$, $u\in(\boldsymbol{a},\boldsymbol{b})$, and accordingly we can consider the derivative $T'(u)$, which may or may not exist at a given point $u_0\in(\boldsymbol{a},\boldsymbol{b})$.

Lemma 11. Assume $\boldsymbol{a}<\boldsymbol{b}$. Consider a point $u_0\in(\boldsymbol{a},\boldsymbol{b})$, such that the derivative $T'(u_0)$ exists and is different from zero. Then (see **II.3.7**)

$$\operatorname{sgn} T'(u_0) = i_{\mathsf{B}}(u_0, T).$$

Proof. The letters a, b will designate two numbers such that

$$\boldsymbol{a} < a < u_0 < b < \boldsymbol{b}. \tag{45}$$

In view of **V.1.1**, lemma 13, the existence of $T'(u_0)$ implies the relations

$$\frac{T(b)-T(u_0)}{b-u_0} \to T'(u_0) \quad \text{for} \quad b\to u_0, \tag{46}$$

$$\frac{T(a)-T(u_0)}{a-u_0} \to T'(u_0) \quad \text{for} \quad a\to u_0, \tag{47}$$

$$\frac{T(b)-T(a)}{b-a} \to T'(u_0) \quad \text{for} \quad a\to u_0, \quad b\to u_0. \tag{48}$$

It is convenient to consider the cases $T'(u_0)>0$ and $T'(u_0)<0$ separately. Since the argument is entirely analogous in both cases, we consider explicitly only the case when

$$T'(u_0)<0. \tag{49}$$

We have to show then that

$$i_{\mathsf{B}}(u_0, T) = -1. \tag{50}$$

From (45) to (49) we infer the existence of a number $\eta>0$ such that

$$T(b)-T(u_0)<0 \quad \text{if} \quad u_0<b<u_0+\eta, \tag{51}$$

$$T(a)-T(u_0)>0 \quad \text{if} \quad u_0-\eta<a<u_0, \tag{52}$$

$$T(b)-T(a)<0 \quad \text{if} \quad u_0-\eta<a<u_0<b<u_0+\eta. \tag{53}$$

From (51) and (52) it follows that $T(u)\neq T(u_0)$ if $0<|u-u_0|<\eta$. Thus u_0 belongs to the set of those points $\mathbf{I}=\mathbf{I}(T,(\boldsymbol{a},\boldsymbol{b}))$ which possess a neighborhood clear of relatives (see **II.3.7**, definition 2). As explained in **II.3.7**, u_0 by itself is therefore an m.m.c. for $(T(u_0), T, (\boldsymbol{a},\boldsymbol{b}))$. From lemma 1 it follows further that if a, b satisfy the inequalities

$$u_0-\eta<a<u_0<b<u_0+\eta, \tag{54}$$

then $\mu(T(u_0), T, (a, b))$ is defined. Since (51) and (52) imply that $T(u_0)\in(T(a), T(b))$, lemma 2 yields [see (53)]

$$\mu(T(u_0), T, (a,b)) = \operatorname{sgn}(T(b)-T(a)) = -1. \tag{55}$$

Thus in every neighborhood of u_0 there are indicator domains (a, b) for $(T(u_0), T, (\boldsymbol{a}, \boldsymbol{b}))$ such that $a < u_0 < b$. Hence u_0 is an e.m.m.c. for $(T(u_0), T, (\boldsymbol{a}, \boldsymbol{b}))$. Therefore (see **II.3.7**)

$$i_{\mathsf{B}}(u_0, T) = i_e(u_0, T). \tag{56}$$

On the other hand, $i_e(u_0, T)$ is equal to $\mu(T(u_0), T, (a, b))$ if (a, b) is an indicator domain for $(T(u_0), T, (\boldsymbol{a}, \boldsymbol{b}))$ such that $a < u_0 < b$ and $[a, b]$ is contained in a characteristic neighborhood of u_0 (see **II.3.4**, definition 1). Clearly, the open interval $(u_0 - \eta, u_0 + \eta)$ is a characteristic neighborhood of u_0, since $T(u) \neq T(u_0)$ if $u \neq u_0$, $u \in (u_0 - \eta, u_0 + \eta)$, as we noted above. Thus (54) and (55) yield $i_e(u_0, T) = -1$, and (50) follows in view of (56).

V.1.3. Bounded variation. The bounded continuous transformation $T: (\boldsymbol{a}, \boldsymbol{b}) \to R^1$, given as in **V.1.2** (1), is considered fixed throughout this section.

Lemma 1. In R^1 the properties $BV\mathsf{B}$ and $sBV\mathsf{B}$ (see **IV.5.1**, definition 1 and **IV.5.3**, definition 1) are equivalent. Explicitly: T is $BV\mathsf{B}$ in $(\boldsymbol{a}, \boldsymbol{b})$ if and only if it is $sBV\mathsf{B}$ in $(\boldsymbol{a}, \boldsymbol{b})$.

Proof. If T is $sBV\mathsf{B}$ in $(\boldsymbol{a}, \boldsymbol{b})$, then it is also $BV\mathsf{B}$ in $(\boldsymbol{a}, \boldsymbol{b})$ by definition. Assume, conversely, that T is $BV\mathsf{B}$ in $(\boldsymbol{a}, \boldsymbol{b})$. Then (see **IV.5.1**, theorem 3) the derivative $D_{\mathsf{B}}(u, T)$ exists a.e. in $(\boldsymbol{a}, \boldsymbol{b})$ and is L-summable there. Now since

$$J_{\mathsf{B}}(u, T) = i_{\mathsf{B}}(u, T)\, D_{\mathsf{B}}(u, T) \quad \text{a.e. in } (\boldsymbol{a}, \boldsymbol{b})$$

by definition, and $|i_{\mathsf{B}}(u, T)| \leq 1$ by **V.1.2**, lemma 6, it follows that $J_{\mathsf{B}}(u, T)$ is L-summable in $(\boldsymbol{a}, \boldsymbol{b})$. Thus T is $sBV\mathsf{B}$ in $(\boldsymbol{a}, \boldsymbol{b})$, and the lemma is proved.

Lemma 2. If (a, b) is any non-degenerate open interval in $(\boldsymbol{a}, \boldsymbol{b})$, then [see **IV.5.1** (1) and **V.1.1**, definition 3]

$$G(a, b) = \omega(T, (a, b)). \tag{1}$$

Proof. In (1), T is interpreted (in connection with ω) as a real-valued function $T(u)$, $u \in (\boldsymbol{a}, \boldsymbol{b})$. Recall that

$$G(a, b) = \mathsf{L}\, T(a, b). \tag{2}$$

Let us put

$$\lambda = \text{gr.l.b. } T(u), \quad \Lambda = \text{l.u.b. } T(u), \quad u \in (a, b). \tag{3}$$

Since T is bounded, clearly λ and Λ are finite, and $\lambda \leq \Lambda$. We first verify that

$$G(a, b) \leq \Lambda - \lambda. \tag{4}$$

Indeed, if $u \in (a, b)$, then $\lambda \leq T(u) \leq \Lambda$ by (3). Thus $T(a, b) \subset [\lambda, \Lambda]$, and (4) follows in view of (2). Next we verify that

$$\omega(T, (a, b)) \leq G(a, b). \tag{5}$$

Assign any $\varepsilon > 0$. By the definition of ω, there exist numbers α, β such that

$$\alpha, \beta \in (a, b), \quad |T(\beta) - T(\alpha)| > \omega(T, (a, b)) - \varepsilon. \tag{6}$$

It follows that $T(\alpha), T(\beta) \in T(a, b)$. Furthermore, the continuity of T implies that if $x \in [T(\alpha), T(\beta)]$, then $x = T(u)$ for some $u \in [\alpha, \beta]$. Thus

$$[T(\alpha), T(\beta)] \subset T[\alpha, \beta] \subset T(a, b).$$

In view of (6) we conclude that

$$|T(\beta) - T(\alpha)| \leq G(a, b). \tag{7}$$

From (6) and (7) we see that

$$\omega(T, (a, b)) < G(a, b) + \varepsilon.$$

As $\varepsilon > 0$ was arbitrary, (5) follows. Finally we verify that

$$\Lambda - \lambda \leq \omega(T, (a, b)). \tag{8}$$

Assign again any $\varepsilon > 0$. In view of (3) there exist numbers u_1, u_2 such that $u_1, u_2 \in (a, b)$, $T(u_1) < \lambda + \varepsilon$, $T(u_2) > \Lambda - \varepsilon$, and hence

$$\Lambda - \lambda - 2\varepsilon < T(u_2) - T(u_1).$$

Since $T(u_2) - T(u_1) \leq \omega(T, (a, b))$, we conclude that $\Lambda - \lambda - 2\varepsilon < \omega(T, (a, b))$. As $\varepsilon > 0$ was arbitrary, (8) follows. Clearly, (4), (5), (8) imply that

$$\Lambda - \lambda = \omega(T, (a, b)) = G(a, b),$$

and the lemma is proved.

Theorem 1. $V(T, (\boldsymbol{a}, \boldsymbol{b})) = V_{\mathsf{B}}(T, (\boldsymbol{a}, \boldsymbol{b})) = V_e(T, (\boldsymbol{a}, \boldsymbol{b}))$.

Proof. By **V.1.1**, lemma 11 we have

$$V(T, (\boldsymbol{a}, \boldsymbol{b})) = V(\omega, (\boldsymbol{a}, \boldsymbol{b})), \tag{9}$$

where ω stands now for the oscillation $\omega(T, (a, b))$ of T in $(a, b) \subset (\boldsymbol{a}, \boldsymbol{b})$. By definition (see **IV.5.1**, definition 2)

$$V_{\mathsf{B}}(T, (\boldsymbol{a}, \boldsymbol{b})) = V(G, (\boldsymbol{a}, \boldsymbol{b})). \tag{10}$$

Now since

$$G(a, b) = \omega(T, (a, b)) \quad \text{for} \quad (a, b) \subset (\boldsymbol{a}, \boldsymbol{b})$$

by lemma 2, (9) and (10) imply that

$$V(T, (\boldsymbol{a}, \boldsymbol{b})) = V_{\mathsf{B}}(T, (\boldsymbol{a}, \boldsymbol{b})). \tag{11}$$

Recall that

$$K(x, T, (\boldsymbol{a}, \boldsymbol{b})) = N(x, T, (\boldsymbol{a}, \boldsymbol{b})) \quad \text{a.e. in } R^1 \tag{12}$$

by **V.1.2**, lemma 1o. Thus $K(x, T, (\boldsymbol{a}, \boldsymbol{b}))$ is L-summable in R^1 if and only if $N(x, T, (\boldsymbol{a}, \boldsymbol{b}))$ is L-summable in R^1. If both are L-summable in R^1, then

$$V_{\mathsf{B}}(T, (\boldsymbol{a}, \boldsymbol{b})) = \int N(x, T, (\boldsymbol{a}, \boldsymbol{b}))\, d\mathsf{L} \tag{13}$$

by **IV.5.1**, lemma 2, and

$$V_e(T, (\boldsymbol{a}, \boldsymbol{b})) = \int K(x, T, (\boldsymbol{a}, \boldsymbol{b}))\, d\mathsf{L} \tag{14}$$

by **IV.4.1**, lemma 3 and definitions 1 and 2. In view of (12), we see from (13) and (14) that

$$V_{\mathsf{B}}(T, (\boldsymbol{a}, \boldsymbol{b})) = V_e(T, (\boldsymbol{a}, \boldsymbol{b})) \tag{15}$$

in this case. On the other hand, if $K(x, T, (\boldsymbol{a}, \boldsymbol{b}))$ and $N(x, T, (\boldsymbol{a}, \boldsymbol{b}))$ are not L-summable in R^1, then

$$V_{\mathsf{B}}(T, (\boldsymbol{a}, \boldsymbol{b})) = \infty = V_e(T, (\boldsymbol{a}, \boldsymbol{b}))$$

by the same references. Thus (15) holds always, and hence in view of (11) the theorem is proved.

Theorem 2. In R^1 the properties BV, $BV\mathsf{B}$, $sBV\mathsf{B}$, eBV are equivalent. Explicitly, if T possesses in $(\boldsymbol{a}, \boldsymbol{b})$ one of these four properties, then it also possesses the other three.

Proof. BV is equivalent to the condition $V(T, (\boldsymbol{a}, \boldsymbol{b})) < \infty$ by **V.1.1**, definition 1; eBV is equivalent to the condition $V_e(T, (\boldsymbol{a}, \boldsymbol{b})) < \infty$ by **IV.4.1**, lemma 3; and $BV\mathsf{B}$ is equivalent to the condition $V_{\mathsf{B}}(T, (\boldsymbol{a}, \boldsymbol{b})) < \infty$ by **III.2.1**, lemma 1 and **IV.5.1**, definition 1. In view of theorem 1 it follows that the properties BV, eBV, $BV\mathsf{B}$ are equivalent. Since $BV\mathsf{B}$ and $sBV\mathsf{B}$ are equivalent by lemma 1, the theorem follows.

V.1.4. Absolute continuity. Again, the bounded continuous transformation $T: (\boldsymbol{a}, \boldsymbol{b}) \to R^1$, given as in **V.1.2** (1), is considered fixed.

Lemma 1. If T is $AC\mathsf{B}$ in $(\boldsymbol{a}, \boldsymbol{b})$, then T is also $sAC\mathsf{B}$ in $(\boldsymbol{a}, \boldsymbol{b})$ (see **IV.5.2** and **IV.5.3**, definition 2).

Proof. Assume that T is $AC\mathsf{B}$ in $(\boldsymbol{a}, \boldsymbol{b})$. By **IV.5.2**, theorem 1, T is then also $BV\mathsf{B}$ in $(\boldsymbol{a}, \boldsymbol{b})$. By **V.1.3**, lemma 1, it follows that the generalized Jacobian $J_{\mathsf{B}}(u, T)$ is L-summable in $(\boldsymbol{a}, \boldsymbol{b})$, and hence T is $sAC\mathsf{B}$ in $(\boldsymbol{a}, \boldsymbol{b})$.

Lemma 2. Suppose that T [considered as a function $T(u)$, $u \in (\boldsymbol{a}, \boldsymbol{b})$] is AC in $(\boldsymbol{a}, \boldsymbol{b})$, in the sense of **V.1.1**, definition 2. Then T is $AC\mathsf{B}$ in $(\boldsymbol{a}, \boldsymbol{b})$.

Proof. Consider any non-degenerate open interval $(a, b) \subset (\boldsymbol{a}, \boldsymbol{b})$. Since T is bounded, its oscillation $\omega(T, (a, b))$ is clearly finite. If $\sigma > 0$ is arbitrarily assigned, then by the definition of $\omega(T, (a, b))$ there exist numbers α, β such that

$$\alpha, \beta \in (a, b), \quad |T(\beta) - T(\alpha)| > \omega(T, (a, b)) - \sigma.$$

Assign now $\varepsilon > 0$. Consider any finite system of non-degenerate open intervals $(a_1, b_1), \ldots, (a_m, b_m)$ such that

$$(a_i, b_i) \subset (\boldsymbol{a}, \boldsymbol{b}), \quad i = 1, \ldots, m, \tag{1}$$

$$(a_i, b_i) \cap (a_j, b_j) = \emptyset \quad \text{if} \quad i \neq j. \tag{2}$$

By the preceding remarks, there exist numbers α_i, β_i such that

$$\alpha_i, \beta_i \in (a_i, b_i), \quad |T(\beta_i) - T(\alpha_i)| > \omega(T, (a_i, b_i)) - \frac{\varepsilon}{2m},$$

for $i = 1, \ldots, m$. Observing that

$$\omega(T, (a_i, b_i)) = G(a_i, b_i)$$

by **V.1.3**, lemma 2, we obtain the inequality

$$\sum_{i=1}^{m} G(a_i, b_i) < \frac{\varepsilon}{2} + \sum_{i=1}^{m} |T(\beta_i) - T(\alpha_i)|. \tag{3}$$

Now since T is AC in $(\boldsymbol{a}, \boldsymbol{b})$ by assumption, we shall have

$$\sum_{i=1}^{m} |T(\beta_i) - T(\alpha_i)| < \frac{\varepsilon}{2}, \tag{4}$$

provided only that

$$\sum_{i=1}^{m} |\beta_i - \alpha_i| < \eta = \eta\left(\frac{\varepsilon}{2}\right), \tag{5}$$

where η corresponds to $\varepsilon/2$ in the sense of the definition of absolute continuity. Since $\alpha_i, \beta_i \in (a_i, b_i)$, (5) will certainly hold if

$$\sum_{i=1}^{m} |b_i - a_i| < \eta. \tag{6}$$

Accordingly, if (1), (2), (6) are satisfied, then from (3) and (4) it follows that

$$\sum_{i=1}^{m} G(a_i, b_i) < \varepsilon.$$

Thus the function $G(a, b)$ of open intervals $(a, b) \subset (\boldsymbol{a}, \boldsymbol{b})$ is absolutely continuous in $(\boldsymbol{a}, \boldsymbol{b})$. Hence, by definition, T is ACB in $(\boldsymbol{a}, \boldsymbol{b})$, and the lemma is proved.

Lemma 3. If T is eAC in $(\boldsymbol{a}, \boldsymbol{b})$, then it is also AC in $(\boldsymbol{a}, \boldsymbol{b})$.

Proof. Consider any non-degenerate closed interval $[a, b] \subset (\boldsymbol{a}, \boldsymbol{b})$, $a < b$. By **IV.4.4** (17) and the *remark* at the end of **IV.4.4**, we have the formula

$$\int_a^b J_e(u, T)\, d\mathsf{L} = \int \mu(x, T, (a, b))\, d\mathsf{L}.$$

On the other hand, **V.1.2**, lemma 3 yields

$$\int \mu(x, T, (a, b))\, d\mathsf{L} = T(b) - T(a),$$

and hence

$$\int_a^b J_e(u, T)\, d\mathsf{L} = T(b) - T(a). \tag{7}$$

Note that since T is eAC in $(\boldsymbol{a}, \boldsymbol{b})$, it is also eBV there. Thus $J_e(u, T)$ is L-summable in $(\boldsymbol{a}, \boldsymbol{b})$ by **IV.4.3**, theorem 1. Hence (see **III.1.1**, lemma 33) for every $\varepsilon > 0$ there exists an $\eta = \eta(\varepsilon) > 0$ such that

$$\int_S |J_e(u, T)|\, d\mathsf{L} < \varepsilon$$

whenever S is an L-measurable set satisfying the conditions $S \subset (\boldsymbol{a}, \boldsymbol{b})$, $\mathsf{L}S < \eta$. Assign now $\varepsilon > 0$, and consider any finite system of non-degenerate closed intervals $[a_1, b_1], \ldots, [a_m, b_m]$ such that

$$a_i < b_i, \quad [a_i, b_i] \subset (\boldsymbol{a}, \boldsymbol{b}), \quad i = 1, \ldots, m, \tag{8}$$

$$(a_i, b_i) \cap (a_j, b_j) = \emptyset \quad \text{if} \quad i \neq j, \tag{9}$$

$$\sum_{i=1}^{m} |b_i - a_i| < \eta. \tag{10}$$

Put $S = \bigcup (a_i, b_i)$, $i = 1, \ldots, m$. Then $\mathsf{L}S < \eta$, and hence

$$\int_S |J_e(u, T)|\, d\mathsf{L} = \sum_{i=1}^{m} \int_{a_i}^{b_i} |J_e(u, T)\, d\mathsf{L} < \varepsilon.$$

In view of (7) it follows that

$$\sum_{i=1}^{m} |T(b_i) - T(a_i)| < \varepsilon$$

whenever (8), (9), (10) hold. Thus T is AC in $(\boldsymbol{a}, \boldsymbol{b})$, and the lemma is proved.

Theorem. In R^1 the properties AC, $AC\mathsf{B}$, $sAC\mathsf{B}$, eAC are equivalent. Explicitly, if T possesses in $(\boldsymbol{a}, \boldsymbol{b})$ one of these four properties, then it also possesses the other three.

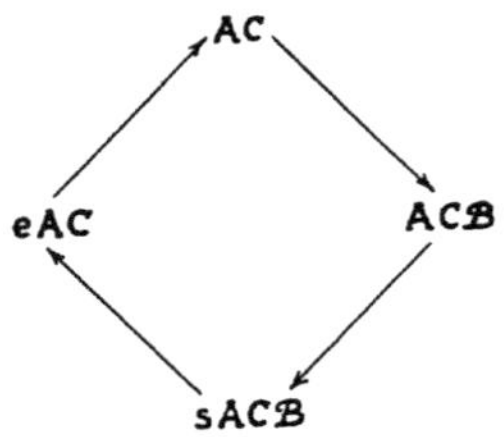

Fig. 43.

Proof. The diagram in Fig. 43, in which the arrows stand for logical implication, summarizes the available relevant information.

The arrows from $AC\mathsf{B}$ to $sAC\mathsf{B}$, from AC to $AC\mathsf{B}$, and from eAC to AC represent the lemmas 1, 2, and 3 respectively. Finally, the arrow from $sAC\mathsf{B}$ to eAC represents **IV.5.3**, theorem 1, according to which the $sAC\mathsf{B}$ property implies the eAC property. It is clear from the diagram that any one of the properties AC, $AC\mathsf{B}$, $sAC\mathsf{B}$, eAC implies the other three, and theorem follows.

V.1.5. Derivatives. The bounded continuous transformation $T\colon(\boldsymbol{a}, \boldsymbol{b}) \to R^1$, given as in **V.1.2**, is considered fixed throughout this section. As noted in **V.1.1**, T may be thought of as a real-valued function $x = T(u)$, $u \in (\boldsymbol{a}, \boldsymbol{b})$. Accordingly, we can speak of the derivative $T'(u)$, which may or may not exist at a given point $u \in (\boldsymbol{a}, \boldsymbol{b})$. Our objective in this section is to compare the derivative $T'(u)$ with the generalized Jacobians $J_{\mathsf{B}}(u, T)$ and $J_e(u, T)$ (see **IV.5.3** and **IV.4.3**). We assume throughout that $\boldsymbol{a} < \boldsymbol{b}$.

Lemma. Assume that T is eBV in $(\boldsymbol{a}, \boldsymbol{b})$, and consider a point $u_0 \in (\boldsymbol{a}, \boldsymbol{b})$. Then $T'(u_0)$ exists if and only if $J_e(u_0, T)$ exists, and

$$T'(u_0) = J_e(u_0, T) \tag{1}$$

if both exist.

Proof. Take any two numbers a, b such that

$$\boldsymbol{a} < a < u_0 < b < \boldsymbol{b}. \tag{2}$$

By **IV.4.3**, lemma 4 we have then

$$\mathfrak{A}_e((a, b)) = \int \mu(x, T, (a, b))\, d\mathsf{L},$$

while **V.1.2**, lemma 3 yields

$$\int \mu(x, T, (a, b))\, d\mathsf{L} = T(b) - T(a).$$

Hence

$$\mathfrak{A}_e((a, b)) = T(b) - T(a). \tag{3}$$

By definition (see **IV.4.3**), the existence of $J_e(u_0, T)$ is equivalent to the relation

$$\frac{\mathfrak{A}_e((a,b))}{b-a} \to J_e(u_0, T) \quad \text{for} \quad a \to u_0, \quad b \to u_0. \tag{4}$$

On the other hand (see **V.1.1**, lemma 13), the existence of $T'(u_0)$ is equivalent to the relation

$$\frac{T(b)-T(a)}{b-a} \to T'(u_0) \quad \text{for} \quad a \to u_0, \quad b \to u_0. \tag{5}$$

In either case, a and b are subjected to the conditions (2). In view of (3), the equivalence of (4) and (5) as well as the validity of (1) is obvious.

Theorem. Assume that T is eBV in $(\boldsymbol{a}, \boldsymbol{b})$. Then the derivative $T'(u)$ and the generalized Jacobians $J_{\mathsf{B}}(u, T)$ and $J_e(u, T)$ exist a.e. in $(\boldsymbol{a}, \boldsymbol{b})$, and

$$T'(u) = J_e(u, T) = J_{\mathsf{B}}(u, T) \quad \text{a.e. in } (\boldsymbol{a}, \boldsymbol{b}). \tag{6}$$

Proof. $J_e(u, T)$ exists a.e. in $(\boldsymbol{a}, \boldsymbol{b})$ by **IV.4.3**, theorem 1. In view of the preceding lemma it follows that $T'(u)$ exists a.e. in $(\boldsymbol{a}, \boldsymbol{b})$ and

$$T'(u) = J_e(u, T) \quad \text{a.e. in } (\boldsymbol{a}, \boldsymbol{b}). \tag{7}$$

Note that T is $BV\mathsf{B}$ in $(\boldsymbol{a}, \boldsymbol{b})$ by **V.1.3**, theorem 2. Hence $J_{\mathsf{B}}(u, T)$ exists a.e. in $(\boldsymbol{a}, \boldsymbol{b})$ by **IV.5.3**. So there remains to show that

$$J_e(u, T) = J_{\mathsf{B}}(u, T) \quad \text{a.e. in } (\boldsymbol{a}, \boldsymbol{b}). \tag{8}$$

Now we have

$$|J_e(u, T)| = D_e(u, T) \quad \text{a.e. in } (\boldsymbol{a}, \boldsymbol{b}) \tag{9}$$

by **IV.4.3**, theorem 2, and

$$J_{\mathsf{B}}(u, T) = i_{\mathsf{B}}(u, T)\, D_{\mathsf{B}}(u, T) \quad \text{a.e. in } (\boldsymbol{a}, \boldsymbol{b}) \tag{10}$$

by **IV.5.3** (1). Since $|i_{\mathsf{B}}(u, T)| \leq 1$ by **V.1.2**, lemma 6, we infer from (10) that

$$|J_{\mathsf{B}}(u, T)| \leq D_{\mathsf{B}}(u, T) \quad \text{a.e. in } (\boldsymbol{a}, \boldsymbol{b}). \tag{11}$$

By definition [see **IV.5.1** (7), (8)], $D_{\mathsf{B}}(u, T)$ is the derivative of the interval function

$$\nu_{\mathsf{B}}\big((a, b)\big) = \int N\big(x, T, (a, b)\big)\, d\mathsf{L}, \quad (a, b) \subset (\boldsymbol{a}, \boldsymbol{b}),$$

and [see **IV.4.3** (13), (15)] $D_e(u, T)$ is the derivative of the interval function

$$\mathfrak{K}((a, b)) = \int K(x, T, (a, b))\, d\mathsf{L}, \quad (a, b) < (\boldsymbol{a}, \boldsymbol{b}).$$

By **V.1.2**, lemma 10 we have

$$K(x, T, (a, b)) = N(x, T, (a, b)) \quad \text{a.e. in } R^1,$$

and hence

$$\mathfrak{K}((a, b)) = v_{\mathsf{B}}((a, b)), \quad (a, b) < (\boldsymbol{a}, \boldsymbol{b}).$$

Thus clearly

$$D_e(u, T) = D_{\mathsf{B}}(u, T) \quad \text{a.e. in } (\boldsymbol{a}, \boldsymbol{b}). \tag{12}$$

From (11), (12), (9) we obtain

$$|J_{\mathsf{B}}(u, T)| \leq |J_e(u, T)| \quad \text{a.e. in } (\boldsymbol{a}, \boldsymbol{b}). \tag{13}$$

Now denote by U the set of those points $u \in (\boldsymbol{a}, \boldsymbol{b})$ where $J_e(u, T)$ exists and is different from zero. From (13) we see that

$$J_{\mathsf{B}}(u, T) = 0 = J_e(u, T) \quad \text{a.e. on} \quad (\boldsymbol{a}, \boldsymbol{b}) - U. \tag{14}$$

Consider now a point $u_0 \in U$. Then $J_e(u_0, T)$ exists and is different from zero. By the preceding lemma, it follows that $T'(u_0)$ exists and

$$T'(u_0) = J_e(u_0, T) \neq 0. \tag{15}$$

From **V.1.2**, lemma 11 we conclude now, in view of (15), that

$$i_{\mathsf{B}}(u_0, T) = \operatorname{sgn} T'(u_0) = \operatorname{sgn} J_e(u_0, T) \quad \text{if} \quad u_0 \in U. \tag{16}$$

By (10), (12), (9), (16) it follows that

$$\begin{aligned} J_{\mathsf{B}}(u, T) &= D_{\mathsf{B}}(u, T)\, i_{\mathsf{B}}(u, T) = D_e(u, T)\, i_{\mathsf{B}}(u, T) \\ &= |J_e(u, T)| \operatorname{sgn} J_e(u, T) = J_e(u, T) \end{aligned}$$

a.e. on U. Thus (8) follows in view of (14), and the theorem is proved.

V.1.6. Applications to real-valued functions of a real variable. Let there be given a real-valued, bounded, continuous function $T(u)$, $u \in (\boldsymbol{a}, \boldsymbol{b})$, where $\boldsymbol{a} < \boldsymbol{b}$. On assigning to each point $u \in (\boldsymbol{a}, \boldsymbol{b})$ the point $x = T(u)$ of R^1 as its image, we obtain a bounded continuous transformation $T: (\boldsymbol{a}, \boldsymbol{b}) \to R^1$. The terms *transformation* T and *function* T will be used interchangeably in this section, and T will be considered as fixed. The general theory of continuous transformations in R^n, developed in Part IV, yields in the case $n = 1$ a series of fundamental classical theorems for real-valued functions of a real variable, as we shall see presently. It should be noted that these theorems can be derived far more economically in a direct manner. Furthermore, the assumption of continuity (necessitated by the formulation of the general theory in Part IV) could be dispensed with in some cases. Yet the discussion of these applications to functions of a real variable serves a useful purpose, since additional light is thrown upon the basic concepts and the scope of the general theory. Furthermore, the *equivalence theorems* discussed in **V.1.2** to **V.1.5** yield a variety of illuminating geometrical interpretations for the concepts of bounded variation, absolute continuity, and derivative.

Theorem 1. T is BV in $(\boldsymbol{a}, \boldsymbol{b})$ if and only if the crude multiplicity function $N(x, T, (\boldsymbol{a}, \boldsymbol{b}))$ is L-summable in R^1.

Proof. By **V.1.3**, theorem 2, T is BV in $(\boldsymbol{a}, \boldsymbol{b})$ if and only if it is $BV\mathsf{B}$ there. By **IV.5.1**, theorem 1, T is $BV\mathsf{B}$ in $(\boldsymbol{a}, \boldsymbol{b})$ if and only if $N(x, T, (\boldsymbol{a}, \boldsymbol{b}))$ is L-summable in R^1, and the present theorem follows.

Theorem 2. If T is BV in $(\boldsymbol{a}, \boldsymbol{b})$, then

$$V(T, (\boldsymbol{a}, \boldsymbol{b})) = \int N(x, T, (\boldsymbol{a}, \boldsymbol{b}))\, d\mathsf{L}.$$

Proof. By assumption, $V(T, (\boldsymbol{a}, \boldsymbol{b})) < \infty$, and thus

$$V_{\mathsf{B}}(T, (\boldsymbol{a}, \boldsymbol{b})) = V(T, (\boldsymbol{a}, \boldsymbol{b})) < \infty$$

in view of **V.1.3**, theorem 1. Hence

$$V_{\mathsf{B}}(T, (\boldsymbol{a}, \boldsymbol{b})) = \int N(x, T, (\boldsymbol{a}, \boldsymbol{b}))\, d\mathsf{L}$$

by **IV.5.1**, lemma 2, and the theorem follows.

Theorem 3. Assume that T is BV in $(\boldsymbol{a}, \boldsymbol{b})$. Then the derivative $T'(u)$ exists a.e. in $(\boldsymbol{a}, \boldsymbol{b})$ and is L-summable there. Furthermore

$$\int_{\boldsymbol{a}}^{\boldsymbol{b}} |T'(u)|\, d\mathsf{L} \leqq V(T, (\boldsymbol{a}, \boldsymbol{b})),$$

where the sign of equality holds if and only if T is AC in $(\boldsymbol{a}, \boldsymbol{b})$.

Proof. By theorem 2 in **V.1.3**, T is eBV in $(\boldsymbol{a}, \boldsymbol{b})$. Hence (see **IV.4.3**, theorem 3 and **IV.4.1**, definition 2), $J_e(u, T)$ exists a.e. in $(\boldsymbol{a}, \boldsymbol{b})$ and is L-summable there, and

$$\int_{\boldsymbol{a}}^{\boldsymbol{b}} |J_e(u, T)|\, d\mathsf{L} \leqq V_e(T, (\boldsymbol{a}, \boldsymbol{b})),$$

where the sign of equality holds if and only if T is eAC in $(\boldsymbol{a}, \boldsymbol{b})$. By the theorem in **V.1.5** it follows that $T'(u)$ exists and is equal to $J_e(u, T)$

a.e. in $(\boldsymbol{a}, \boldsymbol{b})$. Thus

$$\int_{\boldsymbol{a}}^{\boldsymbol{b}} |J_e(u, T)| \, d\mathsf{L} = \int_{\boldsymbol{a}}^{\boldsymbol{b}} |T'(u)| \, d\mathsf{L}.$$

Furthermore

$$V_e(T, (\boldsymbol{a}, \boldsymbol{b})) = V(T, (\boldsymbol{a}, \boldsymbol{b}))$$

by **V.1.3**, theorem 1. Finally, the property eAC is equivalent to the property AC by the theorem in **V.1.4**, and thus the present theorem follows.

Theorem 4. Assume that T is BV in $(\boldsymbol{a}, \boldsymbol{b})$. Then T is AC in $(\boldsymbol{a}, \boldsymbol{b})$ if and only if it satisfies there the condition (N) (see **IV.1.4**).

Proof. T is $BV\mathsf{B}$ in $(\boldsymbol{a}, \boldsymbol{b})$ by **V.1.3**, theorem 2. By **IV.5.2**, theorem 3 it follows that T is $AC\mathsf{B}$ in $(\boldsymbol{a}, \boldsymbol{b})$ if and only if it satisfies there the condition (N). Since the properties $AC\mathsf{B}$ and AC are equivalent by the theorem in **V.1.4**, the present theorem follows.

The preceding theorems are concerned with a function $T(u)$ defined in an *open* interval. In Analysis, it is more customary to consider functions given in a *closed* interval. Let us note that the passage to this latter situation is practically trivial. Consider, indeed, a real-valued function $T(u)$ which is defined and continuous in a non-degenerate closed interval $[\boldsymbol{a}, \boldsymbol{b}]$, where $\boldsymbol{a} < \boldsymbol{b}$. Make the following observations.

(i) By **V.1.1**, lemmas 7 and 8, T is BV (or AC) in $[\boldsymbol{a}, \boldsymbol{b}]$ if and only if it is BV (or AC) in $(\boldsymbol{a}, \boldsymbol{b})$.

(ii) By **V.1.1**, lemma 6, we have $V(T, [\boldsymbol{a}, \boldsymbol{b}]) = V(T, (\boldsymbol{a}, \boldsymbol{b}))$.

(iii) Since $[\boldsymbol{a}, \boldsymbol{b}] - (\boldsymbol{a}, \boldsymbol{b})$ consists of just the two points $\boldsymbol{a}$ and $\boldsymbol{b}$, it is clear that the terms a.e. in $[\boldsymbol{a}, \boldsymbol{b}]$ and a.e. in $(\boldsymbol{a}, \boldsymbol{b})$ are equivalent. It is also clear that T satisfies the condition (N) in $[\boldsymbol{a}, \boldsymbol{b}]$ if and only if it satisfies the condition (N) in $(\boldsymbol{a}, \boldsymbol{b})$. Finally,

$$N(x, T, [\boldsymbol{a}, \boldsymbol{b}]) = N(x, T, (\boldsymbol{a}, \boldsymbol{b})) \quad \text{a.e. in } R^1,$$

since this relation clearly holds if $x \neq T(\boldsymbol{a}), T(\boldsymbol{b})$.

In view of these remarks, the following theorem appears as an immediate corollary of the preceding theorems [parts (a), (b), (c), (d) corresponding to the theorems 1, 2, 3, 4 respectively].

Theorem 5. Let $T(u)$ be a real-valued continuous function in the closed interval $[\boldsymbol{a}, \boldsymbol{b}]$, where $\boldsymbol{a} < \boldsymbol{b}$. Then the following holds.

(a) T is BV in $[\boldsymbol{a}, \boldsymbol{b}]$ if and only if $N(x, T, [\boldsymbol{a}, \boldsymbol{b}])$ is L-summable in R^1.

(b) If T is BV in $[\boldsymbol{a}, \boldsymbol{b}]$, then

$$V(T, [\boldsymbol{a}, \boldsymbol{b}]) = \int N(x, T, [\boldsymbol{a}, \boldsymbol{b}]) \, d\mathsf{L}.$$

(c) Assume that T is BV in $[\boldsymbol{a}, \boldsymbol{b}]$. Then the derivative $T'(u)$ exists a.e. in $[\boldsymbol{a}, \boldsymbol{b}]$ and is L-summable there. Furthermore

$$\int_{\boldsymbol{a}}^{\boldsymbol{b}} |T'(u)|\, d\mathsf{L} \leqq V(T, [\boldsymbol{a}, \boldsymbol{b}]),$$

where the sign of equality holds if and only if T is AC in $[\boldsymbol{a}, \boldsymbol{b}]$.

(d) Assume that T is BV in $[\boldsymbol{a}, \boldsymbol{b}]$. Then T is AC in $[\boldsymbol{a}, \boldsymbol{b}]$ if and only if it satisfies there the condition (N).

We now proceed to discuss *transformation formulas* for definite integrals in R^1. In this connection, the bounded continuous transformation $T: (\boldsymbol{a}, \boldsymbol{b}) \to R^1$ will be assumed to be AC in $(\boldsymbol{a}, \boldsymbol{b})$. By **V.1.1**, remark 1, T can then be extended continuously to the closed interval $[\boldsymbol{a}, \boldsymbol{b}]$. Accordingly, we can assume without loss of generality that T is defined, continuous, and AC in the closed interval $[\boldsymbol{a}, \boldsymbol{b}]$ (see **V.1.1**, remark 1). To conform to general usage in the theory of functions of a single real variable, we shall write dx (or du) instead of $d\mathsf{L}$ in formulas involving definite integrals.

Theorem 6. Assume that the real-valued continuous function $T(u)$ is defined and AC in the closed interval $[\boldsymbol{a}, \boldsymbol{b}]$, where $\boldsymbol{a} < \boldsymbol{b}$. Let $H(x)$ be a real-valued, finite-valued, L-measurable function in R^1. Then

$$\int_{\boldsymbol{a}}^{\boldsymbol{b}} H(T(u))\, |T'(u)|\, du = \int H(x)\, N(x, T, [\boldsymbol{a}, \boldsymbol{b}])\, dx,$$

as soon as one of the two integrals involved exists.

Proof. T is eAC in $(\boldsymbol{a}, \boldsymbol{b})$ by **V.1.1**, lemma 2 and **V.1.4**. Observe that

$$J_e(u, T) = T'(u) \quad \text{a.e. in} \quad (\boldsymbol{a}, \boldsymbol{b})$$

by **V.1.5**, and

$$K(x, T, (\boldsymbol{a}, \boldsymbol{b})) = N(x, T, (\boldsymbol{a}, \boldsymbol{b})) \ \text{a.e. in} \ R^1$$

by **V.1.2**, lemma 10. We noted above that

$$N(x, T, (\boldsymbol{a}, \boldsymbol{b})) = N(x, T, [\boldsymbol{a}, \boldsymbol{b}]) \quad \text{a.e. in } R^1.$$

In view of these remarks, the present theorem appears as a direct consequence of **IV.4.4**, theorem 1, according to which

$$\int_{\boldsymbol{a}}^{\boldsymbol{b}} H(T(u))\, |J_e(u, T)|\, du = \int H(x)\, K(x, T, (\boldsymbol{a}, \boldsymbol{b}))\, dx,$$

as soon as one of the two integrals involved exists.

Theorem 7. Assume that the real-valued continuous function $T(u)$ is defined and AC in the closed interval $[\boldsymbol{a}, \boldsymbol{b}]$, where $\boldsymbol{a} < \boldsymbol{b}$. Let $H(x)$

be a real-valued, finite-valued, L-measurable function in R^1. Then

$$\int_{\boldsymbol{a}}^{\boldsymbol{b}} H(T(u))\, T'(u)\, du = \int_{T(\boldsymbol{a})}^{T(\boldsymbol{b})} H(x)\, dx,$$

as soon as the integral on the left exists.

Proof. T is eAC in $(\boldsymbol{a}, \boldsymbol{b})$ and $T'(u) = J_e(u, T)$ a.e. in $(\boldsymbol{a}, \boldsymbol{b})$, as it has been noted in the proof of theorem 6. Thus if $H(T(u))\, T'(u)$ is L-summable in $[\boldsymbol{a}, \boldsymbol{b}]$ and hence also in $(\boldsymbol{a}, \boldsymbol{b})$, then $H(T(u))\, J_e(u, T)$ is also L-summable there, and

$$\int_{\boldsymbol{a}}^{\boldsymbol{b}} H(T(u))\, T'(u)\, du = \int_{\boldsymbol{a}}^{\boldsymbol{b}} H(T(u))\, J_e(u, T)\, du = \int H(x)\, \mu(x, T, (\boldsymbol{a}, \boldsymbol{b}))\, dx$$

by **IV.4.4**, theorem 3. By **V.1.2**, lemma 1 and lemma 2 we have

$$\mu(x, T, (\boldsymbol{a}, \boldsymbol{b})) = \operatorname{sgn}(T(\boldsymbol{b}) - T(\boldsymbol{a})) \quad \text{if} \quad x \in (T(\boldsymbol{a}), T(\boldsymbol{b})),$$

$$\mu(x, T, (\boldsymbol{a}, \boldsymbol{b})) = 0 \quad \text{if} \quad x \in \mathsf{C}[T(\boldsymbol{a}), T(\boldsymbol{b})].$$

Thus obviously

$$\int H(x)\, \mu(x, T, (\boldsymbol{a}, \boldsymbol{b}))\, dx = \int_{T(\boldsymbol{a})}^{T(\boldsymbol{b})} H(x)\, dx,$$

and the theorem follows.

If T is AC in $[\boldsymbol{a}, \boldsymbol{b}]$, then it is also AC in every closed interval $[a, b]$ in $[\boldsymbol{a}, \boldsymbol{b}]$. Accordingly, the preceding two theorems remain valid if $[\boldsymbol{a}, \boldsymbol{b}]$ is replaced by any closed interval $[a, b] \subset [\boldsymbol{a}, \boldsymbol{b}]$.

A noteworthy special case arises when $H(x) \equiv 1$. Since $T'(u)$ is L-summable in $(\boldsymbol{a}, \boldsymbol{b})$ by theorem 3, we obtain for $H(x) \equiv 1$ the following statements [see also part (c) of theorem 5 and **V.1.4** (7)].

Theorem 8. Assume that the real-valued continuous function $T(u)$ is defined and AC in the closed interval $[\boldsymbol{a}, \boldsymbol{b}]$, where $\boldsymbol{a} < \boldsymbol{b}$. Let a, b be two numbers such that $\boldsymbol{a} < a < b < \boldsymbol{b}$. Then

$$\int_a^b |T'(u)|\, du = V(T, [a, b]), \quad \int_a^b T'(u)\, du = T(b) - T(a).$$

Observe that theorem 7 presents a lack of symmetry with respect to the two integrals involved (in contradistinction with theorem 6 which is symmetric from this point of view). Examples show that this discrepancy is unavoidable in the general case. On the other hand, a symmetric form of theorem 7 is valid in the special case when $T(u)$ is *monotone*. Let us recall the following definitions. If $T(u)$ is a real-valued function in $[\boldsymbol{a}, \boldsymbol{b}]$, where $\boldsymbol{a} < \boldsymbol{b}$, then $T(u)$ is termed *strictly monotone* in $[\boldsymbol{a}, \boldsymbol{b}]$ if either always $T(a) < T(b)$ whenever $\boldsymbol{a} \leq a < b \leq \boldsymbol{b}$, or always $T(a) > T(b)$ whenever $\boldsymbol{a} \leq a < b \leq \boldsymbol{b}$. Furthermore, $T(u)$ is said to be *monotone in the large sense* in $[\boldsymbol{a}, \boldsymbol{b}]$ if either always $T(a) \leq T(b)$ whenever $\boldsymbol{a} \leq a < b \leq \boldsymbol{b}$, or always $T(a) \geq T(b)$ whenever $\boldsymbol{a} \leq a < b \leq \boldsymbol{b}$; in the first case, T is said to be non-decreasing in $[\boldsymbol{a}, \boldsymbol{b}]$, while in the second case T is said to be non-increasing in $[\boldsymbol{a}, \boldsymbol{b}]$. Now if T is strictly monotone in $[\boldsymbol{a}, \boldsymbol{b}]$, then (considered as a transformation) it is one-to-one there, and accordingly **IV.4.6**, theorem 2 yields a symmetric form of theorem 7 for this case. We shall show presently that a similar conclusion holds if T is monotone in the large sense.

Lemma. Assume that the real-valued function $T(u)$ is continuous and monotone in the large sense in $[\boldsymbol{a}, \boldsymbol{b}]$, where $\boldsymbol{a} < \boldsymbol{b}$. Then

$$N(x, T, [\boldsymbol{a}, \boldsymbol{b}]) = 0 \quad \text{if} \quad x \notin [T(\boldsymbol{a}), T(\boldsymbol{b})], \tag{1}$$

$$N(x, T, [\boldsymbol{a}, \boldsymbol{b}]) = 1 \quad \text{a.e. in} \quad [T(\boldsymbol{a}), T(\boldsymbol{b})]. \tag{2}$$

Proof. Suppose, for definiteness, that T is non-decreasing in $[\boldsymbol{a}, \boldsymbol{b}]$ (the case when T is non-increasing can be settled in an entirely similar manner). If $u \in [\boldsymbol{a}, \boldsymbol{b}]$, then $\boldsymbol{a} \leq u \leq \boldsymbol{b}$ and hence $T(\boldsymbol{a}) \leq T(u) \leq T(\boldsymbol{b})$. Thus clearly

$$T[\boldsymbol{a}, \boldsymbol{b}] \subset [T(\boldsymbol{a}), T(\boldsymbol{b})], \tag{3}$$

and (1) follows. If $x \in [T(\boldsymbol{a}), T(\boldsymbol{b})]$, then $T(\boldsymbol{a}) \leq x \leq T(\boldsymbol{b})$. Since T is continuous, it follows that $x = T(u)$ for some $u \in [\boldsymbol{a}, \boldsymbol{b}]$. Hence

$$N(x, T, [\boldsymbol{a}, \boldsymbol{b}]) \geq 1 \quad \text{if} \quad x \in [T(\boldsymbol{a}), T(\boldsymbol{b})]. \tag{4}$$

Consider now the auxiliary set

$$F = \{x \mid N(x, T, [\boldsymbol{a}, \boldsymbol{b}]) > 1\}. \tag{5}$$

We assert that

$$F \subset T S_{\max}, \tag{6}$$

where $S_{\max}$ is the set of those points $u \in (\boldsymbol{a}, \boldsymbol{b})$ where $T(u)$ has a weak local maximum (see **V.1.2**, lemma 8). To verify (6), note that if $x_0 \in F$, then there exist (at least) two points u' and $u'' > u'$ in $[\boldsymbol{a}, \boldsymbol{b}]$ such that $T(u') = T(u'') = x_0$. Since T is non-decreasing, it follows that $T(u) = x_0$ for $u' < u < u''$. Thus if $u_0 \in (u', u'')$, then clearly T has a weak local maximum (and indeed also a weak local minimum) at u_0. Hence $u_0 \in S_{\max}$ and $T(u_0) = x_0$, and thus $x_0 \in T S_{\max}$. So (6) is verified. In view of (1), (4), (6) we have

$$N(x, T, [\boldsymbol{a}, \boldsymbol{b}]) = 1 \quad \text{if} \quad x \in [T(\boldsymbol{a}), T(\boldsymbol{b})] - T S_{\max}. \tag{7}$$

Now the set $T S_{\max}$ is countable (and hence of L-measure zero) by **V.1.2**, lemma 8. Thus (7) implies (2), and the lemma is proved.

Theorem 9. Assume that the real-valued function $T(u)$ is continuous, AC, and monotone in the large sense in the closed interval $[\boldsymbol{a}, \boldsymbol{b}]$, where $\boldsymbol{a} < \boldsymbol{b}$. Let $H(x)$ be a real-valued, finite-valued, L-measurable function in R^1. Then

$$\int_{\boldsymbol{a}}^{\boldsymbol{b}} H(T(u))\, T'(u)\, du = \int_{T(\boldsymbol{a})}^{T(\boldsymbol{b})} H(x)\, dx, \tag{8}$$

as soon as one of the two integrals involved exists.

Proof. The two cases when T is non-decreasing and non-increasing respectively can be treated in an entirely similar manner, and so we consider explicitly only the case when T is non-increasing in $[\boldsymbol{a}, \boldsymbol{b}]$. If the integral on the left in (8) exists, then (8) holds by theorem 7. There remains to verify (8) for the case when $H(x)$ is known to be L-summable in $[T(\boldsymbol{a}), T(\boldsymbol{b})]$. In view of the preceding lemma, we have

$$H(x) = H(x)\, N(x, T, [\boldsymbol{a}, \boldsymbol{b}]) \quad \text{a.e. in} \quad [T(\boldsymbol{a}), T(\boldsymbol{b})] \tag{9}$$

and thus $H(x)\, N(x, T, [\boldsymbol{a}, \boldsymbol{b}])$ is also L-summable in $[T(\boldsymbol{a}), T(\boldsymbol{b})]$. By theorem 6 we have therefore

$$\int_{\boldsymbol{b}}^{\boldsymbol{a}} H(T(u))\, |T'(u)|\, du = \int H(x)\, N(x, T, [\boldsymbol{a}, \boldsymbol{b}])\, dx. \tag{10}$$

Now since T is non-increasing, clearly

$$T'(u) \leq 0 \quad \text{a.e. in } [\boldsymbol{a}, \boldsymbol{b}], \tag{11}$$

and furthermore

$$T(\boldsymbol{b}) \leq T(\boldsymbol{a}). \tag{12}$$

Since $N(x, T, [\boldsymbol{a}, \boldsymbol{b}]) = 0$ for $x \notin [T(\boldsymbol{a}), T(\boldsymbol{b})]$ by the preceding lemma, we have

$$\int H(x)\, N(x, T, [\boldsymbol{a}, \boldsymbol{b}])\, dx = -\int_{T(\boldsymbol{a})}^{T(\boldsymbol{b})} H(x)\, dx \tag{13}$$

in view of (9) and (12), and

$$\int_{\boldsymbol{a}}^{\boldsymbol{b}} H(T(u))\, |T'(u)|\, du = -\int_{\boldsymbol{a}}^{\boldsymbol{b}} H(T(u))\, T'(u)\, du \tag{14}$$

in view of (11). The formula (8) now follows from (10), (13), (14).

In the special case when T is strictly monotone, theorem 9 yields the classical transformation formula for single integrals. Theorem 7 may be now interpreted as revealing the fact that the classical transformation formula remains valid even if T is not monotone, provided that the integral obtained by the substitution $x = T(u)$ exists. The assumption of absolute continuity on the part of T is essential, as simple examples show.

§ V.2. Local approximations in R^n.

V.2.1. Additional comments on linear transformations in R^n. Various basic facts relating to linear transformations in R^n have been already discussed in **II.2.5**. According to the terminology adopted there, a linear transformation $\mathfrak{L}$ from R^n into R^n is given by equations of the form

$$\mathfrak{L}: x^i = b^i + \sum_{j=1}^{n} a_{ij} u^j, \quad i = 1, \ldots, n, \tag{1}$$

where b^i, a_{ij} are real constants and $x = (x^1, \ldots, x^n)$ is the image of $u = (u^1, \ldots, u^n)$ under $\mathfrak{L}$. The determinant of the matrix (a_{ij}) is termed the determinant of $\mathfrak{L}$ and is denoted by $\det \mathfrak{L}$. Thus

$$\det \mathfrak{L} = \det (a_{ij}). \tag{2}$$

Recall that $\mathfrak{L}$ is termed non-singular or singular according as $\det \mathfrak{L} \neq 0$ or $\det \mathfrak{L} = 0$. If $\mathfrak{L}$ is non-singular, then (see **II.2.5**) $\mathfrak{L}$ is a homeomorphism from R^n onto R^n, and its inverse $\mathfrak{L}^{-1}$ is again a non-singular linear transformation. In the following discussion, it will be convenient to use the vector notation explained in **I.2.1**. Given $\mathfrak{L}$ as in (1), put

$$\Lambda(\mathfrak{L}) = 1 + n^2 \max |a_{ij}|, \tag{3}$$

where $\max |a_{ij}|$ denotes the largest one of the numbers $|a_{ij}|$.

Lemma 1. Let $\mathfrak{L}$ be given as in (1). If

$$u_1 = (u_1^1, \ldots, u_1^n), \quad u_2 = (u_2^1, \ldots, u_2^n)$$

are any two points in R^n, then

$$\|\mathfrak{L} u_2 - \mathfrak{L} u_1\| \leqq \Lambda(\mathfrak{L}) \|u_2 - u_1\|. \tag{4}$$

Proof. Put

$$\mathfrak{L} u_2 = x_2 = (x_2^1, \ldots, x_2^n),$$

$$\mathfrak{L} u_1 = x_1 = (x_1^1, \ldots, x_1^n).$$

Then (1) yields

$$x_2^i - x_1^i = \sum_{j=1}^{n} a_{ij} (u_2^j - u_1^j),$$

and hence

$$|x_2^i - x_1^i| \leqq \|u_2 - u_1\| \, n \max |a_{ij}|,$$

$$\|x_2 - x_1\| \leqq \sum_{i=1}^{n} |x_2^i - x_1^i| \leqq \|u_2 - u_1\| \, n^2 \max |a_{ij}|$$

and (4) follows in view of (3).

The preceding lemma shows that $\mathfrak{L}$ is Lipschitzian in R^n (see **V.2.3**, definition 1). The constant $\Lambda(\mathfrak{L})$ could be obviously replaced by a smaller one. However, the advantage of the choice (3) for $\Lambda(\mathfrak{L})$ is that this constant is always positive.

If $\mathfrak{L}$ is non-singular and

$$\mathfrak{L}^{-1}: u^i = \tilde{b}^i + \sum_{j=1}^{n} \tilde{a}_{ij} x^j, \quad i = 1, \ldots, n,$$

is the inverse of $\mathfrak{L}$, then we put [in conformity with (3)]

$$\Lambda(\mathfrak{L}^{-1}) = 1 + n^2 \max |\tilde{a}_{ij}|. \tag{5}$$

Application of lemma 1 to $\mathfrak{L}^{-1}$ yields the following statement.

Lemma 2. Assume that $\mathfrak{L}$ is non-singular, and let u_1, u_2 be any two points in R^n. Then

$$||u_2 - u_1|| \leq \Lambda(\mathfrak{L}^{-1}) \, ||\mathfrak{L} u_2 - \mathfrak{L} u_1||. \tag{6}$$

Given $\mathfrak{L}$ as in (1), select a point $u_0 = (u_0^1, \ldots, u_0^n)$ and put

$$\mathfrak{L} u_0 = x_0 = (x_0^1, \ldots, x_0^n). \tag{7}$$

Let $u = (u^1, \ldots, u^n)$ be a variable point in R^n. Put

$$\mathfrak{L} u = x = (x^1, \ldots, x^n). \tag{8}$$

The point u_0 is thought of as fixed. From (1), (7), (8) we obtain for $\mathfrak{L}$ the representation

$$\mathfrak{L}: x^i - x_0^i = \sum_{j=1}^{n} a_{ij}(u^j - u_0^j), \quad i = 1, \ldots, n, \tag{9}$$

which is convenient in studying $\mathfrak{L}$ in the vicinity of an assigned point u_0.

Lemma 3. Assume that $\mathfrak{L}$, given as in (9), is singular. Then there exist (constant) vectors

$$b_j = (b_j^1, \ldots, b_j^n), \quad j = 1, \ldots, n, \tag{10}$$

such that

(i) $||b_j|| = 1, \quad j = 1, \ldots, n,$
(ii) $b_i \cdot b_j = 0 \quad \text{for} \quad i \neq j,$
(iii) $(\mathfrak{L} u - x_0) \cdot b_n = 0 \quad \text{for every} \quad u \in R^n.$

Proof. In (ii) and (iii), the indicated products are scalar products (see **I.2.1**), and in (iii) the point $\mathfrak{L} u - x_0$ is interpreted as a vector. Consider the system of linear equations

$$\sum_{i=1}^{n} a_{ij} \beta^i = 0, \quad j = 1, \ldots, n, \tag{11}$$

where the coefficients a_{ij} are taken from (9), and $\beta^1, \ldots, \beta^n$ are the unknowns. Since $\mathfrak{L}$ is singular, the determinant of the system vanishes. Thus the system has a non-trivial solution, and hence it also has a solution $\beta^1, \ldots, \beta^n$ such that the sum of the squares of $\beta^1, \ldots, \beta^n$ is equal to 1. For the vector

$$b_n = (\beta^1, \ldots, \beta^n)$$

we have then $\|b_n\| = 1$, and from (11), (9), (8), (7) it follows by direct calculation that (iii) holds. Since b_n is a unit vector in R^n, we can select (see **I.2.1**, exercise 11) vectors $b_1, \ldots, b_{n-1}$ such that (i) and (ii) hold, and the lemma is proved.

Lemma 4. Assume that $\mathfrak{L}$, given as in (9), is non-singular. Let $a_1, \ldots, a_n$ be the vectors given by the formulas

$$a_j = (a_{1j}, \ldots, a_{nj}), \quad j = 1, \ldots, n, \tag{12}$$

where the components a_{ij} are taken from (9). For each positive number β, denote by $Q(\beta)$ the oriented n-cube with center at u_0 and side length 2β (see **I.2.2**). Put

$$P(\beta) = \mathfrak{L}\, Q(\beta). \tag{13}$$

Then the following holds.

(i) $P(\beta)$ is the n-parallelotope with center $x_0 = \mathfrak{L} u_0$ and edge vectors $2\beta a_1, \ldots, 2\beta a_n$ (see **I.2.2**).

(ii) $\mathsf{L}\, P(\beta) = |\det \mathfrak{L}|\, \mathsf{L}\, Q(\beta)$.

(iii) The parameter of regularity $\pi\big(P(\beta)\big)$ of $P(\beta)$ (see **III.1.1**, definition 3) satisfies the inequality

$$\pi\big(P(\beta)\big) \geq \frac{|\det \mathfrak{L}|}{n^n \Lambda(\mathfrak{L})^n}.$$

Proof. The points $u = (u^1, \ldots, u^n) \in Q(\beta)$ are characterized by the fact (see **I.2.2**, exercise 2) that there exist real numbers $\lambda_1, \ldots, \lambda_n$ such that

$$u^j - u_0^j = 2\lambda_j \beta, \quad j = 1, \ldots, n,$$

$$|\lambda_j| \leq 1/2, \quad j = 1, \ldots, n.$$

In view of (9) and (13) it follows that the points

$$x = (x^1, \ldots, x^n) \in P(\beta)$$

are characterized by the fact that there exist real numbers $\lambda_1, \ldots, \lambda_n$ such that

$$x^i = x_0^i + \sum_{j=1}^{n} 2\lambda_j \beta\, a_{ij}, \quad i = 1, \ldots, n, \tag{14}$$

$$|\lambda_j| \leq 1/2, \quad j = 1, \ldots, n. \tag{15}$$

In view of (12), the summation in (14) is equal to the i-th component of the vector

$$2\lambda_1 \beta a_1 + \cdots + 2\lambda_n \beta a_n.$$

Thus the equations (14) are equivalent to the vector equation

$$x = x_0 + \sum_{j=1}^{n} 2\lambda_j \beta a_j. \tag{16}$$

By definition (see **I.2.2**, exercise 4), the relations (16) and (15) yield precisely the points x of the n-parallelotope with center x_0 and edge vectors $2\beta a_1, \ldots, 2\beta a_n$, and thus (i) is proved. From (i) it follows that the matrix of the components of the edge vectors of the parallelotope $P(\beta)$ coincides with the matrix $(2\beta a_{ij})$. Hence (see **III.1.1**, lemma 16)

$$\mathsf{L}\, P(\beta) = |\det(a_{ij})|\, 2^n \beta^n. \tag{17}$$

Now since

$$\mathsf{L}\, Q(\beta) = 2^n \beta^n,$$

and $\det(a_{ij}) = \det \mathfrak{L}$ by (2), the relation (ii) follows from (17). Consider now any point

$$x = (x^1, \ldots, x^n) \in P(\beta).$$

In view of (13) there exists a point

$$u = (u^1, \ldots, u^n) \in Q(\beta)$$

such that $x = \mathfrak{L} u$. From (9) and (3) it is obvious that

$$|x^i - x_0^i| \leq n \beta \Lambda(\mathfrak{L}), \quad i = 1, \ldots, n.$$

Thus $P(\beta)$ is contained in the oriented n-cube with center x_0 and side length $2n\beta\Lambda(\mathfrak{L})$. The volume of this cube being equal to

$$n^n \Lambda(\mathfrak{L})^n 2^n \beta^n = n^n \Lambda(\mathfrak{L})^n \mathsf{L}\, Q(\beta),$$

we conclude from (ii) and **III.1.1**, definition 3 that

$$\pi\big(P(\beta)\big) \geq \frac{\mathsf{L}\, P(\beta)}{n^n \Lambda(\mathfrak{L})^n \mathsf{L}\, Q(\beta)} = \frac{|\det \mathfrak{L}|}{n^n \Lambda(\mathfrak{L})^n},$$

and the proof is complete.

V.2.2. Local linear approximations in R^n †. Let there be given a bounded continuous transformation

$$T: \boldsymbol{D} \to R^n, \tag{1}$$

† The first systematic use of the topological index in conjunction with total differentials has been made by H. Rademacher, Über partielle und totale Differenzierbarkeit von Funktionen mehrerer Variabeln [Math. Ann. **79**, 340–359 (1919) and **81**, 52–63 (1920)].

where $\boldsymbol{D}$ is a bounded domain in R^n. We can also represent T in the form (see **II.2.5**)

$$T: x^i = f_i(u^1, \ldots, u^n), \quad i = 1, \ldots, n, \tag{2}$$

where $f_1, \ldots, f_n$ are real-valued, bounded, continuous functions in $\boldsymbol{D}$ of the coordinates $u^1, \ldots, u^n$ of the point

$$u = (u^1, \ldots, u^n) \in \boldsymbol{D},$$

and $x^1, \ldots, x^n$ are the coordinates of the image point

$$T u = x = (x^1, \ldots, x^n) \in R^n.$$

Select a point

$$u_0 = (u_0^1, \ldots, u_0^n) \in \boldsymbol{D}, \tag{3}$$

and put

$$T u_0 = x_0 = (x_0^1, \ldots, x_0^n). \tag{4}$$

We shall discuss presently certain rather elementary facts which relate to the behavior of T in the vicinity of the assigned point $u_0 \in \boldsymbol{D}$. To avoid trivialities, we assume throughout this section that $n > 1$ (note that the case of continuous transformations in R^1 has been fully discussed in § **V.1**). Assume that the first partial derivatives of the coordinate functions f_i of T [see (2)] exist at u_0, and put

$$a_{ij} = \frac{\partial f_i(u_0)}{\partial u^j}, \quad i, j = 1, \ldots, n. \tag{5}$$

Introduce the linear transformation [see **V.2.1** (9)]

$$\mathfrak{L}: x^i - x_0^i = \sum_{j=1}^{n} a_{ij}(u^j - u_0^j), \quad i = 1, \ldots, n. \tag{6}$$

The determinant of the matrix

$$(a_{ij}) = \left(\frac{\partial f_i(u_0)}{\partial u^j}\right)$$

is termed the *ordinary Jacobian* $J(u_0, T)$ of T at u_0. Thus the ordinary Jacobian exists at a point $u_0 \in \boldsymbol{D}$ if and only if the first partial derivatives of the coordinate functions of T exist at that point. In case $J(u_0, T)$ exists, we have [see **V.2.1** (2)]

$$J(u_0, T) = \det \mathfrak{L}. \tag{7}$$

As stated above, we assume now that $J(u_0, T)$ exists at the assigned point $u_0 \in \boldsymbol{D}$. Observe that the summation in (6) represents the *differential* of the coordinate function f_i at u_0, in the sense of **III.1.3**. Accordingly, the vector whose i-th component is equal to the summation in (6) may be considered as the differential of T at u_0. Let us introduce an auxiliary vector-valued function $f u$ of u by the formulas

$$f u = \frac{T u - \mathfrak{L} u}{\|u - u_0\|} \quad \text{for} \quad u \in \boldsymbol{D}, \quad u \neq u_0, \tag{8}$$

$$f u_0 = 0, \tag{9}$$

where the 0 in (9) stands for the zero vector $(0, \ldots, 0)$. From (8) and (9) there follows the formula

$$T u = \mathfrak{L} u + \|u - u_0\| f u, \quad u \in \boldsymbol{D}. \tag{10}$$

If α is any positive number, then $Q(\alpha)$ will denote the oriented n-cube with center at u_0 and side length 2α (see **V.2.1**). Thus the relation

$$u = (u^1, \ldots, u^n) \in Q(\alpha) \tag{11}$$

is equivalent to the relations

$$|u^i - u_0^i| \leq \alpha, \quad i = 1, \ldots, n. \tag{12}$$

Since $u_0 \in \boldsymbol{D}$ and $\boldsymbol{D}$ is open, we can select a number $\alpha_0 > 0$ such that

$$Q(\alpha) \subset \boldsymbol{D} \quad \text{if} \quad 0 < \alpha < \alpha_0. \tag{13}$$

In the sequel it will be assumed throughout that α is so chosen that (13) holds. We define

$$\gamma(\alpha) = \max \|f u\|, \quad u \in fr\, Q(\alpha), \tag{14}$$

$$\Gamma(\alpha) = \text{l.u.b.} \|f u\|, \quad u \in Q(\alpha). \tag{15}$$

Since T is continuous in $\boldsymbol{D}$, clearly $\|f u\|$ is continuous at every point $u \neq u_0$ in $\boldsymbol{D}$. Thus the use of $\max \|f u\|$ in (14) is legitimate. Clearly $\gamma(\alpha)$ is finite and non-negative. On the other hand, $\|f u\|$ is generally not continuous at u_0, and $\Gamma(\alpha)$ may be equal to $+\infty$. Obviously

$$0 \leq \gamma(\alpha) \leq \Gamma(\alpha). \tag{16}$$

Observe that $\gamma(\alpha)$, for example, depends also upon T and u_0. However, T and u_0 are thought of as fixed, and thus it is unnecessary to use more involved notations like $\gamma(\alpha, T, u_0)$.

Definition 1. T is said to possess a *total differential* at the point $u_0 \in \boldsymbol{D}$ if (a) the first partial derivatives of the coordinate functions of T exist at u_0 and (b) $\Gamma(\alpha) \to 0$ for $\alpha \to 0$.

Definition 2. T is said to possess a *weak total differential* at the point $u_0 \in \boldsymbol{D}$ if (a) the first partial derivatives of the coordinate functions of T exist at u_0 and (b) there exists a sequence

$$\alpha_1 > \cdots > \alpha_m > \cdots > 0, \quad \alpha_m \to 0,$$

such that $\gamma(\alpha_m) \to 0$.

Note that, as a direct consequence of (16), if T possesses a total differential at u_0, then T also possesses a weak total differential at u_0.

Our objective in this section is to study the kind and degree of approximation to T by $\mathfrak{L}$ in the vicinity of u_0 [see (5), (6)]. The following situations are relevant for subsequent applications.

Situation 1. T possesses a total differential at u_0, and $J(u_0, T) = 0$. Under these conditions, we assert that

$$\lim_{\alpha \to 0} \frac{\mathsf{L}\, T Q(\alpha)}{\mathsf{L}\, Q(\alpha)} = 0. \tag{17}$$

To verify this relation, note that the assumption $J(u_0, T) = 0$ implies, by (7), that $\mathfrak{L}$ is singular. By **V.2.1**, lemma 3 we can therefore select (constant) vectors $b_1, \ldots, b_n$ such that

$$\|b_j\| = 1, \quad j = 1, \ldots, n, \tag{18}$$

$$b_i \cdot b_j = 0 \quad \text{for} \quad i \neq j \tag{19}$$

$$(\mathfrak{L} u - x_0) \cdot b_n = 0 \quad \text{for} \quad u \in R^n. \tag{20}$$

Let $b_j^1, \ldots, b_j^n$ be the components of b_j, $j = 1, \ldots, n$. The relations (18) and (19) imply (see **I.2.1**, exercise 10) that

$$\det(b_j^i) = \pm 1. \tag{21}$$

Thus the vectors $b_1, \ldots, b_n$ are linearly independent, and hence every vector v in R^n can be represented in the form

$$v = \sum_{j=1}^{n} \lambda_j b_j. \tag{22}$$

where $\lambda_1, \ldots, \lambda_n$ are real numbers. From (19) and (22) it follows that $\lambda_j = v \cdot b_j$, and hence (see **I.2.1**, exercise 12)

$$|\lambda_j| \leq \|v\| \, \|b_j\|.$$

Since $\|b_j\| = 1$, there follows the inequality

$$|\lambda_j| \leq \|v\|, \quad j = 1, \ldots, n. \tag{23}$$

Consider now any point

$$u \in Q(\alpha). \tag{24}$$

For the vector $Tu - \mathfrak{L}u$ we have a representation of the form

$$Tu - \mathfrak{L}u = \sum_{j=1}^{n} \lambda_j' b_j. \tag{25}$$

From (10) and (15) it follows [in view of (23)] that

$$|\lambda_j'| \leq ||T u - \mathfrak{L} u|| \leq ||u - u_0|| \, \Gamma(\alpha).$$

Since $||u - u_0|| \leq n\alpha$ by (24), we conclude that

$$|\lambda_j'| \leq n \, \alpha \, \Gamma(\alpha), \quad j = 1, \ldots, n. \tag{26}$$

Similarly, for the vector $\mathfrak{L} u - x_0$ we have a representation of the form

$$\mathfrak{L} u - x_0 = \sum_{j=1}^{n} \lambda_j'' \, b_j.$$

From (20) and (19) it follows that

$$\lambda_n'' = (\mathfrak{L} u - x_0) \cdot b_n = 0,$$

and hence

$$\mathfrak{L} u - x_0 = \sum_{j=1}^{n-1} \lambda_j'' \, b_j. \tag{27}$$

Since $x_0 = \mathfrak{L} u_0$, lemma 1 in **V.2.1** yields

$$||\mathfrak{L} u - x_0|| \leq \Lambda(\mathfrak{L}) \, ||u - u_0||.$$

As $||u - u_0|| \leq n\alpha$ by (24), we obtain in view of (23) the inequalities

$$|\lambda_j''| \leq n \, \Lambda(\mathfrak{L}) \, \alpha, \quad j = 1, \ldots, n-1. \tag{28}$$

From (25) and (27) we obtain, by addition,

$$T u - x_0 = \sum_{j=1}^{n-1} (\lambda_j' + \lambda_j'') \, b_j + \lambda_n' \, b_n. \tag{29}$$

By (26) and (28) we have

$$|\lambda_j' + \lambda_j''| \leq n \, \alpha \left(\Lambda(\mathfrak{L}) + \Gamma(\alpha)\right), \quad j = 1, \ldots, n-1, \tag{30}$$

$$|\lambda_n'| \leq n \, \alpha \, \Gamma(\alpha). \tag{31}$$

Assign $\varepsilon > 0$. Denote by $P_\varepsilon^*(\alpha)$ the parallelotope with center at x_0 and edge vectors

$$2n\alpha\left(\Lambda(\mathfrak{L}) + \Gamma(\alpha)\right) b_1, \ldots, \; 2n\alpha\left(\Lambda(\mathfrak{L}) + \Gamma(\alpha)\right) b_{n-1}, \; 2n\alpha\left(\varepsilon + \Gamma(\alpha)\right) b_n.$$

From (29), (30), (31) we conclude that $T u \in P_\varepsilon^*(\alpha)$. Since u was an arbitrary point of $Q(\alpha)$, it follows that

$$T \, Q(\alpha) \subset P_\varepsilon^*(\alpha),$$

and hence

$$\mathsf{L} \, T \, Q(\alpha) \leq \mathsf{L} P_\varepsilon^*(\alpha). \tag{32}$$

In view of (21) and **III.1.1**, lemma 16, we have

$$\mathsf{L}P_\varepsilon^*(\alpha) = 2^n \alpha^n \left(\Lambda(\mathfrak{L}) + \Gamma(\alpha)\right)^{n-1} \left(\varepsilon + \Gamma(\alpha)\right) n^n. \tag{33}$$

Since $L\,Q(\alpha) = 2^n\alpha^n$, and $\Gamma(\alpha)\to 0$ for $\alpha\to 0$, (32) and (33) yield

$$\limsup_{\alpha\to 0} \frac{\mathsf{L}\,T\,Q(\alpha)}{\mathsf{L}\,Q(\alpha)} \leq n^n \Lambda(\mathfrak{L})^{n-1}\varepsilon.$$

As $\varepsilon > 0$ was arbitrary, (17) follows.

Situation 2. T possesses a weak total differential at u_0 and $J(u_0, T) \neq 0$. Then there exists (see definition 2) a sequence $\{\alpha_m\}$ such that

$$\alpha_1 > \cdots > \alpha_m > \cdots > 0, \quad \alpha_m \to 0, \quad \gamma(\alpha_m) \to 0. \tag{34}$$

Consider *any* sequence $\{\alpha_m\}$ satisfying (34). Choose a sequence $\{\eta_m\}$ such that

$$\eta_m > 0, \quad \eta_m \to 0, \tag{35}$$

and put

$$\lambda_m = n\Lambda(\mathfrak{L}^{-1})\left(\gamma(\alpha_m) + \eta_m\right). \tag{36}$$

Note that the assumption $J(u_0, T) \neq 0$ implies, in view of (7), that $\mathfrak{L}$ is non-singular, and hence the positive constant $\Lambda(\mathfrak{L}^{-1})$ is available [see **V.2.1** (5)]. In view of (34) and (35) the inequalities

$$\gamma(\alpha_m) + \eta_m < \frac{1}{n\Lambda(\mathfrak{L}^{-1})}, \quad \lambda_m < 1, \tag{37}$$

will hold for m sufficiently large. Discarding, if necessary, a finite number of terms in the sequences (34) and (35), we can therefore assume that (37) holds for every integer $m \geq 1$. Put

$$\alpha'_m = (1 - \lambda_m)\,\alpha_m, \tag{38}$$

$$\alpha''_m = (1 + \lambda_m)\,\alpha_m. \tag{39}$$

Consider the oriented n-cubes $Q(\alpha'_m)$, $Q(\alpha_m)$, $Q(\alpha''_m)$ with center at u_0 and side length $2\alpha'_m$, $2\alpha_m$, $2\alpha''_m$ respectively. Since $\alpha_m \to 0$ and $\lambda_m \to 0$, these cubes will be contained in $\boldsymbol{D}$ for m sufficiently large. Discarding, if necessary, a finite number of these cubes, we can therefore assume that this is the case for every integer $m \geq 1$. We have then the inclusions (see **I.1.5**, exercise 10)

$$\left.\begin{aligned} &u_0 \in \operatorname{int} Q(\alpha'_m), \quad Q(\alpha'_m) \subset \operatorname{int} Q(\alpha_m), \\ &Q(\alpha_m) \subset \operatorname{int} Q(\alpha''_m), \quad Q(\alpha''_m) \subset \boldsymbol{D}. \end{aligned}\right\} \tag{40}$$

Let us put

$$\mathfrak{L}\,Q(\alpha_m') = P_m', \quad \mathfrak{L}\,Q(\alpha_m) = P_m, \quad \mathfrak{L}\,Q(\alpha_m'') = P_m''. \tag{41}$$

The following diagram, which illustrates the case $n=2$, may help the reader in understanding the discussion.

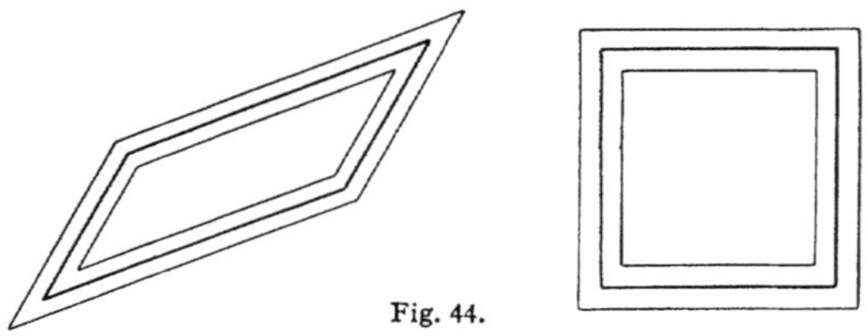

Fig. 44.

In this diagram, the three concentric squares and the three concentric parallelograms correspond to $Q(\alpha_m')$, $Q(\alpha_m)$, $Q(\alpha_m'')$ and P_m', P_m, P_m'' respectively. The following discussion is concerned, of course, with the situation for the case of an arbitrary n. From **V.2.1**, lemma 4 we infer the following facts.

(i) P_m', P_m, P_m'' are parallelotopes with center x_0.

(ii) The volumes $\mathsf{L}P_m'$, $\mathsf{L}P_m$, $\mathsf{L}P_m''$ of these parallelotopes satisfy the relations [see (7)]

$$\frac{\mathsf{L}P_m'}{\mathsf{L}Q(\alpha_m')} = \frac{\mathsf{L}P_m}{\mathsf{L}Q(\alpha_m)} = \frac{\mathsf{L}P_m''}{\mathsf{L}Q(\alpha_m'')} = |J(u_0, T)|\,.$$

(iii) The parameters of regularity of P_m', P_m'' satisfy the inequality [see (7)]

$$\pi(P_m'),\ \pi(P_m'') \geqq \frac{|J(u_0, T)|}{n^n \Lambda(\mathfrak{L})^n}\,.$$

Since $\mathfrak{L}$ is non-singular as a consequence of the assumption $J(u_0, T) \neq 0$ [see (7)], $\mathfrak{L}$ is a homeomorphism from R^n onto R^n, and hence (see **I.1.5**, exercise 10) the inclusions (40) imply that

$$x_0 \in \operatorname{int} P_m', \quad P_m' \subset \operatorname{int} P_m, \quad P_m \subset \operatorname{int} P_m''. \tag{42}$$

We assert that

$$\frac{\mathsf{L}P_m'}{\mathsf{L}Q(\alpha_m)} \to |J(u_0, T)|, \qquad \frac{\mathsf{L}P_m''}{\mathsf{L}Q(\alpha_m)} \to |J(u_0, T)|, \tag{43}$$

for $m \to \infty$. Indeed, by (ii) we have

$$\frac{\mathsf{L}P_m'}{\mathsf{L}Q(\alpha_m)} = |J(u_0, T)|\,\frac{\mathsf{L}Q(\alpha_m')}{\mathsf{L}Q(\alpha_m)}\,.$$

Since [see (38)]

$$\mathsf{L}Q(\alpha_m') = 2^n \alpha_m^n (1-\lambda_m)^n = (1-\lambda_m)^n\, \mathsf{L}Q(\alpha_m),$$

and $\lambda_m \to 0$, the first relation in (43) follows. The second relation in (43) is verified in a similar manner, using (39). Next we assert that

$$\delta P_m' \to 0, \quad \delta P_m'' \to 0 \tag{44}$$

for $m \to \infty$. Consider any two points $x_1'', x_2'' \in P_m''$. In view of (41), there exist points u_1'', u_2'' such that

$$x_1'' = \mathfrak{L} u_1'', \quad x_2'' = \mathfrak{L} u_2'', \quad u_1'', u_2'' \in Q(\alpha_m'').$$

From **V.2.1**, lemma 1 we infer that

$$\|x_2'' - x_1''\| \leq \Lambda(\mathfrak{L}) \|u_2'' - u_1''\| \leq 2n\alpha_m'' \Lambda(\mathfrak{L}).$$

As x_1'', x_2'' were arbitrary points of P_m'', it follows that

$$\delta P_m'' \leq 2n\alpha_m'' \Lambda(\mathfrak{L}),$$

and (since $\alpha_m'' \to 0$) the second relation in (44) is established. The first relation in (44) is verified in a similar manner. We proceed to discuss the topological index $\mu(x, T, \operatorname{int} Q(\alpha_m))$ (see **II.2.2**). We shall establish the fundamental relations

$$\mu(x, T, \operatorname{int} Q(\alpha_m)) = \left\{ \begin{array}{ll} \operatorname{sgn} J(u_0, T) = \pm 1 & \text{if} \quad x \in P_m', \\ 0 & \text{if} \quad x \in \complement P_m''. \end{array} \right\} \tag{45}$$

The proof of (45) requires several steps. First we verify that

$$T\, fr\, Q(\alpha_m) \subset P_m'' - P_m'. \tag{46}$$

Since $\mathfrak{L}$ is a homeomorphism, the inclusion (46) is equivalent [in view of (41)] to the inclusion

$$\mathfrak{L}^{-1} T\, fr\, Q(\alpha_m) \subset Q(\alpha_m'') - Q(\alpha_m'). \tag{47}$$

In turn, the inclusion (47) is implied, in view of (38) and (39), by the following statement: if u is any point such that

$$u \in fr\, Q(\alpha_m), \tag{48}$$

then

$$\|\mathfrak{L}^{-1} T u - u\| < \lambda_m \alpha_m. \tag{49}$$

This latter statement is proved as follows. From **V.2.1**, lemma 2 we obtain the inequality

$$\|\mathfrak{L}^{-1} T u - u\| \leq \Lambda(\mathfrak{L}^{-1}) \|T u - \mathfrak{L} u\|.$$

Since [in view of (10), (48), (14)]

$$\|T u - \mathfrak{L} u\| \leq \|u - u_0\| \|f u\| \leq n \alpha_m \gamma(\alpha_m),$$

by (36) it follows that

$$\|\mathfrak{L}^{-1} T u - u\| \leq n \Lambda(\mathfrak{L}^{-1}) \gamma(\alpha_m) \alpha_m < \lambda_m \alpha_m.$$

Thus (49) holds, and hence (46) is established. From (46) we conclude that

$$\complement P_m'' \subset \complement T\, fr\, Q(\alpha_m), \tag{50}$$

$$P_m' \subset \complement T\, fr\, Q(\alpha_m). \tag{51}$$

These inclusions imply (see **II.2.2**) that $\mu(x, T, \operatorname{int} Q(\alpha_m))$ is actually defined on $\complement P_m''$ and on P_m'. Observe now that the sets $\complement Q(\alpha_m'')$ and $Q(\alpha_m')$ are connected. Hence, by **I.1.5**, exercise 8, the sets

$$\complement P_m'' = \mathfrak{L} \complement Q(\alpha_m''), \quad P_m' = \mathfrak{L} Q(\alpha_m')$$

are also connected. Accordingly, by **II.2.3**, remark 1, the inclusions (50) and (51) imply that

$$\mu(x, T, \operatorname{int} Q(\alpha_m)) \quad \text{is constant on } \complement P_m'', \tag{52}$$

$$\mu(x, T, \operatorname{int} Q(\alpha_m)) \quad \text{is constant on } P_m'. \tag{53}$$

Since the set $\complement P_m''$ is unbounded, from (52) and **II.2.3**, remark 3 we conclude that

$$\mu(x, T, \operatorname{int} Q(\alpha_m)) = 0 \quad \text{for} \quad x \in \complement P_m''.$$

To establish (45), there remains to show that

$$\mu(x, T, \operatorname{int} Q(\alpha_m)) = \operatorname{sgn} J(u_0, T) = \pm 1 \quad \text{if} \quad x \in P_m'. \tag{54}$$

Note that since $x_0 \in P_m'$, we have by (53) the relation

$$\mu(x, T, \operatorname{int} Q(\alpha_m)) = \mu(x_0, T, \operatorname{int} Q(\alpha_m)) \quad \text{for} \quad x \in P_m'. \tag{55}$$

Since $\mathfrak{L}$ is a linear transformation and [see (41), (42)]

$$x_0 \in \operatorname{int} P_m' = \mathfrak{L} \operatorname{int} Q(\alpha_m') \subset \mathfrak{L} \operatorname{int} Q(\alpha_m),$$

we conclude from (7) and **II.2.5**, theorem 4 that

$$\mu(x_0, \mathfrak{L}, \operatorname{int} Q(\alpha_m)) = \operatorname{sgn} \det \mathfrak{L} = \operatorname{sgn} J(u_0, T) = \pm 1, \tag{56}$$

where we used the fact that $J(u_0, T) \neq 0$ by assumption. We shall now verify the relation

$$\mu(x_0, T, \operatorname{int} Q(\alpha_m)) = \mu(x_0, \mathfrak{L}, \operatorname{int} Q(\alpha_m)). \tag{57}$$

For $u \in fr\, Q(\alpha_m)$ we have [see (10) and (14)]

$$\|T u - \mathfrak{L} u\| \leq \|u - u_0\| \gamma(\alpha_m) \leq n\, \alpha_m\, \gamma(\alpha_m).$$

Hence (see **I.1.5**, definition 5)

$$\varrho\left(\mathfrak{L}, T, fr\, Q(\alpha_m)\right) \leq n\, \alpha_m\, \gamma(\alpha_m). \tag{58}$$

Consider now any point $x \in \mathfrak{L}\, fr\, Q(\alpha_m)$. Then there exists a point u such that

$$u \in fr\, Q(\alpha_m), \quad \mathfrak{L} u = x. \tag{59}$$

Lemma 2 in **V.2.1** yields

$$\|u - u_0\| \leq \Lambda(\mathfrak{L}^{-1})\, \|\mathfrak{L} u - \mathfrak{L} u_0\| = \Lambda(\mathfrak{L}^{-1})\, \|x - x_0\|. \tag{60}$$

On the other hand,

$$\|u - u_0\| \geq \alpha_m, \tag{61}$$

since $u \in fr\, Q(\alpha_m)$ and u_0 is the center of $Q(\alpha_m)$. From (60) and (61) we conclude that

$$\|x - x_0\| \geq \frac{\alpha_m}{\Lambda(\mathfrak{L}^{-1})}.$$

Since x was an arbitrary point of the set $\mathfrak{L}\, fr\, Q(\alpha_m)$, there follows the inequality (see **I.1.4**, definition 7)

$$e\left(x_0, \mathfrak{L}\, fr\, Q(\alpha_m)\right) \geq \frac{\alpha_m}{\Lambda(\mathfrak{L}^{-1})}. \tag{62}$$

Comparison of (62) and (58) yields, in view of (37),

$$\varrho\left(\mathfrak{L}, T, fr\, Q(\alpha_m)\right) < e\left(x_0, \mathfrak{L}\, fr\, Q(\alpha_m)\right). \tag{63}$$

By **II.2.3**, remark 8 the inequality (63) implies (57). Finally, (55), (57) and (56) imply (54), and thus (45) is proved. An important inference is the inclusion

$$P'_m \subset T\left(\operatorname{int} Q(\alpha_m)\right), \tag{64}$$

which follows from (54) by **II.2.3**, theorem 2. Next we assert that u_0 by itself is an m.m.c. for $(x_0, T, \boldsymbol{D})$ (see **II.3.1**, definition 1). Indeed, since $T u_0 = x_0$, there exists a component C of $T^{-1} x_0$ such that $u_0 \in C$. We must have

$$C \subset \operatorname{int} Q(\alpha_m). \tag{65}$$

Otherwise, we should have (see **I.1.3**, exercise 17) $C \cap fr\, Q(\alpha_m) \neq \emptyset$, and hence also

$$(T C) \cap \left(T\, fr\, Q(\alpha_m)\right) \neq \emptyset. \tag{66}$$

However [see (46)]

$$T C = x_0 \in P'_m, \quad T\, fr\, Q(\alpha_m) \subset \complement P'_m.$$

Thus (66) cannot hold, and hence (65) is established. As $\alpha_m \to 0$, clearly (65) implies that C reduces to the single point u_0. Accordingly, u_0 by itself is an m.m.c. for $(x_0, T, \boldsymbol{D})$. Consider now any open set O such that $u_0 \in O \subset \boldsymbol{D}$. As $\alpha_m \to 0$, the inclusion $Q(\alpha_m) \subset O$ will hold for m large enough. Now since

$$\mu(x_0, T, \operatorname{int} Q(\alpha_m)) = \operatorname{sgn} J(u_0, T) = \pm 1, \tag{67}$$

by (45), the domain $\operatorname{int} Q(\alpha_m)$ is an indicator domain for $(x_0, T, \boldsymbol{D})$, in the sense of **II.3.2**. In view of **II.3.3**, definition 1 it follows that u_0 by itself is an e.m.m.c. for $(x_0, T, \boldsymbol{D})$. Note that u_0 may or may not be essentially isolated in the sense of **II.3.3**, definition 2. Assume that u_0 is essentially isolated. We have just seen that if any open set O containing u_0 is assigned, then for m sufficiently large $Q(\alpha_m) \subset O$. In view of (67) and **II.3.4** it follows that

$$i_e(u_0, T) = \operatorname{sgn} J(u_0, T) = \pm 1, \tag{68}$$

provided that u_0 is essentially isolated. Note next that u_0 may or may not possess a neighborhood clear of relatives (see **II.3.7**, definition 2). Assume that u_0 does possess such a neighborhood. Since we have already shown that u_0 by itself is an e.m.m.c. for $(x_0, T, \boldsymbol{D})$, it follows from **II.3.7** (3) that $i_B(u_0, T) = i_e(u_0, T)$. In view of (68) there follows the formula

$$i_B(u_0, T) = \operatorname{sgn} J(u_0, T) = \pm 1. \tag{69}$$

Recall that in deriving (69), we had to assume that u_0 possesses a neighborhood clear of relatives. If this condition is not fulfilled, then (69) is false, since then $i_B(u_0, T) = 0$ by **II.3.7** (1).

Situation 3. T possesses a total differential at u_0 and $J(u_0, T) \neq 0$. Consider any sequence

$$\alpha_1 > \cdots > \alpha_m > \cdots > 0, \quad \alpha_m \to 0,$$

subject only to the condition that α_1 is so small that $Q(\alpha_1)$ is contained in $\boldsymbol{D}$ [see (11), (12), (13)]. Then (see definition 1)

$$\Gamma(\alpha_m) \to 0 \quad \text{for} \quad m \to \infty. \tag{70}$$

In view of (16) it follows that

$$\Gamma(\alpha_m) - \gamma(\alpha_m) \geq 0, \tag{71}$$

$$\gamma(\alpha_m) \to 0 \quad \text{for} \quad m \to \infty. \tag{72}$$

Let us put

$$\eta_m = \frac{1}{m} + \Gamma(\alpha_m) - \gamma(\alpha_m), \quad m = 1, 2, \ldots. \tag{73}$$

Then, by (70), (71), (72), $\eta_m > 0$, $\eta_m \to 0$. Thus all the assumptions used in the study of *Situation 2* are fulfilled, and hence the terminology employed and the results derived there are again available. The numbers λ_m [see (36)] are now given, in view of (73), by the formula

$$\lambda_m = n\,\Lambda(\mathfrak{L}^{-1})\left(\frac{1}{m} + \Gamma(\alpha_m)\right). \tag{74}$$

The cubes

$$Q(\alpha_m'), \quad Q(\alpha_m), \quad Q(\alpha_m''),$$

and the parallelotopes P_m', P_m, P_m'' are defined as in *Situation 2*. The new feature in the present *Situation 3* is the validity of (70). On this basis we shall derive presently the following relations

$$P_m' \subset T(\operatorname{int} Q(\alpha_m)) \subset P_m'', \tag{75}$$

$$i_B(u_0, T) = \operatorname{sgn} J(u_0, T) = \pm 1. \tag{76}$$

The first inclusion in (75) is merely a re-statement of (64). The second inclusion in (75) is equivalent to the statement that

$$u \in \operatorname{int} Q(\alpha_m) \quad \text{implies} \quad T u \in P_m''. \tag{77}$$

In turn, (77) is equivalent to the statement that

$$u \in \operatorname{int} Q(\alpha_m) \quad \text{implies} \quad \mathfrak{L}^{-1} T u \in \mathfrak{L}^{-1} P_m'' = Q(\alpha_m'').$$

In view of (39), this last statement will be established if we show that

$$\|\mathfrak{L}^{-1} T u - u\| \leq \lambda_m \alpha_m \quad \text{if} \quad u \in \operatorname{int} Q(\alpha_m). \tag{78}$$

To verify (78), note that

$$\|\mathfrak{L}^{-1} T u - u\| \leq \Lambda(\mathfrak{L}^{-1})\,\|T u - \mathfrak{L} u\|$$

by **V.2.1**, lemma 2. Since $u \in \operatorname{int} Q(\alpha_m)$, we infer from (10) and (15) the inequality

$$\|T u - \mathfrak{L} u\| \leq \|u - u_0\|\,\Gamma(\alpha_m) \leq n\,\alpha_m\,\Gamma(\alpha_m).$$

Hence, in view of (74)

$$\|\mathfrak{L}^{-1} T u - u\| \leq n\,\Lambda(\mathfrak{L}^{-1})\,\Gamma(\alpha_m)\,\alpha_m < \lambda_m\,\alpha_m.$$

Thus (78) is verified, and hence (75) is proved. As regards (76), in view of the comments made in deriving (69) we have only to show that u_0 possesses a neighborhood clear of relatives. In turn, this fact will be established if we show that if $\alpha > 0$ is sufficiently small, then

$$T u \neq x_0 \quad \text{for} \quad u \in Q(\alpha), \quad u \neq u_0. \tag{79}$$

Observe that $\Gamma(\alpha)\to 0$ for $\alpha\to 0$, since by assumption T possesses a total differential at u_0. Accordingly, we can choose $\alpha>0$ so that

$$\Gamma(\alpha)<\frac{1}{\Lambda(\mathfrak{L}^{-1})}, \quad Q(\alpha)\subset \boldsymbol{D}. \tag{80}$$

Consider any point u such that

$$u\in Q(\alpha), \quad u\neq u_0. \tag{81}$$

Then, by (10), (15), and (80),

$$\|Tu-\mathfrak{L}u\|\leq\|u-u_0\|\,\Gamma(\alpha)<\frac{\|u-u_0\|}{\Lambda(\mathfrak{L}^{-1})}.$$

On the other hand, by **V.2.1**, lemma 2,

$$\|u-u_0\|\leq\Lambda(\mathfrak{L}^{-1})\,\|\mathfrak{L}u-x_0\|.$$

There follows the inequality $\|\mathfrak{L}u-x_0\|>\|Tu-\mathfrak{L}u\|$. Hence, by the triangle inequality,

$$\|Tu-x_0\|\geq\|\mathfrak{L}u-x_0\|-\|Tu-\mathfrak{L}u\|>0.$$

Thus (81) implies that $Tu\neq x_0$, and the proof of (76) is complete. We noted above that all the results derived for *Situation 2* are available in the present *Situation 3*. In particular [see (43)], we have again

$$\lim_{m\to\infty}\frac{\mathsf{L}P'_m}{\mathsf{L}Q(\alpha_m)}=|J(u_0,T)|=\lim_{m\to\infty}\frac{\mathsf{L}P''_m}{\mathsf{L}Q(\alpha_m)}. \tag{82}$$

V.2.3. Preliminary study of Lipschitzian transformations in $\boldsymbol{R}^n$. Let $U\neq\emptyset$ be an arbitrary subset of R^n. Consider a transformation

$$T:U\to R^n. \tag{1}$$

We shall also use for T the representation

$$T:x^i=f_i(u^1,\dots,u^n),\quad i=1,\dots,n, \tag{2}$$

where $f_1,\dots,f_n$ are real-valued functions of the coordinates $u^1,\dots,u^n$ of the point $u\in U$, and $x^1,\dots,x^n$ are the coordinates of the image point $x=Tu$.

Definition 1. T is said to be Lipschitzian on U if there exists a (finite) constant $M\geq 0$ such that

$$\|Tu_2-Tu_1\|\leq M\,\|u_2-u_1\| \tag{3}$$

for every pair of points $u_1,u_2\in U$.

Lemma 1. T is Lipschitzian on U if and only if the coordinate functions $f_1, \ldots, f_n$ of T are Lipschitzian on U in the sense of **III.1.3.**

Proof. Consider any two points

$$u_1 = (u_1^1, \ldots, u_1^n), \quad u_2 = (u_2^1, \ldots, u_2^n)$$

of U, and let

$$T u_1 = x_1 = (x_1^1, \ldots, x_1^n), \quad T u_2 = x_2 = (x_2^1, \ldots, x_2^n)$$

be their images under T. Then clearly

$$|x_2^i - x_1^i| \leq ||T u_2 - T u_1||, \quad i = 1, \ldots, n, \tag{4}$$

$$||T u_2 - T u_1|| \leq \sum_{i=1}^{n} |x_2^i - x_1^i|. \tag{5}$$

Assume first that T is Lipschitzian on U, and let M be a constant such that (3) holds for every pair of points $u_1, u_2 \in U$. From (2), (3), (4) it follows then that

$$|f_i(u_2) - f_i(u_1)| \leq M\, ||u_2 - u_1||, \quad i = 1, \ldots, n,$$

and thus $f_1, \ldots, f_n$ are Lipschitzian on U. Assume, conversely, that $f_1, \ldots, f_n$ are Lipschitzian on U. Then there exist (finite) constants $M_i \geq 0$ such that

$$|f_i(u_2) - f_i(u_1)| \leq M_i\, ||u_2 - u_1||, \quad i = 1, \ldots, n, \tag{6}$$

for every pair of points $u_1, u_2 \in U$. On setting

$$M = M_1 + \cdots + M_n,$$

there follows from (5) and (6) the inequality

$$||T u_2 - T u_1|| \leq M\, ||u_2 - u_1||,$$

and thus T is Lipschitzian on U.

Let us now consider a bounded continuous transformation

$$T: \boldsymbol{D} \to R^n, \tag{7}$$

where $\boldsymbol{D}$ is a bounded domain in R^n. Corresponding to (2) we have the alternative representation

$$T: x^i = f_i(u^1, \ldots, u^n), \quad i = 1, \ldots, n, \tag{8}$$

where $f_1, \ldots, f_n$ are now real-valued, bounded, continuous functions of the coordinates $u^1, \ldots, u^n$ of the point $u \in \boldsymbol{D}$.

Lemma 2. Assume that T, given as in (7) and (8), is Lipschitzian in $\boldsymbol{D}$. Then the first partial derivatives of the coordinate functions $f_1, \ldots, f_n$ of T exist a.e. in $\boldsymbol{D}$, and T possesses a total differential a.e. in $\boldsymbol{D}$.

Proof. By lemma 1 the coordinate functions of T are Lipschitzian in $\boldsymbol{D}$. Accordingly, by **III.1.3** the first partial derivatives of the coordinate functions exist a.e. in $\boldsymbol{D}$, and each one of the coordinate functions possesses a total differential a.e. in $\boldsymbol{D}$. Thus the lemma will be proved if we verify the following statement: if $u_0 = (u_0^1, \ldots, u_0^n)$ is a point in $\boldsymbol{D}$ such that each one of the coordinate functions of T possesses a total differential at u_0, then T also possesses a total differential at u_0. To prove this statement, put

$$a_{ij} = \frac{\partial f_i(u_0)}{\partial u^j}, \quad i, j = 1, \ldots, n.$$

For each integer $i = 1, \ldots, n$, put

$$\varphi_i(u) = \frac{1}{\|u - u_0\|} \left[f_i(u) - f_i(u_0) - \sum_{j=1}^{n} a_{ij}(u^j - u_0^j) \right] \tag{9}$$

for $u = (u^1, \ldots, u^n) \in \boldsymbol{D}$, $u \neq u_0$, and let

$$\varphi_i(u_0) = 0. \tag{10}$$

Assign $\varepsilon > 0$. Since f_i possesses a total differential at u_0, there exists (see **III.1.3**) a number $\eta_i = \eta_i(\varepsilon) > 0$ such that

$$|\varphi_i(u)| < \varepsilon \quad \text{if} \quad u \in \boldsymbol{D}, \quad \|u - u_0\| < \eta_i. \tag{11}$$

Define η^* by the formula

$$\eta^* = \frac{1}{n} \min(\eta_1, \ldots, \eta_n).$$

Then clearly

$$0 < \eta^* \leq \frac{\eta_i}{n}, \quad i = 1, \ldots, n.$$

For each positive number α, denote by $Q(\alpha)$ the oriented n-cube with center at u_0 and side length 2α. Choose $\alpha_0 > 0$ so small that

$$0 < \alpha_0 < \eta^*, \quad Q(\alpha) \subset \boldsymbol{D} \quad \text{if} \quad 0 < \alpha < \alpha_0.$$

Consider the linear transformation [see **V.2.2** (6)]

$$\mathfrak{L}: x^i - x_0^i = \sum_{j=1}^{n} a_{ij}(u^j - u_0^j), \quad i = 1, \ldots, n.$$

Assume that $0<\alpha<\alpha_0$, and let $u \neq u_0$ be any point in $Q(\alpha)$. Then

$$\|u - u_0\| \leqq n\,\alpha < n\,\alpha_0 < n\,\eta^* \leqq \eta_i, \quad i = 1, \ldots, n. \tag{12}$$

Clearly the i-th component of the vector [see **V.2.2** (8)]

$$f\,u = \frac{T\,u - \mathfrak{L}\,u}{\|u - u_0\|}$$

is equal to $\varphi_i(u)$. Hence, by (11) and (12),

$$\|f\,u\| \leqq \sum_{i=1}^{n} |\varphi_i(u)| < n\,\varepsilon.$$

In view of **V.2.2** (15) there follows the inequality

$$\Gamma(\alpha) \leqq n\,\varepsilon \quad \text{if} \quad 0<\alpha<\alpha_0.$$

As n is fixed and $\varepsilon>0$ was arbitrary, it follows that $\Gamma(\alpha)\to 0$ if $\alpha\to 0$, and thus T possesses a total differential at u_0 (see **V.2.2**, definition 1).

Lemma 3. Assume that T, given as in (7), is Lipschitzian in $\boldsymbol{D}$. Then the following holds.

(i) T satisfies the condition (N) in $\boldsymbol{D}$ (see **IV.1.4**).

(ii) If S is any L-measurable set in $\boldsymbol{D}$, then the image set TS is L-measurable.

(iii) There exists a (finite) constant $K \geqq 0$ such that

$$\mathsf{L}\,TS \leqq K\,\mathsf{L}\,S \tag{13}$$

for every L-measurable set $S \subset \boldsymbol{D}$.

(iv) T is $AC\mathsf{B}$ in $\boldsymbol{D}$ (see **IV.5.2**).

Proof. Take any L-measurable set $S \subset \boldsymbol{D}$. Assign $\varepsilon>0$. By **III.1.1**, lemma 14 there exists a sequence of oriented n-cubes $Q_1, \ldots, Q_k, \ldots$ such that

$$S \subset \bigcup Q_k \subset \boldsymbol{D}, \quad \sum \mathsf{L}\,Q_k < \mathsf{L}\,S + \varepsilon. \tag{14}$$

Since T is Lipschitzian in $\boldsymbol{D}$, there exists a (finite) constant $M \geqq 0$ such that

$$\|T\,u_2 - T\,u_1\| \leqq M\,\|u_2 - u_1\| \tag{15}$$

for every pair of points u_1, u_2 in $\boldsymbol{D}$. Let u_k be the center and $2\alpha_k$ the side length of Q_k. Then

$$\|u - u_k\| \leqq n\,\alpha_k \quad \text{for} \quad u \in Q_k,$$

and hence, by (15),

$$\|T\,u - T\,u_k\| \leqq n\,M\,\alpha_k \quad \text{for} \quad u \in Q_k.$$

Thus clearly $T Q_k$ is contained in the oriented n-cube with center at $T u_k$ and side length $2nM\alpha_k$. Hence

$$\mathsf{L} T Q_k \leq (2nM\alpha_k)^n = n^n M^n \mathsf{L} Q_k. \tag{16}$$

From (14) and (16) it follows that

$$T S \subset \bigcup T Q_k, \tag{17}$$

$$\sum \mathsf{L} T Q_k \leq n^n M^n (\mathsf{L} S + \varepsilon). \tag{18}$$

From (17) and (18) we conclude that

$$\mathsf{L}^* T S \leq n^n M^n (\mathsf{L} S + \varepsilon),$$

where L^* denotes n-dimensional exterior measure in R^n (see **III.1.1**). Since $\varepsilon > 0$ was arbitrary, it follows that

$$\mathsf{L}^* T S \leq n^n M^n \mathsf{L} S. \tag{19}$$

To prove (i), assume that $\mathsf{L} S = 0$. Then $T S$ is of L-measure zero by (19), and (i) follows. Once we know that (i) holds, (ii) follows directly from **IV.1.4**, corollary 2. Let now S be any L-measurable set in $\boldsymbol{D}$. Then, since $T S$ is L-measurable by (ii), we can replace (see **III.1.1**) L^* by L in (19), and (13) follows. To prove (iv), consider any finite system of n-intervals $I_1, \ldots, I_m$ in $\boldsymbol{D}$ such that

$$\operatorname{int} I_j \cap \operatorname{int} I_k = \emptyset \quad \text{for} \quad j \neq k.$$

By (13) we have then

$$\sum \mathsf{L} T(\operatorname{int} I_k) \leq K \sum \mathsf{L}(\operatorname{int} I_k) \leq K \mathsf{L} \boldsymbol{D} < \infty,$$

and thus T is $BV\mathsf{B}$ in $\boldsymbol{D}$ (see **IV.5.1**, definition 1). Since T satisfies in $\boldsymbol{D}$ the condition (N), as we already proved, it follows by **IV.5.2**, theorem 3 that T is $AC\mathsf{B}$ in $\boldsymbol{D}$.

Lemma 4. Assume that T, given as in (7), is Lipschitzian in $\boldsymbol{D}$. Then

$$|J(u, T)| \leq D_{\mathsf{B}}(u, T) \quad \text{a.e. in } \boldsymbol{D}.$$

Proof. Note first that the ordinary Jacobian $J(u, T)$ exists a.e. in $\boldsymbol{D}$ and T possesses a total differential a.e. in $\boldsymbol{D}$ by lemma 2. Note also that the derivative $D_{\mathsf{B}}(u, T)$ exists a.e. in $\boldsymbol{D}$ by part (iv) of lemma 3 and **IV.5.1**, theorem 3 (see also **IV.5.2**, theorem 1). Accordingly, we have a decomposition $\boldsymbol{D} = s \cup S$, such that $\mathsf{L} s = 0$, and at every point $u \in S$ the ordinary Jacobian $J(u, T)$ exists, the derivative $D_{\mathsf{B}}(u, T)$ exists, and T possesses a total differential. Thus it is sufficient to show that

$$|J(u_0, T)| \leq D_{\mathsf{B}}(u_0, T) \quad \text{if} \quad u_0 \in S. \tag{20}$$

Now (20) is obvious if $J(u_0, T) = 0$. So we can assume that $J(u_0, T) \neq 0$. Then we have the *Situation 3*, studied in **V.2.2**, before us. Consider the sequence of oriented n-cubes $Q(\alpha_m)$ and the parallelotopes P'_m described there. According to **V.2.2** (75) we have the inclusion

$$P'_m \subset T(\text{int}\, Q(\alpha_m)).$$

For the crude multiplicity function $N(x, T, \text{int}\, Q(\alpha_m))$ there follows the inequality

$$N(x, T, \text{int}\, Q(\alpha_m)) \geq 1 \quad \text{for} \quad x \in P'_m.$$

Hence

$$\frac{1}{\mathsf{L}\, Q(\alpha_m)} \int N(x, T, \text{int}\, Q(\alpha_m))\, d\mathsf{L} \geq \frac{\mathsf{L}\, P'_m}{\mathsf{L}\, Q(\alpha_m)}. \tag{21}$$

Since $D_{\mathsf{B}}(u_0, T)$ exists, the term on the left in (21) converges to $D_{\mathsf{B}}(u_0, T)$ for $m \to \infty$ [see **IV.5.1** (8)]. By **V.2.2** (82) the term on the right in (21) converges to $|J(u_0, T)|$ for $m \to \infty$. Thus (21) yields, for $m \to \infty$, the inequality (20), and the lemma is proved.

Lemma 5. Assume that T, given as in (7), is Lipschitzian in $\boldsymbol{D}$. Let S be any L-measurable set in $\boldsymbol{D}$. Then the relation

$$\mathsf{L}\, T S = 0 \tag{22}$$

holds if and only if

$$J(u, T) = 0 \quad \text{a.e. on } S. \tag{23}$$

Proof. Note that T is $AC\mathsf{B}$ in $\boldsymbol{D}$ by lemma 3. Hence T is also $BV\mathsf{B}$ in $\boldsymbol{D}$ by **IV.5.2**, theorem 1. Assume now first that (22) holds. Then

$$D_{\mathsf{B}}(u, T) = 0 \quad \text{a.e. on } S$$

by **IV.5.1**, theorem 5, and (23) follows by lemma 4. Assume, conversely, that (23) holds. Since T possesses a total differential a.e. in $\boldsymbol{D}$ by lemma 2, we have a decomposition

$$S = s \cup S', \tag{24}$$

such that

$$\mathsf{L}\, s = 0, \tag{25}$$

and $J(u, T)$ vanishes and T possesses a total differential at every point $u \in S'$. Thus we have before us the *Situation 1*, described in **V.2.2**, at every point $u \in S'$. Denote by $Q(\alpha, u)$ the oriented n-cube with center at $u \in S'$ and side length 2α. Then (see **V.2.2**, *Situation 1*)

$$\lim_{\alpha \to 0} \frac{\mathsf{L}\, T\, Q(\alpha, u)}{\mathsf{L}\, Q(\alpha, u)} = 0 \quad \text{for} \quad u \in S'.$$

Accordingly, for every point $u \in S'$ we can select an oriented n-cube Q_u such that

$$u \in \operatorname{int} Q_u, \quad Q_u \subset \boldsymbol{D}, \quad \frac{\mathsf{L}\, T\, Q_u}{\mathsf{L}\, Q_u} < \varepsilon,$$

where $\varepsilon > 0$ is arbitrarily assigned. For fixed $\varepsilon > 0$, denote by F_ε the family of all those oriented n-cubes Q which satisfy the conditions

$$Q \subset \boldsymbol{D}, \quad \frac{\mathsf{L}\, T\, Q}{\mathsf{L}\, Q} < \varepsilon.$$

From the preceding remarks we conclude that the cubes $Q \in F_\varepsilon$ cover S' in the manner required in the VITALI covering theorem. Accordingly (see **III.1.1**, lemma 6), we can select from F_ε a sequence of cubes $Q_1, \ldots, Q_k, \ldots$, such that

$$\operatorname{int} Q_j \cap \operatorname{int} Q_k = \emptyset \quad \text{for} \quad j \neq k,$$
$$\mathsf{L}\,(S' - \bigcup \operatorname{int} Q_k) = 0.$$

Thus we have a decomposition

$$S' = s' \cup S'', \tag{26}$$

such that

$$\mathsf{L}\, s' = 0 \tag{27}$$

and

$$S'' \subset \bigcup Q_k. \tag{28}$$

Note that

$$\mathsf{L}\, T\, Q_k < \varepsilon\, \mathsf{L}\, Q_k, \tag{29}$$

since $Q_k \in F_\varepsilon$. Furthermore, since T satisfies the condition (N) in $\boldsymbol{D}$ by lemma 3, the relations (25) and (27) imply that

$$\mathsf{L}\, T\, s = 0, \quad \mathsf{L}\, T\, s' = 0. \tag{30}$$

Combining (28), (30), (26), (24), (29), we obtain

$$\mathsf{L}\, T\, S \leq \sum \mathsf{L}\, T\, Q_k \leq \varepsilon \sum \mathsf{L}\, Q_k \leq \varepsilon\, \mathsf{L}\, \boldsymbol{D}.$$

As $\varepsilon > 0$ was arbitrary, (22) follows, and the lemma is proved.

V.2.4. Local Lipschitzian approximations in $\boldsymbol{R}^n$. In **V.2.2** we have shown that if the transformation $T: \boldsymbol{D} \to R^n$ satisfies certain conditions at a point $u_0 \in \boldsymbol{D}$, then there exists a linear transformation $\mathfrak{L}$ (which depends upon both T and u_0) which yields a useful approximation to T. In the present section we shall show that if T satisfies a certain condition on a set $S \subset \boldsymbol{D}$, then a useful approximation to T may be obtained in terms of a Lipschitzian transformation.

Lemma 1. Let $U \neq \emptyset$ be a subset of R^n, and let $g(u)$ be a real-valued function defined for $u \in U$. Assume that g is Lipschitzian on U (see **III.1.3**). Then there exists a real-valued function $g_*(x)$ with the following properties.

(i) $g_*(x)$ is defined and finite-valued for every point $x \in R^n$.

(ii) $g_*(x)$ is Lipschitzian in R^n.

(iii) $g_*(u) = g(u)$ for $u \in U$.

Briefly, if g is Lipschitzian on U, then there exists a Lipschitzian extension g_* of g to all of R^n.

Proof[1]. By assumption, there exists a (finite) constant $\lambda \geqq 0$ such that

$$|g(u_2) - g(u_1)| \leqq \lambda \|u_2 - u_1\| \tag{1}$$

for every pair of points $u_1, u_2 \in U$. Define, for $x \in R^n$,

$$g_*(x) = \underset{u \in U}{\text{l.u.b.}} \left[g(u) - \lambda \|u - x\|\right]. \tag{2}$$

We proceed to verify that g_* satisfies the requirements (i), (ii), (iii). Choose a point $u_0 \in U$. Let x be any point in R^n. For $u \in U$ we have then [see (1)]

$$\begin{aligned} g(u) - \lambda \|u - x\| = g(u_0) + (g(u) - g(u_0)) - \lambda \|u - x\| \leqq \\ \leqq g(u_0) + \lambda \|u - u_0\| - \lambda \|u - x\|. \end{aligned}$$

Since $\lambda \geqq 0$ and

$$\|u - u_0\| - \|u - x\| \leqq \|x - u_0\|$$

by the triangle inequality, it follows that

$$g(u) - \lambda \|u - x\| \leqq g(u_0) + \lambda \|x - u_0\| \quad \text{for} \quad u \in U. \tag{3}$$

From (2) and (3) we conclude that

$$g_*(x) \leqq g(u_0) + \lambda \|x - u_0\|. \tag{4}$$

On the other hand, (2) yields

$$g_*(x) \geqq g(u_0) - \lambda \|u_0 - x\|. \tag{5}$$

---

[1] This method of proof is due to E. J. McShane, Extension of range of functions [Bull. Amer. Math. Soc. **40**, 837–842 (1934)]. The paper just referred to contains comments on further literature.

From (4) and (5) we see that $g_*(x)$ has a finite value at each point $x \in R^n$, and (i) follows. Next we verify (ii) by showing that

$$|g_*(x_2) - g_*(x_1)| \leq \lambda \|x_2 - x_1\| \tag{6}$$

for every pair of points $x_1, x_2 \in R^n$. Select any two points $x_1, x_2 \in R^n$. Let the notation be so chosen that

$$g_*(x_1) - g_*(x_2) \geq 0. \tag{7}$$

Assign $\varepsilon > 0$. In view of the defining formula (2), there exists a point $u_* \in U$ such that

$$g_*(x_1) \leq g(u_*) - \lambda \|u_* - x_1\| + \varepsilon. \tag{8}$$

On the other hand, (2) yields

$$g_*(x_2) \geq g(u_*) - \lambda \|u_* - x_2\|. \tag{9}$$

From (7), (8), (9) we obtain the inequalities

$$0 \leq g_*(x_1) - g_*(x_2) \leq \varepsilon + \lambda (\|u_* - x_2\| - \|u_* - x_1\|).$$

Observe that

$$\|u_* - x_2\| - \|u_* - x_1\| \leq \|x_1 - x_2\|$$

by the triangle inequality. Hence, since $\lambda \geq 0$,

$$0 \leq g_*(x_1) - g_*(x_2) \leq \varepsilon + \lambda \|x_1 - x_2\|. \tag{10}$$

As $\varepsilon > 0$ was arbitrary, clearly (10) implies (6), and thus (ii) is proved. Consider now any point $u_0 \in U$. Replacing x by u_0 in (4) and (5), we obtain the inequalities

$$g(u_0) \leq g_*(u_0) \leq g(u_0),$$

and (iii) follows.

Definition 1. Let there be given a bounded continuous transformation

$$T: \boldsymbol{D} \to R^n, \tag{11}$$

where $\boldsymbol{D}$ is a bounded domain in R^n. Take a point $u \in \boldsymbol{D}$ and a (finite) constant $M \geq 0$. Then $E(u, M)$ denotes the subset of $\boldsymbol{D}$ defined by the formula

$$E(u, M) = \{u^* | u^* \in \boldsymbol{D},\ \|T u_* - T u\| \leq M \|u^* - u\|\}. \tag{12}$$

Note that $E(u, M)$ depends upon T also. However, T is thought of as fixed, and thus the notation $E(u, M)$ is adequate.

Definition 2. Given T as in (11), we shall say that T is of *approximately bounded linear distortion* at a point $u \in \boldsymbol{D}$ if there exists a (finite) constant $M \geq 0$ such that u is a point of density (see **III.1.1**, definition 1) of the corresponding set $E(u, M)$.

Lemma 2[1]. Given T as in (11), let S be a subset of $\boldsymbol{D}$ such that T is of approximately bounded linear distortion at every point $u \in S$. Then there exists a sequence of sets S_k, $k = 1, 2, \ldots$, such that $S = \bigcup S_k$ and T is Lipschitzian on each of the sets S_k.

Proof. Let us recall that if u is a point in R^n and r is a positive number, then $\bar{\Delta}(u, r)$ denotes the closed spherical neighborhood of u with radius r. Explicitly:

$$\bar{\Delta}(u, r) = \{x \mid x \in R^n, \ \|x - u\| \leq r\}. \tag{13}$$

Let us also recall (see **III.1.1**, lemma 15) that if u_1, u_2 are any two distinct points in R^n, then the number

$$\alpha = \frac{\mathsf{L}[\bar{\Delta}(u_1, \|u_2 - u_1\|) \cap \bar{\Delta}(u_2, \|u_2 - u_1\|)]}{\mathsf{L}\bar{\Delta}(u_1, \|u_2 - u_1\|) + \mathsf{L}\bar{\Delta}(u_2, \|u_2 - u_1\|)} \tag{14}$$

is independent of the choice of u_1, u_2 and satisfies the inequalities

$$0 < \alpha < 1. \tag{15}$$

Let u be a point in $\boldsymbol{D}$, and let M, r be two numbers such that $0 \leq M < \infty$, $0 < r < \infty$. Let us put [see (13) and definition 1]

$$E(u, M, r) = E(u, M) \cap \bar{\Delta}(u, r). \tag{16}$$

Observe that since T is continuous, $E(u, M)$ is closed relative to $\boldsymbol{D}$. As $\boldsymbol{D}$ is open, clearly $E(u, M)$ is a BOREL set and hence it is L-measurable. Hence $E(u, M, r)$ is also L-measurable. For each positive integer k, let S_k be the set of those points u which satisfy the following conditions.

(i) $u \in S$.

(ii) $\|Tu\| \leq k$.

(iii) If r is any number such that $0 < r \leq 1/k$, then [see (14)]

$$\mathsf{L}\, E(u, k, r) > (1 - \alpha)\, \mathsf{L}\, \bar{\Delta}(u, r).$$

We proceed to verify that the sets S_k so defined possess the desired properties. Consider any point $u_0 \in S$. Since T is of approximately bounded linear distortion at u_0, there exists a (finite) constant $M \geq 0$ such that u_0 is a point of density of the corresponding set $E(u_0, M)$. Accordingly (see **III.1.1**, definition 1), in view of (15) and (16) there

[1] See the proof of theorem **5.2** in the paper by H. FEDERER, Surface area II. [Trans. Amer. Math. Soc. **55**, 438–456 (1944)]. This paper of FEDERER contains many important results concerning Lipschitzian transformations.

exists a number $r_0 > 0$ such that

$$\mathsf{L}\, E(u_0, M, r) > (1-\alpha)\, \mathsf{L}\, \bar{\Delta}(u_0, r) \quad \text{if} \quad 0 < r \leqq r_0. \tag{17}$$

of k is a sufficiently large positive integer, then

$$\|T u_0\| < k, \quad M < k, \quad 1/k \leqq r_0. \tag{18}$$

Let k be so selected that (18) holds, and let r be any number such that $0 < r \leqq 1/k$. Then clearly

$$E(u_0, k, r) \supset E(u_0, M, r) \quad \text{and} \quad 0 < r \leqq r_0,$$

and hence by (17) we have the inequalities

$$\mathsf{L}\, E(u_0, k, r) \geqq \mathsf{L}\, E(u_0, M, r) > (1-\alpha)\, \mathsf{L}\, \bar{\Delta}(u_0, r).$$

Thus it is established [see (i), (ii), (iii)] that $u_0 \in S_k$. As u_0 was an arbitrary point of S, it follows that

$$S = \bigcup S_k. \tag{19}$$

We shall now verify that T is Lipschitzian on each one of the sets S_k. This fact will be proved if we show that for any two points $u_1, u_2 \in S_k$ we have the inequality

$$\|T u_2 - T u_1\| \leqq 2k^2 \|u_2 - u_1\|. \tag{20}$$

Since (20) is obvious if $u_1 = u_2$, we can assume that $u_1 \neq u_2$. If

$$\|u_2 - u_1\| > \frac{1}{k},$$

then (since $\|Tu\| \leqq k$ for $u \in S_k$)

$$\|T u_2 - T u_1\| \leqq 2k = 2k^2 \cdot \frac{1}{k} < 2k^2 \|u_2 - u_1\|,$$

and thus (20) holds in this case. So we can assume that

$$0 < \|u_2 - u_1\| \leqq \frac{1}{k}. \tag{21}$$

Let us put

$$E_1 = E(u_1, k, \|u_2 - u_1\|), \quad E_2 = E(u_2, k, \|u_2 - u_1\|), \tag{22}$$

$$\bar{\Delta}_1 = \bar{\Delta}(u_1, \|u_2 - u_1\|), \quad \bar{\Delta}_2 = \bar{\Delta}(u_2, \|u_2 - u_1\|). \tag{23}$$

In view of (14), (16), (21) and (iii) we have then the relations

$$\mathsf{L}\,E_1 > (1-\alpha)\,\mathsf{L}\,\bar{\Delta}_1, \quad \mathsf{L}\,E_2 > (1-\alpha)\,\mathsf{L}\,\bar{\Delta}_2,$$

$$\mathsf{L}\,(E_1 \cup E_2) \leq \mathsf{L}\,(\bar{\Delta}_1 \cup \bar{\Delta}_2), \quad \alpha = \frac{\mathsf{L}\,(\bar{\Delta}_1 \cap \bar{\Delta}_2)}{\mathsf{L}\,\bar{\Delta}_1 + \mathsf{L}\,\bar{\Delta}_2}.$$

On the other hand, by **III.1.1**, lemma 2 we have

$$\mathsf{L}\,(E_1 \cap E_2) + \mathsf{L}\,(E_1 \cup E_2) = \mathsf{L}\,E_1 + \mathsf{L}\,E_2,$$

$$\mathsf{L}\,(\bar{\Delta}_1 \cap \bar{\Delta}_2) + \mathsf{L}\,(\bar{\Delta}_1 \cup \bar{\Delta}_2) = \mathsf{L}\,\bar{\Delta}_1 + \mathsf{L}\,\bar{\Delta}_2.$$

Hence

$$\begin{aligned}\mathsf{L}\,(E_1 \cap E_2) &= \mathsf{L}\,E_1 + \mathsf{L}\,E_2 - \mathsf{L}\,(E_1 \cup E_2) > (1-\alpha)\,(\mathsf{L}\,\bar{\Delta}_1 + \mathsf{L}\,\bar{\Delta}_2) - \mathsf{L}\,(\bar{\Delta}_1 \cup \bar{\Delta}_2) \\ &= (1-\alpha)\,(\mathsf{L}\,\bar{\Delta}_1 + \mathsf{L}\,\bar{\Delta}_2) - [\mathsf{L}\,\bar{\Delta}_1 + \mathsf{L}\,\bar{\Delta}_2 - \mathsf{L}\,(\bar{\Delta}_1 \cap \bar{\Delta}_2)] \\ &= \mathsf{L}\,(\bar{\Delta}_1 \cap \bar{\Delta}_2) - \alpha\,(\mathsf{L}\,\bar{\Delta}_1 + \mathsf{L}\,\bar{\Delta}_2) = 0.\end{aligned}$$

Thus $\mathsf{L}\,(E_1 \cap E_2) > 0$, and hence $E_1 \cap E_2 \neq \emptyset$. So we can select a point $u^* \in E_1 \cap E_2$. Then [see (12), (16), (22), (23)]

$$||T\,u^* - T\,u_1|| \leq k\,||u^* - u_1||, \quad ||T\,u^* - T\,u_2|| \leq k\,||u^* - u_2||,$$

$$||u^* - u_1|| \leq ||u_2 - u_1||, \quad ||u^* - u_2|| \leq ||u_2 - u_1||,$$

and hence

$$\begin{aligned}||T\,u_2 - T\,u_1|| &\leq ||T\,u_2 - T\,u^*|| + ||T\,u^* - T\,u_1|| \leq \\ &\leq k(||u_2 - u^*|| + ||u_1 - u^*||) \leq 2k\,||u_2 - u_1|| \leq 2k^2\,||u_2 - u_1||.\end{aligned}$$

Thus (20) holds, and the proof is complete.

Assume now that the set S, occurring in the statement of lemma 2, is L-measurable. Using the sets S_k whose existence has been just established, we define

$$S'_k = S \cap \bar{S}_k, \quad k = 1, 2, \ldots.$$

Note that since S is L-measurable and $\bar{S}_k$ is closed, clearly S'_k is L-measurable. We assert that T is Lipschitzian on each one of the sets S'_k. Indeed, let u, v be any two points of S'_k. Since $S'_k \subset \bar{S}_k$, we can select two sequences of points $u_m, v_m \in S_k$ such that $u_m \to u$, $v_m \to v$. By (20), we have then

$$||T\,u_m - T\,v_m|| \leq 2k^2\,||u_m - v_m||.$$

For $m \to \infty$ there follows (in view of the continuity of T) the inequality

$$||T\,u - T\,v|| \leq 2k^2\,||u - v||,$$

and thus T is Lipschitzian on S'_k. Consider now the sets

$$S_1^* = S'_1, \ldots, S_k^* = S'_k - \bigcup_{j=1}^{k-1} S_j^*, \ldots.$$

Clearly, $S_1^*, \ldots, S_k^*, \ldots$ are pair-wise disjoint L-measurable sets whose union is S. Furthermore, since $S_k^* \subset S'_k$ and T is Lipschitzian on S'_k, clearly T is Lipschitzian on S_k^* (a vacuous statement if $S_k^* = \emptyset$). These remarks yield the following result.

Lemma 3. Given T as in (11), let S be an L-measurable subset of $\boldsymbol{D}$ such that T is of approximately bounded linear distortion at every point $u \in S$. Then there exists a decomposition

$$S = \bigcup S_k, \quad k = 1, 2, \ldots,$$

where the sets S_k are pair-wise disjoint L-measurable sets and T is Lipschitzian on each one of the sets S_k.

In the following series of lemmas, we shall make use of the representation

$$T: x^i = f_i(u^1, \ldots, u^n), \quad i = 1, \ldots, n, \tag{24}$$

for the bounded continuous transformation given as in (11) (see **V.2.2**).

Lemma 4. Let T be given as in (11) and (24). Assume that the first partial derivatives of the coordinate functions of T exist a.e. in $\boldsymbol{D}$. Let S be an L-measurable subset of $\boldsymbol{D}$ such that T is Lipschitzian on S. Then there exists a transformation $T^*: \boldsymbol{D} \to R^n$ such that the following holds.

(i) T^* is Lipschitzian in $\boldsymbol{D}$.

(ii) $T^* u = T u$ for $u \in S$.

(iii) $J(u, T^*) = J(u, T)$ a.e. on S.

Proof. By **V.2.3**, lemma 1 the coordinate functions $f_1, \ldots, f_n$ of T are Lipschitzian on S. Accordingly, by lemma 1 each one of the functions f_i admits of a Lipschitzian extension to R^n and hence *a fortiori* to $\boldsymbol{D}$. Let f_i^* be a Lipschitzian extension of f_i to $\boldsymbol{D}$, and let $T^*: \boldsymbol{D} \to R^n$ be the transformation defined by the equations

$$T^*: x^i = f_i^*(u^1, \ldots, u^n), \quad i = 1, \ldots, n.$$

Then (ii) is obvious, and (i) holds by **V.2.3**, lemma 1. Observe now that the ordinary Jacobian $J(u, T)$ exists a.e. in $\boldsymbol{D}$ by assumption, while the ordinary Jacobian $J(u, T^*)$ exists a.e. in $\boldsymbol{D}$ by **V.2.3**, lemma 2. Take a point

$$u_0 = (u_0^1, \ldots, u_0^n) \in S,$$

such that $J(u_0, T)$, $J(u_0, T^*)$ both exist and u_0 is a point of linear density of S in the direction of each one of the coordinate axes (note that these conditions are satisfied a.e. on S in view of **III.1.1**, lemma 10). For each integer $j = 1, \ldots, n$ there exists then a sequence of points

$$u_{jk} = (u_{jk}^1, \ldots, u_{jk}^n) \in S, \quad k = 1, 2, \ldots,$$

such that

$$u_{jk}^i = u_0^i \quad \text{if} \quad i \neq j, \quad u_{jk}^j \neq u_0^j,$$

$$\lim_{k \to \infty} u_{jk}^j = u_0^j.$$

In view of (ii) it follows that

$$\frac{\partial f_i(u_0)}{\partial u^j} = \frac{\partial f_i^*(u_0)}{\partial u^j}, \quad i, j = 1, \ldots, n.$$

Hence $J(u_0, T) = J(u_0, T^*)$, and the lemma is proved.

Lemma 5. Given T as in (11) and (24), assume that the first partial derivatives of the coordinate functions of T exist a.e. in $\boldsymbol{D}$. Then T is of approximately bounded linear distortion a.e. in $\boldsymbol{D}$.

Proof. Let $u_0 = (u_0^1, \ldots, u_0^n)$ be a point in $\boldsymbol{D}$ where (a) the first partial derivatives of the coordinate functions of T exist, and (b) each one of the coordinate functions of T possesses an approximate total differential. Since in view of **III.1.3** these conditions are satisfied a.e. in $\boldsymbol{D}$, it is sufficient to show that T is of approximately bounded linear distortion at such a point u_0. For each integer $i = 1, \ldots, n$, denote by E_i the set of those points

$$u = (u^1, \ldots, u^n) \in \boldsymbol{D}$$

for which

$$\left| f_i(u) - f_i(u_0) - \sum_{j=1}^{n} \frac{\partial f_i(u_0)}{\partial u^j} (u^j - u_0^j) \right| \leq \|u - u_0\|. \tag{25}$$

Then, since f_i possesses an approximate total differential at u_0 by assumption, u_0 is a point of density of E_i, $i = 1, \ldots, n$ (see **III.1.3**). Hence (see **III.1.1**, lemma 11) u_0 is also a point of density of the set

$$E = \cap E_i. \tag{26}$$

Accordingly, in view of definition 2, the lemma will be proved if we can exhibit a (finite) constant $M \geq 0$ such that

$$\|T u - T u_0\| \leq M \|u - u_0\| \quad \text{for} \quad u \in E. \tag{27}$$

Let us put

$$a = \max \left| \frac{\partial f_i(u_0)}{\partial u^j} \right|, \quad i, j = 1, \ldots, n.$$

From (25) and (26) we obtain the inequality

$$|f_i(u) - f_i(u_0)| \leq ||u - u_0|| + \sum_{j=1}^{n} \left| \frac{\partial f_i(u_0)}{\partial u^j} \right| |u^j - u_0^j| \leq$$
$$\leq ||u - u_0|| + n\,a\,||u - u_0|| = (1 + n\,a)\,||u - u_0||,$$

for $u \in E$. Hence, for $u \in E$,

$$||T\,u - T\,u_0|| \leq \sum_{i=1}^{n} |f_i(u) - f_i(u_0)| \leq n(1 + n\,a)\,||u - u_0||.$$

Thus (27) holds with $M = n(1 + n a)$, and the lemma is proved.

Lemma 6. Given T as in (11) and (24), assume that the first partial derivatives of the coordinate functions of T exist a.e. in $\boldsymbol{D}$. Let S be an L-measurable subset of $\boldsymbol{D}$. Then there exist pair-wise disjoint L-measurable sets $s, S_1, \ldots, S_k, \ldots$ in $\boldsymbol{D}$ such that the following holds.

(i) $S = s \cup S_1 \cup \cdots \cup S_k \cup \cdots$.

(ii) $\mathsf{L}s = 0$.

(iii) T is Lipschitzian on each one of the sets S_k.

Proof. From lemma 5 there follows the existence of a decomposition $S = s \cup S^*$, where s, S^* are disjoint L-measurable sets such that $\mathsf{L}s = 0$ and T is of approximately bounded linear distortion at every point $u \in S^*$. By lemma 3 there exists a decomposition of S^* into pair-wise disjoint L-measurable sets S_k, $k = 1, 2, \ldots$, such that T is Lipschitzian on each one of the sets S_k, and the present lemma follows.

Lemma 7. Given T as in (11) and (24), assume that the first partial derivatives of the coordinate functions of T exist a.e. in $\boldsymbol{D}$. Let S be an L-measurable subset of $\boldsymbol{D}$ such that

$$\mathsf{L}\,T S = 0. \tag{28}$$

Then

$$J(u, T) = 0 \quad \text{a.e. on } S. \tag{29}$$

Proof. By lemma 6 we have a decomposition

$$S = s \cup S_1 \cup \cdots \cup S_k \cup \cdots$$

with the properties stated there. Thus clearly it is sufficient to show that

$$J(u, T) = 0 \quad \text{a.e. on } S_k \tag{30}$$

for each positive integer k. Now since T is Lipschitzian on S_k, by lemma 4 (applied to S_k) there exists a Lipschitzian transformation $T_k: \boldsymbol{D} \to R^n$ such that

$$T_k u = T u \quad \text{for} \quad u \in S_k, \tag{31}$$

$$J(u, T_k) = J(u, T) \quad \text{a.e. on } S_k. \tag{32}$$

Since $S_k \subset S$, it follows from (28) that $\mathsf{L}\, T S_k = 0$. Since $T_k S_k = T S_k$ by (31), we conclude that $\mathsf{L}\, T_k S_k = 0$. As T_k is Lipschitzian in $\boldsymbol{D}$, by **V.2.3**, lemma 5 (applied to T_k and S_k) we infer that $J(u, T_k) = 0$ a.e. on S_k, and (30) follows in view of (32).

Lemma 8. Given T as in (11) and (24), assume that the first partial derivatives of the coordinate functions of T exist a.e. in $\boldsymbol{D}$. Let S be an L-measurable subset of $\boldsymbol{D}$ such that

$$J(u, T) = 0 \quad \text{a.e. on } S. \tag{33}$$

Then there exists a decomposition $S = s \cup S'$ where s and S' are disjoint L-measurable sets such that

$$\mathsf{L}\, s = 0, \quad \mathsf{L}\, T S' = 0. \tag{34}$$

Proof. By lemma 6 we have a decomposition

$$S = s \cup S_1 \cup \cdots \cup S_k \cup \cdots$$

with the properties stated there. Thus clearly the present lemma will be proved if we show that $\mathsf{L}\, T \cup S_k = 0$. In turn, this last relation will be proved if we show that

$$\mathsf{L}\, T S_k = 0 \tag{35}$$

for every positive integer k. Now since T is Lipschitzian on S_k, by lemma 4 (applied to S_k) there exists a Lipschitzian transformation $T_k: \boldsymbol{D} \to R^n$ such that (31) and (32) hold. Since $S_k \subset S$, (33) and (32) imply that $J(u, T_k) = 0$ a.e. on S_k. As T_k is Lipschitzian in $\boldsymbol{D}$, by **V.2.3**, lemma 5 (applied to T_k and S_k) we conclude that $\mathsf{L}\, T_k S_k = 0$. As $T_k S_k = T S_k$ by (31), the relation (35) follows, and the lemma is proved.

§ V.3. Special classes of differentiable transformations in $\boldsymbol{R}^n$.

V.3.1. Preliminary comments. For continuous transformations in R^n we have developed in § **IV.4** and § **IV.5** two general theories which may be referred to as the *e-theory* and the B-*theory* respectively. In the present § **V.3**, we shall study various relationships between these two theories for the case of a general Euclidean space R^n. The case $n=1$ has been already discussed in detail in § **V.1**. Special features of the case $n=2$ will be studied in Part **VI**.

The results to be derived in the present § **V.3** may be classified as *inclusion theorems* and *equivalence theorems*. An *inclusion theorem* is a statement to the effect that a certain class of transformations is included in some other class of transformations. For example, theorem 1 in **IV.5.3** is an inclusion theorem stating that the class of $sAC\mathsf{B}$ transformations is included in the class of eAC transformations. An *equivalence theorem* is a statement to the effect that under certain conditions corresponding concepts in the e-theory and the B-theory become equivalent to each other or to an analogous concept in classical Analysis. For example, in classical Analysis one operates with the ordinary Jacobian $J(u, T)$, while in the e-theory and in the B-theory one uses the generalized Jacobians $J_e(u, T)$ and $J_{\mathsf{B}}(u, T)$ respectively. Each one of these three Jacobians (see **V.2.2**, **IV.4.3**, **IV.5.3** for the explicit definitions) is meant to represent the *local rate of change of volume* under the transformation T. Accordingly, one may expect that under certain appropriate conditions these three Jacobians will have the same value. In **V.3.2** we shall derive two fundamental equivalence theorems which confirm this expectation. Similarly, the e-theory and the B-theory give rise to the concepts eBV, $BV\mathsf{B}$, $sBV\mathsf{B}$ and eAC, $AC\mathsf{B}$, $sAC\mathsf{B}$ relating to bounded variation and absolute continuity respectively, and in turn these two sets of concepts depend upon two different multiplicity functions. Inclusion and equivalence theorems concerning these basic concepts will be discussed in **V.3.3**.

As regards the scope of results relating to equivalence theorems, it is natural to expect that the e-theory and the B-theory will yield equivalent information if applied to transformations of a more or less elementary character. Section **V.3.4** contains results of interest from this point of view. The term *elementary* is of course a vague one, but it may be assumed that the class of *generalized Lipschitzian transformations*, defined and discussed in **V.3.4**, is sufficiently comprehensive to account for all those transformations which should be thought of as constituting the *elementary range* in the present context.

V.3.2. Comparison of the generalized Jacobians $J_e(u, T)$, $J_{\mathsf{B}}(u, T)$ with the ordinary Jacobian $J(u, T)$. We shall use for the bounded continuous transformation T the alternative representations

$$T: \boldsymbol{D} \rightarrow R^n, \tag{1}$$

$$T: x^i = f_i(u^1, \ldots, u^n), \quad i = 1, \ldots, n, \tag{2}$$

as explained in **V.2.2**.

Theorem 1. Given T as in (1) and (2), assume that (a) T is eBV in $\boldsymbol{D}$, and (b) T possesses a weak total differential a.e. in $\boldsymbol{D}$ (see **V.2.2**, definition 2). Then the essential generalized Jacobian $J_e(u, T)$ is equal to the ordinary Jacobian $J(u,T)$ a.e. in $\boldsymbol{D}$.

Proof. The assumption (a) implies that the essential multiplicity function $K(x,T,\boldsymbol{D})$ is L-summable in R^n (see **IV.4.1**). Hence (see **III.1.1**, lemmas 32, 43) there exists a BOREL set X in R^n such that the following holds.

(α) $\mathsf{L}X=0$.

(β) $K(x,T,\boldsymbol{D})<\infty$ if $x\in \mathsf{C}X$.

(γ) If $x_0\in\mathsf{C}X$ and $\{G_m\}$ is any regular sequence of closed sets such that $x_0\in G_m$, $m=1, 2, \ldots$, and $\delta G_m\to 0$, then

$$\lim_{m\to\infty}\frac{1}{\mathsf{L}\,G_m}\int\limits_{G_m} K(x, T, \boldsymbol{D})\, d\,\mathsf{L} = K(x_0, T, \boldsymbol{D}) < \infty.$$

We first verify the relation

$$J_e(u, T) = 0 = J(u, T) \quad \text{a.e. on } T^{-1}X. \tag{3}$$

Note that since X is a BOREL set, $T^{-1}X$ is also a BOREL set, and hence $T^{-1}X$ is L-measurable (see **IV.1.1**, theorem 1, and **III.1.1**, lemma 19). Furthermore, $TT^{-1}X\subset X$, and thus $\mathsf{L}TT^{-1}X=0$ by (α). Accordingly, $J(u,T)=0$ a.e. on $T^{-1}X$ by **V.2.4**, lemma 7, while $J_e(u,T)=0$ a.e. on $T^{-1}X$ by **IV.4.3**, lemma 5, and (3) follows. Denote now by S_0 the set of those points $u\in\boldsymbol{D}$ where $J(u,T)$ exists and is equal to zero. Thus

$$S_0=\{u|u\in \boldsymbol{D},\ J(u, T) \text{ exists},\ J(u, T) = 0\}. \tag{4}$$

We assert that

$$J_e(u, T) = 0 = J(u, T) \quad \text{a.e. on } S_0. \tag{5}$$

Indeed, by **V.2.4**, lemma 8, we have a decomposition $S_0=s_0\cup S_0'$, where s_0, S_0' are L-measurable sets such that $\mathsf{L}s_0=0$ and $\mathsf{L}TS_0'=0$. This last relation implies (by **IV.4.3**, lemma 5) that $J_e(u,T)=0$ a.e. on S_0', and (5) follows in view of (4). Let us now denote by S^* the set of those points $u\in\boldsymbol{D}$ where one at least of $J(u,T)$, $J_e(u,T)$ fails to exist or T fails to possess a weak total differential. In view of the assumptions (a), (b) and **IV.4.3**, theorem 1, we have then

$$\mathsf{L}\,S^* = 0. \tag{6}$$

In view of (3), (5), (6) the theorem will be proved if we show that

$$J_e(u_0, T) = J(u_0, T) \tag{7}$$

provided that

$$u_0 \in \boldsymbol{D} - (S^* \cup S_0 \cup T^{-1} X). \tag{8}$$

Let us first observe that if (8) holds, then the following conditions are satisfied at u_0.

(i) $J(u_0, T)$ and $J_e(u_0, T)$ both exist.

(ii) $J(u_0, T) \neq 0$ and T possesses a weak total differential at u_0.

(iii) $T u_0 \in \complement X$.

Let us put $T u_0 = x_0$. In view of (ii), we have before us the *Situation 2* discussed in **V.2.2**. Using the terminology employed there, let us consider the sequence of oriented n-cubes $Q(\alpha_m)$ with center at u_0 and side length $2\alpha_m$, where $\alpha_m \to 0$. By **V.2.2** (45) we have

$$\mu(x, T, \operatorname{int} Q(\alpha_m)) = \left\{ \begin{array}{ccc} \operatorname{sgn} J(u_0, T) & \text{if} & x \in P_m', \\ 0 & \text{if} & x \in \complement P_m''. \end{array} \right\} \tag{9}$$

Observe that since

$$K(x, T, \operatorname{int} Q(\alpha_m)) \leq K(x, T, \boldsymbol{D}) < \infty \quad \text{a.e. in } R^n,$$

we have

$$K(x, T, \operatorname{int} Q(\alpha_m)) < \infty \quad \text{a.e. in } R^n. \tag{10}$$

Furthermore [see **V.2.2** (46)]

$$T \, fr \, Q(\alpha_m) \subset P_m'' - P_m'. \tag{11}$$

In view of (10) and (11) we conclude from **II.3.4**, theorem 2 [applied to the domain $\operatorname{int} Q(\alpha_m)$]

$$\mu_e(x, T, \operatorname{int} Q(\alpha_m)) = \mu(x, T, \operatorname{int} Q(\alpha_m)) \quad \text{a.e. on } P_m', \tag{12}$$

$$\mu_e(x, T, \operatorname{int} Q(\alpha_m)) = \mu(x, T, \operatorname{int} Q(\alpha_m)) \quad \text{a.e. on } \complement P_m''. \tag{13}$$

From (9), (12), (13) it follows that

$$\mu_e(x, T, \operatorname{int} Q(\alpha_m)) = \left\{ \begin{array}{cc} \operatorname{sgn} J(u_0, T) & \text{a.e. on } P_m', \\ 0 & \text{a.e. on } \complement P_m''. \end{array} \right\} \tag{14}$$

Consider now a point $x \in R^n$ where

$$K(x, T, \operatorname{int} Q(\alpha_m)) < \infty.$$

From **II.3.2**, theorem 1 and **II.3.4**, theorem 4 [applied to the domain $\operatorname{int} Q(\alpha_m)$] we obtain

$$|\mu_e(x, T, \operatorname{int} Q(\alpha_m)| = |K^+(x, T, \operatorname{int} Q(\alpha_m)) - K^-(x, T, \operatorname{int} Q(\alpha_m))| \leq$$
$$\leq K^+(x, T, \operatorname{int} Q(\alpha_m)) + K^-(x, T, \operatorname{int} Q(\alpha_m)) = K(x, T, \operatorname{int} Q(\alpha_m)).$$

By (10) there follows the inequality

$$|\mu_e(x, T, \operatorname{int} Q(\alpha_m))| \leq K(x, T, \operatorname{int} Q(\alpha_m)) \quad \text{a.e. in } R^n.$$

Since

$$K(x, T, \operatorname{int} Q(\alpha_m)) \leq K(x, T, \boldsymbol{D}),$$

we conclude that

$$|\mu_e(x, T, \operatorname{int} Q(\alpha_m))| \leq K(x, T, \boldsymbol{D}) \quad \text{a.e. in } R^n. \tag{15}$$

Let us put

$$\int_{P'_m} \mu_e(x, T, \operatorname{int} Q(\alpha_m))\, d\mathsf{L} = \lambda'_m, \tag{16}$$

$$\int_{P''_m} \mu_e(x, T, \operatorname{int} Q(\alpha_m))\, d\mathsf{L} = \lambda''_m, \tag{17}$$

$$\int_{P'_m} K(x, T, \boldsymbol{D})\, d\mathsf{L} = \sigma'_m, \tag{18}$$

$$\int_{P''_m} K(x, T, \boldsymbol{D})\, d\mathsf{L} = \sigma''_m. \tag{19}$$

Note that since $J_e(u_0, T)$ exists, we have

$$\lim_{m\to\infty} \frac{1}{\mathsf{L}\, Q(\alpha_m)} \int \mu_e(x, T, \operatorname{int} Q(\alpha_m))\, d\mathsf{L} = J_e(u_0, T) \tag{20}$$

by (4), (18), (19), in **IV.4.3**. From (14), (16), (17) it follows that

$$\int \mu_e(x, T, \operatorname{int} Q(\alpha_m))\, d\mathsf{L} = \lambda''_m, \tag{21}$$

$$\lambda''_m = \lambda'_m + (\lambda''_m - \lambda'_m) = (\mathsf{L}\, P'_m) \operatorname{sgn} J(u_0, T) + \lambda''_m - \lambda'_m. \tag{22}$$

From (15) to (19) we obtain

$$|\lambda''_m - \lambda'_m| \leq \sigma''_m - \sigma'_m. \tag{23}$$

In view of **V.2.2** (iii) and (44), the sequences of parallelotopes P'_m, P''_m constitute regular sequences of closed sets, containing the point $x_0 = T u_0$, such that

$$\delta P'_m \to 0, \quad \delta P''_m \to 0.$$

Hence, by (18), (19), (iii), (β), (γ) above,

$$\frac{\sigma''_m}{\mathsf{L}\, P''_m} \to K(x_0, T, \boldsymbol{D}) < \infty, \quad \frac{\sigma'_m}{\mathsf{L}\, P'_m} \to K(x_0, T, \boldsymbol{D}) < \infty. \tag{24}$$

By **V.2.2** (43) we have

$$\frac{\mathsf{L}\, P'_m}{\mathsf{L}\, Q(\alpha_m)} \to |J(u_0, T)|, \quad \frac{\mathsf{L}\, P''_m}{\mathsf{L}\, Q(\alpha_m)} \to |J(u_0, T)|. \tag{25}$$

From (24) and (25) it follows that

$$\frac{\sigma_m'' - \sigma_m'}{\mathsf{L}\,Q(\alpha_m)} = \frac{\mathsf{L}\,P_m''}{\mathsf{L}\,Q(\alpha_m)}\,\frac{\sigma_m''}{\mathsf{L}\,P_m''} - \frac{\mathsf{L}\,P_m'}{\mathsf{L}\,Q(\alpha_m)}\,\frac{\sigma_m'}{\mathsf{L}\,P_m'} \to 0.$$

Hence, by (23),

$$\frac{\lambda_m'' - \lambda_m'}{\mathsf{L}\,Q(\alpha_m)} \to 0. \tag{26}$$

Combining (20), (21), (22), (25), (26), we obtain

$$\begin{aligned} J_e(u_0, T) = \lim \frac{\lambda_m''}{\mathsf{L}\,Q(\alpha_m)} &= (\operatorname{sgn} J(u_0, T)) \lim \frac{\mathsf{L}\,P_m'}{\mathsf{L}\,Q(\alpha_m)} \\ &= (\operatorname{sgn} J(u_0, T))\,|J(u_0, T)| = J(u_0, T), \end{aligned}$$

and the proof is complete.

Theorem 2. Given T as in (1) and (2), assume that (a) T is $BV\mathsf{B}$ in $\boldsymbol{D}$, and (b) T possesses a total differential a.e. in $\boldsymbol{D}$ (see **V.2.2**, definition 1). Then the generalized Jacobian $J_{\mathsf{B}}(u, T)$ is equal to the ordinary Jacobian $J(u, T)$ a.e. in $\boldsymbol{D}$.

Proof. Let a point $u_0 \in \boldsymbol{D}$ be assigned. We have to show that

$$J_{\mathsf{B}}(u_0, T) = J(u_0, T), \tag{27}$$

except perhaps for certain exceptional points $u_0 \in \boldsymbol{D}$ which constitute a set of L-measure zero. Note that condition (a) implies (see **IV.5.1**, theorem 3) that the derivative $D_{\mathsf{B}}(u, T)$ exists a.e. in $\boldsymbol{D}$. Furthermore (see **IV.5.3**)

$$J_{\mathsf{B}}(u, T) = i_{\mathsf{B}}(u, T)\, D_{\mathsf{B}}(u, T) \quad \text{a.e. in } \boldsymbol{D}. \tag{28}$$

By **IV.5.1**, theorem 4, we have $D_{\mathsf{B}}(u, T) = D(u, G)$ a.e. in $\boldsymbol{D}$, where G stands for the function of open intervals

$$G(\operatorname{int} I) = \mathsf{L}\,T(\operatorname{int} I), \quad \operatorname{int} I \subset \boldsymbol{D}, \tag{29}$$

and $D(u, G)$ denotes the derivative of G. In view of condition (b) we can therefore assume, in proving (27), that the following holds.

(i) $J(u_0, T), D_{\mathsf{B}}(u_0, T), D(u_0, G), J_{\mathsf{B}}(u_0, T)$ exist.

(ii) $D_{\mathsf{B}}(u_0, T) = D(u_0, G)$.

(iii) $J_{\mathsf{B}}(u_0, T) = i_{\mathsf{B}}(u_0, T)\, D_{\mathsf{B}}(u_0, T)$.

(iv) T possesses a total differential at u_0.

Let us introduce the set

$$S_0 = \{u \mid u \in \boldsymbol{D},\ J(u, T) \text{ exists},\ J(u, T) = 0\}. \tag{30}$$

We assert that

$$J_B(u, T) = 0 = J(u, T) \quad \text{a.e. on } S_0. \tag{31}$$

Indeed, in view of condition (b), we can apply **V.2.4**, lemma 8, obtaining a decomposition $S_0 = s \cup S'$, where s, S' are L-measurable sets such that $\mathsf{L}s = 0$ and

$$\mathsf{L}\, T S' = 0. \tag{32}$$

Thus it is sufficient to show that

$$J_B(u, T) = 0 = J(u, T) \quad \text{a.e. on } S'. \tag{33}$$

Now (32) implies (see **IV.5.1**, theorem 5) that $D_B(u, T) = 0$ a.e. on S'. Thus (33) follows in view of (28) and (30), and (31) is established. From (31) we conclude that in proving (27) we can assume that

(v) $J(u_0, T) \neq 0$.

In view of (i), (iv), (v) we have before us the *Situation 3* discussed in **V.2.2**. Using the terminology employed there, consider the sequence $Q(\alpha_m)$ of oriented n-cubes with center at u_0 and side length $2\alpha_m$, where $\alpha_m \to 0$. By **V.2.2** (75) and (76) we have the relations

$$P'_m \subset T \operatorname{int} Q(\alpha_m) \subset P''_m, \tag{34}$$

$$i_B(u_0, T) = \operatorname{sgn} J(u_0, T). \tag{35}$$

From (29) and (34) it follows that

$$\frac{\mathsf{L}\, P'_m}{\mathsf{L}\, Q(\alpha_m)} \leq \frac{G(\operatorname{int} Q(\alpha_m))}{\mathsf{L}\, Q(\alpha_m)} \leq \frac{\mathsf{L}\, P''_m}{\mathsf{L}\, Q(\alpha_m)}. \tag{36}$$

Since $D(u_0, G)$ exists [see (i)], we have the relation

$$\lim \frac{G(\operatorname{int} Q(\alpha_m))}{\mathsf{L}\, Q(\alpha_m)} = D(u_0, G). \tag{37}$$

By **V.2.2** (82) we have

$$\lim \frac{\mathsf{L}\, P'_m}{\mathsf{L}\, Q(\alpha_m)} = \lim \frac{\mathsf{L}\, P''_m}{\mathsf{L}\, Q(\alpha_m)} = |J(u_0, T)|. \tag{38}$$

From (36), (37), (38) we see that

$$D(u_0, G) = |J(u_0, T)|. \tag{39}$$

The relations (ii), (iii), (39), (35) yield

$$J_B(u_0, T) = i_B(u_0, T)\, D(u_0, G) = (\operatorname{sgn} J(u_0, T))\, |J(u_0, T)| = J(u_0, T),$$

and the proof is complete.

V.3.3. Miscellaneous inclusion and equivalence theorems. As in **V.3.2**, we shall use the alternative representations

$$T: \boldsymbol{D} \to R^n, \tag{1}$$

$$T: x^i = f_i(u^1, \ldots, u^n), \quad i = 1, \ldots, n, \tag{2}$$

for the bounded continuous transformation T.

Lemma 1. Given T as in (1) and (2), assume that T possesses a weak total differential a.e. in $\boldsymbol{D}$. Then there exist three pair-wise disjoint L-measurable sets s, S', S^* in $\boldsymbol{D}$ such that the following holds.

(i) $\boldsymbol{D} = s \cup S' \cup S^*$.

(ii) $\mathsf{L}s = 0$.

(iii) $\mathsf{L}\,T S' = 0$.

(iv) The ordinary Jacobian exists and is different from zero at every point $u \in S^*$, and T possesses a weak total differential at every point $u \in S^*$.

(v) $N(x, T, \boldsymbol{D}) = K(x, T, \boldsymbol{D})$ for $x \notin T(s \cup S')$, where N and K are the multiplicity functions defined in **I.1.2** and **II.3.2** respectively.

Proof. By assumption, there exists a decomposition

$$\boldsymbol{D} = s_1 \cup S_1, \tag{3}$$

where s_1, S_1 are L-measurable sets such that

$$\mathsf{L}\, s_1 = 0, \quad s_1 \cap S_1 = \emptyset, \tag{4}$$

and T possesses a weak total differential at every point $u \in S_1$. Let us put

$$S_2 = \{u \mid u \in S_1,\ J(u, T) = 0\}, \tag{5}$$

$$S^* = \{u \mid u \in S_1,\ J(u, T) \neq 0\}. \tag{6}$$

Since $J(u, T)$ is L-measurable in $\boldsymbol{D}$ (see **III.1.3**), it is clear that S_2, S^* are L-measurable sets, and

$$S_2 \cap S^* = \emptyset. \tag{7}$$

Furthermore, the following holds.

(α) T possesses a weak total differential at every point $u \in S^*$.

(β) $J(u, T)$ exists and is different from zero at every point $u \in S^*$.

In view of (5) and **V.2.4**, lemma 8 (applied to S_2), we have a decomposition

$$S_2 = s_2 \cup S', \tag{8}$$

where s_2, S' are L-measurable sets such that

$$\mathsf{L}\, s_2 = 0, \quad s_2 \cap S' = \emptyset, \tag{9}$$

$$\mathsf{L}\, T S' = 0. \tag{10}$$

Setting $s = s_1 \cup s_2$, we assert that the sets s, S', S^* possess the desired properties. Obviously s, S', S^* are pair-wise disjoint L-measurable sets such that (i), (ii), (iii) hold [see (3), (4), (5), (6), (7), (8), (9), (10)]. Furthermore, (iv) holds in view of (α) and (β). To verify (v), consider any point

$$x_0 \notin T(s \cup S'). \tag{11}$$

Then clearly

$$T^{-1} x_0 \subset S^*. \tag{12}$$

If $T^{-1} x_0 = \emptyset$, then (v) is obvious (see **II.3.2**, remark 2). Hence we can assume that $T^{-1} x_0 \neq \emptyset$. Consider any point $u_0 \in T^{-1} x_0$. Then $u_0 \in S^*$ by (12), and in view of (α) and (β) it follows that at u_0 we have before us the *Situation 2* discussed in **V.2.2**. Accordingly, as we have shown there, u_0 by itself is an e.m.m.c. for $(x_0, T, \boldsymbol{D})$. Thus *every* point of $T^{-1} x_0$ is an e.m.m.c. for $(x_0, T, \boldsymbol{D})$, and hence (see **II.3.3**)

$$N(x_0, T, \boldsymbol{D}) = k(x_0, T, \boldsymbol{D}). \tag{13}$$

It is convenient to divide the rest of the proof into two parts.

Case 1. $k(x_0, T, \boldsymbol{D}) = \infty$. Then

$$K(x_0, T, \boldsymbol{D}) = k(x_0, T, \boldsymbol{D})$$

by **II.3.4**, theorem 3, and (v) follows in view of (13).

Case 2. $k(x_0, T, \boldsymbol{D}) < \infty$. Note that $N(x_0, T, \boldsymbol{D}) > 0$ since $T^{-1} x_0 \neq \emptyset$, and hence [in view of (13)]

$$0 < k(x_0, T, \boldsymbol{D}) < \infty. \tag{14}$$

Applying theorem 4, remark 2, and remark 5 in **II.3.4**, we obtain the relation

$$K(x_0, T, \boldsymbol{D}) = \sum |i_e(C, T)|, \quad C \in \mathfrak{E}(x_0, T, \boldsymbol{D}). \tag{15}$$

We observed above that each point of the set $T^{-1} x_0$ is an e.m.m.c. for $(x_0, T, \boldsymbol{D})$. Thus the elements of the class $\mathfrak{E}(x_0, T, \boldsymbol{D})$ are now the individual points of $T^{-1} x_0$. In view of (13) and (14) it follows that each point of $T^{-1} x_0$ is an essentially isolated e.m.m.c. for $(x_0, T, \boldsymbol{D})$. Since $J(u, T)$ exists and is different from zero and T possesses a weak total differential at every point $u \in T^{-1} x_0$ [see (12), (α), (β)], we can apply **V.2.2** (68), obtaining the formula

$$i_e(u, T) = \operatorname{sgn} J(u, T) \quad \text{for} \quad u \in T^{-1} x_0.$$

It is now clear that the summation in (15) contains $N(x_0, T, \boldsymbol{D})$ terms, each of which has the value 1. Thus (15) yields $K(x_0, T, \boldsymbol{D}) = N(x_0, T, \boldsymbol{D})$, and the lemma is proved.

Theorem 1. Given T as in (1) and (2), assume that (a) T possesses a weak total differential a.e. in $\boldsymbol{D}$, and (b) T is $BV\mathsf{B}$ in $\boldsymbol{D}$. Then

(α) $|J_{\mathsf{B}}(u, T)| = D_{\mathsf{B}}(u, T)$ a.e. in $\boldsymbol{D}$ (see **IV.5.1**, **IV.5.3**),

(β) T is $sBV\mathsf{B}$ in $\boldsymbol{D}$ (see **IV.5.3**, definition 1).

Proof. By **IV.5.1**, theorem 3, the derivative $D_{\mathsf{B}}(u, T)$ exists a.e. in $\boldsymbol{D}$ and is L-summable there. In view of the assumption (a) we can apply lemma 1, and thus we have at our disposal the sets s, S', S^* with the properties (i) to (v) described there. Since $\mathsf{L}s = 0$, (α) will be verified if we show that

$$|J_{\mathsf{B}}(u, T)| = 0 = D_{\mathsf{B}}(u, T) \quad \text{a.e. on } S', \tag{16}$$

$$|J_{\mathsf{B}}(u, T)| = D_{\mathsf{B}}(u, T) \quad \text{a.e. on } S^*. \tag{17}$$

Recall that (see **IV.5.3**)

$$J_{\mathsf{B}}(u, T) = i_{\mathsf{B}}(u, T)\, D_{\mathsf{B}}(u, T) \quad \text{a.e. in } \boldsymbol{D}. \tag{18}$$

Now since $\mathsf{L}T S' = 0$, we have $D_{\mathsf{B}}(u, T) = 0$ a.e. on S' by **IV.5.1**, theorem 5, and (16) follows in view of (18). To prove (17), we make use of the set $\mathbf{I} = \mathbf{I}(T, \boldsymbol{D})$ of those points $u \in \boldsymbol{D}$ which have a neighborhood clear of relatives (see **II.3.7**). Let us put

$$S_1^* = S^* \cap \mathbf{I}, \quad S_2^* = S^* \cap (\boldsymbol{D} - \mathbf{I}). \tag{19}$$

Since S^* is L-measurable and $\mathbf{I}$ is a Borel set (see **II.3.7**, lemma 1), S_1^* and S_2^* are L-measurable. As $D_{\mathsf{B}}(u, T) = 0$ a.e. on $\boldsymbol{D} - \mathbf{I}$ (see **IV.5.1**, theorem 6) and $S_2^* \subset \boldsymbol{D} - \mathbf{I}$, we conclude by (18) that

$$|J_{\mathsf{B}}(u, T)| = 0 = D_{\mathsf{B}}(u, T) \quad \text{a.e. on } S_2^*. \tag{20}$$

Observe now that since $S_1^* \subset S^*$, the ordinary Jacobian exists and is different from zero for $u \in S_1^*$ and T possesses a weak total differential at every point $u \in S_1^*$. Thus we have before us, at every point $u \in S_1^*$, the *Situation* 2 discussed in **V.2.2**. Also, since $S_1^* \subset \mathbf{I}$, every point $u \in S_1^*$ has a neighborhood clear of relatives. By **V.2.2** (69) it follows that

$$i_{\mathsf{B}}(u, T) = \operatorname{sgn} J(u, T) = \pm 1 \quad \text{for} \quad u \in S_1^*. \tag{21}$$

From (18) and (21) we conclude that

$$|J_{\mathsf{B}}(u, T)| = D_{\mathsf{B}}(u, T) \quad \text{a.e. on } S_1^*. \tag{22}$$

As (19), (20), (22) imply (17), part (α) of the theorem is proved. Since (as we noted above) $D_{\mathsf{B}}(u,T)$ is L-summable in $\boldsymbol{D}$, (α) implies that $J_{\mathsf{B}}(u,T)$ is also L-summable in $\boldsymbol{D}$. As T is $BV\mathsf{B}$ in $\boldsymbol{D}$ by assumption, part (β) of theorem follows (see **IV.5.3**, definition 1).

Theorem 2. Given T as in (1) and (2), assume that (a) T possesses a weak total differential a.e. in $\boldsymbol{D}$, and (b) T satisfies the condition (N) in $\boldsymbol{D}$ (see **IV.1.4**). Then

$$N(x, T, \boldsymbol{D}) = K(x, T, \boldsymbol{D}) \quad \text{a.e. in } R^n. \tag{23}$$

Proof. Lemma 1 yields the relation

$$N(x, T, \boldsymbol{D}) = K(x, T, \boldsymbol{D}) \quad \text{for} \quad x \notin T(s \cup S'), \tag{24}$$

where s, S' are L-measurable subsets of $\boldsymbol{D}$ such that

$$\mathsf{L}\, s = 0, \tag{25}$$

$$\mathsf{L}\, T S' = 0. \tag{26}$$

In view of the assumption (b), (25) implies that $\mathsf{L}\, Ts = 0$. Hence $\mathsf{L}\, T(s \cup S') = 0$ by (26), and (23) follows now by (24).

Theorem 3. Given T as in (1) and (2), assume that (a) T possesses a weak total differential a.e. in $\boldsymbol{D}$, and (b) T is $AC\mathsf{B}$ in $\boldsymbol{D}$. Then the following holds.

(i) T is both $sAC\mathsf{B}$ and eAC in $\boldsymbol{D}$.

(ii) $J_{\mathsf{B}}(u,T) = J_e(u,T) = J(u,T)$ a.e. in $\boldsymbol{D}$.

(iii) $N(x,T,\boldsymbol{D}) = K(x,T,\boldsymbol{D})$ a.e. in R^n.

Proof. By **IV.5.2**, theorem 1, T is $BV\mathsf{B}$ in $\boldsymbol{D}$. Thus the assumptions of theorem 1 are fulfilled, and it follows that T is $sBV\mathsf{B}$ in $\boldsymbol{D}$. Since T is $AC\mathsf{B}$ in $\boldsymbol{D}$ by assumption, it follows further that T is $sAC\mathsf{B}$ in $\boldsymbol{D}$ (see **IV.5.3**, definition 1 and definition 2). By **IV.5.3**, theorem 1 we conclude that T is eAC (and hence also eBV) in $\boldsymbol{D}$, and

$$J_{\mathsf{B}}(u, T) = J_e(u, T) \quad \text{a.e. in } \boldsymbol{D}.$$

In view of the assumption (a) we can apply **V.3.2**, theorem 1, obtaining

$$J_e(u, T) = J(u, T) \quad \text{a.e. in } \boldsymbol{D}.$$

Thus (i) and (ii) are proved. Observe now that T satisfies the condition (N) in $\boldsymbol{D}$ by **IV.5.2**, theorem 3. Hence (iii) follows directly from theorem 2, and the proof is complete.

Lemma 2. Given T as in (1) and (2), assume that the first partial derivatives of the coordinate functions of T exist a.e. in $\boldsymbol{D}$. Let S be an L-measurable subset of $\boldsymbol{D}$ such that T is Lipschitzian on S. Then

$$\int_S |J(u, T)|\, d\mathsf{L} = \int N(x, T, S)\, d\mathsf{L}. \tag{27}$$

Proof. By **V.2.4**, lemma 4 there exists a transformation

$$T^*: \boldsymbol{D} \to R^n,$$

such that the following holds.

(i) T^* is Lipschitzian in $\boldsymbol{D}$.

(ii) $T^* u = T u$ for $u \in S$.

(iii) $J(u, T^*) = J(u, T)$ a.e. on S.

By **V.2.3**, lemma 3, T^* is $\mathrm{A}C\mathsf{B}$ in $\boldsymbol{D}$. Accordingly, by **IV.5.2**, theorem 5 (applied to T^* and S) we have the formula

$$\int_S D_{\mathsf{B}}(u, T^*)\, d\mathsf{L} = \int N(x, T^*, S) d\mathsf{L}. \tag{28}$$

Since (ii) implies that $N(x, T^*, S) = N(x, T, S)$, we have

$$\int N(x, T^*, S)\, d\mathsf{L} = \int N(x, T, S)\, d\mathsf{L}. \tag{29}$$

Note that T^* possesses a total differential, and hence also a weak total differential, a.e. in $\boldsymbol{D}$ by **V.2.3**, lemma 2. Thus we can apply theorem 3 and theorem 1 to T^*, obtaining

$$J_{\mathsf{B}}(u, T^*) = J(u, T^*) \quad \text{a.e. in } \boldsymbol{D}, \tag{30}$$

$$|J_{\mathsf{B}}(u, T^*)| = D_{\mathsf{B}}(u, T^*) \quad \text{a.e. in } \boldsymbol{D}. \tag{31}$$

Clearly the relations (iii), (28) to (31) imply (27).

Lemma 3. Given T as in (1) and (2), assume that (a) the first partial derivatives of the coordinate functions of T exist a.e. in $\boldsymbol{D}$, and (b) T satisfies the condition (N) in $\boldsymbol{D}$. Let S be any L-measurable subset of $\boldsymbol{D}$. Then

$$\int_S |J(u, T)|\, d\mathsf{L} = \int N(x, T, S)\, d\mathsf{L} \tag{32}$$

as soon as one of the two integrals involved exists.

Proof. The assumption (a) implies (see **V.2.4**, lemma 6) the existence of a decomposition

$$S = s \cup S_1 \cup \cdots \cup S_k \cup \cdots, \tag{33}$$

where s, S_k are L-measurable sets such that

$$\mathsf{L}\, s = 0, \tag{34}$$

$$s \cap S_k = \emptyset, \quad S_j \cap S_k = \emptyset \quad \text{for } j \neq k, \tag{35}$$

and T is Lipschitzian on each one of the sets S_k. Application of lemma 2 (with S replaced by S_k) yields

$$\int_{S_k} |J(u, T)|\, d\mathsf{L} = \int N(x, T, S_k)\, d\mathsf{L}, \quad k = 1, 2, \ldots. \tag{36}$$

Let us put

$$f(x) = \sum N(x, T, S_k). \tag{37}$$

The relations (33), (35), (37) yield

$$N(x, T, S) = f(x) + N(x, T, s). \tag{38}$$

Since T satisfies the condition (N) in $\boldsymbol{D}$, (34) implies that $\mathsf{L}Ts = 0$. Hence $N(x, T, s) = 0$ a.e. in R^n, and thus (38) shows that

$$N(x, T, S) = f(x) \quad \text{a.e. in } R^n. \tag{39}$$

Note that $N(x, T, S)$, $N(x, T, S_k)$ are L-measurable in R^n by **IV.1.4**, corollary 2. Assume now first that $J(u, T)$ is L-summable on S. Then (see **III.1.1**, lemma 37) the relations (33) to (35) yield

$$\sum \int_{S_k} |J(u, T)|\, d\mathsf{L} = \int_S |J(u, T)|\, d\mathsf{L}. \tag{40}$$

From (36) and (40) we conclude that the series

$$\sum \int N(x, T, S_k)\, d\mathsf{L} \tag{41}$$

is convergent. Since $N(x, T, S_k) \geq 0$, the convergence of the series (41) implies, in view of (37) and **III.1.1**, lemma 39, the L-summability in R^n of $f(x)$, as well as the relation

$$\int f(x)\, d\mathsf{L} = \sum \int N(x, T, S_k)\, d\mathsf{L}. \tag{42}$$

Clearly, (40), (36), (42), (39) imply (32). Assume next that $N(x, T, S)$ is L-summable in R^n. By (39) it follows that $f(x)$ is L-summable in R^n. Since $N(x, T, S_k) \geq 0$, (37) implies therefore (by **III.1.1**, lemma 40) that

$$\sum \int N(x, T, S_k)\, d\mathsf{L} = \int f(x)\, d\mathsf{L} < \infty. \tag{43}$$

From (43) and (36) we conclude that the series

$$\sum \int_{S_k} |J(u, T)|\, d\mathsf{L}$$

is convergent. In view of (33), (34), (35) it follows, by **III.1.1**, lemma 39, that $J(u, T)$ is L-summable on S. This fact implies, as we have already shown, that (32) holds, and the proof is complete.

Theorem 4. Given T as in (1) and (2), assume that (a) the first partial derivatives of the coordinate functions of T exist a.e. in $\boldsymbol{D}$, and (b) T satisfies the condition (N) in $\boldsymbol{D}$. Then T is $AC\mathsf{B}$ in $\boldsymbol{D}$ if and only if the ordinary Jacobian $J(u,T)$ is L-summable in $\boldsymbol{D}$.

Proof. Assume first that $J(u,T)$ is L-summable in $\boldsymbol{D}$. By lemma 3 (applied with $S=\boldsymbol{D}$) we conclude that $N(x,T,\boldsymbol{D})$ is L-summable in R^n, and thus T is $BV\mathsf{B}$ in $\boldsymbol{D}$ by **IV.5.1**, theorem 1. Since T satisfies the condition (N) in $\boldsymbol{D}$, it follows that T is $AC\mathsf{B}$ in $\boldsymbol{D}$ (see **IV.5.2**, theorem 3). Assume, conversely, that T is $AC\mathsf{B}$ in $\boldsymbol{D}$. Then T is also $BV\mathsf{B}$ in $\boldsymbol{D}$, and hence $N(x,T,\boldsymbol{D})$ is L-summable in R^n (see **IV.5.2**, theorem 1 and **IV.5.1**, theorem 1). From lemma 3, applied with $S=\boldsymbol{D}$, it follows that $J(u,T)$ is L-summable in $\boldsymbol{D}$, and the proof is complete.

Theorem 5. Given T as in (1) and (2), assume that (a) T possesses a weak total differential a.e. in $\boldsymbol{D}$, and (b) T satisfies the condition (N) in $\boldsymbol{D}$. Then T is eAC in $\boldsymbol{D}$ if and only if the ordinary Jacobian $J(u,T)$ is L-summable in $\boldsymbol{D}$.

Proof. Assume first that $J(u,T)$ is L-summable in $\boldsymbol{D}$. By theorem 4 it follows that T is $AC\mathsf{B}$ in $\boldsymbol{D}$. Thus the assumptions of theorem 3 are satisfied, and hence T is eAC in $\boldsymbol{D}$. Assume, conversely, that T is eAC in $\boldsymbol{D}$. Then $J_e(u,T)$ is L-summable in $\boldsymbol{D}$ by **IV.4.3**, theorem 1. Since $J_e(u,T)=J(u,T)$ a.e. in $\boldsymbol{D}$ by **V.3.2**, theorem 1, it follows that $J(u,T)$ is L-summable in $\boldsymbol{D}$, and the theorem is proved.

Theorem 6. Given T as in (1) and (2), assume that (a) T possesses a weak total differential a.e. in $\boldsymbol{D}$, (b) the ordinary Jacobian $J(u,T)$ is L-summable in $\boldsymbol{D}$, and (c) T satisfies the condition (N) in $\boldsymbol{D}$. Then the following holds.

(i) T is both $sAC\mathsf{B}$ and eAC in $\boldsymbol{D}$.

(ii) $J_{\mathsf{B}}(u,T)=J_e(u,T)=J(u,T)$ a.e. in $\boldsymbol{D}$.

(iii) $D_{\mathsf{B}}(u,T)=D_e(u,T)=|J(u,T)|$ a.e. in $\boldsymbol{D}$.

(iv) $N(x,T,\boldsymbol{D})=K(x,T,\boldsymbol{D})$ a.e. in R^n.

Proof. Recall that the assumption (a) implies the existence of the first partial derivatives of the coordinate functions of T a.e. in $\boldsymbol{D}$ (see **V.2.2**, definition 2). Thus theorem 4 applies, and hence T is $AC\mathsf{B}$ in $\boldsymbol{D}$. By **IV.5.2**, theorem 1 it follows that T is also $BV\mathsf{B}$ in $\boldsymbol{D}$. Hence we can apply theorem 3 and theorem 1. From theorem 3 we infer that (i), (ii), (iv) hold. Theorem 1 yields

$$|J_{\mathsf{B}}(u,T)| = D_{\mathsf{B}}(u,T) \quad \text{a.e. in } \boldsymbol{D},$$

and hence, in view of (ii),

$$D_{\mathsf{B}}(u,T) = |J(u,T)| \quad \text{a.e. in } \boldsymbol{D}.$$

Since T is eAC in $\boldsymbol{D}$, it is also eBV there. Hence (see **IV.4.3**, theorem 2)

$$D_e(u, T) = |J_e(u, T)| \quad \text{a.e. in } \boldsymbol{D}.$$

In view of (ii) it follows that

$$D_e(u, T) = |J(u, T)| \quad \text{a.e. in } \boldsymbol{D},$$

and the theorem is proved.

V.3.4. Transformation formulas for definite integrals in $\boldsymbol{R}^n$ in terms of the ordinary Jacobian $\boldsymbol{J(u, T)}$. We shall use, as in **V.3.2**, the alternative representations

$$T: \boldsymbol{D} \to R^n, \tag{1}$$

$$T: x^i = f_i(u^1, \ldots, u^n), \quad i = 1, \ldots, n, \tag{2}$$

for the bounded continuous transformation T.

Theorem 1. Given T as in (1) and (2), assume that T is eAC in $\boldsymbol{D}$ and possesses a weak total differential a.e. in $\boldsymbol{D}$. Then

$$\int_{\boldsymbol{D}} |J(u, T)| \, d\mathsf{L} = \int K(x, T, \boldsymbol{D}) \, d\mathsf{L}, \tag{3}$$

$$\int_{\boldsymbol{D}} J(u, T) \, d\mathsf{L} = \int \mu_e(x, T, \boldsymbol{D}) \, d\mathsf{L}. \tag{4}$$

More generally, if $H(x)$ is a finite-valued, L-measurable function in R^n, then

$$\int_{\boldsymbol{D}} H(T u) \, |J(u, T)| \, d\mathsf{L} = \int H(x) \, K(x, T, \boldsymbol{D}) \, d\mathsf{L}, \tag{5}$$

as soon as one of the two integrals involved exists, and

$$\int_{\boldsymbol{D}} H(T u) \, J(u, T) \, d\mathsf{L} = \int H(x) \, \mu_e(x, T, \boldsymbol{D}) \, d\mathsf{L}, \tag{6}$$

as soon as the integral on the left exists.

Proof. Note that

$$J_e(u, T) = J(u, T) \quad \text{a.e. in } \boldsymbol{D}$$

by **V.3.2**, theorem 1. We can therefore replace $J_e(u, T)$ by $J(u, T)$ in **IV.4.4**, theorem 1, theorem 2, theorem 4, obtaining the formulas (3), (4), (5), (6).

Theorem 2. Given T as in (1) and (2), assume that (a) T is defined and continuous on $\overline{\boldsymbol{D}}$, (b) $\mathsf{L}T \, fr \boldsymbol{D} = 0$, (c) T is eAC in $\boldsymbol{D}$, and (d) T possesses a weak total differential a.e. in $\boldsymbol{D}$. Then

$$\int_{\boldsymbol{D}} J(u, T) \, d\mathsf{L} = \int \mu(x, T, \boldsymbol{D}) \, d\mathsf{L}. \tag{7}$$

More generally, if $H(x)$ is a finite-valued, L-measurable function in R^n, then

$$\int_{\boldsymbol{D}} H(T u)\, J(u, T)\, d\mathsf{L} = \int H(x)\, \mu(x, T, \boldsymbol{D})\, d\mathsf{L}, \tag{8}$$

as soon as the integral on the left exists.

Proof. We have again $J_e(u, T) = J(u, T)$ a.e. in $\boldsymbol{D}$ by **V.3.2**, theorem 1, and hence (7) and (8) follow from **IV.4.4**, theorem 3, theorem 4 by replacing $J_e(u, T)$ by $J(u, T)$.

The preceding two theorems belong to the *e-theory* (developed in § **IV.4**). To obtain corresponding theorems for the B-*theory* (developed in § **IV.5**), one should replace the assumption that T is eAC in $\boldsymbol{D}$ by the assumption that T is $AC\mathsf{B}$ in $\boldsymbol{D}$. The resulting statements, while true, yield no new information. Indeed, if T is $AC\mathsf{B}$ in $\boldsymbol{D}$ and possesses a weak total differential a.e. in $\boldsymbol{D}$, then T is eAC in $\boldsymbol{D}$ by **V.3.3**, theorem 3, and thus we have before us merely special instances of the situations considered in the preceding theorems 1 and 2.

There arises the question whether the essential multiplicity function K can be replaced by the crude multiplicity function N in the formulas (3) and (5). The following theorem is of interest in this connection.

Theorem 3. Given T as in (1) and (2), assume that (a) T possesses a weak total differential a.e. in $\boldsymbol{D}$, (b) the ordinary Jacobian $J(u, T)$ is L-summable in $\boldsymbol{D}$, and (c) T satisfies the condition (N) in $\boldsymbol{D}$. Then

$$\int_{\boldsymbol{D}} |J(u, T)\, d\mathsf{L} = \int N(x, T, \boldsymbol{D})\, d\mathsf{L}. \tag{9}$$

More generally, if S is an L-measurable subset of $\boldsymbol{D}$ and $H(x)$ is a finite-valued, L-measurable function in R^n, then

$$\int_{S} H(T u)\, |J(u, T)|\, d\mathsf{L} = \int H(x)\, N(x, T, S)\, d\mathsf{L}, \tag{10}$$

as soon as one of the two integrals involved exists. In particular [corresponding to the special case $H(x) \equiv 1$] we have the formula

$$\int_{S} |J(u, T)|\, d\mathsf{L} = \int N(x, T, S)\, d\mathsf{L}. \tag{11}$$

Proof. By **V.3.3**, theorem 6, T is eAC in $\boldsymbol{D}$, and thus (3) holds by theorem 1. Furthermore, by the same theorem 6 in **V.3.3**,

$$K(x, T, \boldsymbol{D}) = N(x, T, \boldsymbol{D}) \quad \text{a.e. in } R^n,$$

and hence (9) follows from (3) by replacing K by N. The formula (11) follows directly from **V.3.3**, lemma 3, since $J(u, T)$ is now L-summable on S as a consequence of the assumption (b). There remains to discuss the formula (10). Observe that, by **V.3.3**, theorem 6, T is $sAC\mathsf{B}$ and hence also ACB in $\boldsymbol{D}$. Accordingly, by **IV.5.2**, theorem 4,

$$\int_S H(T\,u)\, D_{\mathsf{B}}(u, T)\, d\mathsf{L} = \int H(x)\, N(x, T, S)\, d\mathsf{L}, \tag{12}$$

as soon as one of the two integrals involved exists. By **V.3.3**, theorem 6, we have

$$D_{\mathsf{B}}(u, T) = |J(u, T)| \quad \text{a.e. in } \boldsymbol{D},$$

and hence

$$\int_S H(T\,u)\, D_{\mathsf{B}}(u, T)\, d\mathsf{L} = \int_S H(T\,u)\, |J(u, T)|\, d\mathsf{L}, \tag{13}$$

as soon as one of the two integrals involved exists. Combining the preceding remarks, we conclude that if one of the three integrals occurring in (12) and (13) exists, then the other two exist also and the three integrals have then the same value. It follows that (10) holds as soon as one of the two integrals involved exists, and the proof is complete.

The preceding theorem admits of various generalizations. In particular, a similar theorem holds for transformations from subsets of R^n into R^N, where $N \geqq n$. However, the study of these generalizations is beyond the scope of this treatise.

V.3.5. Closure theorems in $\boldsymbol{R}^n$ in terms of the ordinary Jacobian $\boldsymbol{J(u, T)}$. The term *closure theorem* is used here in the sense explained in **IV.4.5**. Accordingly we consider, as in **IV.4.5**, bounded continuous transformations

$$T : \boldsymbol{D} \to R^n, \tag{1}$$

$$T_j : \boldsymbol{D}_j \to R^n, \quad j = 1, 2, \ldots, \tag{2}$$

where $\boldsymbol{D}, \boldsymbol{D}_j$ are bounded domains in R^n. We assume that T, T_j satisfy the following conditions.

(α) The sequence $\{T_j\}$ converges to T uniformly on compact subsets of $\boldsymbol{D}$ (see **II.3.2,** remark 9).

(β) T possesses a weak total differential a.e. in $\boldsymbol{D}$, and the ordinary Jacobian $J(u, T)$ is L-summable in $\boldsymbol{D}$.

(γ) T_j is eAC in $\boldsymbol{D}_j$ and possesses a weak total differential a.e. in $\boldsymbol{D}_j$, $j = 1, 2, \ldots$.

Observe that (γ) implies that T_j is eBV in $\boldsymbol{D}_j$ (see **IV.4.2**). Hence, by **V.3.2**, theorem 1 it follows further that

$$J(u, T_j) = J_e(u, T_j) \quad \text{a.e. in } \boldsymbol{D}_j. \tag{3}$$

From (3) and **IV.4.3**, theorem 1 it follows that

(δ) $J(u, T_j)$ is L-summable in $\boldsymbol{D}_j$.

Consider now a domain D such that $\bar{D} \subset \boldsymbol{D}$. By ($\alpha$) we have then $\bar{D} \subset \boldsymbol{D}_j$ if j is sufficiently large, say $j > j(\bar{D})$. By **IV.4.2**, theorem 1 (applied to T_j) it follows that

(ε) T_j is eAC in D if $\bar{D} \subset \boldsymbol{D}$ and $j > j(\bar{D})$.

Theorem 1. Given T, T_j as in (1) and (2), assume that the following holds in addition to the conditions (α), (β), (γ).

(i) $\liminf_{j\to\infty} \int_{\boldsymbol{D}_j} |J(u, T_j)|\, d\mathsf{L} < \infty$.

(ii) If Q is any oriented n-cube in $\boldsymbol{D}$, then

$$\lim_{j\to\infty} \int_Q |J(u, T_j)|\, d\mathsf{L} = \int_Q |J(u, T)|\, d\mathsf{L}.$$

Then T is eAC in $\boldsymbol{D}$.

Proof. Note that in view of (3) the condition (i) is equivalent to the condition

$$\liminf_{j\to\infty} \int_{\boldsymbol{D}_j} |J_e(u, T_j)|\, d\mathsf{L} < \infty.$$

Hence by **IV.4.5**, lemma 1 it follows that T is eBV in $\boldsymbol{D}$. In view of (β) we can now apply **V.3.2**, theorem 1, obtaining the relation

$$J(u, T) = J_e(u, T) \quad \text{a.e. in } \boldsymbol{D}.$$

Using (3), we can therefore re-write condition (ii) in the form

$$\lim_{j\to\infty} \int_Q |J_e(u, T_j)|\, d\mathsf{L} = \int_Q |J_e(u, T)|\, d\mathsf{L}.$$

Thus the assumptions of **IV.4.5**, theorem 3 are satisfied, and hence T is eAC in $\boldsymbol{D}$.

Theorem 2. Given T, T_j as in (1) and (2), assume that the following holds in addition to the conditions (α), (β), (γ).

(i) If F is any compact subset of $\boldsymbol{D}$, then

$$J(u, T_j) \to J(u, T) \quad \text{a.e. on } F.$$

(ii) If Q is any oriented n-cube in $\boldsymbol{D}$, then

$$\lim_{j\to\infty} \int_Q |J(u, T_j)|\, d\mathsf{L} = \int_Q |J(u, T)|\, d\mathsf{L}.$$

Then T is eAC in $\boldsymbol{D}$.

Proof. Let F be any compact subset of $\boldsymbol{D}$. We assert that

$$\lim_{j\to\infty} \int_F |J(u, T_j)|\, d\mathsf{L} = \int_F |J(u, T)|\, d\mathsf{L}. \tag{4}$$

Note first that (i) implies (see **III.1.1**, lemma 30) the inequality

$$\int_F |J(u, T)|\, d\mathsf{L} \leq \liminf_{j\to\infty} \int_F |J(u, T_j)|\, d\mathsf{L}. \tag{5}$$

Assign now $\varepsilon > 0$. Then (see **III.1.1**, lemma 34) there exists an open set $O = O(\varepsilon)$ such that

$$F \subset O \subset \boldsymbol{D}, \tag{6}$$

$$\int_O |J(u, T)|\, d\mathsf{L} < \int_F |J(u, T)|\, d\mathsf{L} + \varepsilon. \tag{7}$$

We can select a finite system $Q_1, \ldots, Q_m$ of oriented n-cubes such that

$$F \subset \bigcup_{k=1}^{m} Q_k \subset O, \tag{8}$$

$$\operatorname{int} Q_i \cap \operatorname{int} Q_k = \emptyset \quad \text{if} \quad i \neq k. \tag{9}$$

In view of **III.1.1**, lemma 42 the relations (8) and (9) yield

$$\sum_{k=1}^{m} \int_{Q_k} |J(u, T)|\, d\mathsf{L} = \int_{\cup Q_k} |J(u, T)|\, d\mathsf{L} \leq \int_O |J(u, T)|\, d\mathsf{L}.$$

By (7) there follows the inequality

$$\sum_{k=1}^{m} \int_{Q_k} |J(u, T)|\, d\mathsf{L} < \int_F |J(u, T)|\, d\mathsf{L} + \varepsilon. \tag{10}$$

Furthermore, (8) yields

$$\int_F |J(u, T_j)|\, d\mathsf{L} \leq \sum_{k=1}^{m} \int_{Q_k} |J(u\, T_j)|\, d\mathsf{L}. \tag{11}$$

From the assumption (ii) we infer that

$$\lim_{j\to\infty} \sum_{k=1}^{m} \int_{Q_k} |J(u, T_j)|\, d\mathsf{L} = \sum_{k=1}^{m} \int_{Q_k} |J(u, T)|\, d\mathsf{L}. \tag{12}$$

In view of (10) and (12) we obtain from (11) the inequality

$$\limsup_{j\to\infty} \int_F |J(u, T_j)|\, d\mathsf{L} \leq \int_F |J(u, T)|\, d\mathsf{L} + \varepsilon. \tag{13}$$

Since $\varepsilon > 0$ was arbitrary, clearly (13) and (5) imply (4). Consider now any domain D such that

$$\overline{D} \subset \boldsymbol{D}. \tag{14}$$

Applying (4) with $F = \overline{D}$, there follows the relation

$$\lim_{j\to\infty} \int_{\overline{D}} |J(u, T_j)|\, d\mathsf{L} = \int_{\overline{D}} |J(u, T)|\, d\mathsf{L} \leq \int_{\boldsymbol{D}} |J(u, T)|\, d\mathsf{L}.$$

In view of (β) we conclude that

$$\liminf_{j\to\infty} \int_D |J(u, T_j)|\, d\mathsf{L} < \infty. \tag{15}$$

From (ε) and (14) it follows that T_j is eAC in D for $j > j(\bar{D})$. In view of (15) it is now clear that the transformations $T|D$, $T_j|D$, where $j > j(\bar{D})$, satisfy the assumptions of theorem 1, with $\boldsymbol{D}$ and $\boldsymbol{D}_j$ replaced by D. Hence T is eAC in D, and hence also eBV in D (see **IV.4.2**). From (β) and **V.3.2**, theorem 1 it follows now that

$$J(u, T) = J_e(u, T) \quad \text{a.e. in } D.$$

Hence, using again (β),

$$\int_D |J_e(u, T)|\, d\mathsf{L} = \int_D |J(u, T)|\, d\mathsf{L} \le \int_{\boldsymbol{D}} |J(u, T)|\, d\mathsf{L} < \infty. \tag{16}$$

Consider now a sequence of domains D_r, $r = 1, 2, \ldots$, filling up $\boldsymbol{D}$ from the interior (see **II.3.2**, remark 10). Then $\bar{D}_r \subset \boldsymbol{D}$, $r = 1, 2, \ldots$, and thus the preceding remarks can be applied with $D = D_r$. It follows that T is eAC in D_r and [by (16)] that the sequence of integrals

$$\int_{D_r} |J_e(u, T)|\, d\mathsf{L}, \quad r = 1, 2, \ldots,$$

is bounded. Hence T is eAC in $\boldsymbol{D}$ by **IV.4.5**, theorem 2.

Theorem 3. Given T, T_j as in (1) and (2), assume that the following holds in addition to the conditions (α), (β), (γ).

(i) If F is any compact set in $\boldsymbol{D}$, then

$$J(u, T_j) \to J(u, T) \quad \text{a.e. on } F.$$

(ii) If Q is any oriented n-cube in $\boldsymbol{D}$, then the sequence $\{J(u, T_j)\}$ satisfies the condition (V) on Q (see **III.1.1**, definition 6).

Then T is eAC in $\boldsymbol{D}$.

Proof. The assumptions (i) and (ii) imply (see **III.1.1**, lemma 36) that

$$\lim_{j\to\infty} \int_Q |J(u, T_j)|\, d\mathsf{L} = \int_Q |J(u, T)|\, d\mathsf{L},$$

and thus the present theorem follows from the preceding theorem 2.

V.3.6. Generalized Lipschitzian transformations in $\boldsymbol{R}^n$ †. We shall use, as in **V.3.2**, the alternative representations

$$T: \boldsymbol{D} \to R^n, \tag{1}$$

$$T: x^i = f_i(u^1, \ldots, u^n), \quad i = 1, \ldots, n, \tag{2}$$

for the bounded continuous transformation T.

† This class of transformations has been introduced by T. Radó and P. V. Reichelderfer, On generalized Lipschitzian transformations [Riv. Mat. Univ. Parma 2, 289–301 (1951)].

Definition 1. Given T as in (1) and (2), we shall say that T satisfies the condition (Φ, D) if the following holds.

(α) D is a domain in $\boldsymbol{D}$.

(β) $\Phi = \Phi(u)$ is a real-valued, non-negative, L-summable function in D.

(γ) If Q is any oriented n-cube in D, then

$$\delta T Q \leqq \Bigl(\int_Q \Phi \, d\mathsf{L}\Bigr)^{1/n}, \tag{3}$$

where $\delta T Q$ denotes the diameter of the image TQ of Q (see **I.1.4**, definition 4).

Note that the domain D, occurring in the preceding definition, may coincide with $\boldsymbol{D}$.

Lemma 1. Given T as in (1) and (2), assume that T satisfies the condition (Φ, D). Let Q be any oriented n-cube in D. Then

$$\mathsf{L}\, T Q \leqq 2^n \int_Q \Phi \, d\mathsf{L}. \tag{4}$$

Proof. Clearly TQ is contained in an oriented n-cube Q^* with side length $2\delta TQ$. Hence

$$\mathsf{L}\, T Q \leqq \mathsf{L}\, Q^* = 2^n (\delta T Q)^n,$$

and (4) follows in view of (3).

Lemma 2. Given T as in (1) and (2), assume that T satisfies the condition (Φ, D). Then the following holds.

(i) T possesses a total differential a.e. in D. Hence (see **V.2.2**) T also possesses a weak total differential a.e. in D, and the ordinary Jacobian $J(u, T)$ exists a.e. in D.

(ii) $J(u, T)$ is L-summable in D.

(iii) T satisfies the condition (N) in D.

Proof. Since Φ is L-summable in D, there exists (see **III.1.1**, lemma 43) a decomposition

$$D = s \cup S \tag{5}$$

with the following properties.

(a) s, S are disjoint L-measurable sets.

(b) $\mathsf{L}\, s = 0$.

(c) $\Phi(u) < \infty$ on S.

(d) If $u \in S$ and $\{G_m\}$ is any regular sequence of closed sets such that

$$u \in G_m \subset D, \quad m = 1, 2, \ldots, \quad \lim_{m \to \infty} \delta G_m = 0,$$

then

$$\lim_{m\to\infty} \frac{1}{\mathsf{L}\, G_m} \int\limits_{G_m} \Phi\, d\mathsf{L} = \Phi(u) < \infty.$$

In view of (5) and (b), clearly (i) will be proved if we show that T possesses a total differential a.e. on S. Choose a point $u_0 \in S$, and put

$$M = 2\,[\Phi(u_0) + 1]^{1/n}. \tag{6}$$

For $\alpha > 0$, denote by $Q(\alpha)$ the oriented n-cube with center at u_0 and side length 2α. In view of (d) there exists a constant $\alpha_0 > 0$ such that $Q(\alpha) \subset D$ for $0 < \alpha < \alpha_0$ and

$$\int\limits_{Q(\alpha)} \Phi\, d\mathsf{L} < [\Phi(u_0) + 1]\, \mathsf{L}\, Q(\alpha) \quad \text{if} \quad 0 < \alpha < \alpha_0. \tag{7}$$

Consider any point $u \in D$ such that

$$0 < \|u - u_0\| < \alpha_0. \tag{8}$$

Then clearly $u \in Q(\|u - u_0\|) \subset D$. Hence, application of (3) to $Q(\|u - u_0\|)$ yields

$$\|T u - T u_0\| \leq \delta\, T\, Q(\|u - u_0\|) \leq \Big(\int\limits_{Q(\|u - u_0\|)} \Phi\, d\mathsf{L}\Big)^{1/n}. \tag{9}$$

Observe that

$$\mathsf{L}\, Q(\|u - u_0\|) = 2^n\, \|u - u_0\|^n.$$

Thus (9) yields, in view of (8) and (7), the inequality

$$\|T u - T u_0\| \leq [\Phi(u_0) + 1]^{1/n}\, 2\, \|u - u_0\|.$$

Hence, by (6),

$$\|T u - T u_0\| \leq M\, \|u - u_0\| \quad \text{if} \quad 0 < \|u - u_0\| < \alpha_0. \tag{10}$$

Consider now any one of the coordinate functions f_i of T [see (2)]. From (10) it follows that

$$|f_i(u) - f_i(u_0)| \leq M\, \|u - u_0\| \quad \text{if} \quad 0 < \|u - u_0\| < \alpha_0.$$

Hence clearly

$$\limsup_{u\to u_0} \frac{|f_i(u) - f_i(u_0)|}{\|u - u_0\|} < \infty \quad \text{if} \quad u_0 \in S. \tag{11}$$

As $\mathsf{L}(D - S) = 0$ by (a) and (b), (11) implies (see **III.1.3**) that f_i possesses a total differential a.e. in D $(i = 1, \ldots, n)$. In turn, this fact implies (as we have shown in the course of the proof of **V.2.3**, lemma 2) that T

possesses a total differential a.e. in D, and (i) is proved. To verify (iii), observe first that since Φ is L-summable in D, for every $\varepsilon>0$ there exists an $\eta=\eta(\varepsilon)>0$ such that

$$\int_E \Phi\, d\mathsf{L} < \varepsilon \tag{12}$$

if E is any L-measurable set in D which satisfies the inequality

$$\mathsf{L}E < \eta. \tag{13}$$

Consider now any set $F\subset D$ such that

$$\mathsf{L}F = 0. \tag{14}$$

Assign $\varepsilon>0$, and let $\eta=\eta(\varepsilon)>0$ be so chosen that (13) implies (12). By **III.1.1**, lemma 14 the relation (14) implies the existence of a sequence $\{Q_k\}$ of oriented n-cubes such that

$$F\subset \bigcup Q_k \subset D, \tag{15}$$

$$\sum \mathsf{L}Q_k < \eta, \tag{16}$$

$$\operatorname{int} Q_j \cap \operatorname{int} Q_k = \emptyset \quad \text{for} \quad j \neq k. \tag{17}$$

Let us put

$$E = \bigcup Q_k. \tag{18}$$

Then E is an L-measurable set in D, and (16) and (17) imply (see **III.1.1**, lemma 13) that $\mathsf{L}E=\sum \mathsf{L}Q_k<\eta$. Hence [since (13) implies (12) due to the choice of η]

$$\int_E \Phi\, d\mathsf{L} < \varepsilon. \tag{19}$$

On the other hand, (17) and (18) imply (see **III.1.1**, lemma 42) that

$$\int_E \Phi\, d\mathsf{L} = \sum \int_{Q_k} \Phi\, d\mathsf{L}. \tag{20}$$

By lemma 1 we have

$$\mathsf{L}TQ_k \leq 2^n \int_{Q_k} \Phi\, d\mathsf{L}, \quad k=1,2,\ldots. \tag{21}$$

From (15), (18), (21), (20), (19) we conclude that

$$\mathsf{L}TE \leq \sum \mathsf{L}TQ_k \leq 2^n \sum \int_{Q_k} \Phi\, d\mathsf{L} = 2^n \int_E \Phi\, d\mathsf{L} < 2^n\,\varepsilon.$$

As $TF\subset TE$, and $\varepsilon>0$ was arbitrary, it follows that

$$\mathsf{L}TF = 0. \tag{22}$$

Thus (14) implies (22), and (iii) is verified. To prove (ii), consider any point

$$u_0 = (u_0^1, \ldots, u_0^n) \in D$$

such that the first partial derivatives of the coordinate functions of T exist at u_0. Take any sequence of real numbers h_m, $m = 1, 2, \ldots$, such that

$$h_m > 0, \quad h_m \to 0.$$

Let Q_m denote the oriented n-cube determined by the inequalities

$$u_0^i \leq u^i \leq u_0^i + h_m, \quad i = 1, \ldots, n.$$

Let j be any one of the integers $1, \ldots, n$. Consider the point

$$u_{mj} = (u_{mj}^1, \ldots, u_{mj}^n),$$

where

$$u_{mj}^i = u_0^i \quad \text{if} \quad i \neq j, \quad u_{mj}^j = u_0^j + h_m.$$

Clearly $Q_m \subset D$ if m is sufficiently large. Discarding, if necessary, a finite number of the terms of the sequence $\{h_m\}$, we can therefore assume that $Q_m \subset D$ for $m = 1, 2, \ldots$. Since $\partial f_i / \partial u^j$ exists at u_0 by assumption, we have

$$\lim_{m \to \infty} \frac{f_i(u_{mj}) - f_i(u_0)}{h_m} = \frac{\partial f_i(u_0)}{\partial u^j}. \tag{23}$$

Also, since obviously $u_0, u_{mj} \in Q_m$, we have

$$|f_i(u_{mj}) - f_i(u_0)| \leq \|T u_{mj} - T u_0\| \leq \delta T Q_m.$$

By (3) it follows that

$$|f_i(u_{mj}) - f_i(u_0)| \leq \Big(\int_{Q_m} \Phi \, d\mathsf{L}\Big)^{1/n}.$$

Noting that

$$\mathsf{L}\, Q_m = h_m^n,$$

we conclude that

$$\frac{|f_i(u_{mj}) - f_i(u_0)|}{h_m} \leq \left(\frac{1}{\mathsf{L}\, Q_m} \int_{Q_m} \Phi \, d\mathsf{L}\right)^{1/n}.$$

In view of (23) it follows that

$$\left|\frac{\partial f_i(u_0)}{\partial u^j}\right| \leq \limsup_{m \to \infty} \left(\frac{1}{\mathsf{L}\, Q_m} \int_{Q_m} \Phi \, d\mathsf{L}\right)^{1/n}. \tag{24}$$

Observe that in view of (i), which we already proved, the first partial derivatives of the coordinate functions of T exist a.e. in D. Further-

more, since Φ is L-summable in D, by **III.1.1**, lemma 43 we have

$$\lim_{m\to\infty} \frac{1}{\mathsf{L}\, Q_m} \int_{Q_m} \Phi\, d\mathsf{L} = \Phi(u_0) < \infty,$$

except for certain points u_0 which constitute a set of L-measure zero in D. Thus (24) implies that

$$\left|\frac{\partial f_i}{\partial u^j}\right| \leq \Phi^{1/n} \quad \text{a.e. in } D, \quad i, j = 1, \ldots, n.$$

As $J(u, T)$ is a sum of $n!$ terms each of which is a product of n of the first partial derivatives of the coordinate functions of T, it follows that

$$|J(u, T)| \leq n!\, \Phi(u) \quad \text{a.e. in } D. \tag{25}$$

Since Φ is L-summable in D, (25) implies (ii), and the lemma is proved.

Definition 2. Given T as in (1) and (2), T is termed *generalized Lipschitzian* in a domain $D \subset \boldsymbol{D}$ if (a) for every n-interval $I \subset D$ there exists a real-valued, non-negative function $\Phi_I(u)$ such that T satisfies the condition (Φ_I, int I) in the sense of definition 1, and (b) the ordinary Jacobian $J(u, T)$ is L-summable in D.

Note that the domain D, occurring in the preceding definition, may coincide with $\boldsymbol{D}$.

As regards the condition (b) in the preceding definition, observe that by **I.2.2**, exercise 8 there exists a sequence $\{I_k\}$ of n-intervals in D such that $D = \bigcup \operatorname{int} I_k$. The condition (a) implies, in view of lemma 2, that T possesses a total differential a.e. in int I_k, $k = 1, 2, \ldots$. Clearly it follows that T possesses a total differential (and hence also a weak total differential) a.e. in D. Hence the first partial derivatives of the coordinate functions of T exist a.e. in D. Thus $J(u, T)$ exists a.e. in D, and hence the condition (b) in definition 2 is meaningful. Note that the condition (a) in definition 2 implies, by lemma 2, that T satisfies the condition (N) in int I_k, $k = 1, 2, \ldots$. By **IV.1.4**, lemma 1 it follows that T satisfies the condition (N) in D. These remarks yield the following statement.

Theorem 1. Assume that T, given as in (1) and (2), is generalized Lipschitzian in a domain $D \subset \boldsymbol{D}$. Then (a) T possesses a total differential (and hence also a weak total differential) a.e. in D, (b) the ordinary Jacobian $J(u, T)$ is L-summable in D, and (c) T satisfies the condition (N) in D.

In view of this theorem (in which D may coincide with $\boldsymbol{D}$) a number of previously established general results can be applied to generalized Lipschitzian transformations. The following statement summarizes the principal conclusions obtained in this manner.

Theorem 2. Assume that T, given as in (1) and (2), is generalized Lipschitzian in $\boldsymbol{D}$. Then the following holds.

(i) T is both $sAC\mathsf{B}$ and eAC in $\boldsymbol{D}$.

(ii) $J_{\mathsf{B}}(u,T) = J_e(u,T) = J(u,T)$ a.e. in $\boldsymbol{D}$.

(iii) $D_{\mathsf{B}}(u,T) = D_e(u,T) = |J(u,T)|$ a.e. in $\boldsymbol{D}$.

(iv) $N(x,T,\boldsymbol{D}) = K(x,T,\boldsymbol{D})$ a.e. in R^n.

(v) If S is an L-measurable subset of $\boldsymbol{D}$, then

$$\int_S |J(u,T)|\, d\mathsf{L} = \int N(x,T,S)\, d\mathsf{L}.$$

(vi) More generally, if S is an L-measurable subset of $\boldsymbol{D}$ and $H(x)$ is a finite-valued, L-measurable function in R^n, then

$$\int_S H(T\,u)\, |J(u,T)|\, d\mathsf{L} = \int H(x)\, N(x,T,S)\, d\mathsf{L},$$

as soon as one of the two integrals involved exists.

(vii) $\int_{\boldsymbol{D}} J(u,T)\, d\mathsf{L} = \int \mu_e(x,T,\boldsymbol{D})\, d\mathsf{L}$.

(viii) More generally, if $H(x)$ is a finite-valued, L-measurable function in R^n, then

$$\int_{\boldsymbol{D}} H(T\,u)\, J(u,T)\, d\mathsf{L} = \int H(x)\, \mu_e(x,T,\boldsymbol{D})\, d\mathsf{L},$$

as soon as the integral on the left exists.

(ix) If T is defined and continuous on $\bar{\boldsymbol{D}}$ and $\mathsf{L}\,T\,fr\,\boldsymbol{D} = 0$, then

$$\int_{\boldsymbol{D}} J(u,T)\, d\mathsf{L} = \int \mu(x,T,\boldsymbol{D})\, d\mathsf{L}.$$

More generally, if $H(x)$ is a finite-valued, L-measurable function in R^n, then

$$\int_{\boldsymbol{D}} H(T\,u)\, J(u,T)\, d\mathsf{L} = \int H(x)\, \mu(x,T,\boldsymbol{D})\, d\mathsf{L},$$

as soon as the integral on the left exists.

Proof. In view of theorem 1, the statements (i), (ii), (iii), (iv) follow directly from **V.3.3**, theorem 6, and the statements (v) and (vi) follow directly from **V.3.4**, theorem 3. Since T is eAC in $\boldsymbol{D}$ by (i), the statements (vii), (viii), (ix) follow (in view of theorem 1) directly from **V.3.4**, theorem 1 and **V.3.4**, theorem 2.

We proceed to discuss some significant special instances of generalized Lipschitzian transformations in R^n (a more detailed study of this topic will be made in Part VI for the case $n = 2$).

Lemma 3. Given T as in (1) and (2), assume that T satisfies the condition $(\Phi, \boldsymbol{D})$ in the sense of definition 1. Then T is generalized Lipschitzian in $\boldsymbol{D}$.

Proof. The assumption implies (see lemma 2) that $J(u, T)$ exists a.e. in $\boldsymbol{D}$ and is L-sumn able in $\boldsymbol{D}$. Furthermore, if I is any n-interval in $\boldsymbol{D}$, then clearly T satisfies the condition $(\Phi, \operatorname{int} I)$, and the lemma follows.

Lemma 4. Given T as in (1) and (2), assume that T is Lipschitzian in a domain $D \subset \boldsymbol{D}$. Then there exists a (finite) constant $K \geqq 0$ such that on setting $\Phi(u) \equiv K$, T satisfies the condition (Φ, D).

Proof. By assumption there exists a (finite) constant $M \geqq 0$ such that

$$\|T u_2 - T u_1\| \leqq M \|u_2 - u_1\| \tag{26}$$

for every pair of points $u_1, u_2 \in D$. Define

$$\Phi(u) \equiv M^n n^n \quad \text{for} \quad u \in D. \tag{27}$$

Consider any oriented n-cube $Q \subset D$, and let h be the side length of Q. Then [see (27)]

$$\int_Q \Phi \, d\mathsf{L} = M^n n^n \, \mathsf{L} Q = M^n n^n h^n. \tag{28}$$

If u_1, u_2 are any two points in Q, then clearly

$$\|u_2 - u_1\| \leqq n h,$$

and hence, by (26) and (28),

$$\|T u_2 - T u_1\| \leqq M n h = \Big(\int_Q \Phi \, d\mathsf{L}\Big)^{1/n}.$$

Since u_1, u_2 were arbitrary points of Q, it follows (see **I.1.4**, definition 4) that

$$\delta T Q \leqq \Big(\int_Q \Phi \, d\mathsf{L}\Big)^{1/n},$$

and the lemma is proved.

Lemma 5. Assume that T, given as in (1) and (2), is Lipschitzian in $\boldsymbol{D}$. Then T is also generalized Lipschitzian in $\boldsymbol{D}$.

Proof. By lemma 4 (applied with $D = \boldsymbol{D}$) T satisfies the condition $(\Phi, \boldsymbol{D})$ for an appropriately chosen Φ. Hence T is generalized Lipschitzian in $\boldsymbol{D}$ by lemma 3.

Lemma 6. Assume that T, given as in (1) and (2), satisfies the following conditions. (a) The first partial derivatives of the coordinate functions of T exist a.e. in $\boldsymbol{D}$, (b) the ordinary Jacobian $J(u, T)$ is L-summable in $\boldsymbol{D}$, and (c) T is Lipschitzian in every n-interval $I \subset \boldsymbol{D}$. Then T is generalized Lipschitzian in $\boldsymbol{D}$.

Proof. Consider any n-interval $I \subset \boldsymbol{D}$. The condition (c) implies that T is Lipschitzian in $\operatorname{int} I$. Hence by lemma 4 (applied with $D = \operatorname{int} I$) T satisfies the condition $(\Phi_I, \operatorname{int} I)$ for an appropriately chosen Φ_I. Since $J(u, T)$ is L-summable in $\boldsymbol{D}$ by assumption, the lemma follows.

Lemma 7. Assume that T, given as in (1) and (2), satisfies the following conditions. (a) The first partial derivatives of the coordinate functions of T exist and are continuous in $\boldsymbol{D}$, and (b) $J(u, T)$ is L-summable in $\boldsymbol{D}$. Then T is generalized Lipschitzian in $\boldsymbol{D}$.

Proof. In view of lemma 6 it is sufficient to show that T is Lipschitzian on every n-interval $I \subset \boldsymbol{D}$. Let such an interval I be assigned. Since I is compact (see **I.2.2**, exercise 3), and the first partial derivatives of the coordinate functions $f_1, \ldots, f_n$ of T exist and are continuous on I, by **I.1.3**, exercise 38 there follows the existence of a (finite) constant $K = K(I) \geq 0$ such that

$$\left|\frac{\partial f_i}{\partial u^j}\right| \leq K \text{ on } I, \quad i, j = 1, \ldots, n. \tag{29}$$

Take any two points

$$u_1 = (u_1^1, \ldots, u_1^n), \quad u_2 = (u_2^1, \ldots, u_2^n)$$

in I, and let i be one of the integers $1, \ldots, n$. For $0 \leq t \leq 1$, the point

$$u(t) = [t\, u_2^1 + (1-t)\, u_1^1, \ldots, t\, u_2^n + (1-t)\, u_1^n]$$

lies in I (see **I.2.3**, exercise 3). Accordingly the function

$$\varphi_i(t) = f_i(u(t)), \quad 0 \leq t \leq 1,$$

is well defined. Since the first partial derivatives of f_i exist and are continuous in $\boldsymbol{D}$, we can evaluate the difference

$$\varphi_i(1) - \varphi_i(0) = f_i(u_2) - f_i(u_1)$$

by the rules of elementary Calculus, obtaining

$$f_i(u_2) - f_i(u_1) = \int_0^1 \varphi_i'(t)\, dt = \sum_{j=1}^n (u_2^j - u_1^j) \int_0^1 \frac{\partial f_i(u(t))}{\partial u^j}\, dt.$$

In view of (29) there follows the inequality

$$|f_i(u_2) - f_i(u_1)| \leq K \sum_{j=1}^n |u_2^j - u_1^j| \leq n\, K\, \|u_2 - u_1\|.$$

Hence clearly

$$\|T\, u_2 - T\, u_1\| \leq \sum_{i=1}^n |f_i(u_2) - f_i(u_1)| \leq n^2 K\, \|u_2 - u_1\|,$$

and thus T is Lipschitzian in I.

The lemmas 5, 6, 7 yield examples of generalized Lipschitzian transformations. As noted above, more detailed information on this point will be provided in Part VI for transformations in R^2. Let us observe that in view of lemma 7 we can apply theorem 2 to the class of the so-called *transformations of class C'*, obtaining relevant information about the transformation of multiple integrals and the geometrical meaning of the ordinary Jacobian $J(u, T)$ for this classical elementary case.

Part VI. Continuous transformations in R^2 †.

§ VI.1. The topological index in R^2.

VI.1.1. Preliminaries. Throughout Part VI, we shall be concerned with bounded continuous transformations

$$T: \boldsymbol{D} \to R^2, \tag{1}$$

where $\boldsymbol{D}$ is a bounded domain in R^2. It will be convenient to use certain alternative representations for T. Note that the points of R^2 are ordered pairs (x, y) of real numbers x, y. Hence we can associate with the point (x, y) of R^2 the complex number $z = x + iy$. In the sequel, the terms "complex number" and "point of R^2" will be used interchangeably. Using complex numbers, the transformation (1) can be represented in the form

$$T: z = T(w), \quad w \in \boldsymbol{D}, \tag{2}$$

where z is the image point of w under T. Thus T is thought of now as a bounded, continuous, real or complex-valued function of the complex number $w \in \boldsymbol{D}$. If we set $w = u + iv$, $z = x + iy$, where u, v, x, y are real numbers, then we obtain for T the representation

$$T: x = x(u, v), \quad y = y(u, v), \quad (u, v) \in \boldsymbol{D}, \tag{3}$$

where the coordinate functions $x(u, v)$, $y(u, v)$ are bounded, continuous, real-valued functions in $\boldsymbol{D}$.

† For historical comments and further details (including applications in Surface Area Theory and Calculus of Variations) see the treatise on *Length and Area* listed in the *Bibliography*, and the expository paper by L. Cesari and T. Radó, Applications of Area Theory in Analysis, Proc. Internat. Congr. of Math. **2**, 174–179 (1950). For the equivalence (in the case $n = 2$) of the basic concepts used in this volume with those occurring in alternative approaches, see T. Radó, Two-dimensional concepts of bounded variation and absolute continuity [Duke Math. J. **14**, 587–608 (1947)]. It is easy to see that the essential multiplicity function K used in this volume corresponds to the multiplicity function Ψ^* occurring in the paper just referred to.

The two-dimensional Euclidean space R^2 possesses various properties not shared by the general Euclidean spaces R^n. Furthermore, the field of complex numbers possesses various special algebraic properties which play an important role in the sequel. Due to these circumstances, the theory of continuous transformations in R^2 presents several remarkable special features as compared with the general theory developed in Part IV and Part V. Our main objective in Part VI is to discuss some of these special features.

In view of the availability of several excellent and completely rigorous treatises on the topology of R^2 it seems unnecessary to give a detailed presentation of the topological theorems for R^2 which we shall have occasion to use in the sequel (even though many such theorems follow directly from the general results derived in § **I.7**). Accordingly, we shall merely state here some basic definitions and theorems concerning R^2, and throughout Part VI we shall make free use of figures to illustrate statements that are both intuitive and true.

A *simple arc* γ in R^2 is a set which is the homeomorphic image of the closed unit interval $0 \leqq u \leqq 1$. A simple arc is compact and connected, and hence it is a continuum. A *simple closed curve* (or JORDAN *curve*) in R^2 is a set which is the homeomorphic image of the unit circle $u^2 + v^2 = 1$. A simple closed curve is also a continuum. A bounded, finitely connected JORDAN region R in R^2 is obtained as follows (see the figure). One starts with a simple closed curve C_0, and one selects a finite number $m \geqq 0$ of simple closed curves C_j, $j = 1, \ldots, m$, such that each C_j is interior to C_0, and C_j and C_k are exterior to each other if $j \neq k$. Then $C_0, C_1, \ldots, C_m$ constitute the frontier of a uniquely determined bounded domain D. The set

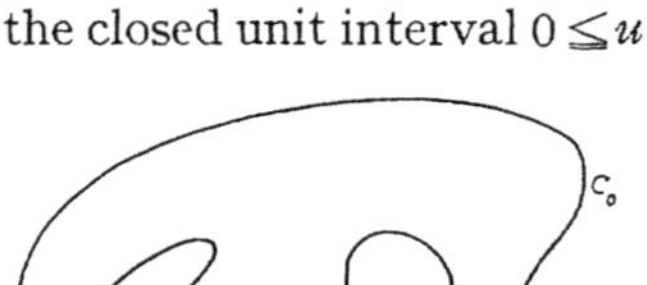

Fig. 45.

$$D \cup C_0 \cup \cdots \cup C_m$$

is termed a bounded, finitely connected JORDAN region R, and the set $B = \cup C_k$, $k = 0, \ldots, m$, is termed the boundary of R. A bounded, finitely connected JORDAN region R is a continuum, and we have the relations

$$fr\, \mathsf{R} = fr\, \mathrm{int}\, \mathsf{R} = B = \bigcup_{k=0}^{m} C_k,$$

where the symbols *fr*, *int* are used in the sense of **I.1.3**, definitions 4, 5. According as $m = 0, 1, 2, \ldots$, the region R is said to be simply, doubly, triply, ... connected. Any two finitely connected, bounded JORDAN regions R_1, R_2 with the same number of boundary curves are homeomorphic. In fact, if w_1, w_2 are arbitrarily assigned points in R_1, R_2

respectively, such that either $w_j \in \text{int}\, \mathsf{R}_j$, $j=1, 2$, or $w_j \in fr\, \mathsf{R}_j$, $j=1, 2$, then there exists a homeomorphism h from R_1 onto R_2 which carries w_1 into w_2.

If we have occasion to consider a continuous transformation T from a bounded, finitely connected JORDAN region R into R^2, then we use [in analogy with (1), (2) (3)] the alternative representations

$$T: \mathsf{R} \to R^2, \tag{4}$$

$$T: z = T(w), \quad w \in \mathsf{R}, \tag{5}$$

$$T: x = x(u, v), \quad y = y(u, v), \quad (u, v) \in \mathsf{R}. \tag{6}$$

If f is a real or complex-valued function on a set $S \subset R^2$, then we shall use any one of the alternative notations

$$f(u, v), \quad f(w), \quad f(x, y), \quad f(z),$$

where $w = u + iv$, $z = x + iy$ (and of course u, v, x, y designate real numbers). Similarly, definite integrals will be written in one of the forms

$$\iint f(u, v)\, du\, dv, \quad \iint f(w)\, du\, dv, \quad \iint f(x, y)\, dx\, dy, \quad \iint f(z)\, dx\, dy.$$

Thus we write now $du\,dv$ or $dx\,dy$ instead of $d\mathsf{L}$, in conformity with general usage.

As we noted above, a point of R^2 is an ordered pair (x, y) of real numbers x, y. In conformity with the practice in Analytic Geometry, we interpret x and y as the first and the second coordinates of the point (x, y) with respect to a Cartesian coordinate system.

The matter of *orientation* enters the picture in this connection. Let us observe that we have at our disposal a *positive orientation* in R^2 according to the abstract pattern developed in § **II.1**. On the other hand, we have at our disposal the intuitive conception of orientation in R^2. According to this intuitive conception (which of course can be formulated in an entirely rigorous manner), one speaks of a *counter-clockwise* orientation, while the opposite orientation is referred to as the *clock-wise* orientation. To avoid circularity, we shall use the term *positive orientation* solely to refer to the abstractly defined orientation described in **II.1.4**.

Let us recall a few definitions and facts concerning complex numbers. If x and y are real numbers and $z = x + iy$, then x and y are termed the real part and the imaginary part respectively of the complex number z. In symbols

$$x = \Re z, \quad y = \Im z.$$

The absolute value of z, denoted by $|z|$, is given by the formula

$$|z| = (x^2 + y^2)^{\frac{1}{2}}.$$

If $z \neq 0$, then one has for z the *trigonometric representation*

$$z = |z| (\cos\varphi + i \sin\varphi), \tag{7}$$

where φ is a real number which represents the radian measure of a certain angle in a well-known manner.

Definition 1. If $z \neq 0$ is a complex number, then every real number φ which satisfies (7) is termed an *argument* of z. No arguments are assigned to $z = 0$.

Lemma 1. If $z \neq 0$ is a complex number, and $r > 0$ and φ are real numbers such that

$$z = r (\cos\varphi + i \sin\varphi),$$

then $r = |z|$ and φ is an argument of z.

Lemma 2. If $z_1 \neq 0, \ldots, z_n \neq 0$ are complex numbers and $\varphi_1, \ldots, \varphi_n$ are arguments of $z_1, \ldots, z_n$ respectively, then $\varphi_1 + \cdots + \varphi_n$ is an argument of the product $z_1 \ldots z_n$.

Lemma 3. If z is a complex number such that $|z - 1| < 1$, then $\Re z > 0$.

Lemma 4. If $z \neq 0$ is a complex number and φ, ψ are any two arguments of z, then $\psi = \varphi + 2n\pi$, where n is an integer.

These lemmas (which state well-known elementary facts) are listed merely for convenient reference.

Consider now the *half-plane* $\Re z > 0$ (that is, the set of those complex numbers z whose real part is positive). If $\Re z > 0$, then clearly z has a unique argument $\alpha(z)$ such that

$$-\frac{\pi}{2} < \alpha(z) < \frac{\pi}{2}. \tag{8}$$

Clearly $\alpha(z)$ is a continuous function of z in the half-plane $\Re z > 0$.

Definition 2. For complex numbers z such that $\Re z > 0$, the uniquely determined argument satisfying (8) will be denoted by $\arg z$.

Thus $\arg z$ is a single-valued, continuous function of z which is defined only in the half-plane $\Re z > 0$.

VI.1.2. The argument of a complex-valued function. Let Γ be a continuum in R^2 (see **I.1.3**, definition 15). Consider a continuous (real or complex-valued) function $f(w)$ on Γ. Put

$$M(f, \Gamma) = \max |f(w)|, \quad w \in \Gamma, \tag{1}$$

$$m(f, \Gamma) = \min |f(w)|, \quad w \in \Gamma, \tag{2}$$

$$\omega(f, \Gamma) = \max |f(w_2) - f(w_1)|, \quad w_1, w_2 \in \Gamma. \tag{3}$$

Since Γ is compact and f is continuous on Γ, the use of the *max* and *min* in (1), (2), (3) is justified in view of **I.1.3**, exercises 38 and 39.

Assume now that $f(w) \neq 0$ on Γ. We have then (see **VI.1.1**, definition 1) for each point $w \in \Gamma$ infinitely many arguments for the complex number $f(w) \neq 0$. If for each $w \in \Gamma$ we select a definite one of the arguments of $f(w)$, then we obtain a single-valued function $\varphi(w)$ on Γ which is termed a single-valued argument for $f(w)$ on Γ. It may or may not be possible to make the selection in such a manner that $\varphi(w)$ is continuous on Γ.

Definition 1. Let $f(w)$ be a continuous, real or complex-valued function on a continuum $\Gamma \subset R^2$, such that $f(w) \neq 0$ on Γ. Then a function $\varphi(w)$ which is real-valued and continuous on Γ is termed a *single-valued continuous argument* of $f(w)$ on Γ if

$$f(w) = |f(w)| \left(\cos\varphi(w) + i \sin\varphi(w)\right), \quad w \in \Gamma.$$

Definition 2. Let $f(w)$ be a continuous, real or complex-valued function on a continuum $\Gamma \subset R^2$ such that $f(w) \neq 0$ on Γ. If there exists a single-valued continuous argument of $f(w)$ on Γ, then f is said to satisfy the condition (arg, Γ).

Lemma 1. Assume that f satisfies the condition (arg, Γ). Then the following holds.

(i) If $\varphi(w)$ is a single-valued continuous argument of $f(w)$ on Γ and n is an integer, then $\varphi(w) + 2n\pi$ is also a single-valued continuous argument of $f(w)$ on Γ.

(ii) Conversely, if $\varphi(w)$ and $\psi(w)$ are any two single-valued continuous arguments of $f(w)$ on Γ, then

$$\psi(w) = \varphi(w) + 2n\pi, \quad w \in \Gamma,$$

where n is an integer independent of w. In particular, if $\psi(w_0) = \varphi(w_0)$ for some point $w_0 \in \Gamma$, then $\psi(w) \equiv \varphi(w)$ on Γ.

Proof. (i) is obvious. To prove (ii), set

$$\lambda(w) = \frac{\psi(w) - \varphi(w)}{2\pi}, \quad w \in \Gamma.$$

Clearly $\lambda(w)$ is continuous on Γ, and from **VI.1.1**, lemma 4 we conclude that $\lambda(w)$ is integral-valued. Since Γ is connected, by **I.1.3**, exercise 37 it follows that $\lambda(w)$ is constant on Γ. Hence $\lambda(w) \equiv n$, where n is a constant integer, and (ii) follows.

Lemma 2. Let $f(w)$ be a continuous, real or complex-valued function on a continuum $\Gamma \subset R^2$, such that $\Re f(w) > 0$ on Γ. Then f satisfies the condition (arg, Γ).

Proof. Using the function arg z introduced in **VI.1.1**, definition 2, we note that the function

$$\varphi(w) = \arg f(w), \quad w \in \Gamma,$$

is clearly a single-valued continuous argument of f on Γ.

Lemma 3. Let $f(w)$ be a continuous, real or complex-valued function on a continuum $\Gamma \subset R^2$, such that [see (2), (3)]

$$\omega(f, \Gamma) < m(f, \Gamma). \tag{4}$$

Then f satisfies the condition $(\arg, \Gamma)$.

Proof. Note that the assumption implies that $f \neq 0$ on Γ. Select a point $w_0 \in \Gamma$ and let φ_0 be an argument of $f(w_0)$. Then

$$f(w_0) = |f(w_0)| (\cos \varphi_0 + i \sin \varphi_0). \tag{5}$$

Introduce the auxiliary function

$$g(w) = \frac{f(w)}{f(w_0)}, \quad w \in \Gamma. \tag{6}$$

Then $g(w)$ is continuous on Γ, and [by (4)]

$$|g(w) - 1| = \frac{|f(w) - f(w_0)|}{|f(w_0)|} \leq \frac{\omega(f, \Gamma)}{m(f, \Gamma)} < 1, \quad w \in \Gamma.$$

By **VI.1.1**, lemma 3 it follows that $\Re g(w) > 0$ on Γ. Hence, by lemma 2 (applied to g) there exists a single-valued continuous argument $\psi(w)$ of g on Γ. In view of (5) and (6) it is clear that $\varphi_0 + \psi(w)$ is a real-valued continuous argument of $f(w)$ on Γ, and the lemma is proved.

Lemma 4. Let $f_1(w), \ldots, f_n(w)$ be continuous, real or complex-valued functions on a continuum $\Gamma \subset R^2$, such that $f_j \neq 0$ on Γ and f_j satisfies the condition $(\arg, \Gamma)$, $j = 1, \ldots, n$. Then the product $f = f_1 \ldots f_n$ also satisfies the condition $(\arg, \Gamma)$. Furthermore, if $\varphi_1(w), \ldots, \varphi_n(w)$ are single-valued continuous arguments of $f_1(w), \ldots, f_n(w)$ respectively on Γ, then $\varphi_1 + \cdots + \varphi_n$ is a single-valued continuous argument of $f = f_1 \ldots f_n$ on Γ.

Proof. By assumption, there exists a single-valued continuous argument $\varphi_j(w)$ of $f_j(w)$ on Γ, $j = 1, \ldots, n$. Then $\varphi_1 + \cdots + \varphi_n$ is a single-valued continuous argument of $f = f_1 \ldots f_n$ on Γ in view of **VI.1.1**, lemma 2, and the present lemma follows.

Lemma 5. Let Γ_1, Γ_2 be two continua in R^2 such that there exists a homeomorphism $h: \Gamma_1 \to \Gamma_2$ from Γ_1 onto Γ_2. Let $f_1(w)$ be a continuous, real or complex-valued function on Γ_1, such that $f_1(w) \neq 0$ on Γ_1 and there exists a single-valued continuous argument $\varphi_1(w)$ of $f_1(w)$ on Γ_1.

Put

$$f_2(w) = f_1(h^{-1}(w)), \quad w \in \Gamma_2, \tag{7}$$

$$\varphi_2(w) = \varphi_1(h^{-1}(w)), \quad w \in \Gamma_2. \tag{8}$$

Then $\varphi_2(w)$ is a single-valued continuous argument of $f_2(w)$ on Γ_2.

Proof. Clearly $\varphi_2(w)$ is a well-defined, real-valued, continuous function on Γ_2. If $w \in \Gamma_2$, then $h^{-1}(w) \in \Gamma_1$, and by assumption

$$f_1(h^{-1}(w)) = |f_1(h^{-1}(w))| \left[\cos \varphi_1(h^{-1}(w)) + i \sin \varphi_1(h^{-1}(w))\right].$$

In view of (7) and (8) this last formula yields

$$f_2(w) = |f_2(w)| \left(\cos \varphi_2(w) + i \sin \varphi_2(w)\right), \quad w \in \Gamma_2,$$

and the lemma follows.

Lemma 6. Let Γ, Γ_1, Γ_2 be three continua in R^2 such that $\Gamma = \Gamma_1 \cup \Gamma_2$ and $\Gamma_1 \cap \Gamma_2$ is non-empty and connected. Let $f(w)$ be a continuous, real or complex-valued function on Γ, such that $f(w) \neq 0$ on Γ and $f(w)$ satisfies the conditions (arg, Γ_1) and (arg, Γ_2). Then $f(w)$ satisfies the condition (arg, Γ).

Proof. Take a point $w_0 \in \Gamma_1 \cap \Gamma_2$, and let β_0 be an argument of $f(w_0)$. By assumption, there exists a single-valued continuous argument $\varphi_1^*(w)$ of $f(w)$ on Γ_1 and also a single-valued continuous argument $\varphi_2^*(w)$ of $f(w)$ on Γ_2. By **VI.1.1**, lemma 4 we have

$$\varphi_1^*(w_0) = \beta_0 + 2n_1\pi, \quad \varphi_2^*(w_0) = \beta_0 + 2n_2\pi,$$

where n_1, n_2 are integers. On setting

$$\varphi_1(w) = \varphi_1^*(w) - 2n_1\pi, \quad \varphi_2(w) = \varphi_2^*(w) - 2n_2\pi,$$

by lemma 1 it follows that $\varphi_1(w), \varphi_2(w)$ are single-valued continuous arguments of $f(w)$ on Γ_1 and Γ_2 respectively. Clearly

$$\varphi_1(w_0) = \varphi_2(w_0) = \beta_0. \tag{9}$$

Furthermore, since $\Gamma_1 \cap \Gamma_2$ is a subset of both Γ_1 and Γ_2, it is clear that $\varphi_1(w)$ and $\varphi_2(w)$ are single-valued continuous arguments of $f(w)$ on $\Gamma_1 \cap \Gamma_2$. Now since $\Gamma_1 \cap \Gamma_2$ is connected, by lemma 1 (applied to $\Gamma_1 \cap \Gamma_2$) it follows that $\varphi_2(w) - \varphi_1(w)$ is constant on $\Gamma_1 \cap \Gamma_2$. Hence, in view of (9),

$$\varphi_1(w) = \varphi_2(w) \quad \text{for} \quad w \in \Gamma_1 \cap \Gamma_2. \tag{10}$$

Define now a function $\varphi(w)$ on Γ by the formulas

$$\varphi(w) = \begin{cases} \varphi_1(w) & \text{if} \quad w \in \Gamma_1, \\ \varphi_2(w) & \text{if} \quad w \in \Gamma_2. \end{cases}$$

Then $\varphi(w)$ is single-valued on Γ as a consequence of (10). Obviously $\varphi(w)$ is an argument of $f(w)$ for every $w \in \Gamma$. Since Γ_1 and Γ_2 are both closed, it follows readily that the continuity of $\varphi_1(w), \varphi_2(w)$ on Γ_1 and Γ_2 respectively implies the continuity of $\varphi(w)$ on $\Gamma = \Gamma_1 \cup \Gamma_2$. Thus $\varphi(w)$ is a single-valued continuous argument of $f(w)$ on Γ, and the lemma is proved.

Definition 3. A continuum $\Gamma \subset R^2$ is said to satisfy the condition (arg) if *every* continuous, real or complex-valued function $f(w)$ on Γ which is different from zero on Γ satisfies the condition (arg, Γ).

Lemma 7. Let Q be the unit square $0 \leq u \leq 1$, $0 \leq v \leq 1$ in R^2. Then Q satisfies the condition (arg).

Proof. Consider any continuous, real or complex-valued function $f(w)$ in Q which is different from zero in Q. We have to show that $f(w)$ has a single-valued continuous argument in Q. Deny this assertion. Divide Q into two congruent rectangles R_1, R_1' by a vertical segment. If $f(w)$ would satisfy both of the conditions (arg, R_1) and (arg, R_1'), then by lemma 6 it would follow that $f(w)$ satisfies the condition (arg, Q), which we denied. Hence $f(w)$ must fail to satisfy one of the conditions (arg, R_1), (arg, R_1'), say (arg, R_1). On subdividing R_1 into two congruent rectangles R_2, R_2' by a horizontal segment, we obtain in the same manner the conclusion that $f(w)$ must fail to satisfy one of the conditions (arg, R_2), (arg, R_2'), say (arg, R_2). Repeating this argument, we obtain an infinite sequence $\{R_n\}$ of oriented rectangles in Q such that

(i) $R_1 \supset \cdots \supset R_n \supset \cdots$,

(ii) $\delta R_n \to 0$, and

(iii) $f(w)$ fails to satisfy the condition (arg, R_n), $n = 1, 2, \ldots$.

Now since $f(w) \neq 0$ in Q, we have [see (2)] $m(f, Q) > 0$, and hence

$$m(f, R_n) \geq m(f, Q) > 0, \quad n = 1, 2, \ldots. \tag{11}$$

Since $f(w)$ is continuous in Q, and Q is compact, $f(w)$ is uniformly continuous in Q. Hence (ii) implies [see (3)] that

$$\omega(f, R_n) \to 0. \tag{12}$$

From (11) and (12) we conclude that if N is a sufficiently large positive integer, then

$$\omega(f, R_N) < m(f, R_N).$$

But then, by lemma 3, $f(w)$ satisfies the condition (arg, R_N), in contradiction with (iii). Thus denial of the lemma leads to a contradiction, and the proof is complete.

Lemma 8. If I is the unit interval $0 \leqq u \leqq 1$ on the number-line, then I satisfies the condition (arg).

The proof, based on successive subdivisions of I, is entirely analogous to that of lemma 7.

Lemma 9. If R is a bounded, simply connected JORDAN region in R^2, then R satisfies the condition (arg).

Proof. Let Q be the unit square $0 \leqq u \leqq 1, 0 \leqq v \leqq 1$ in R^2. Then there exists (see **VI.1.1**) a homeomorphism $h: Q \to \mathsf{R}$ from Q onto R. Assign now any continuous, real or complex-valued function $g(w)$ in R which is different from zero in R. We have to show that there exists a single-valued continuous argument of $g(w)$ in R. Define

$$f(w) = g(h(w)), \quad w \in Q.$$

By lemma 7, there exists a single-valued continuous argument $\varphi(w)$ of $f(w)$ in Q. Since

$$g(w) = f(h^{-1}(w)), \quad w \in \mathsf{R},$$

we conclude from lemma 5 that $\varphi(h^{-1}(w))$ is a single-valued continuous argument of $g(w)$ in R, and the proof is complete.

Lemma 10. If γ is a simple arc in R^2, then γ satisfies the condition (arg).

Noting that γ is homeomorphic to the unit interval $0 \leqq u \leqq 1$, the proof is made in the same manner as in lemma 9, except that lemma 8 is now used instead of lemma 7.

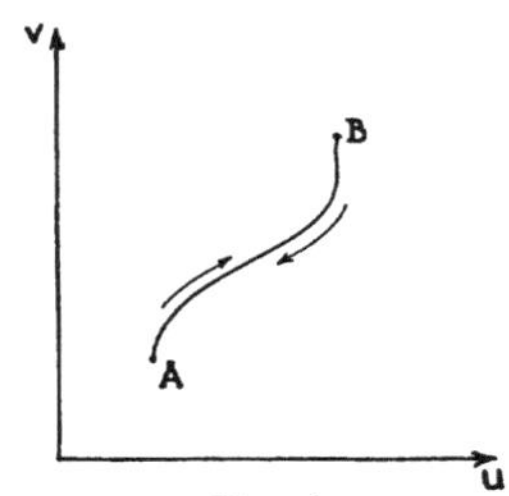

Fig. 46.

Consider now a simple arc in R^2 with end-points A, B. Such an arc admits of two orientations which are indicated by the arrows in the figure. By an *oriented simple arc* γ we mean a simple arc with assigned orientation. In the figure, if one assigns the orientation indicated by the upper arrow, then A is the first end-point and B is the second end-point of the oriented simple arc γ. The same simple arc with the opposite orientation is denoted by $-\gamma$. The first end-point of $-\gamma$ is then B and the second end-point of $-\gamma$ is A.

Definition 4. Let γ be an oriented simple arc in R^2 with first end-point w_1 and second end-point w_2. Let $f(w)$ be a continuous, real or complex-valued function on γ which is different from zero on γ. By lemma 10 there exists a single-valued continuous argument $\varphi(w)$ of $f(w)$ on γ. The quantity

$$V_\gamma[\arg f(w)] = \varphi(w_2) - \varphi(w_1) \tag{13}$$

is termed the *variation of the argument* of $f(w)$ on the oriented simple arc γ.

To justify this definition, we have to show that if $\psi(w)$ is any other single-valued continuous argument of $f(w)$ on γ, then

$$\psi(w_2) - \psi(w_1) = \varphi(w_2) - \varphi(w_1). \tag{14}$$

Now by lemma 1 we have

$$\psi(w) = \varphi(w) + 2n\pi, \quad w \in \gamma,$$

where n is a constant integer, and thus (14) is obvious.

Lemma 11. Let γ be an oriented simple arc in R^2 with first end-point w_1 and second end-point w_2. Then the following holds.

(i) If $f(w)$ is a continuous, real or complex-valued function on γ which is different from zero on γ, then

$$V_\gamma[\arg f(w)] = -V_{-\gamma}[\arg f(w)]. \tag{15}$$

(ii) Given $f(w)$ as under (i), select a point w^* of γ which is different from both w_1 and w_2, and denote by γ_1, γ_2 the oriented sub-arcs of γ with first and second end-points w_1, w^* and w^*, w_2 respectively. Then

$$V_\gamma[\arg f(w)] = V_{\gamma_1}[\arg f(w)] + V_{\gamma_2}[\arg f(w)]. \tag{16}$$

(iii) If $f_1(w), \ldots, f_m(w)$ are continuous, real or complex-valued functions on γ such that $f_j(w) \neq 0$ on γ, $j = 1, \ldots, m$, then

$$V_\gamma\left[\arg \prod_{j=1}^{m} f_j(w)\right] = \sum_{j=1}^{m} V_\gamma[\arg f_j(w)]. \tag{17}$$

Proof. To verify (i) and (ii), select a single-valued continuous argument $\varphi(w)$ of $f(w)$ on γ. Noting that w_2 is the first end-point and w_1 is the second end-point of $-\gamma$, we have by definition the formula

$$V_{-\gamma}[\arg f(w)] = \varphi(w_1) - \varphi(w_2),$$

and (15) follows in view of (13). Furthermore, by definition,

$$V_{\gamma_1}[\arg f(w)] = \varphi(w^*) - \varphi(w_1),$$
$$V_{\gamma_2}[\arg f(w)] = \varphi(w_2) - \varphi(w^*),$$

and (16) follows in view of (13). To prove (17), let $\varphi_j(w)$ be a single-valued continuous argument of $f_j(w)$ on γ, $j = 1, \ldots, m$. By definition

$$V_\gamma[\arg f_j(w)] = \varphi_j(w_2) - \varphi_j(w_1), \quad j = 1, \ldots, m. \tag{18}$$

On the other hand, the function

$$\varphi(w) = \sum_{j=1}^{m} \varphi_j(w), \quad w \in \gamma, \tag{19}$$

is a single-valued continuous argument of $f_1 \ldots f_m$ on γ (see lemma 4). Hence by definition

$$V_\gamma\left[\arg \prod_{j=1}^{m} f_j(w)\right] = \varphi(w_2) - \varphi(w_1),$$

and (17) follows in view of (18) and (19).

Lemma 12. Let w_0, w', w'' be the vertices of a non-degenerate triangle τ in R^2. Denote by λ the radian measure of the angle of τ at w_0 and by γ the oriented segment with first end-point w' and second end-point w''. Then

$$V_\gamma[\arg(w - w_0)] = \lambda$$

if the arrangement w_0, w', w'' corresponds to the counter-clockwise orientation of the perimeter of τ, and

$$V_\gamma[\arg(w - w_0)] = -\lambda$$

if the arrangement w_0, w', w'' corresponds to the clock-wise orientation of the perimeter of τ.

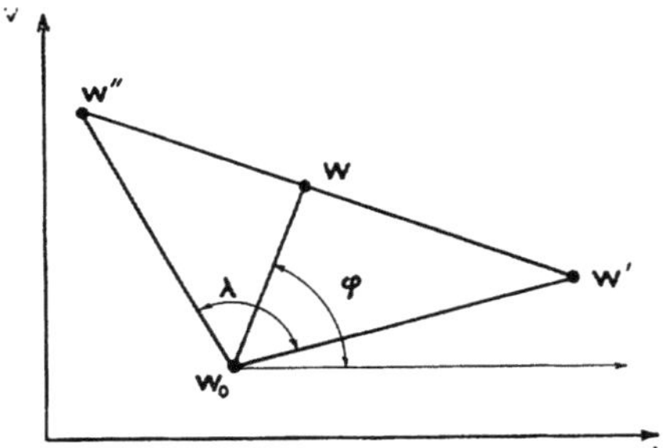

Fig. 47.

Proof. The reasoning is indicated in the figure which illustrates the case when the arrangement w_0, w', w'' corresponds to the counter-clockwise orientation of the perimeter of τ. The radian measure of the angle $\varphi = \varphi(w)$ is selected as a single-valued continuous argument of $w - w_0$ on the oriented segment γ with first end-point w' and second end-point w''. Clearly

$$V_\gamma[\arg(w - w_0)] = \varphi(w'') - \varphi(w') = \lambda.$$

Consider now a simple closed curve in R^2. Such a curve admits of a counter-clockwise orientation and of a clock-wise orientation (in the figure, these orientations are indicated by the upper and the lower arrow respectively). By an oriented simple closed curve C we mean a simple closed curve with assigned orientation. The same simple closed curve with the opposite orientation is then denoted by $-C$.

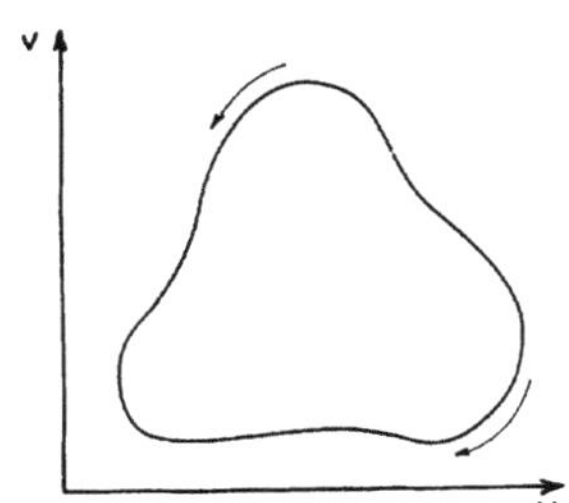

Fig. 48.

Definition 5. Let C be an oriented simple closed curve in R^2. Consider a continuous, real or complex-valued function $f(w)$ on C which is different from zero on C. Take a finite number of distinct points $w_0, w_1, \ldots, w_n$ on C (where $n \geq 1$) which follow upon each other in conformity with the orientation of C. Let γ_j, $j = 0, \ldots, n-1$, be the oriented sub-arc of C with first end-point w_j and second end-point w_{j+1}, and let γ_n be the oriented sub-arc of C with first end-point w_n and second end-point w_0, where these sub-arcs are oriented in conformity with the assigned orientation of C (see the figure). Then the quantity

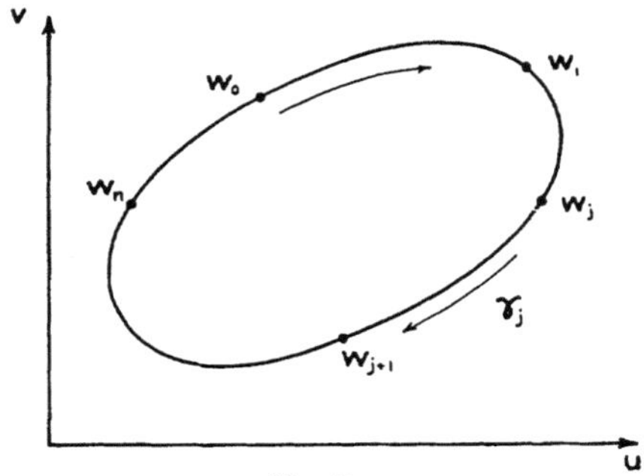

Fig. 49.

$$\left.\begin{aligned} &V_C[\arg f(w)] \\ &\quad = \sum_{j=0}^{n} V_{\gamma_j}[\arg f(w)] \end{aligned}\right\} \tag{20}$$

is termed the variation of the argument of $f(w)$ on the oriented simple closed curve C.

To justify this definition, it must be shown that the value of the summation in (20) is independent of the particular choice of the points of division on C. However, this fact follows readily by repeated application of part (ii) of lemma 11.

Lemma 13. Let C be an oriented simple closed curve in R^2. Then the following holds.

(a) If $f(w)$ is a continuous, real or complex-valued function which is different from zero on C, then

$$V_C[\arg f(w)] = -V_{-C}[\arg f(w)].$$

(b) If $f_1(w), \ldots, f_m(w)$ are continuous, real or complex-valued functions on C such that $f_j(w) \neq 0$ on C, $j = 1, \ldots, m$, then

$$V_C\left[\arg \prod_{j=1}^{m} f_j(w)\right] = \sum_{j=1}^{m} V_C[\arg f_j(w)].$$

(c) Given $f(w)$ as under (a), the quantity

$$\frac{1}{2\pi} V_C[\arg f(w)]$$

is equal to an integer.

(d) Given $f(w)$ as under (a), assume that $f(w)$ satisfies the condition (arg, C). Then

$$V_C[\arg f(w)] = 0. \tag{21}$$

(e) Given $f(w)$ as under (a), select another oriented simple closed curve C^* in R^2, and let h be a homeomorphism from C onto C^*, such that the assigned orientation of C^* corresponds under h to the assigned orientation of C. Define

$$f^*(w) = f\big(h^{-1}(w)\big), \quad w \in C^*.$$

Then

$$V_C[\arg f(w)] = V_{C^*}[\arg f^*(w)].$$

Proof. The statements (a), (b), (e) follow, in view of definition 5, readily from parts (i) and (iii) of lemma 11 and from lemma 5 respectively. To prove (c), select a single-valued continuous argument $\varphi_j(w)$ of $f(w)$ on each one of the oriented arcs γ_j, $j = 0, \ldots, n$, occurring in definition 5. Then definition 4 and definition 5 yield [after re-arranging the summation in (20)] the formula

$$\left.\begin{aligned} V_C[\arg f(w)] = {} & [\varphi_0(w_1) - \varphi_1(w_1)] + \cdots \\ & + [\varphi_{n-1}(w_n) - \varphi_n(w_n)] + [\varphi_n(w_0) - \varphi_0(w_0)]. \end{aligned}\right\} \quad (22)$$

Each one of the differences appearing in (22) is the difference of two arguments of the same complex number. Thus, by **VI.1.1**, lemma 4, each one of these differences is equal to 2π times some integer, and (c) follows. Assume now that there exists a single-valued continuous argument $\varphi(w)$ of $f(w)$ on C. Then we can choose

$$\varphi_j(w) = \varphi(w), \quad w \in \gamma_j, \quad j = 0, \ldots, n,$$

and (21) follows from (22).

Lemma 14. Given an oriented simple closed curve C in R^2, let $F(w)$, $G(w)$, $g(w)$ be three continuous, real or complex-valued functions on C such that

$$F(w) = G(w) + g(w), \quad |G(w)| > |g(w)| \quad \text{on } C. \quad (23)$$

Then

$$V_C[\arg F(w)] = V_C[\arg G(w)]. \quad (24)$$

Proof. Note that the assumptions imply that $F(w) \neq 0$, $G(w) \neq 0$ on C. Consider the auxiliary function

$$f(w) = \frac{F(w)}{G(w)}, \quad w \in C. \quad (25)$$

Clearly $f(w)$ is continuous and different from zero on C. From part (b) of lemma 13 we conclude that

$$V_C[\arg F(w)] = V_C[\arg G(w)] + V_C[\arg f(w)]. \quad (26)$$

On the other hand, (23) and (25) yield

$$|f(w) - 1| = \frac{|F(w) - G(w)|}{|G(w)|} = \frac{|g(w)|}{|G(w)|} < 1, \quad w \in C.$$

Hence, by **VI.1.1**, lemma 3, $\Re f(w) > 0$ on C. By lemma 2 it follows that $f(w)$ satisfies the condition (arg, C), and hence by part (d) of lemma 13 we have

$$V_C[\arg f(w)] = 0. \tag{27}$$

Since (26) and (27) imply (24), the lemma is proved.

Lemma 15. Given an oriented simple closed curve C in R^2, let $f(w)$, $f_n(w)$ be continuous, real or complex-valued functions on C such that $f(w) \neq 0$, $f_n(w) \neq 0$ on C, $n = 1, 2, \ldots$, and

$$f_n(w) \to f(w) \quad \text{uniformly on } C. \tag{28}$$

Then there exists an integer N such that

$$V_C[\arg f(w)] = V_C[\arg f_n(w)] \quad \text{for} \quad n > N. \tag{29}$$

Proof. Since $f(w)$ is continuous and different from zero on C, we have [see (2)] $m(f, C) > 0$. In view of (28) there follows the existence of an integer N such that [see (1)]

$$M(f - f_n, C) < m(f, C) \quad \text{for} \quad n > N. \tag{30}$$

On setting, for fixed $n > N$,

$$G = f, \quad g = f_n - f,$$

we have [in view of (30)]

$$f_n = G + g, \quad |G| > |g| \quad \text{on } C,$$

and thus (29) follows from lemma 14 (applied with $F = f_n$, $n > N$).

Lemma 16. Given a bounded, simply connected Jordan region R in R^2, let C be the arbitrarily oriented boundary curve of R, and let $f(w)$ be a continuous, real or complex-valued function in R which is different from zero in R. Then

$$V_C[\arg f(w)] = 0. \tag{31}$$

Proof. By lemma 9 there exists a single-valued continuous argument $\varphi(w)$ of $f(w)$ in R. Clearly $\varphi(w)$ yields a single-valued continuous argument of $f(w)$ on C, and thus (31) follows from part (d) of lemma 13.

Definition 6. Let R be a bounded, finitely connected Jordan region in R^2. By the *oriented boundary* B of R we mean the boundary of R with the *standard orientation* obtained as follows: the exterior boundary curve of R is oriented counter-clockwise, while the other boundary curves (if present) are oriented clock-wise. We shall write $C \in B$ to state the fact that C is a boundary curve of R, oriented in the manner just described.

Definition 7. Let B be the oriented boundary of the bounded, finitely connected JORDAN region $\mathsf{R} \subset R^2$ (see definition 6), and let $f(w)$ be a continuous, real or complex-valued function in R such that $f(w) \neq 0$ on B. Then the quantity

$$V_B[\arg f(w)] = \sum V_C[\arg f(w)], \quad C \in B, \tag{32}$$

is termed the variation of the argument of $f(w)$ on the oriented boundary B of R.

Lemma 17. Let B be the oriented boundary of the bounded, finitely connected JORDAN region R in R^2, and let $f(w)$ be a continuous, real or complex-valued function in R such that $f(w) \neq 0$ on B. Then the following holds.

(a) If $f(w)$ is different from zero in R, then

$$V_B[\arg f(w)] = 0. \tag{33}$$

(b) If $G(w)$, $g(w)$ are continuous, real or complex-valued functions in R such that $f(w) = G(w) + g(w)$ on B and $|G(w)| > |g(w)|$ on B, then

$$V_B[\arg f(w)] = V_B[\arg G(w)].$$

(c) If $\{f_n(w)\}$ is a sequence of continuous, real or complex-valued functions in R such that $f_n(w) \to f(w)$ uniformly on B, then there exists an integer N such that

$$V_B[\arg f_n(w)] = V_B[\arg f(w)] \quad \text{for} \quad n > N.$$

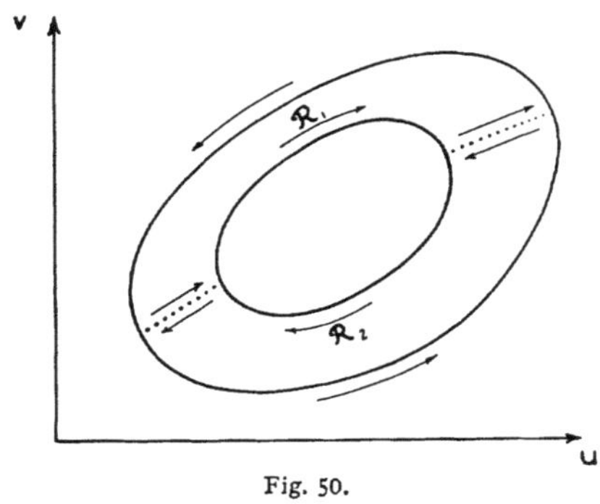

Fig. 50.

Proof. (b) and (c) follow, in view of (32), directly from lemma 14 and lemma 15 respectively. The proof of (a) follows readily by considering the figure (which illustrates the case when R is doubly connected). The auxiliary dotted arcs decompose R into two simply connected JORDAN regions R_1, R_2. On denoting by Γ_1, Γ_2 the counter-clockwise oriented boundary curves of R_1, R_2 respectively, we have

$$V_{\Gamma_1}[\arg f(w)] = 0, \quad V_{\Gamma_2}[\arg f(w)] = 0 \tag{34}$$

by lemma 16. Clearly

$$V_B[\arg f(w)] = V_{\Gamma_1}[\arg f(w)] + V_{\Gamma_2}[\arg f(w)], \tag{35}$$

since the contributions of the dotted arcs cancel out by part (i) of lemma 11, and thus (33) follows from (34) and (35).

Lemma 18. Let z_1, z_2, z_3 be the vertices of a non-degenerate triangle Δ in R^2. Denote by C the perimeter of Δ, oriented in conformity with the arrangement z_1, z_2, z_3 of the vertices (see the figure). If z_0 is an interior point of Δ, then

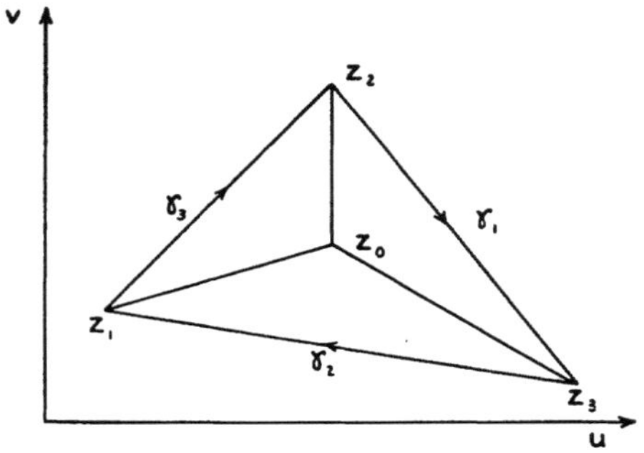

Fig. 51.

$$V_C[\arg(z - z_0)] = 2\pi$$

if the orientation of C is counter-clockwise, and

$$V_C[\arg(z - z_0)] = -2\pi,$$

if the orientation of C is clockwise.

Proof. On denoting by $\gamma_1, \gamma_2, \gamma_3$ the oriented sides of Δ (see the figure, which illustrates the case when the orientation of C is clockwise), we have in view of definition 5

$$V_C[\arg(z - z_0)] = \sum_{i=1}^{3} V_{\gamma_i}[\arg(z - z_0)]. \tag{36}$$

Let $\lambda_1, \lambda_2, \lambda_3$ be the radian measures of the angles at the vertex z_0 in the triangles $z_2 z_0 z_3$, $z_3 z_0 z_1$, $z_1 z_0 z_2$ respectively. Then by lemma 12

$$V_{\gamma_i}[\arg(z - z_0)] = -\lambda_i, \quad i = 1, 2, 3. \tag{37}$$

Since $\lambda_1 + \lambda_2 + \lambda_3 = 2\pi$, it follows from (36) and (37) that

$$V_C[\arg(z - z_0)] = -2\pi.$$

A similar argument applies in the case when the orientation of C is counter-clockwise.

Consider now the following situation. Let w_1, w_2, w_3 be the vertices of a non-degenerate triangle t in R^2, such that the arrangement w_1, w_2, w_3 of the vertices corresponds to the counter-clockwise orientation of the perimeter Γ of t. Let $\mathfrak{L}$ be a non-singular linear transformation from R^2 into R^2 (see **II.2.5**). On setting $z_i = \mathfrak{L}(w_i)$, $i = 1, 2, 3$, the points z_1, z_2, z_3 are then the vertices of a non-degenerate triangle Δ in R^2. Denote by C the perimeter of Δ, oriented in conformity with the arrangement z_1, z_2, z_3 of the vertices. Let z_0 be a point in R^2 which does not lie on C.

Lemma 19. Under the conditions just described, we have

$$V_\Gamma[\arg(\mathfrak{L}(w) - z_0)] = 2\pi \operatorname{sgn} \det \mathfrak{L} \tag{38}$$

if z_0 is interior to Δ, and

$$V_\Gamma[\arg(\mathfrak{L}(w) - z_0)] = 0 \tag{39}$$

if z_0 is exterior to Δ.

Proof. Assume first that z_0 is interior to Δ, and consider for instance the case when $\det \mathfrak{L} > 0$. Then the arrangement z_1, z_2, z_3 corresponds to the counter-clockwise orientation of the perimeter of Δ, and hence by lemma 18

$$V_C[\arg(z - z_0)] = 2\pi = 2\pi \operatorname{sgn} \det \mathfrak{L}, \tag{40}$$

since $\det \mathfrak{L} > 0$ by assumption. Also, since $\mathfrak{L}$ is a homeomorphism, we have by part (e) of lemma 13

$$V_\Gamma[\arg(\mathfrak{L}(w) - z_0)] = V_C[\arg(z - z_0)],$$

and (38) follows in view of (40). Assume next that z_0 is exterior to Δ. Then $\mathfrak{L}(w) - z_0 \neq 0$ in t, and (39) follows by part (d) of lemma 13.

Lemma 20. Let B be the oriented boundary of the bounded, finitely connected JORDAN region $\mathsf{R} \subset R^2$. Consider a continuous transformation $T: \mathsf{R} \to R^2$. Let z_0 be a point in R^2 such that $z_0 \notin TB$. Then there exists a real number $\eta > 0$ such that

$$V_B[\arg(T(w) - z_0^*)] = V_B[\arg(T(w) - z_0)] \quad \text{if} \quad |z_0 - z_0^*| < \eta. \tag{41}$$

Proof. We assert that the number

$$\eta = \min |T(w) - z_0|, \quad w \in B, \tag{42}$$

is adequate for our purposes. Note first that $\eta > 0$ since $T(w) - z_0 \neq 0$ on B. Take now any point $z_0^* \in R^2$ such that

$$|z_0 - z_0^*| < \eta. \tag{43}$$

Define, for $w \in \mathsf{R}$,

$$f(w) = T(w) - z_0^*, \quad G(w) = T(w) - z_0, \quad g(w) = z_0 - z_0\,.$$

We have then, in view of (42) and (43),

$$f(w) = G(w) + g(w), \quad |G(w)| > |g(w)| \quad \text{on } B,$$

and (41) follows by part (b) of lemma 17.

Lemma 21. Let B be the oriented boundary of the bounded, finitely connected JORDAN region $\mathsf{R} \subset R^2$. Consider a continuous transformation $T: \mathsf{R} \to R^2$. Let z_0 be a point in R^2 such that $z_0 \notin TB$. Take any finitely connected JORDAN region R_* such that

$$\mathsf{R}_* \subset \operatorname{int} \mathsf{R}, \quad T^{-1} z_0 \subset \operatorname{int} \mathsf{R}_* .$$

Denote by B_* the oriented boundary of R_*. Then

$$V_B[\arg(T(w) - z_0)] = V_{B_*}[\arg(T(w) - z_0)]. \tag{44}$$

Proof. The boundary curves of R_* divide R into a finite number of (finitely connected) JORDAN regions $\mathsf{R}_0, \mathsf{R}_1, \ldots, \mathsf{R}_m$, one of which is R_*, say

$$\mathsf{R}_* = \mathsf{R}_0. \tag{45}$$

Let B_j be the oriented boundary of R_j, $j = 0, 1, \ldots, m$ (thus $B_0 = B_*$). If C_* is a (properly oriented) boundary curve of $\mathsf{R}_* = \mathsf{R}_0$, then $-C_*$ is a (properly oriented) boundary curve for precisely one of the regions $\mathsf{R}_1, \ldots, \mathsf{R}_m$. Note also that $z_0 \notin TB_j$, $j = 0, 1, \ldots, m$ since $T^{-1} z_0 \subset \operatorname{int} \mathsf{R}_*$. These remarks yield [in view of (45) and part (a) of lemma 13] the relation

$$V_B[\arg(T(w) - z_0)] = V_{B_*}[\arg(T(w) - z_0)] + \sum_{j=1}^{m} V_{B_j}[\arg(T(w) - z_0)]. \tag{46}$$

Now since $T^{-1} z_0 \subset \operatorname{int} \mathsf{R}_*$, clearly $T(w) - z_0 \neq 0$ for $w \in \mathsf{R}_j$, $j = 1, \ldots, m$. By part (a) of lemma 17 we conclude that

$$V_{B_j}[\arg(T(w) - z_0)] = 0, \quad j = 1, \ldots, m,$$

and (44) follows in view of (46).

VI.1.3. The k-th root of a complex-valued function. Let us recall that if $z \neq 0$ is a complex number and k is a positive integer, then there exist precisely k distinct solutions ζ of the equation $\zeta^k = z$, while the equation $\zeta^k = 0$ has the unique solution $\zeta = 0$.

Definition. Let $f(w)$ be a continuous, real or complex-valued function on a continuum $\Gamma \subset R^2$, and let k be a positive integer. If $g(w)$ is a continuous, real or complex-valued function on Γ such that $g(w)^k = f(w)$ on Γ, then $g(w)$ is termed a single-valued continuous k-th root of $f(w)$ on Γ.

Lemma 1. Let $f(w)$ be a continuous, real or complex-valued function on a continuum $\Gamma \subset R^2$ such that (i) $f(w) \neq 0$ on Γ, and (ii) $f(w)$ satisfies the condition $(\arg, \Gamma)$ (see **VI.1.2**, definition 2). Then for every positive integer k there exists a single-valued continuous k-th root of $f(w)$ on Γ.

Proof. By assumption there exists a single-valued continuous argument $\varphi(w)$ of $f(w)$ on Γ. Then

$$f(w) = |f(w)|\,[\cos\varphi(w) + i\sin\varphi(w)], \quad w\in\Gamma.$$

Thus clearly

$$g(w) = |f(w)|^{1/k}\left(\cos\frac{\varphi(w)}{k} + i\sin\frac{\varphi(w)}{k}\right), \quad w\in\Gamma,$$

is a single-valued continuous k-th root of $f(w)$ on Γ.

Lemma 2. Let $f(w)$ be a continuous, real or complex-valued function on a continuum $\Gamma\subset R^2$, and let k be a positive integer. Assume that there exists a single-valued continuous k-th root $g_0(w)$ of $f(w)$ on Γ. Then the following holds.

(i) If λ is a k-th root of unity (that is, $\lambda^k=1$), then $\lambda g_0(w)$ is also a single-valued continuous k-th root of $f(w)$ on Γ.

(ii) If $f(w)\neq 0$ on Γ and $g(w)$ is any other single-valued continuous k-th root of $f(w)$ on Γ, then $g(w)\equiv\lambda g_0(w)$, where λ is a k-th root of unity independent of w. In particular, if $g(w_*)=g_0(w_*)$ at some point $w_*\in\Gamma$, then $g(w)\equiv g_0(w)$ on Γ.

Proof. (i) is obvious. To verify (ii), note first that $g_0(w)\neq 0$ on Γ [since $f(w)\neq 0$ on Γ by assumption]. Hence, if $g(w)$ is any single-valued continuous k-th root of $f(w)$ on Γ, then we can consider the auxiliary function $h(w)=g(w)/g_0(w)$ on Γ. Since $h(w)^k\equiv 1$ on Γ, it follows that each value of $h(w)$ is a k-th root of unity. Thus $h(w)$ is a continuous function on Γ which takes on only a finite number of distinct values on Γ. Since Γ is connected, it follows (see **I.1.3**, exercise 37) that $h(w)$ is constant on Γ. As $h(w)^k\equiv 1$, the constant value of $h(w)$ must be a k-th root of unity, and (ii) follows.

Lemma 3. Let $f(w)$ be a continuous, real or complex-valued function on a continuum $\Gamma\subset R^2$, and let k be a positive integer. Assume that there exists a single-valued continuous k-th root $g(w)$ of $f(w)$ on Γ. Let $h:\Gamma\to\Gamma^*$ be a homeomorphism from Γ onto a continuum $\Gamma^*\subset R^2$. Define

$$f^*(w) = f\big(h^{-1}(w)\big), \quad g^*(w) = g\big(h^{-1}(w)\big), \quad w\in\Gamma^*.$$

Then $f^*(w)$ is continuous on Γ^*, and $g^*(w)$ is a single-valued continuous k-th root of $f^*(w)$ on Γ^*.

This statement is an obvious consequence of the definitions involved.

Lemma 4. Let R be a bounded, simply connected JORDAN region in R^2, and let w_0 be a point on the boundary curve of R. Consider a continuous, real or complex-valued function $f(w)$ in R such that

(i) $f(w_0)=0$, and (ii) $f(w) \neq 0$ if $w_0 \neq w \in \mathsf{R}$. Then for every positive integer k there exists a single-valued continuous k-th root of $f(w)$ in R.

Proof. Without loss of generality we can assume (in view of lemma 3) that R is the region

$$0 \leq u^2 + v^2 \leq 1, \quad v \geq 0,$$

where $u + iv = w$, and $w_0 = 0$. For each positive integer $n \geq 2$, let us denote by R_n the region

$$\frac{1}{n^2} \leq u^2 + v^2 \leq 1, \quad v \geq 0.$$

Clearly R_n is simply connected, and $f(w) \neq 0$ in R_n. Note that the point $w = 1$ lies in R_n. Let α be a k-th root of $f(1)$. Since R_n is simply connected and $f(w) \neq 0$ in R_n, by **VI.1.2**, lemma 9 it follows that $f(w)$ satisfies the condition (arg, R_n). By lemma 1 it follows further that there exists a single-valued continuous k-th root $g_n(w)$ of $f(w)$ in R_n. From lemma 2 we conclude that we can select $g_n(w)$ so that $g_n(1) = \alpha$ for every integer $n \geq 2$. Consider now two positive integers m, n such that $m \leq n$. Then clearly $\mathsf{R}_m \subset \mathsf{R}_n$. We assert that

$$g_n(w) = g_m(w) \quad \text{for} \quad w \in \mathsf{R}_m. \tag{1}$$

Indeed, in R_m the functions $g_n(w)$ and $g_m(w)$ are single-valued continuous k-th roots of $f(w)$, and $g_n(1) = g_m(1) = \alpha$. Thus (1) follows by part (ii) of lemma 2. We define now a function $g(w)$ in R as follows:

$$g(w) = g_n(w) \quad \text{if} \quad w \in \mathsf{R}_n, \quad g(0) = 0.$$

In view of (1) it is clear that $g(w)$ is single-valued and continuous in R, except that the continuity of $g(w)$ at $w = 0$ is perhaps not immediately obvious. To settle this point, note that since $g_n(w)^k = f(w)$ in R_n and $f(0) = 0$, we have [in view of (1)]

$$g(w)^k = f(w) \quad \text{for} \quad w \in \mathsf{R}. \tag{2}$$

Now since $f(w)$ is continuous in R and $f(0) = 0$, we have $f(w) \to 0$ for $w \to 0$, and hence by (2) also $g(w) \to 0$ for $w \to 0$. In view of (2), it is now clear that $g(w)$ is a single-valued continuous k-th root of $f(w)$ in R, and the lemma is proved.

Lemma 5. Let R be a bounded, simply connected JORDAN region in R^2, w_0 an interior point of R, and $j \neq 0$ an integer. Consider a continuous, real or complex-valued function $f(w)$ in R such that (i) $f(w_0) = 0$, (ii) $f(w) \neq 0$ at every point $w \neq w_0$ in R, and (iii) the variation of the argument of $f(w)$ on the counter-clockwise oriented boundary curve C of R is equal to $2\pi j$ [see part (c) of **VI.1.2**, lemma 13]. Then there exists a single-valued continuous $|j|$-th root of $f(w)$ in R.

Proof. In view of lemma 3 and part (e) of **VI.1.2**, lemma 13 we can assume without loss of generality that R coincides with the unit disc $0 \leq u^2 + v^2 \leq 1$ and $w_0 = 0$. Introduce the auxiliary regions

$$R_1 : 0 \leq u^2 + v^2 \leq 1, \quad v \geq 0,$$
$$R_2 : 0 \leq u^2 + v^2 \leq 1, \quad v \leq 0,$$

and let α be an $|j|$-th root of $f(1)$. By lemma 4 there exist single-valued continuous $|j|$-th roots $g_1(w)$, $g_2(w)$ of $f(w)$ in R_1, R_2 respectively, and in view of part (i) of lemma 2 we can select $g_1(w)$, $g_2(w)$ so that

$$g_1(1) = g_2(1) = \alpha. \tag{3}$$

Denote by C_1 the semi-circle $u^2 + v^2 = 1, v \geq 0$, oriented from $w = 1$ toward $w = -1$, and by C_2 the semi-circle $u^2 + v^2 = 1, v \leq 0$, oriented from $w = 1$ toward $w = -1$. By **VI.1.2**, definition 5 and part (i) of **VI.1.2**, lemma 11 we have then

$$V_{C_1}[\arg f(w)] - V_{C_2}[\arg f(w)] = V_C[\arg f(w)] = 2\pi j. \tag{4}$$

Let φ_0 be an argument of α [see (3)]. Since $g_1(w) \neq 0$ on C_1, by **VI.1.2**, lemma 10 there exists a single-valued continuous argument $\varphi_1(w)$ of $g_1(w)$ on C_1. In view of **VI.1.1**, lemma 4 we can select $\varphi_1(w)$ so that $\varphi_1(1) = \varphi_0$. Then (see **VI.1.2**, definition 4)

$$V_{C_1}[\arg g_1(w)] = \varphi_1(-1) - \alpha. \tag{5}$$

Similarly it follows that we can select a single-valued continuous argument $\varphi_2(w)$ of $g_2(w)$ on C_2 so that

$$V_{C_2}[\arg g_2(w)] = \varphi_2(-1) - \alpha. \tag{6}$$

Since $g_1(w)^{|j|} = f(w)$ on C_1 and $g_2(w)^{|j|} = f(w)$ on C_2, we obtain from part (iii) of **VI.1.2**, lemma 11 the relations

$$V_{C_1}[\arg f(w)] = |j|\, V_{C_1}[\arg g_1(w)], \tag{7}$$
$$V_{C_2}[\arg f(w)] = |j|\, V_{C_2}[\arg g_2(w)]. \tag{8}$$

From (4), (7), (8), (5), (6) we conclude that

$$2\pi j = |j| \left(\varphi_1(-1) - \varphi_2(-1)\right),$$

and hence (since $j \neq 0$)

$$\varphi_1(-1) - \varphi_2(-1) = \pm 2\pi. \tag{9}$$

Observe that

$$|g_1(-1)| = |g_2(-1)| = |f(-1)|^{1/|j|}.$$

Thus (9) implies that

$$g_1(-1) = g_2(-1). \tag{10}$$

Consider now any real number u_0 such that $-1 < u_0 < 0$. Then $g_1(w)$, $g_2(w)$ are single-valued continuous $|j|$-th roots of $f(w)$ on the segment $-1 \leq u \leq u_0$, and $f(w) \neq 0$ on this segment. In view of (10) it follows by part (ii) of lemma 2 that $g_1(w) = g_2(w)$ on this segment. Since u_0 was any real number such that $-1 < u_0 < 0$, we conclude [in view of the continuity of $g_1(w)$, $g_2(w)$] that

$$g_1(u) = g_2(u) \quad \text{for} \quad -1 \leq u \leq 0. \tag{11}$$

A similar argument, using (3) instead of (10), yields

$$g_1(u) = g_2(u) \quad \text{for} \quad 0 \leq u \leq 1. \tag{12}$$

If we now define

$$g(w) = \begin{cases} g_1(w) & \text{if} \quad w \in \mathsf{R}_1, \\ g_2(w) & \text{if} \quad w \in \mathsf{R}_2, \end{cases}$$

then [in view of (11) and (12)] it is clear that $g(w)$ is a single-valued continuous $|j|$-th root of $f(w)$ in R, and the lemma is proved.

VI.1.4. The topological index in R^2. We consider in this section a bounded continuous transformation

$$T: \mathsf{R} \to R^2, \tag{1}$$

where R is a bounded, finitely connected JORDAN region in R^2. We shall also use for T the alternative representation (see **VI.1.1**)

$$T: z = T(w), \quad w \in \mathsf{R}. \tag{2}$$

Let B be the oriented boundary of R (see **VI.1.2**, definition 6). Select a point z_0 in R^2 such that

$$z_0 \notin TB. \tag{3}$$

Then $T(w) - z_0$ is continuous in R and different from zero on B. Thus the variation of the argument of $T(w) - z_0$ on B is available (see **VI.1.2**, definition 7). Observe that (3) is equivalent (see **VI.1.1**) to the relation

$$z_0 \notin T\, fr \operatorname{int} \mathsf{R}. \tag{4}$$

Accordingly (see **II.2.2**) the topological index $\mu(z_0, T, \operatorname{int} \mathsf{R})$ is available.

Theorem. Under the conditions just described, we have the relation

$$\mu(z_0, T, \operatorname{int} \mathsf{R}) = \frac{1}{2\pi} V_B[\arg(T(w) - z_0)]. \tag{5}$$

Proof. To simplify notations, let us put

$$\frac{1}{2\pi} V_B[\arg(T(w) - z_0)] = \mu^*(z_0, T, \operatorname{int} \mathsf{R}). \tag{6}$$

We have to show that

$$\mu(z_0, T, \operatorname{int} \mathsf{R}) = \mu^*(z_0, T, \operatorname{int} \mathsf{R}). \tag{7}$$

If $z_0 \notin T\mathsf{R}$, then

$$\mu(z_0, T, \operatorname{int} \mathsf{R}) = 0$$

by **II.2.3**, theorem 2, and

$$\mu^*(z_0, T, \operatorname{int} \mathsf{R}) = 0$$

by part (a) of **VI.1.2**, lemma 17. Thus (7) holds if $z_0 \notin T\mathsf{R}$, and hence we can assume, in proving (7), that $z_0 \in T\mathsf{R}$. This inclusion implies, in view of (4), that

$$z_0 \in T \operatorname{int} \mathsf{R}. \tag{8}$$

It is convenient to divide the proof of (7) into several parts.

Case 1. R is a non-degenerate triangle, and $T = \mathfrak{L}$, a linear transformation from R^2 into R^2 (see **II.2.5**). We assert that $\mathfrak{L}$ is non-singular. Indeed, if $\mathfrak{L}$ were singular, then we would have $\mathfrak{L}\mathsf{R} = \mathfrak{L}B$ and hence, by (8), $z_0 \in \mathfrak{L}B$, in contradiction with (3). Once we know that $\mathfrak{L}$ is non-singular, we can apply **II.2.5**, theorem 4, obtaining in view of (8) the relation

$$\mu(z_0, \mathfrak{L}, \operatorname{int} \mathsf{R}) = \operatorname{sgn} \det \mathfrak{L}. \tag{9}$$

On the other hand, **VI.1.2**, lemma 19 yields

$$V_B[\arg(\mathfrak{L}(w) - z_0)] = 2\pi \operatorname{sgn} \det \mathfrak{L},$$

and (7) follows in view of (9) and (6).

Case 2. R is a polygonal region (that is, the boundary curves of R are simple closed polygons), and T is *quasi-linear* in R. This means that there exists a triangulation of R, consisting of a finite number of triangles $t_1, \ldots, t_m$ (see the figure), such that T coincides with a linear transformation $\mathfrak{L}_j$ in t_j, $j = 1, \ldots, m$. Let S be the union of all the edges of $t_1, \ldots, t_m$. It is convenient to consider now two sub-cases.

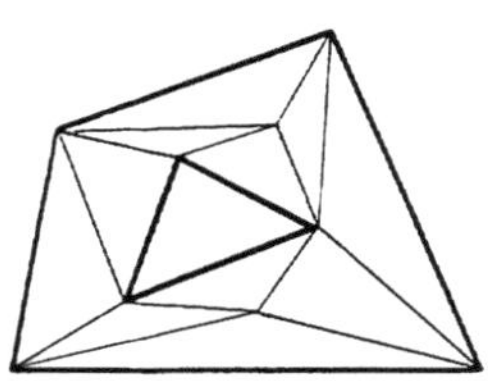

Fig. 52.

Case 2a. $z_0 \notin TS$. Then we have

$$\left.\begin{aligned} \mu(z_0, T, \operatorname{int} t_j) &= \mu^*(z_0, T, \operatorname{int} t_j), \\ j &= 1, \ldots, m. \end{aligned}\right\} \tag{10}$$

Indeed, if $z_0 \in T \operatorname{int} t_j$ for a certain j, then (10) holds by *Case 1.* If $z_0 \notin T \operatorname{int} t_j$ for a certain j, then (since $z_0 \notin TS$) we have $z_0 \notin T t_j$, and hence both sides in (10) are equal to zero, as we noted above. Thus (10) is verified. Denote by B_j the counter-clockwise oriented perimeter of t_j. Then

$$V_B[\arg(T(w) - z_0)] = \sum_{j=1}^{m} V_{B_j}[\arg(T(w) - z_0)], \tag{11}$$

since the contributions of those edges which are not parts of B cancel out by part (i) of **VI.1.2**, lemma 11. From (11) it follows, in view of (6), that

$$\mu^*(z_0, T, \text{int } \mathsf{R}) = \sum_{j=1}^{m} \mu^*(z_0, T, \text{int } t_j). \tag{12}$$

Observe now that the assumption $z_0 \notin TS$ implies that

$$T^{-1} z_0 \subset \bigcup_{j=1}^{m} \text{int } t_j.$$

Thus int $t_1, \ldots$, int t_m is a (finite) sequence of domains in int R which is $(z_0, T, \text{int } \mathsf{R})$-complete (see **II.2.1**, definition 5). Hence

$$\mu(z_0, T, \text{int } \mathsf{R}) = \sum_{j=1}^{m} \mu(z_0, T, \text{int } t_j)$$

by **II.2.3**, theorem 3, and (7) follows in view of (10) and (12).

Case 2b. Assume now that $z_0 \in TS$. In view of (6), (3) and (4) we conclude from **II.2.3**, theorem 1 and **VI.1.2**, lemma 20 that there exists an $\eta > 0$ such that

$$\mu(z_0^*, T, \text{int } \mathsf{R}) = \mu(z_0, T, \text{int } \mathsf{R}) \quad \text{if} \quad |z_0 - z_0^*| < \eta, \tag{13}$$

$$\mu^*(z_0^*, T, \text{int } \mathsf{R}) = \mu^*(z_0, T, \text{int } \mathsf{R}) \quad \text{if} \quad |z_0 - z_0^*| < \eta. \tag{14}$$

Note that the set TS consists of a finite number of straight segments, some of which may reduce to single points. Thus clearly we can select a point z_0^* such that

$$z_0^* \notin TS, \quad |z_0 - z_0^*| < \eta. \tag{15}$$

The point z_0^* satisfies the assumptions used in *Case 2a*, and hence

$$\mu(z_0^*, T, \text{int } \mathsf{R}) = \mu^*(z_0^*, T, \text{int } \mathsf{R}). \tag{16}$$

Clearly (13) to (16) imply (7).

Case 3. R is a polygonal region (as in *Case 2*), but T is not assumed to be quasi-linear. Take a triangulation τ_1 of R (as indicated in the figure used in *Case 2*). Subdividing each triangle t of τ_1 into six smaller triangles by drawing the three medians of t, we obtain a second triangulation τ_2 of R. Repetition of this process yields a sequence $\tau_1, \ldots, \tau_m, \ldots$

of triangulations of R. For each positive integer m denote by T_m the quasi-linear transformation obtained as follows: in each triangle t of τ_m, T_m coincides with the (uniquely determined) linear transformation which agrees with T at the vertices of t. Since T is uniformly continuous in R, clearly

$$T_m \to T \quad \text{uniformly in R}. \tag{17}$$

From (3) and (17) we conclude that $z_0 \notin T_m B$ if m is large enough, say $m > M$. By *Case 2* we have therefore

$$\mu(z_0, T_m, \text{int R}) = \mu^*(z_0, T_m, \text{int R}) \quad \text{for} \quad m > M. \tag{18}$$

Part (c) of **VI.1.2**, lemma 17 yields, in view of (17) and (6),

$$\mu^*(z_0, T, \text{int R}) = \mu^*(z_0, T_m, \text{int R}) \quad \text{for } m \text{ large}. \tag{19}$$

Note now that in view of (4) we have (see **I.1.4**, exercise 3)

$$e(z_0, T\, fr\, \text{int R}) > 0.$$

From (17) it follows therefore that

$$\varrho(T, T_m, fr\, \text{int R}) < e(z_0, T\, fr\, \text{int R}) \quad \text{for } m \text{ large}. \tag{20}$$

By **II.2.3**, remark 8 we conclude from (20) that

$$\mu(z_0, T, \text{int R}) = \mu(z_0, T_m, \text{int R}) \quad \text{for } m \text{ large},$$

and (7) follows in view of (18) and (19).

We are now prepared to deal with the general case. In view of (4) and (8), the set $T^{-1} z_0$ is a non-empty, compact subset of int R. Accordingly, we can select a polygonal region R_* such that

$$T^{-1} z_0 \subset \text{int } \mathsf{R}_*, \quad \mathsf{R}_* \subset \text{int R}. \tag{21}$$

Then R_* presents *Case 3*, and hence

$$\mu(z_0, T, \text{int } \mathsf{R}_*) = \mu^*(z_0, T, \text{int } \mathsf{R}_*). \tag{22}$$

By **II.2.3**, remark 4 it follows from (21) that

$$\mu(z_0, T, \text{int R}) = \mu(z_0, T, \text{int } \mathsf{R}_*). \tag{23}$$

Denote by B_* the oriented boundary of R_*. From **VI.1.2**, lemma 21 we infer [in view of (21) and (6)] that

$$\mu^*(z_0, T, \text{int R}) = \mu^*(z_0, T, \text{int } \mathsf{R}_*),$$

and (7) follows by (22) and (23).

§ VI.2. Special features of continuous transformations in R^2.

VI.2.1. A countability theorem. The validity of the principal theorem to be derived in this section is restricted to R^2. Simple examples show that the corresponding statement for R^n is generally false if $n > 2$.

Our first lemma is concerned with a continuous transformation T from a finitely connected, bounded JORDAN region $\mathsf{R} \subset R^2$ into R^2. We shall use for T the alternative representations

$$T: \mathsf{R} \to R^2, \tag{1}$$

$$T: z = T(w), \quad w \in \mathsf{R}, \tag{2}$$

in the sense of **VI.1.1** (4), (5).

Lemma 1. Given T as in (1) and (2), and given a point $w_0 \in \operatorname{int} \mathsf{R}$, assume that the following holds.

(i) R is simply connected.
(ii) $T(w) \neq T(w_0)$ for $w \in \mathsf{R}$, $w \neq w_0$.
(iii) $|\mu(T w_0, T, \operatorname{int} \mathsf{R})| \geq 1$.

Then there exists a number $\eta > 0$ such that

$$N(z, T, \operatorname{int} \mathsf{R}) \geq |\mu(T w_0, T, \operatorname{int} \mathsf{R})| \quad \text{for} \quad 0 < |z - T w_0| < \eta. \tag{3}$$

Proof. The assumption (ii) implies that $T(w_0) \notin T\,fr \operatorname{int} \mathsf{R}$, and thus $\mu(T w_0, T, \operatorname{int} R)$ is defined (see **II.2.2**). Let us put

$$|\mu(T w_0, T, \operatorname{int} R)| = k. \tag{4}$$

Then by (iii)

$$k \geq 1. \tag{5}$$

Let C be the counter-clockwise oriented boundary curve of R. By **VI.1.4** (5) we have then

$$2\pi \mu(T w_0, T, \operatorname{int} \mathsf{R}) = V_C[\arg(T(w) - T(w_0))]. \tag{6}$$

Introduce the auxiliary function

$$f(w) = T(w) - T(w_0), \quad w \in \mathsf{R}. \tag{7}$$

From (i), (ii), (iii), (4), (5), (6) we conclude that $f(w)$ satisfies the assumptions of **VI.1.3**, lemma 5, and hence $f(w)$ possesses a single-valued, continuous k-th root $T^*(w)$ in R. Thus

$$f(w) = T^*(w)^k, \quad w \in \mathsf{R}. \tag{8}$$

Consider the transformation

$$T^*: z = T^*(w), \quad w \in \mathsf{R}. \tag{9}$$

From (7), (8), (ii) we see that

$$T^*(w_0) = 0, \quad T^*(w) \neq 0 \quad \text{for} \quad w \in C. \tag{10}$$

Thus $0 \notin T^* fr \operatorname{int} \mathsf{R}$, and hence $\mu(0, T^*, \operatorname{int} \mathsf{R})$ is defined (see **II.2.2**). By **VI.1.4** (5) we have

$$\mu(0, T^*, \operatorname{int} \mathsf{R}) = \frac{1}{2\pi} V_C [\arg T^*(w)]. \tag{11}$$

On the other hand, (8) implies (see **VI.1.2**, lemma 13) that

$$V_C [\arg T^*(w)] \doteq \frac{1}{k} V_C [\arg f(w)]. \tag{12}$$

From (11), (12), (7), (6), (4) it follows that

$$\mu(0, T^*, \operatorname{int} \mathsf{R}) = \pm 1 \neq 0. \tag{13}$$

In view of **II.2.3**, theorem 1, the relation (13) implies that $\mu(z, T^*, \operatorname{int} \mathsf{R})$ is different from zero in a certain neighborhood of the point $z = 0$. Thus there exists a number $\delta > 0$ such that

$$\mu(\zeta, T^*, \operatorname{int} \mathsf{R}) \neq 0 \quad \text{for} \quad 0 \leqq |\zeta| < \delta. \tag{14}$$

Let us put

$$\eta = \delta^k. \tag{15}$$

We proceed to verify that this number η has the property required in our lemma. Since $\delta > 0$, clearly $\eta > 0$. Consider any point z such that

$$0 < |z - T(w_0)| < \eta. \tag{16}$$

Then the complex number $z - T(w_0)$ has k distinct k-th roots $z_1, \ldots, z_k$. Thus

$$z_r^k = z - T(w_0), \quad r = 1, \ldots, k, \tag{17}$$

$$z_r \neq z_s \quad \text{if} \quad r \neq s. \tag{18}$$

Furthermore, from (17), (16), (15) it follows that

$$0 < |z_r| < \delta, \quad r = 1, \ldots, k. \tag{19}$$

From (19) and (14) we conclude that

$$\mu(z_r, T^*, \operatorname{int} \mathsf{R}) \neq 0, \quad r = 1, \ldots, k. \tag{20}$$

By **II.2.3**, theorem 2 the relations (20) imply that

$$z_r \in T^* \operatorname{int} \mathsf{R}, \quad r = 1, \ldots, k.$$

Hence there exist points $w_1, \ldots, w_k$ such that

$$w_r \in \operatorname{int} \mathsf{R}, \quad r = 1, \ldots, k, \tag{21}$$

$$T^*(w_r) = z_r, \quad r = 1, \ldots, k. \tag{22}$$

From (22) and (18) it follows that $w_r \neq w_s$ for $r \neq s$. The relations (7), (8), (22), (17) yield

$$T(w_r) = T(w_0) + f(w_r) = T(w_0) + T^*(w_r)^k = T(w_0) + \big(z - T(w_0)\big) = z.$$

Hence the set $(T^{-1}z) \cap \operatorname{int} \mathsf{R}$ contains at least k distinct points, namely, at least the points $w_1, \ldots, w_k$. Thus it is established that (16) implies the inequality $N(z, T, \operatorname{int} \mathsf{R}) \geqq k$, and (3) follows in view of (4).

Lemma 2. Given a bounded continuous transformation

$$T : \boldsymbol{D} \to R^2 \tag{23}$$

as in **VI.1.1** (1), let D be a domain in $\boldsymbol{D}$, and let $k \geqq 2$ be an integer. Denote by $F(k, T, D)$ the set of those points $z_0 \in R^2$ which satisfy the following conditions. (a) There exists precisely one point $w_0 \in D$ such that $T(w_0) = z_0$, and (b) at this point w_0 we have (see **II.3.7**)

$$|i_{\mathsf{B}}(w_0, T)| = k. \tag{24}$$

Then the set $F(k, T, D)$ is countable.

Proof. The assertion is obvious if $F(k, T, D) = \emptyset$. So we can assume that $F(k, T, D) \neq \emptyset$. Consider any point $z_0 \in F(k, T, D)$, and let w_0 be the unique point in D such that $T(w_0) = z_0$ [see condition (a)]. Then, in view of (24),

$$w_0 \in D, \quad |i_{\mathsf{B}}(w_0, T)| \geqq 2. \tag{25}$$

Note that condition (a) implies that D is clear of relatives of w_0 (see **II.3.7**). If R is a circular disc with center at w_0 and a sufficiently small radius, then we have $\mathsf{R} \subset D$, and hence $\operatorname{int} \mathsf{R}$ is also clear of relatives of w_0. Therefore [see **II.3.7** (3) and **II.3.4**, definition 1]

$$i_{\mathsf{B}}(w_0, T) = \mu(z_0, T, \operatorname{int} \mathsf{R}). \tag{26}$$

From (25) and (26) we conclude that the transformation $T|\mathsf{R}$ and the point w_0 satisfy the assumptions of lemma 1. Hence there exists a number $\eta = \eta(z_0) > 0$ such that

$$N(z, T, \operatorname{int} \mathsf{R}) \geqq k \geqq 2 \quad \text{if} \quad 0 < |z - z_0| < \eta(z_0).$$

Since $\mathsf{R} \subset D$, we have *a fortiori*

$$N(z, T, D) \geqq 2 \quad \text{if} \quad 0 < |z - z_0| < \eta(z_0). \tag{27}$$

We assert that

$$0 < |z - z_0| < \eta(z_0) \quad \text{implies} \quad z \notin F(k, T, D). \tag{28}$$

Indeed, if $0 < |z - z_0| < \eta(z_0)$, then by (27) there exist at least two distinct points w', w'' in D such that $T(w') = z$, $T(w'') = z$, and thus z violates the condition (a) required of the points of the set $F(k, T, D)$. From (28) it follows that every point z_0 of $F(k, T, D)$ possesses a neighborhood clear of further points of $F(k, T, D)$. Thus $F(k, T, D)$ is an *isolated set*. Hence (see **I.2.1**, exercise 5) $F(k, T, D)$ is countable, and the lemma is proved.

Theorem. Given T as in (23), let $\tilde{\mathbf{I}}$ be the set of those points $w \in \boldsymbol{D}$ where $|i_{\mathsf{B}}(w, T)| \geqq 2$. Then $\tilde{\mathbf{I}}$ is countable.

Proof. For each integer k let $\mathbf{I}_k$ be the set of those points $w \in \boldsymbol{D}$ where $i_{\mathsf{B}}(w, T) = k$. Then clearly $\tilde{\mathbf{I}} = \bigcup \mathbf{I}_k$, $|k| \geqq 2$, and thus it is sufficient to prove that $\mathbf{I}_k$ is countable if $|k| \geqq 2$. We divide the proof of this fact into two parts.

Part 1. Let k be an integer such that $|k| \geqq 2$. We assert that the set $T\mathbf{I}_k$ is countable. To see this, let $\mathfrak{Q}$ be the class of all those oriented squares $Q \subset \boldsymbol{D}$ which satisfy the following conditions. (a) The coordinates of the center of Q are rational numbers, and (b) the side-length of Q is a rational number. Then clearly $\mathfrak{Q}$ is a countable class. We first verify the inclusion

$$T\,\mathbf{I}_k \subset \bigcup F(|k|, T, \operatorname{int} Q), \quad Q \subset \mathfrak{Q}, \tag{29}$$

where $F(|k|, T, \operatorname{int} Q)$ is defined as in lemma 2. To establish (29), note that if $z_0 \in T\mathbf{I}_k$, then there exists a point w_0 such that

$$w_0 \in \mathbf{I}_k, \quad T\,w_0 = z_0, \quad i_{\mathsf{B}}(w_0, T) = k \neq 0.$$

Since $i_{\mathsf{B}}(w_0, T) \neq 0$, the point w_0 possesses (see **II.3.7**) a neighborhood clear of relatives. Thus there exists an open set O such that

$$w_0 \in O \subset \boldsymbol{D}, \quad T(w) \neq z_0 \quad \text{if} \quad w \in O, \quad w \neq w_0.$$

Since O is open, there follows the existence if a square $Q_0 \in \mathfrak{Q}$ such that $w_0 \in \operatorname{int} Q_0$, $Q_0 \subset O$. Then clearly w_0 is the only point in $\operatorname{int} Q_0$ whose image under T is z_0. Furthermore, $i_{\mathsf{B}}(w_0, T) = k$. Thus, by the definition of the set $F(|k|, T, \operatorname{int} Q_0)$, we have $z_0 \in F(|k|, T, \operatorname{int} Q_0)$. Since z_0 was an arbitrary point of $T\mathbf{I}_k$ and $Q_0 \in \mathfrak{Q}$, (29) follows. As

$\mathfrak{Q}$ is a countable class and $F(|k|, T, \operatorname{int} Q)$ is countable for $|k| \geq 2$ by lemma 2, the inclusion (29) implies that $T\mathbf{I}_k$ is countable for $|k| \geq 2$.

Part 2. We next show that if $|k| \geq 2$, then the set $\mathbf{I}_k$ is countable. Since we already know that $T\mathbf{I}_k$ is countable if $|k| \geq 2$, it is sufficient to verify that for each point $z_0 \in T\mathbf{I}_k$ the set

$$G(z_0) = \mathbf{I}_k \cap T^{-1} z_0 \tag{30}$$

is countable. In turn, the countability of $G(z_0)$ will be established (see **I.2.1**, exercise 5) if we show that each point $w_0 \in G(z_0)$ possesses a neighborhood clear of further points of $G(z_0)$. Consider a point $w_0 \in G(z_0)$. Then $w_0 \in \mathbf{I}_k$, and thus $|i_{\mathsf{B}}(w_0, T)| = |k| \neq 0$. By the definition of the local index i_{B} (see **II.3.7**) it follows that w_0 possesses a neighborhood O clear of relatives. Hence if $w \in O$ and $w \neq w_0$, then $w \notin T^{-1} z_0$ and [see (30)] *a fortiori* $w \notin G(z_0)$. Thus O contains no further point of $G(z_0)$, and the proof is complete.

VI.2.2. Some applications of the countability theorem. We consider a bounded continuous transformation

$$T: z = T(w), \quad w \in \boldsymbol{D}, \tag{1}$$

as in **VI.1.1** (2).

Theorem 1. Given T as in (1), we have

$$K(z, T, \boldsymbol{D}) \leq N(z, T, \boldsymbol{D}) \quad \text{a.e. in } R^2. \tag{2}$$

Proof. Lemma 6 in **IV.5.3** yields

$$K(z, T, \boldsymbol{D}) \leq N(z, T, \boldsymbol{D}) \quad \text{if} \quad z \notin T\tilde{\mathbf{I}}. \tag{3}$$

Since we are now operating in R^2, the set $T\tilde{\mathbf{I}}$ is countable by the *countability theorem* in **VI.2.1**. Hence the set $T\tilde{\mathbf{I}}$ is of L-measure zero, and thus (2) follows from (3).

Theorem 2. In R^2 the concepts $BV\mathsf{B}$ and $sBV\mathsf{B}$ are equivalent. Explicitly, the transformation (1) is $sBV\mathsf{B}$ in $\boldsymbol{D}$ if and only if it is $BV\mathsf{B}$ in $\boldsymbol{D}$.

Proof. Assume that T is $BV\mathsf{B}$ in $\boldsymbol{D}$. Then (see **IV.5.1**, theorem 3) the derivative $D_{\mathsf{B}}(w, T)$ exists a.e. in $\boldsymbol{D}$ and is L-summable in $\boldsymbol{D}$. Furthermore (see **IV.5.3**)

$$J_{\mathsf{B}}(w, T) = i_{\mathsf{B}}(w, T)\, D_{\mathsf{B}}(w, T) \quad \text{a.e. in } \boldsymbol{D}. \tag{4}$$

The countability theorem in **VI.2.1** yields

$$\mathsf{L}\tilde{\mathbf{I}} = 0, \tag{5}$$

$$|i_{\mathsf{B}}(w, T)| \leq 1 \quad \text{for} \quad w \in \boldsymbol{D} - \tilde{\mathbf{I}}. \tag{6}$$

From (4), (5), (6) we conclude that

$$|J_{\mathsf{B}}(w, T)| \leq D_{\mathsf{B}}(w, T) \quad \text{a.e. in } \boldsymbol{D}. \tag{7}$$

Since $D_{\mathsf{B}}(w, T)$ is L-summable in $\boldsymbol{D}$, (7) implies that $J_{\mathsf{B}}(w, T)$ is also L-summable in $\boldsymbol{D}$. In view of **IV.5.3**, definition 1 it follows that T is $sBV\mathsf{B}$ in $\boldsymbol{D}$. Conversely, if T is $sBV\mathsf{B}$ in $\boldsymbol{D}$, then T is $BV\mathsf{B}$ in $\boldsymbol{D}$ by definition.

Theorem 3. In R^2 the concepts $AC\mathsf{B}$ and $sAC\mathsf{B}$ are equivalent. Explicitly, the transformation (1) is $sAC\mathsf{B}$ in $\boldsymbol{D}$ if and only if it is $AC\mathsf{B}$ in $\boldsymbol{D}$.

Proof. Assume that T is $AC\mathsf{B}$ in $\boldsymbol{D}$. Then T is $BV\mathsf{B}$ in $\boldsymbol{D}$ by **IV.5.2**, theorem 1, and hence T is $sBV\mathsf{B}$ in $\boldsymbol{D}$ by the preceding theorem. Accordingly, T is $sAC\mathsf{B}$ in $\boldsymbol{D}$ as a direct consequence of **IV.5.3**, definition 2. Conversely, if T is $sAC\mathsf{B}$ in $\boldsymbol{D}$, then by definition T is also $AC\mathsf{B}$ in $\boldsymbol{D}$.

Theorem 4. In R^2 the class of $BV\mathsf{B}$ transformations is included in the class of eBV transformations. Explicitly, if the transformation (1) is $BV\mathsf{B}$ in $\boldsymbol{D}$, then it is also eBV in $\boldsymbol{D}$.

Proof. If T is $BV\mathsf{B}$ in $\boldsymbol{D}$, then $N(z, T, \boldsymbol{D})$ is L-summable in R^2 by **IV.5.1**, theorem 1. By (2) it follows that $K(z, T, \boldsymbol{D})$ is L-summable in R^2. Hence T is eBV in $\boldsymbol{D}$ in view of **IV.4.1**, definition 1.

Theorem 5. In R^2 the class of $AC\mathsf{B}$ transformations is included in the class of eAC transformations. Explicitly, if the transformation (1) is $AC\mathsf{B}$ in $\boldsymbol{D}$, then it is also eAC in $\boldsymbol{D}$. Furthermore, if the transformation (1) is $AC\mathsf{B}$ in $\boldsymbol{D}$, then

$$J_{\mathsf{B}}(w, T) = J_e(w, T) \quad \text{a.e. in } \boldsymbol{D}. \tag{8}$$

Proof. Assume that T is $AC\mathsf{B}$ in $\boldsymbol{D}$. Then T is $sAC\mathsf{B}$ in $\boldsymbol{D}$ by theorem 3. Hence T is eAC in $\boldsymbol{D}$ and (8) holds by **IV.5.3**, theorem 1.

Remark. According to the preceding results, the concepts $sBV\mathsf{B}$ and $sAC\mathsf{B}$ are irrelevant in R^2, and the B-*theory* is included in the *e-theory* in R^2.

VI.2.3. On the existence of the weak total differential in $\boldsymbol{R^2}$. We shall use in this section the alternative representations

$$T: z = T(w), \quad w \in \boldsymbol{D}, \tag{1}$$

$$T: x = x(u, v), \quad y = y(u, v), \quad (u, v) \in \boldsymbol{D}, \tag{2}$$

for the bounded continuous transformation $T: \boldsymbol{D} \to R^2$, as explained in **VI.1.1**. We shall establish the fact that if the first partial derivatives of the coordinate functions of T exist a.e. in $\boldsymbol{D}$, then T possesses a weak total differential a.e. in $\boldsymbol{D}$, in the sense of **V.2.2**, definition 2. Simple examples show that the analogous statement for R^n is generally false if $n > 2$.

Given T as in (1) and (2), assume that the first partial derivatives x_u, x_v, y_u, y_v exist a.e. in $\boldsymbol{D}$. Let (u_0, v_0) be a point in $\boldsymbol{D}$ where x_u, x_v, y_u, y_v exist. To simplify notations, we put [for $(u, v) \in \boldsymbol{D}$]

$$X(u,v,u_0,v_0) = x(u,v) - x(u_0,v_0) - (u-u_0)\, x_u(u_0,v_0) - (v-v_0)\, x_v(u_0,v_0), \quad (3)$$

$$Y(u,v,u_0,v_0) = y(u,v) - y(u_0,v_0) - (u-u_0)\, y_u(u_0,v_0) - (v-v_0)\, y_v(u_0,v_0). \quad (4)$$

Lemma 1. Given T as in (1) and (2), assume that the first partial derivatives x_u, x_v, y_u, y_v exist a.e. in $\boldsymbol{D}$. Assign $\varepsilon > 0, \tau > 0$. Then there exists a set $E = E(\varepsilon, \tau) \subset \boldsymbol{D}$ and a number $\sigma = \sigma(\varepsilon, \tau) > 0$ such that the following holds.

(i) E is closed.

(ii) $\mathsf{L}(\boldsymbol{D} - E) < \varepsilon$.

(iii) x_u, x_v, y_u, y_v exist at every point $(u, v) \in E$.

(iv) x_u, x_v, y_u, y_v are continuous on E (relative to E).

(v) If (u_0, v_0), (u, v) are any two points such that (a) $(u_0, v_0) \in E$, (b) $(u, v) \in \boldsymbol{D}$, (c) at least one of the inclusions $(u_0, v) \in E$, $(u, v_0) \in E$ holds, and (d) the distance of (u_0, v_0) and (u, v) does not exceed σ, then

$$|X(u, v, u_0, v_0)| \leq \tau [(u - u_0)^2 + (v - v_0)^2]^{\frac{1}{2}}, \quad (5)$$

$$|Y(u, v, u_0, v_0)| \leq \tau [(u - u_0)^2 + (v - v_0)^2]^{\frac{1}{2}}. \quad (6)$$

(vi) Let Q be any oriented square in $\boldsymbol{D}$, with center $(u_0, v_0) \in E$ and side-length $2h \leq \sigma$, such that the points of intersection of the perimeter of Q with the lines $u = u_0$ and $v = v_0$ lie in E. Then (5) and (6) hold for every point (u, v) on the perimeter of Q.

Proof. The first partial derivatives x_u, x_v, y_u, y_v exist a.e. in $\boldsymbol{D}$ by assumption and are L-measurable in $\boldsymbol{D}$ by **III.1.3**. Accordingly (see **III.1.1**, lemma 22), we can select in $\boldsymbol{D}$ a set E_* such that the following holds.

(α) E_* is closed.

(β) $\mathsf{L}(\boldsymbol{D} - E_*) < \varepsilon/2$.

(γ) x_u, x_v, y_u, y_v exist at every point $(u, v) \in E_*$.

(δ) x_u, x_v, y_u, y_v are continuous on E_* (relative to E_*).

Since E_* is compact, (δ) implies that x_u, x_v, y_u, y_v are uniformly continuous on E_* (relative to E_*). Hence we can select a number

$\sigma_* > 0$ such that

$$|x_u(u'', v'') - x_u(u', v')| \leq \tau/3, \quad (7)$$

$$|x_v(u'', v'') - x_v(u', v')| \leq \tau/3, \quad (8)$$

$$|y_u(u'', v'') - y_u(u', v')| \leq \tau/3, \quad (9)$$

$$|y_v(u'', v'') - y_v(u', v')| \leq \tau/3, \quad (10)$$

whenever

$$(u'', v'') \in E_*, \quad (u', v') \in E_*, \quad [(u'' - u')^2 + (v'' - v')^2]^{\frac{1}{2}} \leq \sigma_*. \quad (11)$$

For each positive integer n, denote by E_*^n the set of those points $(u_*, v_*) \in E_*$ for which the following statements (a), (b) hold simultaneously.

(a) If $(u, v_*) \in \boldsymbol{D}$ and $|u - u_*| \leq 1/n$, then

$$|x(u, v_*) - x(u_*, v_*) - (u - u_*)\, x_u(u_*, v_*)| \leq \frac{\tau}{3} |u - u_*|, \quad (12)$$

$$|y(u, v_*) - y(u_*, v_*) - (u - u_*)\, y_u(u_*, v_*)| \leq \frac{\tau}{3} |u - u_*|. \quad (13)$$

(b) If $(u_*, v) \in \boldsymbol{D}$ and $|v - v_*| \leq 1/n$, then

$$|x(u_*, v) - x(u_*, v_*) - (v - v_*)\, x_v(u_*, v_*)| \leq \frac{\tau}{3} |v - v_*|, \quad (14)$$

$$|y(u_*, v) - y(u_*, v_*) - (v - v_*)\, y_v(u_*, v_*)| \leq \frac{\tau}{3} |v - v_*|. \quad (15)$$

Obviously

$$E_*^n \subset E_*^{n+1}. \quad (16)$$

Furthermore, since x_u, x_v, y_u, y_v exist at every point $(u_*, v_*) \in E_*$, clearly

$$E_* = \cup E_*^n, \quad n = 1, 2, \ldots. \quad (17)$$

Finally, since x, y, x_u, x_v, y_u, y_v are continuous on E_*, and E_* is closed, it is easy to see that E_*^n is closed (and hence L-measurable). Hence (see **III.1.1**, lemma 4) the relations (16) and (17) imply that

$$\mathsf{L}(E_* - E_*^n) \to 0 \quad \text{for} \quad n \to \infty.$$

Thus we can select an integer N such that

$$\mathsf{L}(E_* - E_*^N) < \varepsilon/2. \quad (18)$$

Let us put

$$\sigma = \min\left(\sigma_*, \frac{1}{N}\right), \quad E = E_*^N. \quad (19)$$

We assert that σ and E, as defined in (19), satisfy the conditions (i) to (vi). Note first that since E_*^N is closed, the conditions (i) to (iv) are obviously satisfied in view of (α), (β), (γ), (δ), (18), (19). There remains to verify (v) and (vi). Take any two points (u_0, v_0), (u, v) such that

$$(u_0, v_0) \in E, \quad (u, v) \in \boldsymbol{D}, \quad [(u-u_0)^2 + (v-v_0)^2]^{\frac{1}{2}} \leq \sigma, \tag{20}$$

and assume that one at least of the inclusions $(u_0, v) \in E$, $(u, v_0) \in E$ holds. Suppose, for instance, that

$$(u, v_0) \in E. \tag{21}$$

The quantity $X(u, v, u_0, v_0)$, defined by (3), can be re-written as follows.

$$\left.\begin{aligned} X(u, v, u_0, v_0) = {} & [x(u, v) - x(u, v_0) - (v - v_0)\, x_v(u, v_0)] + \\ & + [x(u, v_0) - x(u_0, v_0) - (u - u_0)\, x_u(u_0, v_0)] + \\ & + (v - v_0)\, [x_v(u, v_0) - x_v(u_0, v_0)]. \end{aligned}\right\} \tag{22}$$

Note that

$$|u - u_0|, \quad |v - v_0| \leq [(u - u_0)^2 + (v - v_0)^2]^{\frac{1}{2}} \leq \sigma \leq \sigma_* \tag{23}$$

by (19) and (20). In view of (21), (19), (17), (20), (23), (11) we can apply (8) with

$$(u'', v'') = (u, v_0), \quad (u', v') = (u_0, v_0).$$

It follows that

$$|(v - v_0)\, [x_v(u, v_0) - x_v(u_0, v_0)]| \leq \frac{\tau}{3} [(u - u_0)^2 + (v - v_0)^2]^{\frac{1}{2}}. \tag{24}$$

Next, observe that

$$(u, v_0) \in E_*^N, \quad (u, v) \in \boldsymbol{D}, \quad |v - v_0| \leq 1/N,$$

by (21), (19), (20), (23). Accordingly, we can apply (14) with $(u_*, v_*) = (u, v_0)$, obtaining the inequality

$$|x(u, v) - x(u, v_0) - (v - v_0)\, x_v(u, v_0)| \leq \frac{\tau}{3} [(u - u_0)^2 + (v - v_0)^2]^{\frac{1}{2}}. \tag{25}$$

Similarly, we have

$$(u_0, v_0) \in E_*^N, \quad (u, v_0) \in \boldsymbol{D}, \quad |u - u_0| \leq 1/N,$$

by (21), (19), (20), (23). Hence we can apply (12) with $(u_*, v_*) = (u_0, v_0)$, obtaining the inequality

$$|x(u, v_0) - x(u_0, v_0) - (u - u_0)\, x_u(u_0, v_0)| \leq \frac{\tau}{3} [(u - u_0)^2 + (v - v_0)^2]^{\frac{1}{2}}. \tag{26}$$

Clearly (5) follows from (22), (24), (25), (26) under the assumption (21). An entirely similar reasoning yields the verification of (6) under the conditions stated in (v).

Consider now any square Q with the properties described under (vi). The lines $u=u_0$ and $v=v_0$ intersect the perimeter of Q in the points

$$(u_0+h, v_0), \quad (u_0-h, v_0), \quad (u_0, v_0+h), \quad (u_0, v_0-h). \tag{27}$$

Let (u, v) be any point on the perimeter of Q. Then clearly

$$[(u-u_0)^2+(v-v_0)^2]^{\frac{1}{2}} \leq 2h \leq \sigma. \tag{28}$$

Furthermore, one at least of the relations $u=u_0 \pm h$, $v=v_0 \pm h$ holds. Thus at least one of the points (u, v_0), (u_0, v) is included among the points (27), and hence at least one of the points (u, v_0), (u_0, v) lies in E. In view of (28) it follows that the points (u, v), (u_0, v_0) satisfy the assumptions made under (v), and hence (5) and (6) hold.

Lemma 2. Let S be a bounded L-measurable set in R^2. Then there exists a decomposition

$$S = s \cup S' \tag{29}$$

such that the following holds. (i) $\mathsf{L}s=0$. (ii) If $(u_0, v_0) \in S'$, then there exists a sequence of oriented squares Q_n, with center at (u_0, v_0) and side-length $2h_n \to 0$, such that the points of intersection of the perimeter of Q_n with the lines $u=u_0$ and $v=v_0$ lie in S.

Proof. By **III.1.1**, lemma 10 there exists a decomposition of S of the form (29) such that (i) holds and every point of S' is a point of linear density of S in the direction of both of the coordinate axes. We assert that (ii) also holds for this decomposition of S. Indeed, let (u_0, v_0) be a point of S'. Denote by H_u^+ the set of those real numbers h for which $(u_0+h, v_0) \in S$, and by H_u^- the set of those real numbers h for which $(u_0-h, v_0) \in S$. Similarly, denote by H_v^+, H_v^- the set of those real numbers h for which $(u_0, v_0+h) \in S$ and $(u_0, v_0-h) \in S$ respectively. Since (u_0, v_0) is a point of linear density of S in the direction of both of the coordinate axes, it is clear that the point $h=0$ of the number-line is a point of density of each one of the sets $H_u^+, H_u^-, H_v^+, H_v^-$. Hence (see **III.1.1**, lemma 11) $h=0$ is also a point of density of the set

$$H = H_u^+ \cap H_u^- \cap H_v^+ \cap H_v^-.$$

Accordingly we can select a sequence of real numbers h_n such $h_n \in H$, $h_n \neq 0$, $h_n \to 0$. Now since the inclusion $h_n \in H$ implies the inclusions $h_n \in H_u^+$, $h_n \in H_u^-$, $h_n \in H_v^+$, $h_n \in H_v^-$, it follows that

$$(u_0 \pm h_n, v_0) \in S, \quad (u_0, v_0 \pm h_n) \in S, \quad n = 1, 2, \ldots. \tag{30}$$

Denote by Q_n the oriented square with center at (u_0, v_0) and side-length $2|h_n|$. Then (30) shows that the points of intersection of the perimeter of Q_n with the lines $u=u_0$ and $v=v_0$ lie in S, and thus the lemma is proved.

Theorem 1. Given T as in (1) and (2), assume that the first partial derivatives x_u, x_v, y_u, y_v exist a.e. in $\boldsymbol{D}$. Then T possesses a weak total differential a.e. in $\boldsymbol{D}$, in the sense of **V.2.2**, definition 2.

Proof. Assign a number $\eta>0$. For each positive integer j select numbers $\varepsilon_j>0$, $\tau_j>0$ such that

$$\varepsilon_j \to 0, \quad \tau_j \to 0, \quad \sum \varepsilon_j < \eta. \tag{31}$$

Denote by E_j, σ_j the set E and the number $\sigma>0$ which correspond to $\varepsilon=\varepsilon_j$, $\tau=\tau_j$ in the sense of lemma 1. Put

$$S = \bigcap E_j, \quad j=1,2,\ldots. \tag{32}$$

Then S, as the intersection of closed sets, is closed and hence is L-measurable. Furthermore, since

$$\boldsymbol{D} - S \subset \bigcup (\boldsymbol{D} - E_j),$$

and $\mathsf{L}(\boldsymbol{D}-E_j) < \varepsilon_j$, there follows in view of (31) the inequality

$$\mathsf{L}(\boldsymbol{D} - S) < \eta. \tag{33}$$

On applying lemma 2 to the set S defined by (32), we obtain a decomposition

$$S = s \cup S', \tag{34}$$

such that

$$\mathsf{L}\, s = 0 \tag{35}$$

and for every point $(u_0, v_0) \in S'$ there exists a sequence of oriented squares $Q_n(u_0, v_0)$ satisfying the following conditions. (a) The point (u_0, v_0) is the center of $Q_n(u_0, v_0)$. (b) The side-length $2h_n(u_0, v_0)$ of $Q_n(u_0, v_0)$ converges to zero. (c) The points of intersection of the perimeter of $Q_n(u_0, v_0)$ with the lines $u=u_0$ and $v=v_0$ lie in S.

Consider a point $(u_0, v_0) \in S'$. Keeping (u_0, v_0) fixed, select an integer $j>0$. In view of (32) it follows that $(u_0, v_0) \in E_j$ and that the points of intersection of the perimeter of $Q_n(u_0, v_0)$ with the lines $u=u_0$ and $v=v_0$ lie in E_j. Since $h_n(u_0, v_0) \to 0$, we shall have $2h_n(u_0, v_0) \leq \sigma_j$ for n sufficiently large, say for $n > N(u_0, v_0, j)$. By part (vi) of lemma 1 it follows that

$$|X(u, v, u_0, v_0)| \leq \tau_j [(u-u_0)^2 + (v-v_0)^2]^{\frac{1}{2}}, \tag{36}$$

$$|Y(u, v, u_0, v_0)| \leq \tau_j [(u-u_0)^2 + (v-v_0)^2]^{\frac{1}{2}}, \tag{37}$$

provided that

$$(u, v) \in fr\, Q_n(u_0, v_0), \quad n > N(u_0, v_0, j). \tag{38}$$

In view of (3) and (4) the quantity $\gamma(h_n(u_0, v_0))$, defined in **V.2.2** (14), is now given by the formula

$$\left.\begin{aligned} \gamma(h_n(u_0, v_0)) &= \max \frac{[X(u, v, u_0, v_0)^2 + Y(u, v, u_0, v_0)^2]^{\frac{1}{2}}}{[(u-u_0)^2 + (v-v_0)^2]^{\frac{1}{2}}}, \\ &(u, v) \in fr\, Q_n(u_0, v_0). \end{aligned}\right\} \tag{39}$$

From (36) to (39) we conclude that

$$\gamma(h_n(u_0, v_0)) \leq \sqrt{2}\,\tau_j \quad \text{for} \quad n > N(u_0, v_0, j).$$

For $n \to \infty$ there follows the inequality

$$\limsup_{n\to\infty} \gamma(h_n(u_0, v_0)) \leq \sqrt{2}\,\tau_j.$$

Since the positive integer j was arbitrary and $\tau_j \to 0$, $\gamma(h_n(u_0, v_0)) \geq 0$, it follows finally that

$$\lim_{n\to\infty} \gamma(h_n(u_0, v_0)) = 0.$$

Thus it is established (see **V.2.2**, definition 2) that T possesses a weak total differential at every point $(u_0, v_0) \in S'$. Since $\mathsf{L}(\boldsymbol{D} - S') < \eta$ by (33), (34), (35) and $\eta > 0$ was arbitrary, we conclude that T possesses a weak total differential a.e. in $\boldsymbol{D}$.

In applying the general theorems derived in § **V.3** to differentiable transformations in R^2 we can replace, in view of the preceding theorem 1, the assumption that T possesses a weak total differential a.e. in $\boldsymbol{D}$ by the assumption that the first partial derivatives of the coordinate functions of T exist a.e. in $\boldsymbol{D}$. From **V.3.2**, theorem 1; **V.3.3**, theorem 6; **V.3.4**, theorem 1; **V.3.4**, theorem 2; **V.3.4**, theorem 3 we obtain in this manner the following series of statements for bounded continuous transformations in R^2 (see **VI.1.1** for the special notations adopted in R^2).

Theorem 2. Given T as in (1) and (2), assume that T is eBV in $\boldsymbol{D}$ and the first partial derivatives x_u, x_v, y_u, y_v exist a.e. in $\boldsymbol{D}$. Then the essential generalized Jacobian $J_e(w, T)$ is equal to the ordinary Jacobian $J(w, T)$ a.e. in $\boldsymbol{D}$.

Theorem 3. Given T as in (1) and (2), assume that the first partial derivatives x_u, x_v, y_u, y_v exist a.e. in $\boldsymbol{D}$, the ordinary Jacobian $J(w, T)$ is L-summable in $\boldsymbol{D}$, and T satisfies the condition (N) in $\boldsymbol{D}$. Then the following holds.

(i) T is both $sACB$ and eAC in $\boldsymbol{D}$.
(ii) $J_B(w,T) = J_e(w,T) = J(w,T)$ a.e. in $\boldsymbol{D}$.
(iii) $D_B(w,T) = D_e(w,T) = |J(w,T)|$ a.e. in $\boldsymbol{D}$.
(iv) $N(z,T,\boldsymbol{D}) = K(z,T,\boldsymbol{D})$ a.e. in R^2.

Theorem 4. Given T as in (1) and (2), assume that T is eAC in $\boldsymbol{D}$ and the first partial derivatives x_u, x_v, y_u, y_v exist a.e. in $\boldsymbol{D}$. Then

$$\int_{\boldsymbol{D}} |J(w,T)|\, du\, dv = \int K(z,T,\boldsymbol{D})\, dx\, dy,$$

$$\int_{\boldsymbol{D}} J(w,T)\, du\, dv = \int \mu_e(z,T,\boldsymbol{D})\, dx\, dy.$$

More generally, if $H(z)$ is a finite-valued, L-measurable function in R^2, then

$$\int_{\boldsymbol{D}} H(Tw)\, |J(w,T)|\, du\, dv = \int H(z)\, K(z,T,\boldsymbol{D})\, dx\, dy,$$

as soon as one of the two integrals involved exists, and

$$\int_{\boldsymbol{D}} H(Tw)\, J(w,T)\, du\, dv = \int H(z)\, \mu_e(z,T,\boldsymbol{D})\, dx\, dy,$$

as soon as the integral on the left exists.

Theorem 5. Given T as in (1) and (2), assume that (a) T is defined and continuous in $\bar{\boldsymbol{D}}$, (b) $LT\, fr\, \boldsymbol{D} = 0$, (c) x_u, x_v, y_u, y_v exist a.e. in $\boldsymbol{D}$, and (d) T is eAC in $\boldsymbol{D}$. Then

$$\int_{\boldsymbol{D}} J(w,T)\, du\, dv = \int \mu(z,T,\boldsymbol{D})\, dx\, dy.$$

More generally, if $H(z)$ is a finite-valued, L-measurable function in R^2, then

$$\int_{\boldsymbol{D}} H(Tw)\, J(w,T)\, du\, dv = \int H(z)\, \mu(z,T,\boldsymbol{D})\, dx\, dy,$$

as soon as the integral on the left exists.

Theorem 6. Given T as in (1) and (2), assume that the first partial derivatives x_u, x_v, y_u, y_v exist a.e. in $\boldsymbol{D}$, the ordinary Jacobian $J(w,T)$ is L-summable in $\boldsymbol{D}$, and T satisfies the condition (N) in $\boldsymbol{D}$. Then

$$\int_{\boldsymbol{D}} |J(w,T)|\, du\, dv = \int N(z,T,\boldsymbol{D})\, dx\, dy.$$

More generally, if S is an L-measurable subset of $\boldsymbol{D}$ and $H(z)$ is a finite-valued, L-measurable function in R^2, then

$$\int_{S} H(Tw)\, |J(w,T)|\, du\, dv = \int H(z)\, N(z,T,S)\, dx\, dy,$$

as soon as one of the two integrals involved exists. In particular, corresponding to the special case $H(z) \equiv 1$,

$$\int_S |J(w, T)| \, du \, dv = \int N(z, T, S) \, dx \, dy.$$

The *closure theorems* derived in **V.3.5** can be improved in a similar manner in the special case of continuous transformations in R^2. We state explicitly only the result obtained by combining theorem 1 with **V.3.5**, theorem 3.

Theorem 7. Let there be given bounded continuous transformations [see **VI.1.1** (2)]

$$T: z = T(w), \quad w \in \boldsymbol{D},$$

$$T_j: z = T_j(w), \quad w \in \boldsymbol{D}_j, \quad j = 1, 2, \ldots,$$

such that the following holds.

(i) $\boldsymbol{D}, \boldsymbol{D}_j$ are bounded domains in R^2.

(ii) The sequence $\{T_j\}$ converges to T uniformly on compact subsets of $\boldsymbol{D}$.

(iii) The first partial derivatives of the coordinate functions of T exist a.e. in $\boldsymbol{D}$, and the ordinary Jacobian $J(w, T)$ is L-summable in $\boldsymbol{D}$.

(iv) T_j is eAC in $\boldsymbol{D}_j$, and the first partial derivatives of the coordinate functions of T_j exist a.e. in $\boldsymbol{D}_j$, $j = 1, 2, \ldots$.

(v) If F is any compact subset of $\boldsymbol{D}$, then

$$J(w, T_j) \to J(w, T) \quad \text{a.e. on } F.$$

(vi) If Q is any oriented square in $\boldsymbol{D}$, then the sequence $\{J(w, T_j)\}$ satisfies the condition (V) in Q (see **III.1.1**, definition 6).

Then T is eAC in $\boldsymbol{D}$.

§ VI.3. Special classes of differentiable transformations in $\boldsymbol{R}^2$.

VI.3.1. Some elementary inequalities. We begin with a review of various facts relating to real-valued functions of two real variables. Many of the results discussed in the present § **VI.3** remain valid for functions of an arbitrary number of variables.

Our objective in this section is to prove some inequalities needed later on. Free use will be made of familiar devices in elementary Differential and Integral Calculus.

Lemma 1. Let $g(u, v)$ denote a real-valued continuous function in a bounded domain D in R^2, such that

$$g(u, v) \geq 1 \quad \text{for} \quad (u, v) \in D. \tag{1}$$

Take a square $Q \subset D$ (it is not required that the sides of Q should be parallel to the coordinate axes). Select a point $(u_0, v_0) \in Q$, and put

$$\varrho = [(u - u_0)^2 + (v - v_0)^2]^{\frac{1}{2}}, \quad (u, v) \in R^2.$$

Let p be a real number such that

$$p > 2. \tag{2}$$

Then there exists a (finite) positive constant $\Gamma_1(p)$, which depends solely upon p, such that

$$\iint\limits_Q \frac{g(u, v)}{\varrho} \, du \, dv \leq \Gamma_1(p) \left[\iint\limits_Q g(u, v)^p \, du \, dv\right]^{\frac{1}{p}}. \tag{3}$$

Proof. Let h be the side-length of Q. Denote by K the circular disc with center at (u_0, v_0) and radius $\sqrt{2}h$. Then clearly

$$Q \subset K. \tag{4}$$

Introduce polar coordinates ϱ, α with pole at (u_0, v_0). Define the real number q by the equation

$$\frac{1}{p} + \frac{1}{q} = 1. \tag{5}$$

From (2) and (5) it follows that

$$2 - q = \frac{p - 2}{p - 1} > 0. \tag{6}$$

Using (6) and (4), we obtain the relations

$$\iint\limits_Q \frac{du \, dv}{\varrho^q} < \iint\limits_K \frac{du \, dv}{\varrho^q} = \int\limits_0^{2\pi} \int\limits_0^{\sqrt{2}h} \varrho^{1-q} \, d\varrho \, d\alpha = C_1(p) \, h^{2-q}, \tag{7}$$

where

$$C_1(p) = \frac{\pi \, 2^{\frac{4-q}{2}}}{2 - q}. \tag{8}$$

The inequality of HÖLDER (see **III.1.1**, lemma 29) yields, in view of (5) and (7),

$$\left.\begin{aligned} \iint\limits_Q \frac{g(u, v)}{\varrho} \, du \, dv &\leq \left[\iint\limits_Q g(u, v)^p \, du \, dv\right]^{1/p} \left[\iint\limits_Q \frac{du \, dv}{\varrho^q}\right]^{1/q} \leq \\ &\leq \Gamma_1(p) \, h^{\frac{2-q}{q}} \left[\iint\limits_Q g(u, v)^p \, du \, dv\right]^{1/p}, \end{aligned}\right\} \tag{9}$$

where

$$\Gamma_1(p) = C_1(p)^{1/q}. \tag{10}$$

From (1) it follows that

$$\frac{1}{h^2}\iint\limits_Q g(u, v)^p\, du\, dv \geq 1,$$

and hence, since $p > 2$,

$$\left[\frac{1}{h^2}\iint\limits_Q g(u,v)^p\, du\, dv\right]^{1/p} \leq \left[\frac{1}{h^2}\iint\limits_Q g(u, v)^p\, du\, dv\right]^{\frac{1}{2}}. \tag{11}$$

From (11) we conclude that

$$\left[\iint\limits_Q g(u, v)^p\, du\, dv\right]^{1/p} \leq \left[\iint\limits_Q g(u, v)^p\, du\, dv\right]^{\frac{1}{2}} h^{\frac{2}{p}-1}. \tag{12}$$

The inequalities (9) and (12) imply that

$$\iint\limits_Q \frac{g(u, v)}{\varrho}\, du\, dv \leq \Gamma_1(p)\, h^{\frac{2-q}{q}+\frac{2}{p}-1}\left[\iint\limits_Q g(u, v)^p\, du\, dv\right]^{\frac{1}{2}},$$

and (3) is established, since [in view of (5)]

$$\frac{2-q}{q} + \frac{2}{p} - 1 = 0.$$

Lemma 2. Let $f(u, v)$ be a real-valued continuous function in a bounded domain D in R^2, such that the first partial derivatives f_u, f_v of f exist and are continuous in D. Select a square $Q < D$ (it is not required that the sides of Q should be parallel to the coordinate axes), and denote by P_1, P_2 a pair of opposite vertices of Q. Let a real number $p > 2$ be assigned. Then there exists a (finite) positive constant $\Gamma_2(p)$, which depends solely upon p, such that

$$\left.\begin{aligned}&|f(P_2) - f(P_1)| \leq \Gamma_2(p) \times\\ &\times\left[\iint\limits_Q (1+|f_u|+|f_v|)^p\, du\, dv\right]^{\frac{1}{2}}.\end{aligned}\right\} \tag{13}$$

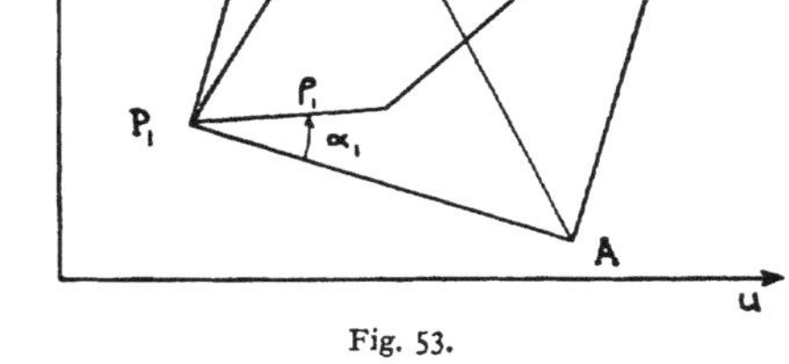

Fig. 53.

Proof. Denote by A, B the remaining two vertices of Q (see the figure). Introduce a first system of polar coordinates ϱ_1, α_1 with pole at P_1, and a second system of polar coordinates ϱ_2, α_2 with pole at P_2, where the polar angles α_1, α_2 are measured in the manner indicated in the figure. Denote by Δ_1, Δ_2 the triangles with the vertices P_1, A, B and P_2, A, B respectively. Applying lemma 1 with

$$g = 1 + |f_u| + |f_v|, \quad (u_0, v_0) = P_1$$

we obtain the relations

$$\iint_{\Delta_1} (1 + |f_u| + |f_v|)\, d\varrho_1\, d\alpha_1 = \iint_{\Delta_1} \frac{1 + |f_u| + |f_v|}{\varrho_1}\, du\, dv \leq$$
$$\leq \iint_{Q} \frac{1 + |f_u| + |f_v|}{\varrho_1}\, du\, dv \leq \Gamma_1(p) \left[\iint_{Q} (1 + |f_u| + |f_v|)^p\, du\, dv\right]^{\frac{1}{p}}.$$

The same argument applies with P_1 replaced by P_2. Accordingly, we have the inequalities

$$\iint_{\Delta_1} (1 + |f_u| + |f_v|)\, d\varrho_1\, d\alpha_1 \leq \Gamma_1(p) \left[\iint_{Q} (1 + |f_u| + |f_v|)^p\, du\, dv\right]^{\frac{1}{p}}, \tag{14}$$
$$\iint_{\Delta_2} (1 + |f_u| + |f_v|)\, d\varrho_2\, d\alpha_2 \leq \Gamma_1(p) \left[\iint_{Q} (1 + |f_u| + |f_v|)^p\, du\, dv\right]^{\frac{1}{p}}. \tag{15}$$

Denote by P a variable point on the diagonal AB of Q. Then

$$|f(P_2) - f(P_1)| \leq |f(P) - f(P_2)| + |f(P) - f(P_1)|. \tag{16}$$

The polar angles α_1, α_2 corresponding to P are clearly equal to each other. On denoting by α their common value, the absolute values on the right-hand side in (16) are continuous functions of α for $0 \leq \alpha \leq \pi/2$, while the absolute value on the left hand side in (16) is independent of α. Hence integration of (16), with respect to $\alpha = \alpha_1 = \alpha_2$ from 0 to $\pi/2$ yields the inequality

$$\frac{\pi}{2} |f(P_2) - f(P_1)| \leq \int_0^{\pi/2} |f(P) - f(P_2)|\, d\alpha_2 + \int_0^{\pi/2} |f(P) - f(P_1)|\, d\alpha_1. \tag{17}$$

For fixed α_1 we have

$$f(P) - f(P_1) = \int_{P_1 P} \frac{df}{d\varrho_1}\, d\varrho_1. \tag{18}$$

Now since

$$\frac{df}{d\varrho_1} = f_u \frac{du}{d\varrho_1} + f_v \frac{dv}{d\varrho_1},$$

and

$$\left|\frac{du}{d\varrho_1}\right| \leq 1, \quad \left|\frac{dv}{d\varrho_1}\right| \leq 1,$$

we have the inequality

$$\left|\frac{df}{d\varrho_1}\right| \leq |f_u| + |f_v|.$$

Hence (18) yields

$$|f(P) - f(P_1)| \leq \int_{P_1 P} (|f_u| + |f_v|)\, d\varrho_1 < \int_{P_1 P} (1 + |f_u| + |f_v|)\, d\varrho_1,$$

and consequently

$$\left.\begin{aligned} \int_0^{\pi/2} |f(P) - f(P_1)|\, d\alpha_1 &\leq \int_0^{\pi/2} \int_{P_1 P} (1 + |f_u| + |f_v|)\, d\varrho_1\, d\alpha_1 \\ &= \iint_{\Delta_1} (1 + |f_u| + |f_v|)\, d\varrho_1\, d\alpha_1. \end{aligned}\right\} \qquad (19)$$

An entirely similar argument yields the inequality

$$\int_0^{\pi/2} |f(P) - f(P_2)|\, d\alpha_2 \leq \iint_{\Delta_2} (1 + |f_u| + |f_v|)\, d\varrho_2\, d\alpha_2. \qquad (20)$$

From (17), (19), (20), (14), (15) we conclude that

$$|f(P_2) - f(P_1)| \leq \frac{4}{\pi} \Gamma_1(p) \left(\iint_Q (1 + |f_u| + |f_v|)^p\, du\, dv\right)^{\frac{1}{p}}.$$

Thus (13) holds with

$$\Gamma_2(p) = \frac{4\,\Gamma_1(p)}{\pi},$$

and the lemma is proved.

Lemma 3. Let $f(u, v)$ be a real-valued, continuous function in a bounded domain $D \subset R^2$, such that the first partial derivatives f_u, f_v of f exist and are continuous in D. Let Q be a square in D, and let $\omega(f, Q)$ be the oscillation of f in Q (see **I.1.2**). Let a real number $p > 2$ be assigned. Then there exists a (finite) positive constant $\Gamma(p)$, which depends solely upon p, such that

$$\omega(f, Q) \leq \Gamma(p) \left[\iint_Q (1 + |f_u| + |f_v|)^p\, du\, dv\right]^{\frac{1}{p}}. \qquad (21)$$

Proof. Denote by P_0 the center of Q. Select any two points P', P'' in Q (if the points P_0, P', P'' are not all distinct, then the following argument becomes simpler in an obvious manner). Denote by Q' the square for which P_0 and P' are opposite vertices. Obviously $Q' \subset Q$. Applying lemma 2 with Q, P_1, P_2 replaced by Q', P_0, P', there follows the inequality

$$|f(P') - f(P_0)| \leq \Gamma_2(p) \left[\iint_{Q'} (1 + |f_u| + |f_v|)^p\, du\, dv\right]^{\frac{1}{p}}.$$

Since $Q' \subset Q$, it follows further that

$$|f(P') - f(P_0)| \leq \Gamma_2(p) \left[\iint_Q (1 + |f_u| + |f_v|)^p\, du\, dv\right]^{\frac{1}{p}}. \qquad (22)$$

An entirely similar argument yields the inequality

$$|f(P'') - f(P_0)| \leq \Gamma_2(p) \left[\iint\limits_Q (1 + |f_u| + |f_v|)^p \, du \, dv\right]^{\frac{1}{p}}. \tag{23}$$

Now since

$$|f(P'') - f(P')| \leq |f(P'') - f(P_0)| + |f(P') - f(P_0)|,$$

we obtain from (22) and (23) the inequality

$$|f(P'') - f(P')| \leq 2\Gamma_2(p) \left[\iint\limits_Q (1 + |f_u| + |f_v|)^p \, du \, dv\right]^{\frac{1}{p}}. \tag{24}$$

Since P', P'' were arbitrary points in Q, (24) implies that (21) holds with $\Gamma(p) = 2\Gamma_2(p)$.

VI.3.2. Approximation by integral means. It frequently happens that a certain statement relating to real-valued functions can be proved readily by elementary methods if the functions involved are sufficiently smooth. The proof of the same statement under more general conditions may then be obtained, in certain cases, by appropriate approximations. The method of *approximation by integral means*, to be discussed briefly in the present section, is often useful in this connection. Let $f(u, v)$ be a real-valued, L-measurable function in a bounded domain $\boldsymbol{D} \subset R^2$. Assume that $f(u, v)$ is L-summable in every oriented rectangle in $\boldsymbol{D}$. Select a domain D such that $\bar{D} \subset \boldsymbol{D}$. Then there exists a number $\delta > 0$, which depends upon D, such that

$$(u, v) \in D \quad \text{implies} \quad \bar{\Delta}(u, v, \delta) \subset \boldsymbol{D}, \tag{1}$$

where $\bar{\Delta}(u, v, \delta)$ is the circular disc (closed spherical neighborhood) with center (u, v) and radius δ. Let h be a real number such that

$$0 < h < \delta/\sqrt{2}. \tag{2}$$

If $Q(u, v, h)$ denotes the oriented square with center (u, v) and side-length $2h$, then (1) implies that

$$Q(u, v, h) \subset \boldsymbol{D} \quad \text{if} \quad (u, v) \in D, \quad 0 < h < \delta/\sqrt{2}. \tag{3}$$

Accordingly, $f(u, v)$ is L-summable in $Q(u, v, h)$ if $(u, v) \in D$ and (2) holds. We can therefore define a function $f_h(u, v)$ in D by the formula

$$f_h(u, v) = \frac{1}{4h^2} \int\limits_{-h}^{h} \int\limits_{-h}^{h} f(u + \alpha, v + \beta) \, d\alpha \, d\beta, \quad (u, v) \in D, \quad 0 < h < \delta/\sqrt{2}. \tag{4}$$

Thus $f_h(u, v)$ is equal to the integral mean of f on $Q(u, v, h)$.

Lemma 1. For fixed h satisfying (2), $f_h(u, v)$ is continuous in D.

Proof. Select a point $(u_0, v_0) \in D$. Let $\{(u_n, v_n)\}$ be any sequence of points in D such that $(u_n, v_n) \to (u_0, v_0)$. Put

$$Q_0 = Q(u_0, v_0, h), \quad Q_n = Q(u_n, v_n, h).$$

Since $Q_0 \subset \boldsymbol{D}$, we can select an oriented square Q_0^* with center at (u_0, v_0) such that $Q_0 \subset \operatorname{int} Q_0^*$, $Q_0^* \subset \boldsymbol{D}$. Since $(u_n, v_n) \to (u_0, v_0)$, we can choose a positive integer N such that $Q_n \subset Q_0^*$ for $n > N$. In view of (4) we have the inequality

$$|f_h(u_0, v_0) - f_h(u_n, v_n)| \leq \frac{1}{4h^2}\left[\iint_{Q_n - Q_0} |f(u, v)|\, du\, dv + \iint_{Q_0 - Q_n} |f(u, v)|\, du\, dv\right]. \quad (5)$$

Now clearly

$$\lim_{n\to\infty} \mathsf{L}(Q_n - Q_0) = 0, \quad \lim_{n\to\infty} \mathsf{L}(Q_0 - Q_n) = 0. \quad (6)$$

Since f is L-summable in Q_0^*, and $Q_n - Q_0$, $Q_0 - Q_n$ are L-measurable subsets of Q_0^* for $n < N$, the relations (6) imply (see **III.1.1**, lemma 33) that the integrals occurring in (5) converge to zero for $n \to \infty$. Thus (5) yields (since h is fixed)

$$\lim_{n\to\infty} |f_h(u_0, v_0) - f_h(u_n, v_n)| = 0,$$

and the lemma is proved.

Lemma 2. If $\{h_n\}$ is any sequence of real numbers such that

$$0 < h_n < \delta/\sqrt{2}, \quad h_n \to 0, \quad (7)$$

then $f_{h_n}(u, v) \to f(u, v)$ a.e. in D.

Proof. Since $\bar{D} \subset \boldsymbol{D}$, it follows that f is L-summable in $\bar{D}$. Indeed, since $\bar{D}$ is compact, we can select a finite sequence $R_1, \ldots, R_m$ of oriented rectangles in $\boldsymbol{D}$ such that $\bar{D} \subset \bigcup R_j$, $j = 1, \ldots, m$. Since f is L-summable by assumption in R_j, $j = 1, \ldots, m$, it follows that f is L-summable in $\bar{D}$, and hence also in D. Accordingly (see **III.1.1**, lemma 43) there exists a decomposition $D = s \cup S$ such that $\mathsf{L}s = 0$ and the following holds for every point $(u_0, v_0) \in S$. If $\{G_n\}$ is any regular sequence of closed sets in D such that $(u_0, v_0) \in G_n$ and the diameter of G_n converges to zero, then

$$\lim_{n\to\infty} \frac{1}{\mathsf{L}G_n} \iint_{G_n} f(u, v)\, du\, dv = f(u_0, v_0).$$

On selecting G_n as the oriented square with center (u_0, v_0) and side-length $2h_n$, in view of (4) it follows that

$$\lim_{n\to\infty} f_{h_n}(u_0, v_0) = f(u_0, v_0) \quad \text{for} \quad (u_0, v_0) \in S,$$

and the lemma is proved.

Lemma 3. Assume that $|f|^p$ is L-summable in every oriented rectangle in $\boldsymbol{D}$, where $p \geq 1$. Then the family of functions

$$|f_h|^p, \quad 0 < h < \delta/\sqrt{2},$$

satisfies the condition (V) in every oriented rectangle in D (see **III.1.1**, definition 6).

Proof. Select an oriented rectangle $R \subset D$. Clearly we can choose an oriented rectangle R_* such that $R \subset \operatorname{int} R_*$, $R_* \subset \boldsymbol{D}$, and such that

$$(u, v) \in R, \quad |\alpha| < \delta/\sqrt{2}, \quad |\beta| < \delta/\sqrt{2} \quad \text{imply} \quad (u+\alpha, v+\beta) \in R_*. \tag{8}$$

Assign now $\varepsilon > 0$. By assumption, $|f|^p$ is L-summable in R_*. Hence (see **III.1.1**, lemma 33) there exists a number $\eta > 0$ such that

$$\iint_{S_*} |f(u, v)|^p \, du\, dv < \varepsilon, \tag{9}$$

whenever S_* is an L-measurable subset of R_* such that

$$\mathsf{L}\, S_* < \eta. \tag{10}$$

Select any L-measurable set S such that

$$S \subset R, \quad \mathsf{L}\, S < \eta. \tag{11}$$

Let α, β be any two real numbers such that

$$|\alpha| < \delta/\sqrt{2}, \quad |\beta| < \delta/\sqrt{2}, \tag{12}$$

and denote by $S_{\alpha\beta}$ the set obtained by subjecting S to the translation by the vector with components α, β. Then clearly $S_{\alpha\beta}$ is an L-measurable set which satisfies [see (8), (11), (12)] the conditions

$$\mathsf{L}\, S_{\alpha\beta} < \eta, \quad S_{\alpha\beta} \subset R_*. \tag{13}$$

From (9) to (13) we conclude that

$$\iint_{S_{\alpha\beta}} |f(u, v)|^p \, du\, dv < \varepsilon. \tag{14}$$

It is convenient to divide the rest of the proof into two parts.

Part 1. Assume that $p>1$. Then using (4) and the inequality of HÖLDER, it follows that

$$|f_h(u,v)| \leq \frac{1}{4h^2} \int_{-h}^{h}\int_{-h}^{h} |f(u+\alpha, v+\beta)|\, d\alpha\, d\beta \leq$$

$$\leq \frac{1}{4h^2}\left[\int_{-h}^{h}\int_{-h}^{h} |f(u+\alpha, v+\beta)|^p\, d\alpha\, d\beta\right]^{1/p}\left[\int_{-h}^{h}\int_{-h}^{h} 1^q\, d\alpha\, d\beta\right]^{1/q}$$

$$= \left[\int_{-h}^{h}\int_{-h}^{h} |f(u+\alpha, v+\beta)|^p\, d\alpha\, d\beta\right]^{1/p} (4h^2)^{\frac{1}{q}-1},$$

where q satisfies the equation

$$\frac{1}{p} + \frac{1}{q} = 1.$$

Thus it follows that

$$|f_h(u,v)|^p \leq \frac{1}{4h^2} \int_{-h}^{h}\int_{-h}^{h} |f(u+\alpha, v+\beta)|^p\, d\alpha\, d\beta.$$

In view of the FUBINI-TONELLI theorem (see the Saks treatise listed in the *Bibliography*) we conclude that

$$\iint_S |f_h(u,v)|^p\, du\, dv \leq \frac{1}{4h^2} \int_{-h}^{h}\int_{-h}^{h}\left[\iint_S |f(u+\alpha, v+\beta)|^p\, du\, dv\right] d\alpha\, d\beta$$

$$= \frac{1}{4h^2} \int_{-h}^{h}\int_{-h}^{h}\left[\iint_{S_{\alpha\beta}} |f(u,v)|^p\, du\, dv\right] d\alpha\, d\beta.$$

Using (14) it follows finally that

$$\iint_S |f_h(u,v)|^p\, du\, dv < \varepsilon. \tag{15}$$

Thus it is established that (11) implies (15), and the lemma is proved for the case $p>1$.

Part 2. Assume that $p=1$. From (4) it follows that

$$\iint_S |f_h(u,v)|\, du\, dv \leq \iint_S \left[\frac{1}{4h^2}\int_{-h}^{h}\int_{-h}^{h} |f(u+\alpha, v+\beta)|\, d\alpha\, d\beta\right] du\, dv$$

$$= \frac{1}{4h^2}\int_{-h}^{h}\int_{-h}^{h}\left[\iint_S |f(u+\alpha, v+\beta)|\, du\, dv\right] d\alpha\, d\beta$$

$$= \frac{1}{4h^2}\int_{-h}^{h}\int_{-h}^{h}\left[\iint_{S_{\alpha\beta}} |f(u,v)|\, du\, dv\right] d\alpha\, d\beta.$$

In view of (14), applied with $p=1$, there follows the inequality

$$\iint_S |f_h(u, v)|\, du\, dv < \varepsilon \quad \text{for} \quad S \subset Q, \quad \mathsf{L} S < \eta,$$

and the lemma is proved.

Returning to the function $f_h(u, v)$ given by (4), let us note that this function possesses various useful properties beyond those explicitly stated in the preceding lemmas. We shall discuss only a few additional properties which will be needed later on. The following preliminary comments will simplify the discussion. Let a, b be two real numbers such that $a<b$, and let $\varphi(u)$ be a real-valued continuous function in the open interval $a-h<u<b+h$ of the real number-line, where $h>0$. Then we can define, in analogy with (4), a function $\varphi_h(u)$ by the formula

$$\varphi_h(u) = \frac{1}{2h} \int_{-h}^{h} \varphi(u+\alpha)\, d\alpha, \quad a<u<b. \tag{16}$$

We shall need only the fact that the derivative $\varphi_h'(u)$ of $\varphi_h(u)$ exists for $a<u<b$ and is given by the formula

$$\varphi_h'(u) = \frac{\varphi(u+h) - \varphi(u-h)}{2h}, \quad a<u<b. \tag{17}$$

To verify (17), take a point u_0 such that $a<u_0<b$, and consider the difference quotient

$$\frac{\varphi_h(u_0+\Delta u) - \varphi_h(u_0)}{\Delta u}.$$

If $|\Delta u|$ is sufficiently small, then this quotient is defined, and from (16) there follows the formula

$$\frac{\varphi_h(u_0+\Delta u)-\varphi_h(u_0)}{\Delta u} = \frac{1}{2h}\left[\frac{1}{\Delta u}\int_{u_0+h}^{u_0+h+\Delta u} \varphi(u)\, du - \frac{1}{\Delta u}\int_{u_0-h}^{u_0-h+\Delta u} \varphi(u)\, du\right]. \tag{18}$$

Since $\varphi(u)$ is continuous, clearly

$$\lim_{\Delta u \to 0} \frac{1}{\Delta u} \int_{u_0+h}^{u_0+h+\Delta u} \varphi(u)\, du = \varphi(u_0+h), \tag{19}$$

$$\lim_{\Delta u \to 0} \frac{1}{\Delta u} \int_{u_0-h}^{u_0-h+\Delta u} \varphi(u)\, du = \varphi(u_0-h). \tag{20}$$

From (18), (19), (20) it follows that $\varphi_h'(u_0)$ exists and is given by the formula

$$\varphi_h'(u_0) = \frac{\varphi(u_0+h)-\varphi(u_0-h)}{2h}.$$

Thus (17) is proved.

Lemma 4. Assume that $f(u, v)$ is continuous in $\boldsymbol{D}$. Then the function $f_h(u, v)$ given by (4) has continuous first partial derivatives

$$f_{hu} = \frac{\partial f_h}{\partial u}, \quad f_{hv} = \frac{\partial f_h}{\partial v}$$

in D, and

$$f_{hu}(u, v) = \frac{1}{4h^2}\int_{-h}^{h} [f(u+h, v+\beta) - f(u-h, v+\beta)]\, d\beta, \tag{21}$$

$$f_{hv}(u, v) = \frac{1}{4h^2}\int_{-h}^{h} [f(u+\alpha, v+h) - f(u+\alpha, v-h)]\, d\alpha. \tag{22}$$

Proof. Take a point $(u_0, v_0) \in D$. Consider the auxiliary function

$$\varphi(u) = \frac{1}{2h}\int_{-h}^{h} f(u, v_0+\beta)\, d\beta. \tag{23}$$

If a, b are two numbers, sufficiently close to u_0, such that $a < u_0 < b$, then clearly $\varphi(u)$ is defined and continuous for $a-h < u < b+h$ (recall that f is continuous by assumption). Accordingly, on defining $\varphi_h(u)$ by (16), the formula (17) will hold. From (16), (23), and (4) it follows that $\varphi_h(u) = f_h(u, v_0)$, and hence, by (17) and (23),

$$\begin{aligned} f_{hu}(u_0, v_0) &= \varphi_h'(u_0) = \frac{\varphi(u_0+h)-\varphi(u_0-h)}{2h} \\ &= \frac{1}{4h^2}\left[\int_{-h}^{h} f(u_0+h, v_0+\beta)\, d\beta - \int_{-h}^{h} f(u_0-h, v_0+\beta)\, d\beta\right]. \end{aligned}$$

Thus (21) is proved. The formula (22) is verified in an entirely similar manner. Since f is continuous, clearly (21) and (22) imply the continuity of f_{hu}, f_{hv} in D.

Lemma 5. Assume that $f(u, v)$ is continuous in $\boldsymbol{D}$. Let D be a domain such that $\overline{D} \subset \boldsymbol{D}$, and consider the function $f_h(u, v)$ given by (4). Assign $\varepsilon > 0$. Then there exists a number $\eta > 0$ such that

$$|f_h(u, v) - f(u, v)| < \varepsilon \quad \text{if} \quad (u, v) \in D, \quad 0 < h < \min\left(\eta, \frac{\delta}{\sqrt{2}}\right). \tag{24}$$

Proof. Since $\bar{D}$ is compact and f is continuous in $\boldsymbol{D}$, we have a number $\eta > 0$ such that

$$|f(u+\alpha, v+\beta) - f(u,v)| < \varepsilon \quad \text{if} \quad (u,v) \in \bar{D}, \quad |\alpha| < \eta, \quad |\beta| < \eta. \tag{25}$$

Let (u, v) be any point in D, and let h satisfy the inequalities

$$0 < h < \min\left(\eta, \frac{\delta}{\sqrt{2}}\right).$$

From (4) and (25) it follows that

$$|f_h(u,v) - f(u,v)| = \left| \frac{1}{4h^2} \int_{-h}^{h} \int_{-h}^{h} [f(u+\alpha, v+\beta) - f(u,v)]\, d\alpha\, d\beta \right| <$$

$$< \frac{1}{4h^2} \int_{-h}^{h} \int_{-h}^{h} \varepsilon\, d\alpha\, d\beta = \varepsilon,$$

and the lemma is proved.

VI.3.3. Bounded variation and absolute continuity in the TONELLI sense. Let $f(u, v)$ be a real-valued continuous function in an oriented rectangle

$$R: a \leq u \leq b, \quad c \leq v \leq d, \tag{1}$$

where $a < b$, $c < d$. For fixed v satisfying the inequalities $c \leq v \leq d$, $f(u, v)$ is then a (continuous) function of u for $a \leq u \leq b$, and accordingly we can consider the total variation of $f(u, v)$, as a function of u for fixed v, in the closed linear interval $a \leq u \leq b$ (see **V.1.1**, definition 1). This quantity will be denoted by

$$V_u(v, f, [a, b]). \tag{2}$$

Thus (2) designates a function of v defined for $c \leq v \leq d$. Similarly, we denote by

$$V_v(u, f, [c, d]) \tag{3}$$

the total variation of $f(u, v)$ with respect to v, for fixed u, in the closed linear interval $c \leq v \leq d$. Note that the total variations (2) and (3) may assume the value $+\infty$.

Lemma 1. $V_u(v, f, [a, b])$ is a lower semi-continuous function of v, and similarly $V_v(u, f, [c, d])$ is a lower semi-continuous function of u. Hence (see **III.1.1**, lemma 20) the total variations (2) and (3) are L-measurable functions for $c \leq v \leq d$ and $a \leq u \leq b$ respectively.

Proof. Since the argument is entirely analogous in either case, we consider explicitly only the total variation (2). Take v_0, v_n such that

$$c \leq v_0 \leq d, \quad c \leq v_n \leq d, \quad v_n \to v_0.$$

To simplify notations, put

$$f(u, v_0) = g_0(u), \quad f(u, v_n) = g_n(u) \quad \text{for} \quad a \leq u \leq b.$$

Since f is continuous in R, clearly $g_n(u) \to g_0(u)$ for $a \leq u \leq b$. Hence (see **V.1.1**, lemma 4)

$$V(g_0, [a, b]) \leq \liminf_{n \to \infty} V(g_n, [a, b]),$$

and the lemma follows, since clearly

$$V(g_0, [a, b]) = V_u(v_0, f, [a, b]), \quad V(g_n, [a, b]) = V_u(v_n, f, [a, b]).$$

Definition 1. If $f(u, v)$ is a real-valued continuous function in an oriented rectangle R given as in (1), then $f(u, v)$ is said to be of bounded variation in R in the TONELLI sense (briefly, BVT in R) provided that the total variations $V_u(v, f, [a, b])$ and $V_v(u, f, [c, d])$ are L-summable in the closed intervals $[c, d]$ and $[a, b]$ respectively.

Lemma 2. If $f(u, v)$ is BVT in R, then the first partial derivatives f_u, f_v exist a.e. in R and are L-summable in R.

Proof. We shall consider explicitly only f_u, since an entirely similar argument applies to f_v. Let E be the set of those points $(u, v) \in R$ where f_u fails to exist. Since E is L-measurable (see **III.1.3**), the fact that $LE = 0$ will be established (see **III.1.1**, lemma 17) if we show that for a.e. $v_0 \in [c, d]$ the intersection of E with the segment $v = v_0$, $a \leq u \leq b$ is of (linear) L-measure zero. Now if v is so chosen that

$$V_u(v, f, [a, b]) < \infty, \tag{4}$$

then $f(u, v)$ is of bounded variation, as a function of u, in the closed interval $[a, b]$. Hence, by theorem 5 in **V.1.6**, $f_u(u, v)$ exists for a.e. $u \in [a, b]$ if (4) holds. Next, since $V_u(v, f, [a, b])$ is L-summable in $[c, d]$, by **III.1.1**, lemma 31 it follows that (4) holds for a.e. $v \in [c, d]$, and thus it is established that f_u exists a.e. in R. Consider again a number $v \in [c, d]$ such that (4) holds. By **V.1.6**, theorem 5 it follows that $f_u(u, v)$, as a function of u, is L-summable in $[a, b]$ and

$$\int_a^b |f_u(u, v)| \, du \leq V_u(v, f, [a, b]). \tag{5}$$

Now since (4) holds for a.e. $v \in [c, d]$ and $V_u(v, f, [a, b])$ is L-summable in $[c, d]$, (5) also holds for a.e. $v \in [c, d]$ and yields the inequality

$$\int_c^d \left[\int_a^b |f_u(u, v)|\, du\right] dv \leq \int_c^d V_u(v, f, [a, b])\, dv < \infty. \tag{6}$$

By the FUBINI-TONELLI theorem, the existence of the iterated integral in (6) implies the L-summability of f_u in R, as well as the inequality

$$\iint_R |f_u(u, v)|\, du\, dv \leq \int_c^d V_u(v, f, [a, b])\, dv. \tag{7}$$

An entirely similar argument yields the L-summability of f_v in R and the inequality

$$\iint_R |f_v(u, v)|\, du\, dv \leq \int_a^b V_v(u, f, [c, d])\, du. \tag{8}$$

Definition 2. If $f(u, v)$ is a real-valued, continuous function in an oriented rectangle R given as in (1), then $f(u, v)$ is said to be absolutely continuous in R in the TONELLI sense (briefly, ACT in R) provided that (i) T is BVT in R, (ii) for a.e. $v \in [c, d]$, $f(u, v)$ is absolutely continuous as a function of u in the closed interval $[a, b]$, and (iii) for a.e. $u \in [a, b]$, $f(u, v)$ is absolutely continuous as a function of v in the closed interval $[c, d]$.

Lemma 3. Let $f(u, v)$ be a real-valued, continuous function in a bounded domain $\boldsymbol{D} \subset R^2$, such that $f(u, v)$ is ACT in every oriented rectangle in $\boldsymbol{D}$. Let D be a domain such that $\overline{D} \subset \boldsymbol{D}$. Consider the function

$$f_h(u, v) = \frac{1}{4h^2} \int_{-h}^{h} \int_{-h}^{h} f(u+\alpha, v+\beta)\, d\alpha\, d\beta, \quad (u, v) \in D, \quad 0 < h < \frac{\delta}{\sqrt{2}}, \tag{9}$$

which has been introduced in **VI.3.2**. Then the following holds for the partial derivatives

$$f_{hu} = \frac{\partial f_h}{\partial u}, \quad f_{hv} = \frac{\partial f_h}{\partial v}.$$

(α) f_{hu}, f_{hv} exist and are continuous in D.

(β) f_{hu}, f_{hv} are given in D by the formulas

$$f_{hu}(u, v) = \frac{1}{4h^2} \int_{-h}^{h} \int_{-h}^{h} f_u(u+\alpha, v+\beta)\, d\alpha\, d\beta, \tag{10}$$

$$f_{hv}(u, v) = \frac{1}{4h^2} \int_{-h}^{h} \int_{-h}^{h} f_v(u+\alpha, v+\beta)\, d\alpha\, d\beta. \tag{11}$$

(γ) If $\{h_n\}$ is any sequence of positive numbers such that

$$0 < h_n < \frac{\delta}{\sqrt{2}}, \quad h_n \to 0,$$

then

$$f_{h_n u} \to f_u, \quad f_{h_n v} \to f_v \quad \text{a.e. in } D.$$

Proof. (α) follows directly from **VI.3.2**, lemma 4, which also yields the formulas

$$f_{hu}(u, v) = \frac{1}{4h^2} \int_{-h}^{h} [f(u+h, v+\beta) - f(u-h, v+\beta)]\, d\beta, \tag{12}$$

$$f_{hv}(u, v) = \frac{1}{4h^2} \int_{-h}^{h} [f(u+\alpha, v+h) - f(u+\alpha, v-h)]\, d\alpha, \tag{13}$$

for $(u, v) \in D$. Consider now a point $(u_0, v_0) \in D$. Denote by Q the oriented square with center at (u_0, v_0) and side-length $2h$. Then $Q \subset \boldsymbol{D}$ [see **VI.3.2** (2), (3)], and hence $f(u, v)$ is ACT in Q by assumption. Accordingly, for a.e. $v \in [v_0 - h, v_0 + h]$, $f(u, v)$ is absolutely continuous as a function of u for $u \in [u_0 - h, u_0 + h]$, or equivalently, for a.e. $\beta \in [-h, h]$, $f(u, v_0 + \beta)$ is absolutely continuous as a function of u for $u \in [u_0 - h, u_0 + h]$. By **V.1.6**, theorem 8 it follows that

$$\left.\begin{aligned} f(u_0+h, v_0+\beta) - f(u_0-h, v_0+\beta) &= \int_{u_0-h}^{u_0+h} f_u(u, v_0+\beta)\, du \\ &= \int_{-h}^{h} f_u(u_0+\alpha, v_0+\beta)\, d\alpha \quad \text{for a.e. } \beta \in [-h, h]. \end{aligned}\right\} \tag{14}$$

Note that f_u is L-summable in Q by lemma 2. Accordingly, in view of the FUBINI-TONELLI theorem there follows from (14) the formula

$$\begin{aligned} &\frac{1}{4h^2} \int_{-h}^{h} [f(u_0 + h, v_0 + \beta) - f(u_0 - h, v_0 + \beta)]\, d\beta \\ &\qquad = \frac{1}{4h^2} \int_{-h}^{h} \int_{-h}^{h} f_u(u_0 + \alpha, v_0 + \beta)\, d\alpha\, d\beta, \end{aligned}$$

and thus (10) is proved. In an entirely similar manner, (11) is seen to follow from (13). There remains to verify (γ). Since $f(u, v)$ is ACT in every oriented rectangle in $\boldsymbol{D}$, by lemma 2 it follows that f_u exists a.e. in every oriented rectangle in $\boldsymbol{D}$. Since $\boldsymbol{D}$ can be represented (see **I.2.2**, exercise 8) as a countable union of oriented rectangles, it follows that f_u exists a.e. in $\boldsymbol{D}$, and by **III.1.3** it follows further that f_u is L-measurable in $\boldsymbol{D}$. Lemma 2 also yields the fact that f_u is L-summable in every oriented rectangle in $\boldsymbol{D}$. Thus we can apply **VI.3.2**, lemma 2 with f replaced by f_u, obtaining the relation

$$(f_u)_{h_n} \to f_u \quad \text{a.e. in } D. \tag{15}$$

On the other hand

$$f_{h_n u} = (f_u)_{h_n} \tag{16}$$

by (10). From (15) and (16) we conclude that $f_{h_n u} \to f_u$ a.e. in D. An entirely similar argument yields the relation $f_{h_n v} \to f_v$ a.e. in D, and (γ) is proved.

Lemma 4. Let $f(u, v)$ be a real-valued, continuous function in a bounded domain $\boldsymbol{D} \subset R^2$, and let p be a real number greater than or equal to one. Assume that $f(u, v)$ is ACT in every oriented rectangle in $\boldsymbol{D}$ and $|f_u|^p$, $|f_v|^p$ are L-summable in every oriented rectangle in $\boldsymbol{D}$. Let D be a domain such that $\overline{D} \subset \boldsymbol{D}$. Consider the functions (see **VI.3.2**)

$$f_h(u, v) = \frac{1}{4h^2} \int_{-h}^{h} \int_{-h}^{h} f(u+\alpha, v+\beta)\, d\alpha\, d\beta, \quad (u, v) \in D, \quad 0 < h < \frac{\delta}{\sqrt{2}}.$$

Then the family of functions

$$|f_{hu}|^p, \quad |f_{hv}|^p, \quad 0 < h < \frac{\delta}{\sqrt{2}},$$

satisfies the condition (V) in every oriented rectangle in D.

Proof. We note again, as in the proof of part (γ) of lemma 3, that f_u exists a.e. in $\boldsymbol{D}$ and is L-measurable in $\boldsymbol{D}$, and f_u is L-summable in every oriented rectangle in $\boldsymbol{D}$. Furthermore, $|f_u|^p$ is L-summable by assumption in every oriented rectangle in $\boldsymbol{D}$. Accordingly, we can apply **VI.3.2**, lemma 3 with f replaced by f_u, obtaining the conclusion that the family of functions

$$|(f_u)_h|^p, \quad 0 < h < \frac{\delta}{\sqrt{2}},$$

satisfies the condition (V) in every oriented rectangle in D. Similarly we find that the family of functions

$$|(f_v)_h|^p, \quad 0 < h < \frac{\delta}{\sqrt{2}},$$

satisfies the condition (V) in every oriented rectangle in D. Since

$$(f_u)_h = f_{hu}, \quad (f_v)_h = f_{hv}$$

by part (β) of lemma 3, the present lemma follows.

Lemma 5[1]. Let $f(u, v)$ be a real-valued, continuous function in a bounded domain $\boldsymbol{D} \subset R^2$, such that (a) $f(u, v)$ is ACT in every oriented rectangle in $\boldsymbol{D}$, and (b) there exists a real number $p>2$ such that $|f_u|^p$, $|f_v|^p$ are L-summable in every oriented rectangle in $\boldsymbol{D}$. Let Q be an oriented square in $\boldsymbol{D}$ and $\omega(f, Q)$ the oscillation of f in Q. Then

$$\omega(f, Q) \leq \Gamma(p) \left[\iint_Q (1 + |f_u| + |f_v|)^p \, du \, dv\right]^{\frac{1}{p}}, \tag{17}$$

where $\Gamma(p)$ is the constant occurring in **VI.3.1**, lemma 3 [thus $\Gamma(p)$ is positive, finite, and depends solely upon p].

Proof. Note that (17) has been already established in **VI.3.1**, lemma 3 for the special case when $f(u, v)$ has continuous first partial derivatives. The more general situation considered in the present lemma will be treated by the *method of approximation by integral means*, as follows. Select an oriented square $Q \subset \boldsymbol{D}$. Since, by assumption, $|f_u|^p$ and $|f_v|^p$ are L-summable in Q, by **III.1.1**, lemma 28 it follows that the integral appearing in (17) exists. Select a domain D such that

$$Q \subset D, \quad \bar{D} \subset \boldsymbol{D}, \tag{18}$$

and consider, as in **VI.3.2**, the functions

$$f_h(u, v) = \frac{1}{4h^2} \int_{-h}^{h} \int_{-h}^{h} f(u+\alpha, v+\beta) \, d\alpha \, d\beta, \quad (u, v) \in D, \quad 0 < h < \frac{\delta}{\sqrt{2}}. \tag{19}$$

Select a sequence of real numbers h_n such that

$$0 < h_n < \frac{\delta}{\sqrt{2}}, \quad h_n \to 0. \tag{20}$$

We have then, by lemma 3,

$$f_{h_n u} \to f_u, \quad f_{h_n v} \to f_v \quad \text{a.e. in } D,$$

and hence also [see (18)]

$$(1 + |f_{h_n u}| + |f_{h_n v}|)^p \to (1 + |f_u| + |f_v|)^p \quad \text{a.e. in } Q. \tag{21}$$

In view of (20), the sequences

$$|f_{h_n u}|^p, \quad |f_{h_n v}|^p, \quad n = 1, 2, \ldots, \tag{22}$$

[1] The inequality (17) is a special case of an inequality due to A. P. CALDERÓN, On the differentiability of absolutely continuous functions [Riv. Mat. Univ. Parma 2, 203–213 (1951)].

satisfy the condition (V) in Q by lemma 4. Take now any L-measurable set $S \subset Q$. By **III.1.1**, lemma 28 we have the inequality

$$\left. \begin{aligned} &\left[\iint_S (1 + |f_{h_n u}| + |f_{h_n v}|)^p \, du \, dv\right]^{1/p} \leq \\ &\leq \left[\iint_S 1^p \, du \, dv\right]^{1/p} + \left[\iint_S |f_{h_n u}|^p \, du \, dv\right]^{1/p} + \left[\iint_S |f_{h_n v}|^p \, du \, dv\right]^{1/p}. \end{aligned} \right\} \quad (23)$$

Since the sequences (22) satisfy the condition (V) in Q, it is clear from (23) that the sequence

$$(1 + |f_{h_n u}| + |f_{h_n v}|)^p, \quad n = 1, 2, \ldots,$$

also satisfies the condition (V) in Q. In view of (21) it follows (see **III.1.1**, lemma 36) that

$$\iint_Q (1 + |f_{h_n u}| + |f_{h_n v}|)^p \, du \, dv \to \iint_Q (1 + |f_u| + |f_v|)^p \, du \, dv. \quad (24)$$

Since $f(u, v)$ is continuous in $\boldsymbol{D}$, from **VI.3.2**, lemma 5 we infer that

$$f_{h_n} \to f \quad \text{uniformly in } Q.$$

Hence clearly

$$\omega(f_{h_n}, Q) \to \omega(f, Q). \quad (25)$$

Observe now that f_{h_n} has continuous first partial derivatives in D by lemma 3. Accordingly, we can apply **VI.3.1**, lemma 3 to f_{h_n}, obtaining the inequality

$$\omega(f_{h_n}, Q) \leq \Gamma(p) \left[\iint_Q (1 + |f_{h_n u}| + |f_{h_n v}|)^p \, du \, dv\right]^{\frac{1}{p}}. \quad (26)$$

Clearly (26), (25), (24) imply (17).

VI.3.4. Special classes of differentiable transformations in $\boldsymbol{R}^2$. In this section we shall use the representation

$$T: x = x(u, v), \quad y = y(u, v), \quad (u, v) \in \boldsymbol{D}, \quad (1)$$

for the bounded continuous transformation $T: \boldsymbol{D} \to R^2$ (see **VI.1.1**).

Definition 1. Given T as in (1), we shall say that T satisfies the hypothesis H_0 in $\boldsymbol{D}$ if (i) the first partial derivatives x_u, x_v, y_u, y_v exist a.e. in $\boldsymbol{D}$, (ii) the ordinary Jacobian $J(w, T)$ is L-summable in $\boldsymbol{D}$, and (iii) $x(u, v)$, $y(u, v)$ are ACT in every oriented rectangle in $\boldsymbol{D}$.

Practically all the special classes of differentiable transformations in R^2 which have been studied in the literature satisfy the hypothesis H_0, and hence the clarification of the implications of this hypothesis is a matter of considerable interest. It has been conjectured that the hypothesis H_0 implies that T is eAC in $\boldsymbol{D}$. Since the validity of this conjecture is still an open question, we shall discuss only a few typical results relating to the hypothesis H_0. The proofs will be based upon the *method of approximation by integral means.*

Given T as in (1), take a sequence $\{D_j\}$ of domains which fill up $\boldsymbol{D}$ from the interior (see **II.3.2**, remark 10). Then $\overline{D}_j \subset \boldsymbol{D}$, $j = 1, 2, \ldots$. Hence (see **VI.3.2**) for every positive integer j we can consider the functions

$$x_h(u, v) = \frac{1}{4h^2} \int_{-h}^{h} \int_{-h}^{h} x(u+\alpha, v+\beta)\, d\alpha\, d\beta, \quad (u, v) \in D_j, \quad 0 < h < \frac{\delta_j}{\sqrt{2}}, \tag{2}$$

$$y_h(u, v) = \frac{1}{4h^2} \int_{-h}^{h} \int_{-h}^{h} y(u+\alpha, v+\beta)\, d\alpha\, d\beta, \quad (u, v) \in D_j, \quad 0 < h < \frac{\delta_j}{\sqrt{2}}, \tag{2*}$$

where δ_j is an appropriately chosen positive constant. By **VI.3.2**, lemma 5 we can select for each j a number h_j such that

$$0 < h_j < \frac{\delta_j}{\sqrt{2}}, \quad h_j \to 0, \quad h_1 > \cdots > h_j > \cdots, \tag{3}$$

and furthermore

$$|x_{h_j}(u, v) - x(u, v)| < \frac{1}{j} \quad \text{for} \quad (u, v) \in D_j, \tag{4}$$

$$|y_{h_j}(u, v) - y(u, v)| < \frac{1}{j} \quad \text{for} \quad (u, v) \in D_j. \tag{5}$$

To simplify notations, we put

$$x_{h_j}(u, v) = X_j(u, v), \quad (u, v) \in D_j, \tag{6}$$

$$y_{h_j}(u, v) = Y_j(u, v), \quad (u, v) \in D_j. \tag{7}$$

Lemma 1. X_j and Y_j are continuous in D_j, and the first partial derivatives $X_{ju}, X_{jv}, Y_{ju}, Y_{jv}$ exist and are continuous in D_j.

Proof. Since $x(u, v)$ and $y(u, v)$ are continuous in $\boldsymbol{D}$, this lemma is a direct consequence of the lemmas 1 and 4 in **VI.3.2.**

Let us introduce now the auxiliary transformations

$$T_j: x = X_j(u, v), \quad y = Y_j(u, v), \quad (u, v) \in D_j.$$

Lemma 2. T_j is eAC in D_j.

Proof. Since the first partial derivatives of the coordinate functions of T_j exist and are continuous in D_j by lemma 1, in view of **V.3.6**, lemma 7 (applied to T_j) it is sufficient to show that the ordinary Jacobian $J(w, T_j)$ is L-summable in D_j. Now since T is bounded in $\boldsymbol{D}$, we have a constant $M > 0$ such that

$$|x(u, v)|, \ |y(u, v)| < M, \quad (u, v) \in \boldsymbol{D}. \tag{8}$$

Furthermore, by **VI.3.2**, lemma 4 we have [in view of (1) and (6)] the formula

$$X_{ju}(u, v) = \frac{1}{4 h_j^2} \int_{-h_j}^{h_j} [x(u + h_j, v + \beta) - x(u - h_j, v + \beta)] \, d\beta, \quad (u, v) \in D_j.$$

By (8) it follows that

$$|X_{ju}(u, v)| < \frac{M}{h_j}, \quad (u, v) \in D_j \quad \text{for } j \text{ large}.$$

A similar argument yields the same bound for $|X_{jv}|$, $|Y_{ju}|$, $|Y_{jv}|$. Hence

$$|J(w, T_j)| < \frac{2M^2}{h_j^2}, \quad w \in D_j \quad \text{for } j \text{ large}.$$

Thus $J(w, T_j)$ is continuous and bounded in D_j. Hence $J(w, T_j)$ is L-summable in D_j, and the lemma is proved.

In the course of the following discussion we shall make statements about the behavior of the sequence $\{T_j\}$ on some compact subset F of $\boldsymbol{D}$. Since the domains D_j fill up $\boldsymbol{D}$ from the interior, there exists an integer $j_0 = j_0(F)$ such that $F \subset D_j$ for $j > j_0$. Then T_j is defined on F for $j > j_0$, and in the sequel it will be understood that our statements are meant to apply for $j > j_0$ only.

Lemma 3. The sequence $\{T_j\}$ converges to T uniformly on compact subsets of $\boldsymbol{D}$, in the sense of **II.3.2**, remark 9.

Proof. Take a compact set $F \subset \boldsymbol{D}$. Since the domains D_j fill up $\boldsymbol{D}$ from the interior, we can select an integer j_0 such that $F \subset D_j$ for $j > j_0$. From (4), (5), (6), (7) we infer that

$$\varrho(T, T_j, F) < \frac{2}{j} \quad \text{for} \quad j > j_0.$$

Thus clearly $\varrho(T, T_j, F) \to 0$ for $j \to \infty$, and the lemma follows.

Theorem 1. Given T as in (1), assume that the following holds. (a) T satisfies the hypothesis H_0 in $\boldsymbol{D}$. (b) If Q is any oriented square in $\boldsymbol{D}$, then the sequence $\{J(w, T_j)\}$ satisfies the condition (V) in Q. Then T is eAC in $\boldsymbol{D}$.

Proof. The assumption (a) implies that the first partial derivatives x_u, x_v, y_u, y_v exist a.e. in $\boldsymbol{D}$ and the ordinary Jacobian $J(w,T)$ is L-summable in $\boldsymbol{D}$. By lemma 3, the sequence $\{T_j\}$ converges to T uniformly on compact subsets of $\boldsymbol{D}$. The first partial derivatives of the coordinate functions of T_j exist and are continuous in D_j by lemma 1, and T_j is eAC in D_j by lemma 2. By assumption, the sequence $\{J(w, T_j)\}$ satisfies the condition (V) in every oriented square $Q \subset \boldsymbol{D}$. Accordingly, the present theorem will appear as a consequence of **VI.2.3**, theorem 7 if we show yet that if F is a compact set in $\boldsymbol{D}$, then

$$J(w, T_j) \to J(w, T) \quad \text{a.e. on } F. \tag{9}$$

Now since the domains D_j fill up $\boldsymbol{D}$ from the interior, we have an integer $j_0 = j_0(F)$ such that $F \subset D_{j_0}$. By (3) we have then $0 < h_j < \delta_{j_0}/\sqrt{2}$ for $j > j_0$. Since $x(u, v)$ and $y(u, v)$ are ACT in every oriented rectangle in $\boldsymbol{D}$, we can apply **VI.3.3**, lemma 3 to the functions $x(u, v)$, $y(u, v)$ and the domain D_{j_0}. In view of (7) and (6) it follows (since $h_j \to 0$) that $X_{ju}, X_{jv}, Y_{ju}, Y_{jv}$ converge to x_u, x_v, y_u, y_v respectively a.e. in D_{j_0}. Hence (9) holds, since $F \subset D_{j_0}$.

We proceed to consider a few of the numerous applications of the preceding theorem.

Theorem 2. Given T as in (1), assume that the following holds. (a) T satisfies the hypothesis H_0 in $\boldsymbol{D}$. (b) There exist two real numbers p, q such that

$$p > 1, \quad q > 1, \quad \frac{1}{p} + \frac{1}{q} = 1, \tag{10}$$

and $|x_u|^p, |x_v|^p, |y_u|^q, |y_v|^q$ are L-summable in every oriented rectangle in $\boldsymbol{D}$. Then T is eAC in $\boldsymbol{D}$.

Proof. In view of theorem 1 it is sufficient to verify that the sequence $\{J(w, T_j)\}$ satisfies the condition (V) in every oriented square $Q \subset \boldsymbol{D}$. In turn, this latter fact will be established if we show that the sequences

$$\{X_{ju} Y_{jv}\}, \quad \{X_{jv} Y_{ju}\} \tag{11}$$

satisfy the condition (V) in every oriented square $Q \subset \boldsymbol{D}$. Select such a square Q. Since the domains D_j fill up $\boldsymbol{D}$ from the interior, we can select an integer j_0 such that $Q \subset D_{j_0}$. By (3) we have then $0 < h_j < \delta_{j_0}/\sqrt{2}$ for $j > j_0$. In view of the assumptions (a) and (b) we conclude that we can apply **VI.3.3**, lemma 4 to the functions $x(u, v)$, $y(u, v)$ and the domain D_{j_0}. It follows that the sequences

$$\{|X_{ju}|^p\}, \quad \{|X_{jv}|^p\}, \quad \{|Y_{ju}|^q\}, \quad \{|Y_{jv}|^q\}, \quad j > j_0,$$

satisfy the condition (V) in Q. Hence if $\varepsilon > 0$ is assigned, then there exists a number $\eta > 0$ such that (for $j > j_0$)

$$\left.\begin{aligned} &\iint_S |X_{ju}|^p\, du\, dv < \varepsilon, \quad \iint_S |X_{jv}|^p\, du\, dv < \varepsilon, \\ &\iint_S |Y_{ju}|^q\, du\, dv < \varepsilon, \quad \iint_S |Y_{jv}|^q\, du\, dv < \varepsilon, \end{aligned}\right\} \tag{12}$$

whenever S is an L-measurable set such that

$$S \subset Q, \quad \mathsf{L}\, S < \eta. \tag{13}$$

Assume that (13) holds. Then the HÖLDER inequality (see **III.1.1**, lemma 29) yields, in view of (12),

$$\left|\iint_S |X_{ju} Y_{jv}\, du\, dv\right| \leq \left[\iint_S |X_{ju}|^p\, du\, dv\right]^{1/p} \left[\iint_S |Y_{jv}|^q\, du\, dv\right]^{1/q} < \varepsilon^{\frac{1}{p}+\frac{1}{q}} = \varepsilon$$

for $j > j_0$. An entirely similar reasoning shows that

$$\left|\iint_S X_{jv} Y_{ju}\, du\, dv\right| < \varepsilon \quad \text{for} \quad j > j_0.$$

Thus the sequences (11) satisfy the condition (V) in Q, and the proof is complete.

Definition 2. A transformation T, given as in (1), is termed a DIRICHLET *transformation* in $\boldsymbol{D}$ if (i) T satisfies the hypothesis H_0 in $\boldsymbol{D}$, and (ii) for every oriented rectangle $R \subset \boldsymbol{D}$ the DIRICHLET integrals

$$\iint_R (x_u^2 + x_v^2)\, du\, dv, \quad \iint_R (y_u^2 + y_v^2)\, du\, dv$$

of the coordinate functions of T exist.

Theorem 3. If T is a DIRICHLET transformation in $\boldsymbol{D}$, then T is eAC in $\boldsymbol{D}$.

This statement is merely a special case of theorem 2, corresponding to $p = q = 2$. However, theorem 3 is of interest for two reasons. First, DIRICHLET transformations play an important role in various applications. Second, relatively simple examples show that a DIRICHLET transformation is generally not absolutely continuous in the BANACH sense. Thus the important class of DIRICHLET transformations is beyond the scope of the B-*theory*, while on the other hand (in view of theorem 3), these transformations constitute a very special sub-class of the class of eAC transformations.

The following theorem is concerned with another very special but quite important class of transformations in R^2.

Theorem 4. Given T as in (1), assume that the following holds. (a) T satisfies the hypothesis H_0 in $\boldsymbol{D}$. (b) One of the coordinate functions $x(u, v)$, $y(u, v)$ of T is Lipschitzian in $\boldsymbol{D}$. Then T is eAC in $\boldsymbol{D}$.

Proof. Assume, for instance, that $y(u, v)$ is Lipschitzian in $\boldsymbol{D}$. Having selected an oriented square $Q < \boldsymbol{D}$ and a number $\varepsilon > 0$, the proof begins as that of theorem 2, the modifications taking place when the inequalities (12) are reached. Instead of (12) we have now the following facts at our disposal. Since $x(u, v)$ is ACT in every oriented rectangle in $\boldsymbol{D}$, the partial derivatives x_u, x_v are L-summable in every oriented rectangle in $\boldsymbol{D}$ by **VI.3.3**, lemma 2. Applying **VI.3.3**, lemma 4 to $x(u, v)$ with $p = 1$, we obtain [corresponding to the first line in (12)] the conclusion that (for $j > j_0$)

$$\iint_S |X_{ju}|\, du\, dv < \varepsilon, \quad \iint_S |X_{jv}|\, du\, dv < \varepsilon, \tag{14}$$

whenever S is an L-measurable set such that

$$S < Q, \quad \mathsf{L}\, S < \eta. \tag{15}$$

Since $y(u, v)$ is Lipschitzian in $\boldsymbol{D}$, there exists a constant $M \geq 0$ such that

$$|y(u_2, v_2) - y(u_1, v_1)| \leq M\,[(u_2 - u_1)^2 + (v_2 - v_1)^2]^{\frac{1}{2}} \tag{16}$$

for any two points (u_2, v_2), (u_1, v_1) in $\boldsymbol{D}$. In view of (7) and **VI.3.2**, lemma 4 we have the formula

$$Y_{ju}(u, v) = \frac{1}{4 h_j^2} \int_{-h_j}^{h_j} [y(u + h_j, v + \beta) - y(u - h_j, v + \beta)]\, d\beta, \quad (u, v) \in D_j.$$

By (16) it follows that

$$|Y_{ju}| \leq M \quad \text{in } D_j. \tag{17}$$

A similar argument yields

$$|Y_{jv}| \leq M \quad \text{in } D_j. \tag{18}$$

Consider now any L-measurable set S such that (15) holds. From (14), (17), (18) we conclude that

$$\left|\iint_S X_{ju} Y_{jv}\, du\, dv\right| \leq M \iint_S |X_{ju}|\, du\, dv \leq M\, \varepsilon \quad \text{for} \quad j > j_0,$$

and similarly

$$\left|\iint_S X_{jv} Y_{ju}\, du\, dv\right| \leq M\, \varepsilon \quad \text{for} \quad j > j_0.$$

Since M is fixed and ε was arbitrary, these inequalities imply that the sequences

$$\{X_{ju} Y_{jv}\}, \quad \{X_{jv} Y_{ju}\}, \quad j > j_0,$$

satisfy the condition (V) in Q. Hence the sequence $\{J(w, T_j)\}, j > j_0$, also satisfies the condition (V) in Q, and thus the present theorem follows from theorem 1.

The next theorem yields an important class of generalized Lipschitzian transformations in R^2.

Theorem 5. Given T as in (1), assume that the following holds. (a) T satisfies the hypothesis H_0 in $\boldsymbol{D}$. (b) There exists a number $p > 2$ such that $|x_u|^p$, $|x_v|^p$, $|y_u|^p$, $|y_v|^p$ are L-summable in every oriented rectangle in $\boldsymbol{D}$.

Then T is generalized Lipschitzian in $\boldsymbol{D}$ (and hence T possesses all the properties stated in **V.3.6**, theorem 2).

Proof. Take an oriented rectangle $R \subset \boldsymbol{D}$ and select an oriented square $Q \subset R$. We can then apply **VI.3.3**, lemma 5 to both $x(u, v)$ and $y(u, v)$, obtaining the inequalities

$$\omega(x, Q) \leq \Gamma(p) \left[\iint_Q (1 + |x_u| + |x_v|)^p \, du \, dv\right]^{\frac{1}{p}}, \tag{19}$$

$$\omega(y, Q) \leq \Gamma(p) \left[\iint_Q (1 + |y_u| + |y_v|)^p \, du \, dv\right]^{\frac{1}{p}}. \tag{20}$$

Let us put

$$\varphi = 1 + |x_u| + |x_v| + |y_u| + |y_v|. \tag{21}$$

The assumption (b) implies, by **III.1.1**, lemma 28, that φ^p is L-summable in R. Since clearly

$$\delta T Q \leq \omega(x, Q) + \omega(y, Q),$$

we obtain from (19), (20), (21) the inequality

$$\delta T Q \leq 2\Gamma(p) \left[\iint_Q \varphi^p \, du \, dv\right]^{\frac{1}{p}}. \tag{22}$$

Hence on setting

$$\Phi(u, v) = 4\Gamma(p)^2 \varphi(u, v)^p, \quad (u, v) \in R,$$

it follows that Φ is a non-negative, L-summable function in R such that T satisfies the condition $(\Phi, \operatorname{int} R)$ [see (22) and **V.3.6**, definition 1]. Since the assumption (a) implies that the ordinary Jacobian $J(w, T)$ is L-summable in $\boldsymbol{D}$, it follows finally (see **V.3.6**, definition 2) that T is generalized Lipschitzian in $\boldsymbol{D}$.

Index.

Bibliography.

ALEXANDROFF, P., and H. HOPF: Topologie. Grundlehren der mathematischen Wissenschaften, Vol. 45. Berlin: Springer 1953.

EILENBERG, S., and N. STEENROD: Foundations of Algebraic Topology. Princeton University Press 1952.

HAHN, H.: Reelle Funktionen, Part I. Leipzig 1932.

HALMOS, P. R.: Measure Theory. New York: D. Van Nostrand Co. 1950.

HAUSDORFF, F.: Mengenlehre. Leipzig 1935.

KURATOWSKI, C.: Topologie, I, II. Monografie Matematyczne, Warszawa, 1948, 1950.

LEFSCHETZ, S.: Algebraic Topology. Amer. Math. Soc. Colloquium Publications, Vol. 27, 1942.

MAYRHOFER, K.: Inhalt und Maß. Wien: Springer 1952.

RADO, T.: Length and Area. Amer. Math. Soc. Colloquium Publications, Vol. 30, 1948.

SAKS, S.: Theory of the Integral. Monografie Matematyczne, Warszawa, 1937.

SIERPINSKI, W.: Introduction to General Topology. University of Toronto Press 1934.

WHYBURN, G. T.: Analytic Topology. Amer. Math. Soc. Colloquium Publications, Vol. 28, 1942.

WILDER, R. L.: Topology of Manifolds. Amer. Math. Soc. Colloquium Publications, Vol. 32, 1949.

MIX
Papier aus verantwortungsvollen Quellen
Paper from responsible sources
FSC® C105338

If you have any concerns about our products,
you can contact us on
ProductSafety@springernature.com

In case Publisher is established outside the EU,
the EU authorized representative is:
Springer Nature Customer Service Center GmbH
Europaplatz 3, 69115 Heidelberg, Germany

Printed by Libri Plureos GmbH
in Hamburg, Germany